Gravitation

Following the Prague Inspiration

A Volume in Celebration of the 60th Birthday of Jiří Bičák

Ten, kdo nám dovedl ukázati některou novou linii, některou netušenou perspektivu z krásy věcí a světa, je naším dobrodincem, jediným a nezapomenutelným. Zdá se nám, jako by nám patřil před svým narozením a celou věčností se k nám blížil a v tomto životě nás nalezl, aby nám řekl, co nám měl říci. Po celý život jsme mu vděčni dle zákona lásky, která nedovede zapomenouti než jen to, co sama rozdala.

Otokar Březina (1868-1929), Hudba pramenů — Krása světa

He who was able to reveal to us a new line, an unforeseen perspective within the beauty of the things and of the world, is our benefactor, unique and unforgettable. It seems to us as if he belonged to us before his birth and was approaching us throughout all eternity, to find us in this life and tell us what he had to tell. We are grateful to him all our lives according to the law of love which is able to forget nothing but that which she herself gave to others.

editors

O Semerák, J Podolský & M Žofka

Charles University Prague, Czech Republic

Gravitation

Following the Prague Inspiration

A Volume in Celebration of the 60th Birthday of Jiří Bičák

World Scientific

New Jersey • London • Singapore • Hong Kong

Published by

World Scientific Publishing Co. Pte. Ltd.
P O Box 128, Farrer Road, Singapore 912805
USA office: Suite 1B, 1060 Main Street, River Edge, NJ 07661
UK office: 57 Shelton Street, Covent Garden, London WC2H 9HE

British Library Cataloguing-in-Publication Data
A catalogue record for this book is available from the British Library.

GRAVITATION: FOLLOWING THE PRAGUE INSPIRATION
A Volume in Celebration of the 60th Birthday of Jiří Bičák

ISBN 981-238-093-0

Printed by FuIsland Offset Printing (S) Pte Ltd, Singapore

PREFACE

This book is dedicated to the worldtube of Professor Jiří Bičák on the occasion of the value $\tau = \tau_{60} = 5.676438379482 \cdot 10^{19}\,\mathrm{cm}$ of his proper time.[a] It has been written by Jiří's ex-students, once dragged into the realm of relativistic physics. The contributions present original results or survey those already published elsewhere. Some of them were directly incited or influenced by the teacher, while others at least felt his inspiration field.

Although the subjects involved range from motion of stars in galactic nuclei to quantum mechanics on a boundary, we still have not been able to cover the full scope of the honoured person. Also, his "it has to be re-formulated" kept running round in our heads during the preparation of this volume, mainly when writing introductory parts of the contributions. At the Prague Relativity Seminar, one manages with mere equations, because after a couple of sentences Jiří takes the floor and explains things better. It is often thanks to these intermezzi that the audience (as well as the speaker) can understand at least something. If not even these help and the person fights 'face redshift', Jiří appeases: "Don't worry, all is vanity of vanities anyway..."

What can be extracted from his ergosphere is, however, quite an opposite attitude — that certain actions are required by the very principle. Some time ago, shortly before midnight, Jiří had a cursory look at a final version of our grant report and declared: "here, in 'spacetimes of type N', N should be in italics." But seriously, in our spacetimes Jiří has been one of the principal rays. Within his "open society" (in the import of K. Popper), we were presented covariant formulae and stories about distant galaxies (and much more), but also a training in balance — between modesty and showiness, between universal devoutness and independence, between abstraction permeating into jungle and wide contexts seen from the heights. A training threaded by the sense of fair play and humour and rooted in general culture.

Certainly, Jiří would rather give money to young teachers or poor people, he would support tigers and elephants. Fortunately, no expensive binding is necessary for this book, since he will not open it wider than a light cone (its future part, hopefully). Jiří's most decent attitude towards literature — and surely towards light cones — has obviously been cultivated by his academic teacher Professor Karel Kuchař,[b] the author of the only hard(-cover) — and still canonical — Czech textbook on general relativity.

No place in Prague is more appropriate for presenting our gift than the salon of Berta Fanta in the Unicorn House ('U jednorožce') at the Old Town Square. The birthday falls on a Monday, but we hope to defer the party for one day. Actually, Jiří will certainly prefer playing his violin in a duet — and we inquired that Prof. Einstein only shows himself there on Tuesdays, on intellectual gatherings. Karlova street being within earshot, Johannes Kepler is also sure to come round: he has never been indifferent to harmonies. Yet if he starts his "Wherever there is matter

[a] Geometrized units are used in which $c = G = 1$.

[b] "I have hardly ever met closer anybody that cultured, educated, sensitive, and rational at once."

there is also geometry", the music will be over, because Jiří will immediately remark that even "the matter *there* governs geometry *here*", not noticing that Kepler in fact contemplates the snowflake on his coat. After such words, Prof. Mach *must* appear, coming in slippers from the house next door. (It is to be hoped he will commend the acoustics!)

According to lady Fanta, her son-in-law announced the work of Bernard Bolzano to be the talking point for January. Not a bad choice with such speakers. However, if the conversation diverts from mathematics to God, we fear it might not be profound enough for everyone. Jiří should take the physicists downstairs and captivate them with Bolzano's cosmological transcendences. Otherwise Kepler will better cross the square to the Church of Our Lady in front of Týn to pray at the tomb of his former employer Tycho Brahe,[c] while Prof. Einstein will quit[d] — followed by Ernst Mach, of course — to discuss the frequency shift with Christian Doppler. Prof. Doppler should be at Bolzano's place in the nearby Celetná street right now (Augustin Cauchy might call in for a dinner as well); he is surely wondering about Einstein's recent results on *gravitational* shift and light bending.

Let us apologize for this childish construction at last. However, it actually only suffers from one flaw: the (partial) de-synchronization of clocks. Had we a time-machine, Jiří would be the right person to really enjoy *inspiring* the atmosphere of such a company. Not only for how much he 'paid in advance' (in Hesiod's sense). *Who else* could so much resonate with (and himself personify) Jiří's favourite ideas of Otokar Březina,[e] Josef Čapek[f] or Albert Schweitzer[g]? With *whom else* could he wish as much to share the delight in omnipresent geometry? Not just any geometry, but — in words of Z. Horský — the geometry that was preferred by Pythagoreans: the one which lends order to beauty and vice versa. The geometry of spacetime, of Bach's fugue, or of snow, perceived from different distances (as H. Moore recommended to S. Chandrasekhar), with new aspects of beauty discovered on every scale. Incidentally, Jiří is often asked about such volatiles: creative inspiration, anthropic principle, ethics of science, the coming millenium... He is not afraid of touching them, but always distinguishing between 'sacral and profane' — always vigilant, together with T. Mann or T. G. Masaryk, whether 'belief', 'beauty' or 'being' are not pronounced as easily as 'beer'.

Much as he looks up to the natural, moral and aesthetic unity, Jiří can't stand any mainstream, but also not any 'right view'. Near Březina's "... so that all the values be laid on places where they belong, and so that there be order in the spiritual world.", one can read "When everyone agrees around here it means that something must be wrong." on his office wall. Jiří can hardly be imagined voting for the election winners; when one professes Schweitzer's respect to life as a creed, he needles "it may not be well to be humble...", and with his almost ostentatiously

[c] "Who holds a telescope in his right hand, isn't he put on a par with a king or a ruler over the work of God?"

[d] ... thinking the circle to be "a troop of unrealistic people, harking back to the Middle Ages".

[e] "... there will always be minds ... who will aspire to create, in a joint might of knowledge and dream, science and poetry, the unified image of universal events that would accord with perpetual yearning of a human spirit for harmony and beauty, as well as with the heart's thirst for justice."

[f] "Having felt oneself into a great unity of the world, you are urged into more moral behaviour."

[g] ... about whom Albert Einstein said: "Man at last in this tragic century!"

fusty Škoda 120 car he would like best to drive in the opposite direction.

Let us return to our celebration, though.[h] Before adjourning to Mach's apartment, everyone will surely wait for the hour outside. It is a special experience to watch the Old Town Astronomical Clock striking at January midnight, in a snowy coat and without crowds. Between strokes of the bellworks, through the depths of centuries, one perceives the brilliance of its designer, Jan Ondřejů called Šindel, and creator Mikuláš of Kadaň. The masterpiece was constructed in 1410, only some 60 years after 'artes reales' (arithmetics, geometry, astronomy — and music, of course) had began to be taught at the newly founded 'Studium generale', later known as Universitas Carolina. Time and again, one's thoughts encounter the personality of its founder, Charles IV. Who was that man, capable of raising the young Czech Crown Lands to a level that would only to be dreamt of later? Who, as a medieval Emperor, let himself be figured on his knees handing over the university foundation charter under the shield of Saint Wenceslas. Who gave the university the same inviolable protection that was enjoyed by the very symbols of his royal power, the Crown Jewels, and who also took under his wing all its teachers and students coming from various parts of Europe. (At this moment, Jiří will definitely interrupt by saying something low:[i] pathos is not his cup of tea. Well, except for that of Březina, Einstein, ...)

Towards the morning, we slowly part at the Jan Hus monument. After such a party, Fermi-Walker transport may be difficult. Everybody grumbles Mach shouldn't have bragged of his swivel chair — now we see Mach's bands everywhere; Einstein says something about an extremal path to the Smíchov district. Prof. Bičák would boost him there, being glad the car has not been stolen, but it does not start. Kepler drops behind a little, his cough has come back in a winter smog. The moon rose; perhaps it could be reached as in Cyrano... The crescent peeks out just over the Fantas'-house pinnacle, with a misty ring around. "Look: God, being too good not to reveal himself, invested his game into symbols and thus sent evidence of his appearance to the world. I therefore dare to think that the whole nature and the beatified heaven are written in the language of geometry and in a simple one."

In a morning newspaper, one reads that intellectuals annoy the society. Mostly they are no longer so untidy, grubby, no-hair, unsightly and barefoot as Socrates. But even nowadays, from time to time, they 'petrify and meditate'. Those whose confession is close to Einstein's or Schweitzer's are called extremists, not mentioning what label Čapek would be given today. Those like Hus, but also Bolzano or Kepler, remain suspect, and fashion dictates to pass off the shared dream of Březina's great unity as a dark relic. In bookshops, various rehashes of old sophists can be found, dubbing a human the gauge of everything and leaving the values to be rated on his or her market.

The crescent sinks in dawn. Our heroes have retired to their centuries. In Prague future, their Platonic beliefs may not have an easy time.

Oldřich Semerák

[h] "It's taking sort of long already..."

[i] ("I tell you, we 'oldies' are only for those quaintnesses.")

Acknowledgements

We thank our colleagues for various help, in particular V. Karas for conceptual discussions and P. Krtouš, T. Ledvinka and V. Pravda for technical assistance. And, of course, the contributors for their papers. Prof. K. Kuchař kindly improved our translation of the preamble excerpt from Březina; some of our Czenglish was also checked by Prof. J. B. Griffiths. M. Wolf and C. Ron helped us to convert the J.B.'s age from Central-European to International-Atomic-Time seconds. The photograph of Jiří Bičák is by courtesy of K. Meister, other photographs are by O.S. Also acknowledged gratefully is a partial support from the Faculty of Mathematics and Physics of the Charles University in Prague, from the Faculty of Philosophy and Science of the Silesian University in Opava, from the Czech grants GACR-202/99/0261, GACR-205/00/1685, GAUK-141/2000, GACR 101-13/201380, GACR-202/00/P030 and GACR-202/00/P031, and from the Slovak grant VEGA-1/7069/20.

List of publications of Jiří Bičák

A. Original publications

1. JB, *On the question of the uniqueness of the energy-momentum complex in the special and general theory of relativity*, Czech. J. Phys. **B15** (1965) 81-94
2. JB, *On the uniqueness of the energy-momentum complex in the special and general relativity*, Wiss. Zeits. der Friedrich-Schiller-Universität Jena **15** (1966) 83-85
3. JB, *On the Rainich geometrization of a vector meson field in the Kibble theory*, Czech. J. Phys. **B16** (1966) 95-98
4. JB, *Gravitational radiation from uniformly accelerated particles in general relativity*, Proc. Roy. Soc. London **A302** (1968) 201-224
5. JB, *A note on relativistic heat engines*, Lett. Nuovo Cim. **1** (1969) 302-304
6. JB, *Selected topics in the problem of energy and radiation*, in *Relativity and gravitation*, eds. Ch. Kuper and A. Peres, Gordon and Breach, New York (1971) 47-67
7. JB, *Gravitational collapse with charge and small asymmetries I. Scalar perturbations*, Gen. Rel. Grav. **3** (1972) 331-349
8. JB, *Collapsing stars and the physics of black holes*, in *Proc. of the Conf. Stellar Astronomy*, Folia Univ. Brunensis **XIV** Physica 13 (1973) 125-130
9. JB, *Perturbations of spherical collapse with charge*, in *Proc. of the Conf. Stellar Astronomy*, Folia Univ. Brunensis **XIV** Physica 13 (1973) 131-132
10. JB, *Nonspherical perturbations of gravitational collapse*, in *Int. School of Cosmology and Gravitation, III. Course*, Ettore Majorana Centre for Scientific Culture (1974) 1-15
11. JB and J. Slavík, *Non-linear electrodynamics in the Newman-Penrose formalism*, Acta Phys. Polon. **B6** (1975) 489-508
12. JB and P. Hadrava, *General-relativistic radiation transfer theory in refractive and dispersive media*, Astron. Astrophys. **44** (1975) 389-399
13. JB, *Rotating black holes*, in *Proc. of the Int. Symp. on Rel. and Grav.*, Scripta Fac. Sci. Univ. Brno (Physica) **5** (1975) 181-183
14. JB, Z. Stuchlík and M. Šob, *On perturbations of collapse and black holes with charge*, in *Proc. of the Int. Symp. on Rel. and Grav.*, Scripta Fac. Sci. Univ. Brno (Physica) **5** (1975) 189-191
15. JB and J. Slavík, *Non-linear electrodynamics in the Newman-Penrose formalism*, in *Proc. of the Int. Symp. on Rel. and Grav.*, Scripta Fac. Sci. Univ. Brno (Physica) **5** (1975) 215-216
16. JB and P. Hadrava, *General-relativistic radiative trasfer theory in refractive and dispersive media*, in *Proc. of the Int. Symp. on Rel and Grav.*, Scripta Fac. Sci. Univ. Brno (Physica) **5** (1975) 191-192
17. JB, *On the present state of cosmology*, in *Memoirs and Observations of the Czechosl. Astron. Society* (The 500th anniversary of the birth of Nicolas Copernicus), No. 15 (1975) 193-239

18. JB and Z. Stuchlík, *On the latitudinal and radial motion in the field of a rotating black hole*, Bull. Astron. Inst. Czechosl. **27** (1976) 129-133
19. JB and Z. Stuchlík, *The fall of the shell of dust onto a rotating black hole*, Mon. Not. R. Astron. Soc. **175** (1976) 381-393
20. JB and L. Dvořák, *Stationary electromagnetic fields around black holes II. General solutions and the fields of some special sources near a Kerr black hole*, Gen. Rel. Grav. **7** (1976) 959-983
21. JB and L. Dvořák, *Stationary electromagnetic fields around black holes I. General solutions and the fields of some special sources near a Schwarzschild black hole*, Czech. J. Phys. **B27** (1977) 127-147
22. JB, *Stationary interacting fields around an extreme Reissner-Nordström black hole*, Phys. Lett. **64A** (1977) 279-281
23. JB, *On the outside of cold black holes*, Czech. J. Phys. **B28** (1978) 1-7
24. JB, Z. Stuchlík and M. Šob, *Scalar fields around a charged, rotating black hole*, Czech. J. Phys. **B28** (1978) 121-124
25. JB, *Einsteinova cesta k obecné teorii relativity* (Einstein's route to the general theory of relativity - in Czech), Čs. čas. fyz. **A29** (1979) 222-243
26. JB, *On the theories of the interacting perturbations of the Reissner-Nordström black hole*, Czech. J. Phys. **B29** (1979) 945-980
27. JB, *Einstein Prágábán* (Einstein in Prague), Fizikai Szemle **29** (1979) 228-231
28. JB, *Gravitational collapse with charge and small asymmetries II. Interacting electromagnetic and gravitational perturbations*, Gen. Rel. Grav. **12** (1980) 195-204
29. JB, *The motion of a charged black hole in an electromagnetic field*, Proc. Roy. Soc. London **A371** (1980) 429-438
30. JB and L. Dvořák, *Stationary electromagnetic fields around black holes III. General solutions and the fields of current loops near the Reissner-Nordström black hole*, Phys. Rev. **D22** (1980) 2933-2940
31. JB, *Mají Bolzanovy myšlenky vztah k nynější fyzice a kosmologii?* (Have the ideas of Bolzano any relation to the present-day physics and cosmology? - in Czech), in *Bernard Bolzano*, Conf. of Czech math., Zvíkovské podhradí (mimeographed), JČSMF, Prague (1981) 12
32. JB, *Perturbations of the Reissner-Nordström black hole*, in *Proc. of the 2nd M. Grossmann Meeting on Gen. Rel.*, ed. R. Ruffini, North-Holland, Amsterdam (1982) 277-286
33. JB and B.G. Schmidt, *Isometries compatible with gravitational radiation*, J. Math. Phys. **25** (1984) 600-606
34. JB, C. Hoenselaers and B.G. Schmidt, *The solutions of the Einstein equations for uniformly accelerated particles without nodal singularities I. Freely falling particles in external fields*, Proc. Roy. Soc. London **A390** (1983) 397-409
35. JB, C. Hoenselaers and B.G. Schmidt, *The solutions of the Einstein equations for uniformly accelerated particles without nodal singularities II. Self-accelerating particles*, Proc. Roy. Soc. London **A390** (1983) 411-419
36. JB, *The penetration of magnetic field into black holes*, in *10th Int. Conf. on Gen. Rel. and Grav.*, eds. B. Bertotti, F. de Felice and A. Pascolini, Padova (1983) Vol. 2, 688-689

37. JB and B.G. Schmidt, *Isometries compatible with gravitational radiation*, in *10th Int. Conf. on Gen. Rel. and Grav.*, eds. B. Bertotti, F. de Felice and A. Pascolini, Padova (1983) Vol. 1, 29-31
38. JB and B.G. Schmidt, *The boost rotation symmetric solutions*, in *10th Int. Conf. on Gen. Rel. and Grav.*, eds. B. Bertotti, F. de Felice and A. Pascaloni, Padova (1983) Vol. 1, 178-179
39. JB, *Einstein's days and works in Prague*, in *Physics and Prague*, eds. J. Janta and J. Niederle, Union of Czechosl. Math. and Phys., Prague (1984) 59-76
40. JB, *The symmetries of an asymptotically flat space-time and gravitational radiation*, in *Group theoretical methods in physics*, eds. M. Markov, I. Manko and V. Shabad, Harwood Academic Publ., London (1985) 87-94
41. JB and C. Hoenselaers, *Two equal Kerr-Newman sources in stationary equilibrium*, Phys. Rev. **D31** (1985) 2476-2479
42. JB and V. Janiš, *Magnetic fluxes across black holes*, Mon. Not. R. Astron. Soc. **212** (1985) 899-915
43. JB, *Einstein's Prague ideas on gravitation: the history and the present*, in *Trends in Physics*, Proc. of the 6th Gen. Conf. of the European Phys. Soc., eds. J. Janta and J. Pantoflíček, Union of Czechosl. Math. and Phys., Prometheus, Prague (1985) 65-75
44. JB, *On exact radiative solutions representing finite sources*, in *Galaxies, axisymmetric systems and relativity* (Essays presented to W.B. Bonnor on his 65th birthday), ed. M.A.H. MacCallum, Cambridge University Press, Cambridge (1985) 91-124
45. JB and Z. Perjés, *Asymptotic behaviour of Robinson-Trautman pure radiation solutions*, Class. Quantum Grav. **4** (1987) 595-597
46. JB, *Radiative properties of space-times with the axial and boost symmetries*, in *Gravitation and geometry* (A volume in honour of I. Robinson), eds. W. Rindler and A. Trautman, Bibliopolis, Naples (1987) 55-69
47. JB, P. Reilly and J. Winicour, *Boost-rotation symmetric gravitational null cone data*, Gen. Rel. Grav. **20** (1988) 171-181
48. JB and J. Podolský, *The exact models of gravitational waves in the inflationary universe*, in *Evolution of galaxies*, Proc. of the 10th European Regional Astron. Meeting of the IAU, ed. J. Palouš, Publ. of the Astron. Inst. no. 69, Prague (1988) 427-430
49. JB, *Three problems in the analysis of radiative spacetimes*, in *Relativity today*, ed. Z. Perjés, World Scientific, Singapore (1988) 84-93
50. JB, *Exact radiative space-times*, in *Proc. of the 5th M. Grossman Meeting on Gen. Rel.*, eds. D.G. Blair and M.J. Buckingham, World Scientific, Singapore (1989) 309-341
51. JB and V. Karas, *The influence of black holes on uniform magnetic fields*, in *Proc. of the 5th M. Grossmann Meeting on Gen. Rel.*, eds. D.G. Blair and M.J. Buckingham, World Scientific, Singapore (1989) 1199-1207
52. JB, *Einstein's Prague articles on gravitation*, in *Proc. of the 5th M. Grossmann Meeting on Gen. Rel.*, eds. D.G. Blair and M.J. Buckingham, World Scientific, Singapore (1989) 1325-1334

53. JB, *Axisymmetric asymptotically flat radiative space-times with another symmetry: the general definition and comments*, in *Conf. on Math. Relativity*, Proc. of the Centre for Mathem. Analysis, ed. R. Bartnik, Vol. 19, Australian Nat. Univ., Canberra (1989) 1-8
54. JB, Z. Stuchlík and V. Balek, *The motion of the charged particles in the field of rotating charged black holes and naked singularities I. The general features of the radial motion and the motion along the axis of symmetry*, Bull. Astron. Inst. Czechosl. **40** (1989) 65-92
55. JB, V. Balek and Z. Stuchlík, *The motion of the charged particles in the field of rotating charged black holes and naked singularities II. The motion in the equatorial plane*, Bull. Astron. Inst. Czechosl. **40** (1989) 133-165
56. JB and B.G. Schmidt, *On the asymptotic structure of axisymmetric radiative space-times*, Class. Quantum Grav. **6** (1989) 1547-1559
57. JB and B.G. Schmidt, *Asymptotically flat radiative space-times with boost-rotation symmetry: the general structure*, Phys. Rev. **D40** (1989) 1827-1853
58. JB and R. Muschall, *Electromagnetic fields and radiation patterns from multipoles in hyperbolic motion*, in *Relativistische Physik* (E. Schmutzer Festschrift), Wiss. Zeits. der Friedrich-Schiller-Universität Jena **39** (1990) 15-20
59. JB, *Recent rigorous studies in gravitational radiation theory*, in *Selected topics in quantum field theory and mathematical physics*, eds. J. Niederle and J. Fischer, World Scientific, Singapore (1990) 37-54
60. JB, *Is there a news function for an infinite cosmic string?*, Astron. Nachr. **311** (1990) 189-192
61. M.A. Abramowicz and JB, *On the interplay between relativistic gravitational, centrifugal and electric forces: a simple example*, Gen. Rel. Grav. **23** (1991) 941-946
62. JB, *Machian aspects of the present-day cosmology*, in *Ernst Mach and the development of physics*, eds. J. Folta and V. Prosser, Karolinum, Prague (1991) 131-149
63. JB, O. Semerák and P. Hadrava, *Could the Kerr geometry contribute to the collimation of cosmic jets?*, in *Proc. of the 6th M. Grossmann Meeting on Gen. Rel.*, eds. H. Sato and T. Nakamura, World Scientific, Singapore (1992) 674-677
64. JB, *Gravitational radiation from axisymmetric systems and infinite cosmic strings*, in *Proc. of the 6th M. Grossmann Meeting on Gen. Rel.*, eds. H. Sato and T. Nakamura, World Scientific, Singapore (1992) 234-237
65. JB and J. Podolský, *Radiative solutions with cosmological constant and cosmic no-hair conjecture*, in *Relativistic astrophysics and cosmology*, eds. S. Gottlöber, J. P. Mücket and V. Müller, World Scientific, Singapore (1992) 167-171
66. JB, O. Semerák and P. Hadrava, *Collimation effects of the Kerr field*, Mon. Not. R. Astron. Soc. **263** (1993) 545-559
67. JB, D. Lynden-Bell and J. Katz, *Relativistic discs as sources of static vacuum spacetimes*, Phys. Rev. **D47** (1993) 4334-4343
68. JB, D. Lynden-Bell and C. Pichon, *Relativistic discs and flat galaxy models*, Mon. Not. R. Astron. Soc. **265** (1993) 126-144

69. JB and T. Ledvinka, *Sources for stationary axisymmetrie gravitational fields*, Max-Planck-Institut für Astrophysik preprint, MPA 726 (1993) pp 13
70. JB and T. Ledvinka, *Relativistic disks as sources of the Kerr metric*, Phys. Rev. Lett. **71** (1993) 1669-1672
71. JB and J.B. Griffiths, *The scattering and collision of gravitational waves in Friedmann-Robertson-Walker open universes*, Phys. Rev. **D49** (1994) 900-906
72. JB and J. Podolský, *The cosmic no-hair conjecture and black-hole formation: an exact model with gravitational radiation*, Phys. Rev. **D52** (1995) 887-895
73. JB, C. Cris, P. Hájíček and A. Higuchi, *Gauge symmetry breakdown at the horizon of extreme black holes*, Class. Quantum Grav. **12** (1995) 479-498
74. J.B. Griffiths and JB, *Singularity-free stiff fluid cosmologies*, Class. Quantum Grav. **12** (1995) L81-L83
75. D. Lynden-Bell, J. Katz and JB, *Mach's priciple from the relativistic constraint equations*, Mon. Not. R. Astron. Soc. **252** (1995) 150-160; Erratum: Mon. Not. R. Astron. Soc. **277** (1995) 1600
76. JB and J.B. Griffiths, *Gravitational waves propagating into Friedmann-Robertson-Walker universes*, Ann. Phys. (N.Y.) **252** (1996) 180-210
77. O. Semerák and JB, *The interplay between forces in the Kerr-Newman field*, Class. Quantum Grav. **14** (1997) 3135-3147
78. A. Ashtekar, JB and B.G. Schmidt, *Asymptotic structure of symmetry-reduced general relativity*, Phys. Rev. **D55** (1997) 669-686
79. A. Ashtekar, JB and B.G. Schmidt, *Behaviour of Einstein-Rosen waves at null infinity*, Phys. Rev. **D55** (1997) 687-694
80. JB, *Radiative spacetimes: exact approaches*, in *Relativistic gravitation and gravitational radiation*, Proc. Les Houches School of Phys., eds. J.-A. Marck and J.-P. Lasota, Cambridge University Press, Cambridge (1997) 67-87
81. JB and K. Kuchař, *Null dust in canonical gravity*, Phys. Rev. **D56** (1997) 4878-4895
82. JB and J. Podolský, *The global structure of Robinson-Trautman radiative space-times with cosmological constant*, Phys. Rev. **D55** (1997) 1985-1993
83. P. Hájíček and JB, *Gauge-invariant Hamiltonian formalism for spherically symmetric gravitating shells*, Phys. Rev. **D56** (1997) 4706-4719
84. J. Katz, JB and D. Lynden-Bell, *Relativistic conservation laws and integral constraints for large cosmological perturbations*, Phys. Rev. **D55** (1997) 5957-5969
85. JB and V. Pravda, *Curvature invariants in type-N spacetimes*, Class. Quantum Grav. **15** (1998) 1539-1555
86. JB and A. Pravdová, *Symmetries of asymptotically flat electrovacuum spacetimes and radiation*, J. Math. Phys. **39** (1998) 6011-6039
87. JB and O. Semerák, *Interplay between forces in Kerr-Newman field*, in *Proc. of the 8th M. Grossmann Meeting on Gen. Rel.*, ed. T. Piran, World Scientific, Singapore (1999) 401-403
88. JB, J. Katz and D. Lynden-Bell, *Conservation laws for cosmological perturbations and Mach's principle*, in *Proc. of the 8th M. Grossmann Meeting on Gen. Rel.*, ed. T. Piran, World Scientific, Singapore (1999) 277-279

89. T. Ledvinka, M. Žofka, and JB, *Relativistic discs as sources of Kerr-Newman fields*, in *Proc. of the 8th M. Grossmann Meeting on Gen. Rel.*, ed. T. Piran, World Scientific, Singapore (1999) 339-341

90. JB and J. Podolský, *Robinson-Trautman radiative space-times with cosmological constant*, in *Proc. of the 8th M. Grossmann Meeting on Gen. Rel.*, ed. T. Piran, World Scientific, Singapore (1999) 310-312

91. D. Lynden-Bell, J. Katz and JB, *Machian thought experiments*, in *Proc. of the 8th M. Grossmann Meeting on Gen. Rel.*, ed. T. Piran, World Scientific, Singapore (1999) 286-287

92. Z. Stuchlík, JB and V. Balek, *The shell of incoherent charged matter falling onto a charged rotating black hole*, Gen. Rel. Grav. **31** (1999) 53-71

93. J. Katz, D. Lynden-Bell and JB, *Instantaneous inertial frames but retarded electromagnetism in rotating relativistic collapse*, Class. Quantum. Grav. **15** (1998) 3177-3194

94. D. Lynden-Bell, JB and J. Katz, *On accelerated inertial frames in gravity and electromagnetism*, Ann. Phys. (N.Y.) **271** (1999) 1-22

95. JB and B.G. Schmidt, *Self-gravitating fluid shells and their nonspherical oscillations in Newtonian theory*, Astrophys. J. **521** (1999) 708-717

96. JB and A. Pravdová, *Axisymmetric electrovacuum spacetimes with a translational Killing vector at null infinity*, Class. Quantum Grav. **16** (1999) 2023-2041

97. JB and V. Pravda, *Spinning C-metric: radiative spacetime with accelerating, rotating black holes*, Phys. Rev. **D60** (1999) 044004 (10 pages)

98. JB and J. Podolský, *Gravitational waves in vacuum spacetimes with cosmological constant. I. Classification and geometrical properties of non-twisting type N solutions*, J. Math. Phys. **40** (1999) 4495-4505

99. JB and J. Podolský, *Gravitational waves in vacuum spacetimes with cosmological constant. II. Deviation of geodesics and interpretation of non-twisting type N solutions*, J. Math. Phys. **40** (1999) 4506-4517

100. J. Katz, JB and D. Lynden-Bell, *Disc sources for conformastationary metrics*, Class. Quantum Grav. **16** (1999) 4023-4034

101. JB, *Selected solutions of Einstein's field equations: their role in general relativity and astrophysics* in *Einstein's field equations and their physical implications* (Selected essays in honour of Jürgen Ehlers), ed. B.G. Schmidt, Lect. Notes Phys. **540**, Springer-Verlag, Berlin Heidelberg (2000) 1-126

102. JB, *Exact radiative spacetimes: some recent developments*, Ann. Phys. (Leipzig) **9** (2000) 207-217

103. T. Doležel, JB and N. Deruelle, *Slowly rotating voids in cosmology*, Class. Quantum Grav. **17** (2000) 2719-2737

104. JB and T. Ledvinka, *Electromagnetic fields around black holes and Meissner effect*, Nuovo Cim. **B115** (2000) 739-750

105. JB, *Black holes under external influence*, Pramana **55** (2000) 481-496

106. A. Pravdová and JB, *Axisymmetric electrovacuum spacetimes with an additional Killing vector*, in *Proc. of the 9th M. Grossmann Meeting on Gen. Rel.*, World Scientific, Singapore (2001) in print

107. V. Pravda and JB, *Curvature invariants in algebraically special spacetimes*, in *Proc. of the 9th M. Grossmann Meeting on Gen. Rel.*, World Scientific, Singapore (2001) in print
108. JB and P. Krtouš, *Accelerated sources in de Sitter spacetime and the insufficiency of retarded fields*, Phys. Rev. D **64** (2001) 124020 (16 pages)
109. JB, *Radiative spacetimes*, in *Recent developments in general relativity* (Genoa, 2000), eds. R. Cianci, R. Collina, M. Francaviglia and P. Fre, Springer-Verlag, Berlin (2002) in print
110. JB, *Exact solutions and their interpretation* (report of the chair of the A.1 workshop), in *16th Int. Conf. on Gen. Rel. and Grav.* (Durban, 2001), eds. N.T. Bishop and S.D. Maharaj, World Scientific, Singapore (2002) in print

B. Conference abstracts

1. JB, *Gravitational waves from accelerated particles*, in *GR-5*, Theses of the 5th Int. Conf. on Grav. and Theory of Rel., Publ. House Tbilisi Univ. (1968) 79
2. JB, *Perturbations of cold black holes*, in *GR-8*, Abstracts of the 8th Int. Conf. on Gen. Rel. and Grav., Univ. of Waterloo Press, Waterloo (1977) 86
3. JB, *On the coupled perturbations of the Reissner-Nordström metric*, in *GR-9*, Abstracts of the 9th Int. Conf. on Gen. Rel. and Grav., F. Schiller Univ., Jena (1980) Vol. 1, 238-239
4. JB, *The boost-rotation symmetric solutions*, in *Proc. of the Seminar on Rel. Theories of Grav.* (mimeographed), F. Schiller Univ., Jena (1982) 35
5. JB, *Radiative space-times with boost-rotation symmetry*, in *GR-11*, Abstracts of the 11th Int. Conf. on Gen. Rel. and Grav., Stockholm (1986) 470
6. JB and V. Karas, *The influence of black holes on uniform magnetic fields*, in *Abstracts of the 5th M. Grossmann Meeting on Gen. Rel.*, eds. B. Lohmann, A. Peakall and P. J. Veitch, Univ. of Western Australia, Perth (1988) 140
7. JB, *Einstein's Prague articles on gravitation*, in *Abstracts of the 5th M. Grossmann Meeting on Gen. Rel.*, eds. B. Lohmann, A. Peakall and P. J. Veitch, Univ. of Western Australia, Perth (1988) 189
8. JB and B.G. Schmidt, *On axisymmetric radiative space-times*, in *GR-12*, Abstracts of the 12th Int. Conf. on Gen. Rel. and Grav., eds. N. Ashby and P. Bender, Univ. of Colorado, Boulder (1989) Vol. 1, 219
9. JB, *From the vicinity of a rotating black hole*, in *GR-12*, Abstracts of the 12th Int. Conf. on Gen. Rel. and Grav., eds. N. Ashby and P. Bender, Univ. of Colorado, Boulder (1989) Vol. 2, 403
10. JB and J. Podolský, *On the interpretation of the Robinson-Trautman solutions with cosmological constant*, in *GR-13*, Abstracts of the 13th Int. Conf. on Gen. Rel. and Grav., eds. P.W. Lamberti and O.E. Otiz, Univ. of Cordoba, Cordoba (1992) 12
11. JB, *CD's as sound sources of the Kerr and other stationary spacetimes*, in *GR-14*, Abstracts of the 14th Int. Conf. on Gen. Rel. and Grav., SIGRAV, Florence (1995) A16

12. JB and J. Podolský, *The cosmic no-hair conjecture and black-hole formation under the presence of gravitational waves*, in *GR-14*, Abstracts of the 14th Int. Conf. on Gen. Rel. and Grav., SIGRAV, Florence (1995) A17
13. JB, J. Katz and D. Lynden-Bell, *Mach's principle in closed universes*, in *GR-14*, Abstracts of the 14th Int. Conf. on Gen. Rel. and Grav., SIGRAV, Florence (1995) B66
14. JB, T. Ledvinka, O. Semerák and Z. Stuchlík, *Some aspects of motion in the Kerr and Kerr-Newman fields*, in *GR-14*, Abstracts of the 14th Int. Conf. on Gen. Rel. and Grav., SIGRAV, Florence (1995) B8
15. JB, J. Katz and D. Lynden-Bell, *Conservation laws and evolution of cosmological perturbations*, in *Abstr. of the 7th JENAM Conf.*, ed. J. Palouš, EAS&CAS, Prague (1998) 272
16. JB, V. Pravda and A. Pravdová, *Boost-rotation symmetric spacetimes as test beds in relativistic astrophysics*, in *Abstr. of the 7th JENAM Conf.*, ed. J. Palouš, EAS&CAS, Prague (1998) 271
17. T. Ledvinka, M. Žofka and JB, *Energy conditions for disk-like sources of Kerr and Kerr-Newman spacetimes*, in *Abstr. of the 7th JENAM Conf.*, ed. J. Palouš, EAS&CAS, Prague (1998) 288
18. D. Lynden-Bell, J. Katz and JB, *Machianly induced inertia within a massive shell*, in *GR-15*, Abstr. of Plenary Lectures and Contrib. Papers of the 15th Int. Conf. on Gen. Rel. and Grav., IUCAA, Pune (1998) 5
19. JB and P. Hájíček, *Hamiltonian formalism for spherical shells collapsing on black holes*, in *GR-15*, Abstr. of Plenary Lectures and Contrib. Papers of the 15th Int. Conf. on Gen. Rel. and Grav., IUCAA, Pune (1998) 244
20. JB and K. Kuchař, *Hamiltonian formalism for null dust in general relativity*, in *GR-15*, Abstr. of Plenary Lectures and Contrib. Papers of the 15th Int. Conf. on Gen. Rel. and Grav., IUCAA, Pune (1998) 241
21. JB, P. Krtouš, V. Pravda and A. Pravdová, *Boost-rotation symmetric radiative spacetimes and fields: post-GR15 developments*, in *GR-16*, Abstr. of Plenary Lectures and Contrib. Papers of the 16th Int. Conf. on Gen. Rel. and Grav., Durban (2001) 19

C. Books, book translations and epilogues

1. JB, *Einstein a Praha* (Einstein and Prague - in Czech, parts in German), JČSMF, Praha (1979) pp 64
2. JB and V.N. Rudenko, *Teorie relativity a gravitační vlny* (The theory of relativity and gravitational waves - in Czech), Universita Karlova, Praha (1986) pp 314
3. JB and V.N. Rudenko, *Gravitacionnyje vol'ny v obshtchej teorii otnositel'nosti i metody ich detekcii* (Gravitational waves in general relativity and the methods of their detection - in Russian), Moscow University Press, Moscow (1987) pp 270

4. JB and J. Klíma, *Pan Tompkins v říši divů*, Czech translation of the book *Mr. Tompkins in paperback* by G. Gamow, with commentaries *Co by mohl profesor dodat dnes* (What could professor add today) and epilogue *George Gamow a jeho pan C.G.H. Tompkins* (George Gamow and his Mr. C.G.H. Tompkins), Mladá fronta, Praha (1986) pp 232
5. JB, *Mr. X*, epilogue of the book *To přece nemyslíte vážně, pane Feynmane!* (Surely you are joking, Mr. Feynman!) by R. Feynman, Mladá fronta, Praha (1989) 311-316
6. JB, *Krátce o Stephenu Hawkingovi a jeho pohledu na svět* (Briefly on Stephen Hawking and his view of the universe - in Czech), epilogue of the book *Stručná historie času* (A brief history of time) by S. Hawking, Mladá fronta, Praha (1991) 181-187; 2nd edtition Praha (1997)
7. JB, *Víra v rozum a spiritus loci* (Trust in reason and spiritus loci), epilogue of the book *Snění o finální teorii* (Dreams of a final theory) by S. Weinberg, Hynek, Praha (1996) 239-247
8. JB, *Prostoročasová setkání Rogera Penrose a Stephena Hawkinga* (Spacetime meetings of Roger Penrose and Stephen Hawking), foreword of the book *Povaha prostoru a času* (The nature of space and time) by R. Penrose and S. Hawking, Academia, Praha (2000) 7-10
9. A. Bičáková and JB, *Moje světočára - neformální autobiografie*, Czech translation of the book *My worldline. An informal autobiography* by G. Gamow, with epilogue *Kolem světočáry George Gamowa* (Around the worldline of George Gamow), Mladá fronta, Praha (2000) pp 160

D. Review and pedagogical publications

1. JB, *Záření černého tělesa z raných stádií ve vývoji vesmíru* (Blackbody radiation from the early stages of the universe - in Czech), Čs. čas. fyz. **A18** (1968) 725-727
2. JB, *Objev černé díry v naší galaxii?* (The discovery of a black hole in our Galaxy? - in Czech), Čs. čas. fyz. **A22** (1972) 107-109
3. JB, *Struktura a vývoj vesmíru* (The structure and evolution of the universe - in Czech), Čs. čas. fyz. **A24** (1974) 425-445
4. JB, *Kosmologické koincidence* (The cosmological coincidences - in Czech), Čs. čas. fyz. **A24** (1974) 507-508
5. JB, *Paradox rychlého muže na mříži* (The paradox of a fast man on a lattice - in Czech), Čs. čas. fyz. **A24** (1974) 519; Čs. čas. fyz. **A25**(1975) 200
6. JB, *Nová relativistická laboratoř - pulsar ve dvojhvězdě* (A new relativistic laboratory - the pulsar in a binary system - in Czech), Čs. čas. fyz. **A25** (1975) 628-629
7. JB, *Gravitace a vesmír* (Gravitation and the universe - in Czech), in *Sborník 12. celostátní konference o matematice na VŠTEZ* (mimeographed), JČSMF, Žilina (1977) 10-12

8. JB, *Fyzikální procesy kolem gravitačně zkolabovaných objektů* (Physical processes around gravitationally collapsed objects - in Czech), in *Sborník referátů k 25. výročí MFF UK*, část II - fyzika, ed. L. Eckertová, Univerzita Karlova, Praha (1979) 145-152
9. JB, *Role Einsteinovy obecné teorie relativity v současné fyzice* (The role of Einstein's general theory of relativity in contemporary physics - in Czech), in *Sborník VI. konference čs. fyziků*, Part 2 - souborné referáty, JČMF, Praha (1980) 265-277
10. L. Dvořák, JB et al., *Výklad musí být především fyzikálně správný* (Explanation must above all be physically correct - in Czech), Pokroky MFA **XXXI** (1986) 52-54
11. JB, *Hmotnost a energie před Einsteinem a po něm* (Mass and energy before and after Einstein - in Czech), in *Pedagogicko-fyzikální dialogy 5*, ed. M. Černohorský, Brno (1986) 16-31
12. JB, *Kosmická cenzura (komenář k problémm B. Simona)* (Cosmic censorship (commentary on the problems by B. Simon) - in Czech), Čs. čas. fyz. **A38** (1988) 300-302
13. JB, *Binární pulzary a základní fyzika* (Binary pulsars and fundamental physics - in Czech), Čs. čas. fyz. **A43** (1993) 224
14. JB, *Kosmologie a vyvratitelnost* (Cosmology and refutability - in Czech), Universum, č. 11 (1993)
15. JB, *Črta k Machovu principu* (Sketch on Mach's principle - in Czech), Čs. čas. fyz. **A48** (1998) 347-348
16. JB, *Poznání a údiv včera, dnes a zítra: očima fyzika* (Knowledge and wonder yesterday, today and tomorrow: through the eyes of a physicist - in Czech), Vesmír **78** (1999) 146-151
17. JB, *Zářivé prostoročasy* (Radiative spacetimes - in Czech), Čs. čas. fyz. **A50** (2000) 249-251

E. Miscellanea

1. Articles on encounters with physicists (e.g. with S. Chandrasekhar, E.M. Lifshitz, K.S. Thorne), interviews on science and related issues (with, e.g., W.B. Bonnor, R. Penrose, A. Trautman, V. Votruba, J.A. Wheeler), and reports on international conferences, published (in Czech) in Čs. čas. fyz. A in the years 1969-89
2. Articles and interviews on science in Czech journals and newspapers, published mostly after 1990
3. More than 50 book reviews published both in the Czech and international journals, including, for example, the review of *The mathematical theory of black holes* by S. Chandrasekhar (Gen. Rel. Grav. **17** (1985) 615-619), of *A first course in general relativity* by B. Schutz (Class. Quantum Grav. **2** (1985) 798-799), or of *Black holes: the membrane paradigm* by K.S. Thorne, R.H. Price and D.A. MacDonald (Class. Quantum Grav. **5** (1988) 533-534)

Contents

SLOWLY ROTATING VOIDS IN COSMOLOGY AND MACH'S PRINCIPLE

TOMÁŠ DOLEŽEL

16, Rue René Jacques, Issy-Les-Moulineaux, 92130, France

In this yet another contribution to the still continuing Machian discussion we first summarize the history and meaning of Mach's principle. Then a specific model allowing to discuss Mach's ideas in a cosmological context is treated: we consider a spacetime consisting of an empty void separated from an almost Friedmann-Lemaître-Robertson-Walker (FLRW) dust universe by a spherically symmetric slowly rotating shell which is comoving with the cosmic dust. We treat in a unified manner all types of the FLRW universes. The metric is expressed in terms of a constant characterizing the angular momentum of the shell and parametrized by the comoving radius of the shell. Assuming the rotation to be a first order perturbation, we compute the dragging of inertial frames as well as the apparent motion of distant stars within the void. Finally, we discuss the Machian features of the model in terms of quantities that are measurable in principle.

1 A brief history of Mach's principle[a]

The debate about motion and its character (absolute or relative?) goes back to the ancient Greeks. In the view of Aristotle bodies remain at rest unless they are forced to move and forces produce velocities according to $F = mv$. The velocities v are measured with respect to the Earth, thus understood as relative. The Earth provides an absolute element in the theory: the fundamental rest frame is attached to the Earth.

Copernicus defined his reference frame in the following way:

The first and highest of all is the sphere of the fixed stars, which contains itself and everything, and is therefore immovable. It is unquestionably the place of the universe, to which the motion and position of all the other heavenly bodies are compared.

Kepler was strongly convinced that the planets could not follow such precise orbits in the empty space if they were not "guided" by the distribution of real masses in the universe, by the Sun, and the continuous sphere of distant stars.

Galileo, on the other hand, did not need to use the notion of distant stars, he believed in motion relative to *space.* In this context he was the first to employ the notion of the *absolute motion.* On the contrary, Descartes defended a strictly 'relativistic' position: all motion is relative, and everything is in motion.

Newton disagreed profoundly with Descartes' rationalism. He was not able to formulate the laws of dynamics without a rigid framework — *the absolute space.* Unlike his contemporary Leibniz who understood absolute space as a mere abstraction or convenience introduced in order to simplify calculations and to determine

[a] This section draws upon the exhaustive book [1] on Mach's principle and its status. The citations from Copernicus, Mach and Einstein (their English translations, respectively) are taken from that work. A brief history of Mach's principle is reviewed, for instance, in the introduction of Lynden-Bell, Katz and Bičák's paper [19].

relative motion of different bodies, Newton thought that absolute space was a part of physical reality. According to Newton, absolute rotation, i.e. rotation with respect to absolute space, and more generally absolute acceleration can be detected without any reference to astronomical observations. Newton argued employing the famous bucket experiment. When a test body rotates it is acted upon by inertial centrifugal forces. The presence of inertial forces seems to be completely independent of whether the test body rotates with respect to bodies placed in the immediate vicinity. This is what Newton observed: the centrifugal forces induce a concavity of the surface of the rotating body of water independently of whether the water rotates with respect to the bucket or not. Hence, it seems to be possible to determine the acceleration absolutely — without any reference to surrounding bodies. Inertial forces act on bodies accelerating with respect to the absolute space.

Mach's role was to point out (looking back to Aristotle, Copernicus and Kepler) that this 'absolute rotation' is in fact relative with respect to other masses in the universe. In 1883 he wrote in his *Mechanik* [3] (p. 216):

Newton's experiment with the rotating bucket teaches only that the rotation of the water relative to the sides of the vessel does not include noticeable centrifugal forces, but that such forces are induced by its rotation relative to the mass of the Earth and other celestial bodies. Nobody can say how the experiment would turn out if the sides of the vessel increased in thickness and mass till they were ultimately several leagues thick. The one experiment lies before us, and our business is to bring it into accord with the other facts known to us, and not with the arbitrary fictions of our imagination.

Another important remark in Mach's *Mechanik* traces back to Descartes stating that the right mechanics should be based on a purely relative description of all physical phenomena in a cosmological context:

The universe is not twice given, with an Earth at rest and an Earth in motion, but only once with its relative motions only determinable. Therefore we cannot say what would happen if the Earth did not rotate. We may interpret the one case that is given us in different ways. If however, we so interpret it that we come into conflict with experience, our interpretation is simply wrong. The principles of mechanics can presumably be so conceived that centrifugal forces arise also for relative rotations...

...No one is competent to predicate things about absolute space and absolute motion; they are pure things of thought, pure mental constructs, that cannot be produced in experience. All our principles of mechanics are, as we have shown in detail, experimental knowledge concerning the relative positions and motions of bodies. Even in the provinces in which they are now recognized as valid, that could not, and were not, admitted without previously being subject to experimental tests. No one is warranted in extending these principles beyond the boundaries of experience. In fact, such an extension is meaningless, as no one possesses the requisite knowledge to make use of it.

These passages form the basis of the discussion that has been ever since continuing under the name of Mach's principle (the name coined first apparently by A. Einstein). They summarize two distinctive themes. The first is that the goal of physical description is the simplest abstract expression of facts coming only from experience (elsewhere Mach says more explicitly that *physics is experience, arranged in economical order* [2]). The second one is that Newton's absolute space, acceleration, time etc. do not correspond to this goal. Newton's bucket experiment does not imply any existence or need of these notions. The only thing that one should conclude from this experiment is a mutual relationship between two observations, the presence of inertial centrifugal forces and rotation with respect to the stars. (Today's observations reveal that, remarkably, local inertial frames near the Earth coincide with the frame of the most distant galaxies and quasars with the accuracy $2.5 \times 10^{-4} arcsec/year$[b].) Hence, according to Mach, Newton inferred far more than he should have done and, on the other hand, an alternative mechanism for the generation of inertial forces, which is the rotation with respect to other bodies (since relative motions are only determinable), is not ruled out by such an experiment.

Nevertheless, 'Mach's principle' has never been clearly stated as a principle either by Mach[c] himself or by others. It was Mach's disciples and his later readers (namely Einstein) who extracted some proposals from his writings, unfortunately rather vague and ambiguous. The only thing that seems to be clearly established is the criticism towards Newton.

Although Einstein's attitude to Mach's ideas changed in the course of years, there is no doubt that he was very influenced by Mach's criticism of Newton while constructing the theory of general relativity. When Einstein arrived at the conclusion that the principle of relativity should be extended to all moving reference frames and that gravitation would be described by the metric tensor of a curved spacetime (1912), he writes: *This makes it plausible that the entire inertia of a point mass is the effect of the presence of all other masses, deriving from a kind of interaction with the latter* [4].

In the following years several different formulations of Mach's principle can be found in the writings of Einstein. First, there are repeated expressions of the type 'The metric $g_{\mu\nu}$ should be completely determined by the material distribution described by the stress-energy tensor $T_{\mu\nu}$'.

In 1916, when Einstein realized that the field equations alone do not satisfy this demand, he reformulated the Machian problem as a problem of boundary conditions. The boundary conditions should complete the field equations in order to exclude the solutions that Einstein felt contradictory to Mach's ideas. For instance, Minkowski spacetime and Schwarzschild's solution both violate Mach's ideas because inertial forces cannot be caused by material content and its relative motion. The boundary conditions imposed in order to derive these solutions Einstein considered as 'absolutist' and 'non-Machian'.

In 1917 Einstein made a further step by replacing the boundary condition formulation by the demand of a *closed universe* [5]. His understanding of Mach can

---

[b]Mentioned in [1], page 325.

[c]Paradoxically enough, Mach would probably find the formulation of such a principle contradictory to his goal of economical description.

be expressed verbally by the following two requirements:

1. The universe should be spatially finite and closed, i.e. have no boundary region. Then the absence of boundary conditions assures that the metric is completely determined by the matter distribution in spacetime.

2. The modified[d] field equations should not allow matter-free, singularity-free solutions. This is necessary to ensure that the theory as a whole (not just a subset of solutions) is of Machian character.

The requirement 1 can only hold if the spacetime is not empty which implies the importance of the requirement 2. However, the introduction of the cosmological constant Λ failed to satisfy this condition. De Sitter found a solution to the modified field equations with $T_{\mu\nu} = 0$ the same year and Einstein spent over a year trying to show that the de Sitter solution was for some reasons physically unacceptable. After the failure of these efforts Einstein gave up the struggle and in the 1918 paper on the general theory of relativity 'Prinzipielles zur algemeinen Relativitätstheorie' [6] he expresses his current understanding of Mach's principle (for the first time he uses this term explicitly) as:

Mach's principle: The $g_{\mu\nu}$ field is without remainder determined by the masses of bodies. Since mass and energy are, according to results of the special theory relativity, the same, and since energy is formally described by the symmetric energy tensor ($T_{\mu\nu}$), this therefore entails that the $g_{\mu\nu}$ field should be conditioned and determined by the energy tensor.

Although Einstein's enthusiasm for Mach's principle had begun to decline, he did not cease to discuss Mach's principle — without any attempt to mathematically formulate or precisely state the principle. A semi-popular discussion in the textbook 'The meaning of relativity' is typical:

The theory of relativity makes it appear probable that Mach was on the right road in his thought that inertia depends upon a mutual action of matter.... What is to be expected along the line of Mach's thought?

1. *The inertia of a body must increase when ponderable masses are piled up in its neighbourhood.*

2. *A body must experience an accelerating force when neighbouring masses are accelerated, and, in fact, the force must be in the same direction as the acceleration.*

3. *A rotating hollow body must generate inside of itself a 'Coriolis field', which deflect moving bodies in the sense of the rotation, and a radial centrifugal forces as well.*[e]

...Although all these effects are inaccessible to experiment, because κ is so small, nevertheless they certainly exist according to the general theory of relativity. We must see in them a strong support for Mach's ideas as to the relativity of all inertial actions. If we think these ideas consistently through to the end we must expect that the whole inertia, that is, the whole $g_{\mu\nu}$ field, to be determined by the matter of the universe, and not mainly by the boundary conditions at infinity.

---

[d] A modification by the cosmological constant term Λ.

[e] Bodies, and more generally inertial frames, subject to these induced 'forces' are said to be dragged. Hence the term 'dragging' for all effects of this type.

Einstein seems to have accepted the attitude that the general theory of relativity fulfills Mach's ideas on the base of intuition without any mathematical criterion.

Due to the fact that Mach does not propose any definite ansatz for the mechanism of inertia induction by cosmic masses, Mach's principle says more about other author's reading of Mach than about Mach's original intention. The following list summarizes (without any pretension to be complete) the different understandings of Mach's principle treated in the literature[f]: Mach's principle was understood as

- problem of defining the velocity (motion)
- proposition that overall rigid rotations and translations of a system are unobservable
- determination of inertial frames from relative motion of masses
- mechanical interaction of masses, in particular through an interactive Machian 'kinetic energy'
- generation of inertial forces in any body accelerated with respect to distant masses
- proposition that an isolated body in an otherwise empty space has no inertia
- requirement that if you take away all matter, there is no more space
- induction of inertial forces by accelerated masses by analogy with electromagnetic induction
- requirement that the metric tensor should be completely determined by matter
- prediction of the future when relational initial data are given (initial-value approach)
- cosmic derivation of the inertial mass
- generalization of the special principle of relativity to nonuniform motions
- requirement of general covariance of the laws of motion
- need for generally covariant boundary conditions in general relativity
- need for nonexistence of boundary conditions (spatial closure of the universe)
- requirement that the total energy, angular and linear momentum of the universe should be zero
- requirement that there should be no matter-free, singularity-free solutions of general relativity.

[f]The list is taken from [1], page 530, and completed by another one in [7].

- appeal to the principle of sufficient reason (observable effects must have observable causes).
- expectation that there will be dragging effects.
- explanation of the fact that the compass of inertia does not rotate relative to distant masses
- proposition that the universe at large influences local physics (Dicke-type approach).
- selection principle of solutions that are intuitively 'Machian'.
- requirement that dynamics should not contain absolute elements.
- proposition that Newton's gravitational constant G is a dynamical field.

Hence the use of the terms 'Machian' and 'anti-Machian' is rather confusing, crucially depending on the formulation adopted. For example, as indicated in [7], the Lense-Thirring effect, mentioned in the next section, can be considered both Machian and anti-Machian. Aware of the controversies connected with the formulation and interpretation of Mach's principle, we apply a kind of 'Occam razor' and return to Mach's observation that the universe, as represented by the average motion of distant galaxies, does not appear to rotate relative to local inertial frames. This means that in any universe model, the totality of matter should not be *observed* by inertial observers to be rotating in a preferred direction. As a consequence, we expect that there will be dragging effects near rotating bodies. More particularly, inertial frames inside a rotating massive shell are expected to be dragged into corotation so that in the event that the shell models all cosmic matter, its rotation would not be *observable.*

2 Dragging effects near and inside rotating shells

The model of an infinitely thin rotating, spherical mass shell was first used by Einstein in his quest for a new type of 'force' that he expected would cause dragging (by analogy with electromagnetic induction). In the framework of a preliminary scalar gravitation theory he first showed that the translational acceleration of the mass shell induces an acceleration (dragging) of a mass point at its centre (1912) [4]. One year later he showed, now within a preliminary tensorial theory, that a rotating mass shell induces a Coriolis-type dragging force on an inertial frame inside the shell.

The first dragging result within general relativity came from Hans Thirring (1918). In his classical papers [8], Thirring reconsidered Einstein's model of the rotating mass shell and calculated dragging effects in Minkowski spacetime. To the first order in the shell mass M (i.e., for shells of small masses), and to the first order in its angular velocity $\mathbf{\Omega}$, Thirring found the Coriolis-type force[g] inside the

[g]Thirring considered also the second order in $\mathbf{\Omega}$ and found a centrifugal-type force. The calculations were done using Einstein's perturbation scheme in Minkowski spacetime.

shell in the form

$$\mathbf{F}_C = -\frac{8GMm}{3R} \left(\mathbf{\Omega} \times \mathbf{v} \right), \tag{1}$$

where m is a test mass passing through the centre of the shell of radius R with velocity $\mathbf{v}$. Inertial effects inside the shell thus behave as R^{-1}. In the following paper [9], Lense and Thirring did similar calculations far outside the shell, $r >> R$, to get the 'Lense-Thirring effect': inertial frames far outside the shell are dragged with an angular velocity (near the axis of rotation of the shell)

$$\omega_{LT} = \frac{4GMR^2}{3r^3} \mathbf{\Omega}. \tag{2}$$

This result has been discussed in connection with various types of experiments over the years — at present a series of LAGEOS experiments involving satellites in suitable orbits around the Earth is in progress (Ciufolini [1], page 386). However, the order of magnitude of Lense-Thirring effect does not allow to reach a sufficient accuracy for a clear positive conclusion of the experiments.

The results of Thirring were generalized to an arbitrary shell mass M by Brill and Cohen (1966) [10], who considered a slowly rotating shell separating an empty void from the Schwarzschild solution to linear order in the rotational perturbation[h]. The rotation of the inertial frames induced by a rotating massive shell is calculated for both small masses (here the result of Thirring is recovered) and large masses. In the limiting case, when the shell radius approaches the Schwarzschild radius, the induced rotation approaches the rotation of the shell and the shell is thus at rest with respect to interior inertial observers[i], i.e., dragging becomes *perfect*. Hence, the shell completely *screens out* the influence of the asymptotic Minkowski frame and inertial properties inside the shell depend only on its matter (in general, behaviour of the interior depends on the features of the asymptotic region through the boundary condition of asymptotic flatness used in the derivation of the induced rotation). Brill and Cohen note that in this sense the result explains why *the 'fixed stars' are indeed fixed*, i.e., do not rotate with respect to our inertial frame. Nevertheless, as pointed out later by Cohen (1968) [11], a stationary mass at the Schwarzschild radius has to be considered unphysical since the stresses in the shell diverge as the shell approaches the horizon.

Lindblom and Brill (1974) [12] extended this work by treating a freely falling, slowly rotating shell of dust. They found that in this dynamic model the induced rotation of inertial frames does also reach the shell rotation (both as seen from infinity) as the shell collapses towards the Schwarzschild radius. However, interior observers would observe a nonzero angular velocity of the shell even in this extreme case. Hence, dragging is never perfect, i.e., the inertial frames inside the shell remain under influence of spacelike infinity. In addition, in order to interpret in physical terms the rotational perturbation inside the shell, they studied the

[h] In the Poisson gauge, this is described by the off-diagonal '$dt\, dx^i$' metric coefficient.

[i] Since the asymptotic flatness provides an absolute fixed point, all rotations are related to observers at infinity. Although the fact that both angular velocities approach the same value as seen from infinity does not imply that the shell is at rest for interior inertial observers, it remains true in the case of a static (non-collapsing) shell treated by Brill and Cohen. The contrary is true for a collapsing shell considered by Lindblom and Brill [12].

apparent motion of a 'searchlight' at the centre of the shell as seen from infinity. They concluded that the rotational perturbation, i.e. the off-diagonal '$dt\,dx^i$' metric coefficient, corresponds to the angular velocity of the searchlight observed from infinity. Dragging thus acquires a kind of absolute character.

More recently, Katz, Lynden-Bell and Bičák (1998) [13] solved the opposite problem (which is more interesting from the Machian point of view) of how the points fixed at infinity ('fixed stars') in the asymptotic Minkowski frame appear to rotate to an inertial observer at the centre of the shell. As expected, if the shell is seen to be rotating *forwards*, the asymptotic flat infinity rotates *backwards*.

An evident drawback of these Schwarzschild-shell models is the asymptotic flatness incorporated as a boundary condition. The whole universe is described by a shell in an asymptotically flat spacetime so that all rotations are related to the inertial frames at infinity. The rotation of all inertial frames inside the shell is therefore determined, at least in part, by these asymptotic frames that play the rôle of something given absolutely, and the shell then cannot be considered to be the only source of inertia in the universe. Hence it is necessary to study dragging effects in a cosmological context — in agreement with Mach's notice ... *The universe is not twice given...* which confirms that Mach expected a purely relative description of all physical phenomena only within a realistic cosmological model.

3 Dragging effects in cosmological models

The mathematically simplest model, the static spatially closed Einstein cosmos, was treated by Lausberg in 1969 [14]. He perturbed the model by the rigid-body rotation of a shell with finite thickness and came to a rather expected result: perfect dragging is achieved when the rigidly rotating shell represents the whole universe.

Lewis (1980) [15] was the first to discuss dragging in a truly cosmological context, in the framework of an expanding, spatially closed $k = +1$ FLRW universe model. A rotating shell separates a Minkowski interior from the outer universe. He gave the solution for the rotational perturbation in the first order and discussed, somewhat vaguely, some Machian features of the model. Dragging of interior inertial frames is found to depend both on the properties of the shell and cosmic matter which is claimed to be in accordance with the spirit of Mach's principle.

Later Klein (1993) [16] analysed in more detail a slowly rotating shell immersed at a given comoving position into a FLRW spacetime with open flat spatial sections ($k = 0$). He discussed dragging of inertial frames inside the shell under the condition that the rotational perturbation is caused by the shell only. He found that perfect dragging is reached only as the radius of the shell goes to infinity when the totality of matter in the universe is redistributed on the shell. In his Ph.D. thesis (1993) [17], Klein gave a solution for the rotational perturbations in the models with all hyperbolic, flat and spherical spatial sections ($k = -1, 0, 1$), and pointed out the non-uniqueness of the boundary conditions determining the motion of cosmic dust in the closed case. In the following paper (1994) [18], Klein also considered some effects caused by slowly rotating shells in $k = 0$ universes to the second order in angular velocity.

A large step from the work of Lindblom and Brill [12] towards a more general cosmological context was done by Lynden-Bell, Katz and Bičák (1995) [19]. They considered a system of freely falling spheres slowly rotating about a common axis[j]. When the continuously distributed spheres fill the whole spacetime one obtains a rotationally perturbed Lemaître-Bondi-Tolman universe. The paper provides a fine Machian discussion with the following results. By analogy with the asymptotically flat models, infinity in *open* universes provides a boundary condition on the rotational perturbation and *gives an absolute rotation.* However, the *closed* model reveals nice Machian features. First, when the universe is closed, dragging of inertial frames at each point of the spacetime is determined *uniquely* by the matter distribution. For FLRW universes this comes from the constraint equations of general relativity. Second, remarkably enough, close spaces with less and less mass do not approach a flat spacetime. Instead, they have smaller and smaller radii of expansion, with shorter and shorter time of existence. In the limit of zero mass, the universe is a point existing for one instant. *Thus the conundrum of inertia in almost empty nearly flat spaces disappears if the space is closed.*

The work presented in the next chapters has been done in collaboration with Jiří Bičák and Nathalie Deruelle [20] (2000). We extend the work of Klein [17], [16] and study some new aspects of the results of Lynden-Bell, Katz and Bičák (1995) [19]. We analyse a rotating shell separating a Minkowski interior from a FLRW dust exterior in a unified manner for all $k = -1, 0, +1$ cases[k]. We give the spacetime metric as well as all angular velocities in terms of two parameters — of a constant characterizing the angular momentum of the shell and of the comoving radius of the shell. We calculate the dragging of inertial frames within the void, determine the conditions for perfect dragging to occur and illustrate that perfect dragging cannot be achieved in a closed universe. In accordance with [19], open universes give an absolute rotation in the sense that one is allowed to define a preferred coordinate system (in which the rotational perturbation vanishes at infinity) with respect to which the dragging of interior inertial frames can be related. In the closed case all motions are relative since there is no such system.

We stress the observational character of Mach's ideas. Our Machian discussion is therefore based on quantities measurable in principle. We analyse, in the spirit of Katz, Lynden-Bell and Bičák (1998) [13], the apparent motion of the shell and of the distant stars as seen by inertial observers inside the void. As far as we are aware, this problem has not been tackled before in a cosmological context. We conclude that cosmic matter (the shell and cosmic dust) of the models considered is indeed never *observed* to be globally (slowly) rotating in a preferred direction. Inertial frames inside the void are in this sense determined by the 'average motion of distant galaxies' - in conformity with Mach's ideas.

4 Spacetime metric

The void is empty so that spacetime inside the void can be represented by a part of Minkowski space. Consider a family of standard clocks at rest with respect to

---

[j] The case of "misaligned rotations" is also treated.

[k] Recent observations of voids with a size not negligible with respect to the Hubble radius [21] suggest that these models do not have to be purely academic.

each other — they define an inertial frame S_{in} with the line element

$$ds^2|_{in} = -d\bar{t}^2 + d\bar{r}^2 + \bar{r}^2(d\theta^2 + \sin^2\theta\, d\bar{\phi}^2), \tag{3}$$

where $\bar{t}$ is the time measured by the clocks and $(\bar{r}, \theta, \bar{\phi})$ their spherical coordinates.

Now one can use another time coordinate t such that $d\bar{t} = \alpha(t)dt$ where the lapse function $\alpha(t)$ is to be specified later. One can also use coordinates which rotate with respect to $(\bar{r}, \theta, \bar{\phi})$ in such a way that $d\bar{\phi} = d\phi - \tilde{\omega}_0(t)dt$ where the angular velocity $\tilde{\omega}_0(t)$ is a function of t. In this rotating frame S_{rot} with coordinates $(t, \bar{r}, \theta, \phi)$ the line element (3) up to the first order in $\tilde{\omega}_0(t)$ reads

$$ds^2|_{in} = -\alpha^2(t)dt^2 + d\bar{r}^2 + \bar{r}^2(d\theta^2 + \sin^2\theta\, d\phi^2) - 2\bar{r}^2\,\tilde{\omega}_0(t)\sin^2\theta\, d\phi\, dt. \tag{4}$$

The shell is supposed to be spherically symmetric. Its equation of motion can thus be written as $\bar{r} = l(t)$; $l(t)$ is for the moment an arbitrary function of time. The metric induced on the shell by (4) is then

$$ds^2|_{in}^{\Sigma} = -[\alpha^2(t) - \dot{l}^2(t)]dt^2 + l^2(t)(d\theta^2 + \sin^2\theta\, d\phi^2 - 2\tilde{\omega}_0(t)\sin^2\theta\, d\phi\, dt), \tag{5}$$

where $\dot{l} \equiv dl/dt$ and Σ denotes the shell.

Outside the void, the universe is supposed to be almost perfectly homogeneous and isotropic up to a rotational perturbation. Therefore, there exists a coordinate system in which the line element reads

$$ds^2|_{out} = -dt^2 + a^2\left[\frac{dr^2}{1-kr^2} + r^2(d\theta^2 + \sin^2\theta\, d\phi^2) - 2r^2\,\omega(t,r)\sin^2\theta\, d\phi\, dt\right]. \tag{6}$$

As usual, $a(t)$ is the scale factor and the function $\omega(t,r)$ is the rotational metric perturbation. For the perturbation theory of the first order to apply it must be 'small', that is $r\,\omega(t,r) \ll 1$.

We will also assume that the shell is comoving with the cosmic dust, thus placed at a constant comoving radius $r = l_0$. The metric induced on such a shell by (6) reads

$$ds^2|_{out}^{\Sigma} = -dt^2 + a^2(t)l_0^2\left[d\theta^2 + \sin^2\theta\, d\phi^2 - 2\omega_0(t)\sin^2\theta\, d\phi\, dt\right], \tag{7}$$

where $\omega_0(t)$ is the value of the rotational perturbation on the shell: $\omega_0(t) \equiv \omega(t, r = l_0)$.

Let us now complete the matching of the two coordinate grids on the shell. The induced metrics (5) and (7) must be continuous [22], [23]. This first gives the radial position of the shell $l(t)$ in terms of its comoving coordinate l_0 and the scale factor $a(t)$ as

$$l(t) = l_0 a(t). \tag{8}$$

Second, it implies that the angular velocity $\tilde{\omega}_0(t)$ and the time shift $\alpha(t)$ of the coordinate system S_{rot} with respect to the system S_{in} are

$$\tilde{\omega}_0(t) = \omega_0(t) \quad , \quad \alpha(t) = \sqrt{1 + l_0^2\dot{a}^2}. \tag{9}$$

Israel's junction conditions [22], [23] give the stress-energy tensor of the shell in terms of the jump in its extrinsic curvature (see appendix A). If we consider the shell to be a rotating 2-dimensional perfect fluid, we can write

$$\mathcal{T}_{ab} = (\rho_\Sigma + p_\Sigma)w_a w_b + p_\Sigma \gamma_{ab}, \quad \text{with} \quad w^a = (1, 0, \Omega_\Sigma), \tag{10}$$

where ρ_Σ, p_Σ are the shell energy density and surface pressure, respectively, and $\Omega_\Sigma \equiv d\phi_\Sigma/dt$ ($\phi_\Sigma(t)$ being the azimuthal coordinate of a fixed point on the shell in the outside coordinates)[l] is the angular velocity of the shell with respect to the outside coordinate grid or, also, with respect to the rotating frame $\mathcal{S}_{rot}$ inside the void. The indices (a, b) stand for (t, θ, ϕ) and are raised and lowered by the induced metric on the shell (7). The matching conditions (appendix A, eq. (46)) result in

$$\rho_\Sigma(t) = \frac{1}{4\pi G}\frac{1}{al_0}\left(\sqrt{1+l_0^2\dot{a}^2} - \sqrt{1-kl_0^2}\right),$$
$$p_\Sigma(t) = -\frac{1}{2}\rho_\Sigma - \frac{1}{8\pi G}\frac{l_0\ddot{a}}{\sqrt{1+l_0^2\dot{a}^2}},$$
$$\Omega_\Sigma(t) = \omega_0(t) - \frac{1}{16\pi G}\frac{\sqrt{1-kl_0^2}}{a(\rho_\Sigma + p_\Sigma)}\frac{\partial\omega}{\partial r}|_0, \tag{11}$$

where we define $\partial\omega/\partial r|_0 \equiv \partial\omega/\partial r|_{r=l_0}$. The energy density and pressure of the shell are determined in terms of l_0 and the scale factor, and its angular velocity will be known once the metric perturbation $\omega(t, r)$ is determined. We note that $\rho_\Sigma(t)$ is always positive and that the equation of state of the shell differs in general from that of the FLRW universe outside.

A number of global quantities can be associated with the shell. For example we can define its total (proper) mass as

$$m_\Sigma(t) = 4\pi a^2 l_0^2 \rho_\Sigma = \frac{al_0}{G}\left(\sqrt{1+l_0^2\dot{a}^2} - \sqrt{1-kl_0^2}\right). \tag{12}$$

Note that this mass is *not* equal to the (proper) mass $M_\Sigma(t)$ of the universe that has to be removed in order to create the void. Indeed

$$M_\Sigma(t) = \frac{a}{2G}\dot{a}^2 l_0^3 \quad \text{for} \quad k=0$$
$$M_\Sigma(t) = \frac{3a}{4G}(1+\dot{a}^2)\left(\arcsin l_0 - l_0\sqrt{1-l_0^2}\right) \quad \text{for} \quad k=1 \tag{13}$$

(and a similar expression for $k = -1$). Only in case of voids small as compared to the Hubble and curvature radii ($l_0\dot{a} \ll 1$, $l_0 \ll 1$) do $m_\Sigma(t)$ and $M_\Sigma(t)$ coincide. Otherwise in general $M_\Sigma(t) > m_\Sigma(t)$. In any case, unless the void is much larger than the Hubble radius, which is unlikely for causality reasons, $m_\Sigma(t)$ represents a sizeable fraction of $M_\Sigma(t)$ and therefore is large which is not surprising since we assumed the void to be totally empty.

More interesting is the angular momentum of the shell

$$j_\Sigma = \int_\Sigma \mathcal{T}^t{}_\phi \sqrt{-\gamma}\, d\phi d\theta, \tag{14}$$

γ being the determinant of the induced metric (7), which is conserved due to the axial symmetry of the problem at hand. Using the decomposition (10-11) it can be rewritten as

$$j_\Sigma = \frac{8\pi}{3}a^4 l_0^4(\Omega_\Sigma - \omega_0)(\rho_\Sigma + p_\Sigma) = -\frac{1}{6G}l_0^4\sqrt{1-kl_0^2}\, a^3\frac{\partial\omega}{\partial r}|_0. \tag{15}$$

[l]To the first order in the rotational perturbations $dt = d\tau$, τ being the proper time on the shell.

The source which governs the outside metric (6) is the stress-energy tensor of cosmic dust

$$T_{\mu\nu} = \rho u_\mu u_\nu, \quad \text{with} \quad u^\mu = (1, 0, 0, \Omega), \tag{16}$$

where $g_{\mu\nu}$ is the metric (6), ρ is the energy density of the dust and $\Omega(t,r)$ is its (small) angular velocity.

Einstein's equations for the rotational metric perturbation $\omega(t,r)$ are solved in appendix B. They give

$$\omega(t,r) = -\frac{1}{a^3}(\beta(r) + g(t)), \tag{17}$$

where [17]

$$\beta(r) = -\frac{2G\sqrt{1-kr^2}}{r^3}(1+2kr^2)j_\Sigma + 6G\int_{l_0}^{r}\frac{j(r')}{r'^4\sqrt{1-kr'^2}}\,dr'. \tag{18}$$

The radially dependent function $j(r)$

$$j(r) \equiv \int_{l_0}^{r} T^t{}_\phi\sqrt{-g}\,d\phi\,d\theta\,dr' = \frac{8\pi}{3}a^5\rho\int_{l_0}^{r}\frac{(\Omega-\omega)}{\sqrt{1-kr'^2}}r'^4\,dr' \tag{19}$$

is the conserved total angular momentum of the dust layer $(l_0, r]$ outside the shell. Hence the first term on the right hand side of (18), proportional to j_Σ, comes from the rotating shell while the second one is the contribution of cosmic matter outside the shell.

As for the function $g(t)$, it is clear that it can be absorbed in the azimuthal coordinate ϕ and hence expresses the option to choose any global (slowly) rotating axes, i.e. the frame $\mathcal{S}_{rot}$, but we keep this freedom for later purposes. Einstein's equations thus yield the metric perturbation $\omega(t,r)$ in terms of the function of time $g(t)$ and the radial dependent function $j(r)$. These will be determined by boundary conditions.

- *Case of open universes*

In open universes, the function $g(t)$ is determined uniquely by the condition $r\,\omega(t,r) \ll 1$ for all radii r so that the perturbation $\omega(t,r)$ must die away at infinity at least as r^{-1}. When $\lim_{r\to\infty} r\int_{l_0}^{r} j(r')dr'/(r'^4\sqrt{1-kr'^2})$ vanishes, as will be the case, this amounts, by (17) and (18), to choosing the coordinate system in such a way that

$$g(t) = 0 \quad \text{for} \quad k = 0; \qquad g(t) = -4Gj_\Sigma \quad \text{for} \quad k = -1. \tag{20}$$

The metric (4) inside the void is then completely known: $\alpha(t)$ and $\tilde{\omega}_0(t)$ are given by (9) and

$$\omega_0(t) = \frac{2Gj_\Sigma}{a^3 l_0^3} \quad \text{for} \quad k = 0$$

$$\omega_0(t) = \frac{2Gj_\Sigma}{a^3 l_0^3}\left[\sqrt{1+l_0^2}(1-2l_0^2) + 2l_0^3\right] \quad \text{for} \quad k = -1. \tag{21}$$

Let us finally turn to the metric outside the shell, which still depends on the (so far unknown) function $j(r)$. The conservation laws for the dust universe imply $\rho \propto a^{-3}$ so that from (19) we deduce that the combination $a^2(\Omega - \omega)$ does not depend on time. Hence we define a function $f(r) \equiv a^2(\Omega - \omega)$ which describes the perturbation of the stress-energy tensor of cosmic dust outside the shell, $T^t{}_\phi = \rho(t) r^2 \sin^2\theta f(r)$, or, equivalently, its angular momentum distribution. In general, it is not determined uniquely just by requiring $\omega \to 0$ at infinity and can in principle be fixed by the initial conditions or deduced from observations. However, if we demand, in accord with Klein [16], that the only source for the rotational perturbation ω at any r is the shell itself (brought in some way into a slow rotation), we eliminate the contribution of cosmic dust in (18) by choosing $f(r) = 0$. Then

$$\Omega(t,r) = \omega(t,r) \quad \text{i.e.} \quad j(r) = 0. \tag{22}$$

With the condition (22), the metric outside the shell is known in terms of a single parameter j_Σ

$$\begin{aligned} \omega(t,r) &= \frac{2Gj_\Sigma}{a^3 r^3} \quad \text{for} \quad k = 0 \\ \omega(t,r) &= \frac{2Gj_\Sigma}{a^3 r^3}\left[\sqrt{1+r^2}(1-2r^2) + 2r^3\right] \quad \text{for} \quad k = -1. \end{aligned} \tag{23}$$

- *Case of closed universes*

In this case the condition $r\,\omega(t,r) \ll 1$ holds for any small $g(t)$ (with $\beta(r)$ small — cf. (17)), and we are left with a freedom to choose (slowly) rotating axes. We can choose the gauge simply as

$$g(t) = 0. \tag{24}$$

The metric inside the void is then known. In particular

$$\omega_0(t) = \frac{2Gj_\Sigma}{a^3 l_0^3}\sqrt{1 - l_0^2}(1 + 2l_0^2). \tag{25}$$

In order to determine the metric outside the void now, we first take into account the fact that the total angular momentum of a *closed* universe must vanish [16], [19]. The choice $f(r) \equiv a^2(\Omega - \omega) = 0$ used in the open cases is not compatible with this constraint. This means that the integral in (18) cannot be eliminated, i.e., the matter outside the shell *does* necessarily influence the rotational perturbation: cosmic dust necessarily possesses some intrinsic angular momentum to counteract the rotation of the shell. Nevertheless even the zero total angular momentum condition does not determine the function $f(r)$ uniquely. We shall, for definiteness, treat an example $f = constant$ where the constant is fixed by

$$j(\pi) = -j_\Sigma = (8\pi f/3)(\rho a^3)\int_{\chi_0}^{\pi} \sin^4\chi d\chi$$

where we put as usual $\sin\chi = r$ ($\sin\chi_0 = l_0$). Then $a^2(\Omega - \omega) \propto j_\Sigma$ and the metric outside the shell is known in terms of the angular momentum of the shell j_Σ. We obtain

$$f = -j_\Sigma \left[\frac{8\pi}{3} (\rho a^3)(\xi(\pi) - \xi(\chi_0)) \right]^{-1},$$

where

$$\xi(\chi) = \frac{1}{32}(12\chi - 8\sin 2\chi + \sin 4\chi), \tag{26}$$

and the analogue of condition (22) reads

$$\Omega(t,r) = \omega(t,r) - j_\Sigma \left[\frac{8\pi}{3} (\rho a^5)(\xi(\pi) - \xi(\chi_0)) \right]^{-1},$$

i.e.

$$j(\chi) = -j_\Sigma \frac{\xi(\chi) - \xi(\chi_0)}{\xi(\pi) - \xi(\chi_0)}. \tag{27}$$

Equation (51) can now be integrated to obtain the rotational perturbation outside the shell in the form

$$\omega(t,\chi) = \frac{3}{8} \frac{G j_\Sigma}{\xi(\pi) - \xi(\chi_0)} \frac{1}{a^3} \times \left\{ \frac{1}{\sin^3 \chi} [2\sin\chi + (\pi - \chi)(3\cos\chi - \cos 3\chi)] - C(\chi_0) \right\}, \tag{28}$$

where $C(\chi_0)$ is given by

$$C(\chi_0) = \frac{1}{\sin^3 \chi_0} \left\{ \sin 2\chi_0 (\frac{1}{2} + \frac{1}{3}\sin^2 \chi_0)(\cos 3\chi_0 - 3\cos\chi_0) + 2\sin\chi_0 \right\}, \tag{29}$$

the comoving radius of the shell $\chi_0 \in (0, \pi)$ being a parameter of the solution. The value of the rotational perturbation at the shell is obtained by the limit $\chi \to \chi_0$ in (28)

$$\omega_0(t) = -\frac{G j_\Sigma}{a^3 \sin^3 \chi_0} [\cos 3\chi_0 - 3\cos\chi_0], \tag{30}$$

which is identical to (25). The function $a^3\omega_0$ decreases monotonously as a function of the comoving radius of the shell χ_0; it vanishes at $\chi_0 = \frac{\pi}{2}$. Fig. 1 shows the rotational perturbation $a^3\omega$ as a function of the radial coordinate χ for several positions χ_0 of the shell. For $\chi_0 < \frac{\pi}{2}$, the rotational perturbation vanishes at some point, $\tilde{\chi} \in [\chi_0, \pi]$ say. For $\chi_0 > \frac{\pi}{2}$, the rotational perturbation is negative in the whole spacetime outside the shell. However, we can always choose the function $g(t)$ in such a way that $a^3\omega$ vanishes at some given $\tilde{\chi}$ whatever the value of χ_0 is — see section 6 for a further discussion of this point.

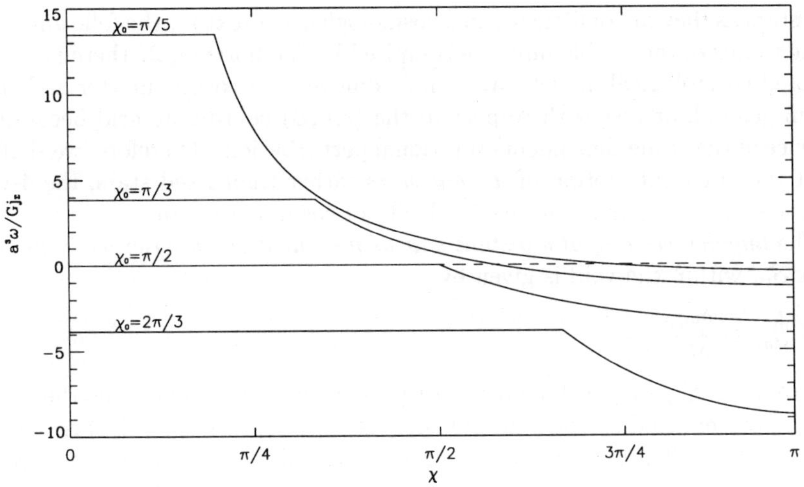

Figure 1. *Rotational perturbation $a^3\omega/Gj_\Sigma$ as a function of the radial coordinate χ for several comoving radii of the shell χ_0 ($k=+1$). Inside the void, $\chi<\chi_0$, the perturbation is constant, given by (30). The curves are plotted in the gauge corresponding to $g(t)=0$; any non-zero $g(t)=constant$ would shift the picture along the y-axis.*

5 Angular velocity of the shell and apparent motion of distant stars

5.1 Proper angular velocity of the shell

Let us now turn to the rotation of the shell. In (10) we defined the shell angular velocity with respect to the outside coordinate grid $\Omega_\Sigma \equiv d\phi_\Sigma/dt$. We can as well define its proper angular velocity $\bar{\Omega}_\tau \equiv d\bar{\phi}_\Sigma/d\tau$, $\bar{\phi}_\Sigma(t)$ being the trajectory of a point of the shell in the inertial frame $\mathcal{S}_{in}$ and τ being its proper time. As the shell is slowly rotating we see to the first order, i.e., for $(l_0 a)\bar{\Omega}_\tau \ll 1$ and $d\tau = dt$, that

$$\bar{\Omega}_\tau = \Omega_\Sigma - \omega_0. \tag{31}$$

Putting together equations (11), (17) and (51), and recalling that $\lim_{r\to l_{0_+}} j(r) = 0$, we find for all $k=-1,0,1$

$$\bar{\Omega}_\tau = \frac{3Gj_\Sigma}{a^3 l_0^3}\frac{\sqrt{1+l_0^2\dot{a}^2}}{1-\sqrt{1+l_0^2\dot{a}^2}\sqrt{1-kl_0^2-l_0^2a^2(\dot{a}/a)^{\cdot}}}\,. \tag{32}$$

Note that in order to keep j_Σ constant the angular velocity of the shell, in general, depends on time through the scale factor $a(t)$.

5.2 Apparent motion of distant stars

A *fixed star*, as defined in [13] in the case of the Schwarzschild shell, denotes a source of light placed at infinity at rest with respect to the asymptotic inertial frame attached to the centre of the shell. In general, we shall not consider such fixed stars as they are of little use in a cosmological context for the following reasons: **1.** the radius of the visible universe is limited by the horizon; **2.** there is no infinity in closed cosmological models; **3.** 'stars' comove with cosmic matter and they are not, in general, at rest with respect to the (r,θ,ϕ) coordinate grid because of the presence of the (time-dependent) rotational perturbation. Therefore, we shall speak about the apparent motion of *distant stars* rather than fixed stars, the 4-velocity of these stars being given by the 4-velocity of cosmic dust (16).

The *angular velocity of a distant star as measured by an observer in the inertial frame S_{in}* within the void is given by

$$\bar{\Omega}^m_{star} \equiv \frac{\delta\bar{\phi}_{rec}}{\delta\bar{t}_{rec}}, \tag{33}$$

where $\delta\bar{t}_{rec} \equiv \bar{t}^2_{rec} - \bar{t}^1_{rec}$ is the observer's proper time interval between the arrivals of two photons emitted by the star and $\delta\bar{\phi}_{rec} \equiv \bar{\phi}_{star}(\bar{t}^2_{rec}) - \bar{\phi}_{star}(\bar{t}^1_{rec})$ is the angular separation travelled by the star during the interval $\delta\bar{t}_{rec}$ as measured in S_{in}. The results of appendix C show that a star comoving with cosmic dust at a given radius χ_*, emitting photons radially inwards in the equatorial plane,[m] rotates apparently with

$$\bar{\Omega}^m_{star} = \frac{1}{[a(\alpha + l_0\dot{a})]_0}\left([a\Omega]_* - [a\omega]_* + \int_{\chi_0}^{\chi_*} \frac{d(a^3\omega)}{d\chi}\frac{1}{a^2}d\chi\right). \tag{34}$$

The subscripts '*' and '0' mean that the quantities are evaluated at the time of emission and at the time when the first photon reaches the shell, respectively.[n]

- *Case of open universes*

The scale factor of an open flat universe is given in terms of the conformal time η and a constant a_M as $a(\eta) = \frac{1}{4}a_M\eta^2$. Using the condition (22) (i.e. the rotational perturbation arises only due to the rotation of the shell, $\Omega(\eta_*, \chi_*) = \omega(\eta_*, \chi_*)$) one obtains from eq. (34) directly

$$\bar{\Omega}^m_{star} = -\frac{384 G j_\Sigma}{a^3_M\eta^2_0}\frac{1}{\sqrt{1 + 4\chi^2_0/\eta^2_0} + 2\chi_0/\eta_0}\int_{\chi_0}^{\chi_*}\frac{d\chi}{\chi^4[\chi_0 + \eta_0 - \chi]^4}, \tag{35}$$

where $\eta_0 = \chi_* + \eta_* - \chi_0$. Since $\bar{\Omega}^m_{star}$ is never positive the distant stars are seen to be rotating *backwards*. Note that the stars placed just behind the shell, $\chi_* \to \chi_0$, do not rotate with respect to inertial observers inside the void.

Interesting from the observational point of view is today's dependence of $\bar{\Omega}^m_{star}/j_\Sigma$ on the cosmological redshift $z(\chi_*)$ of the star. η_0 is given as a function of the conformal time η_{rec} elapsed by the time $\bar{t}_{rec}$ when the first ray reaches the observer at the centre of the void. Between $\bar{t}_0$ and $\bar{t}_{rec}$, when the light travels within the void, the conformal time changes from η_0 to η_{rec} so that

$$a(\eta_0)\chi_0 = \frac{1}{4}a_M\eta^2_0\chi_0 = \int_{\bar{t}_0}^{\bar{t}_{rec}} d\bar{t} = \int_{\eta_0}^{\eta_{rec}} a(\eta)\alpha(\eta)d\eta = \frac{1}{4}a_M\int_{\eta_0}^{\eta_{rec}}\eta^2\sqrt{1 + 4\frac{\chi^2_0}{\eta^2}}d\eta,$$

[m] Putting the observer off the centre complicates the calculation by introducing a radial velocity of the star in which we are not interested here.

[n] We recall that $r = \chi, \sinh\chi, \sin\chi$ for $k = 0, -1, +1$, respectively, with the comoving radius of the shell denoted as $l_0 = \chi_0, \sinh\chi_0$ or $\sin\chi_0$.

which results in

$$(\eta_{rec}^2 + 4\chi_0^2)^{3/2} = 3\chi_0\eta_0^2 + (\eta_0^2 + 4\chi_0^2)^{3/2}. \tag{36}$$

In the above expression, η_{rec} is given as a function of today's estimated age of the universe, $t_{today} = 1.1 \times 10^{10}$ years, as $\eta_{rec} = (12t_{today}/a_M)^{1/3}$. Equation (36) allows to calculate η_0 explicitly as

$$\eta_0^2 = \left\{[X+Y]^{1/3} + [X-Y]^{1/3} - \chi_0\right\}^2 - 4\chi_0^2$$

where

$$X = \frac{1}{2}\left(\eta_{rec}^2 + 4\chi_0^2\right)^{3/2} + 5\chi_0^3\,, \quad Y = \sqrt{\frac{1}{4}\left[(\eta_{rec}^2 + 4\chi_0^2)^{3/2} + 10\chi_0^3\right]^2 - \chi_0^6}.$$

The equality (35) can now be evaluated as a dependence $\bar{\Omega}^m_{star}/j_\Sigma\,(z, t_{today})$, z being the measured redshift

$$z(\chi_*) \equiv \frac{a(\eta_0)}{a(\eta_0 + \chi_0 - \chi_*)} - 1 = \frac{\eta_0^2}{[\eta_0 + \chi_0 - \chi_*]^2} - 1.$$

Several curves illustrating today's angular velocity of a star as a function of z are plotted in fig. 2. Since the measured redshift depends on $a(\eta_0)$ when the light ray leaves the outer universe and not on $a(\eta_{rec})$ at the time of observation, the cosmological redshift does not bring information about the radius of the void.

It is worth noting that in a nonstationary universe distant stars do not rotate uniformly. For χ_0 and χ_* fixed, $\bar{\Omega}^m_{star}$ depends on the time of observation through η_0 and (36). $\bar{\Omega}^m_{star}$ decreases monotonously down to zero as $\eta_0 \to \infty$ (then $a \to \infty$ so that the rotational perturbation vanishes).

The $k = -1$ models give qualitatively the same results.

- *Case of closed universes*

In the closed case, cosmic dust contributes to the rotational perturbation, i.e. $\Omega(\eta_*, \chi_*) - \omega(\eta_*, \chi_*) \neq 0$. Equations (51) and (27) imply

$$\begin{aligned} a(\eta_*)[\Omega(\eta_*, \chi_*) - \omega(\eta_*, \chi_*)] &= -\frac{j_\Sigma G}{[\xi(\pi) - \xi(\chi_0)]}\frac{1}{a_M a(\eta_*)}, \\ \frac{d[a^3\omega](\chi)}{d\chi} &= -\frac{6Gj_\Sigma}{\sin^4\chi}\frac{\xi(\pi) - \xi(\chi)}{\xi(\pi) - \xi(\chi_0)}, \end{aligned} \tag{37}$$

so that eq. (34) becomes

$$\begin{aligned} \bar{\Omega}^m_{star} = &-\frac{4Gj_\Sigma}{[\xi(\pi) - \xi(\chi_0)]a_M^3}\left[\frac{1}{\alpha(\eta_0)(1-\cos\eta_0) + \sin\eta_0\sin\chi_0}\right] \\ &\times \left[\frac{1}{1-\cos[\chi_* - \chi_0 - \eta_0]} + \mathcal{I}\right], \end{aligned} \tag{38}$$

where

$$\mathcal{I} \equiv 12\int_{\chi_0}^{\chi_*}\frac{\xi(\pi) - \xi(\chi)}{\sin^4\chi(1-\cos[\chi - \chi_0 - \eta_0])^2}d\chi\,, \quad \alpha(\eta_0) = \sqrt{1 + \frac{\sin^2\chi_0\sin^2\eta_0}{(1-\cos\eta_0)^2}}.$$

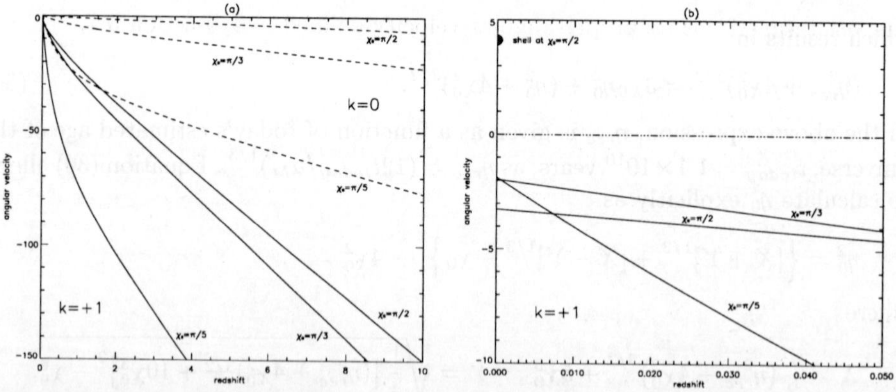

Figure 2. *(a) Observed angular velocity of cosmic matter $\bar{\Omega}^m_{star} a^3_M / G j_\Sigma$ as a function of the redshift and corresponding to today's time of observation, plotted for both open (dashed lines) and closed (solid lines) universes. (b) The near vicinity of the shell for $k = +1$ is shown. While for $k = 0$ all the curves start at the origin, the starting point of each curve is negative when $k = +1$ and depends on the radius of the shell. The 'discontinuity' between the angular velocity of the shell and those of stars just above the shell, $\bar{\Omega}^m_{star} \not\to \bar{\Omega}^m_\Sigma$ in the limit $\chi_* \to \chi_0$, is shown for $\chi_0 = \frac{\pi}{2}$.*

$\bar{\Omega}^m_{star}$ in the above expression is negative for all χ_*, χ_0, η_0. As in the open case, the distant stars are seen to be rotating *backwards*. In contrast with the open case, cosmic dust induces the first term on the right hand side of (38) which does not vanish, unlike the integral $\mathcal{I}$, in the limit $\chi_* \to \chi_0$. Hence, if the dust outside the shell contributes to the rotational perturbation, the stars just above the shell, in general, rotate with respect to the inertial observers inside the void (see fig. 2).

The dependence of η_0 on η_{rec} (obtained similarly as in the case $k = 0$) is given by

$$a(\eta_0) l_0 = \frac{a_M}{2}(1 - \cos\eta_0)\sin\chi_0 = \frac{a_M}{2}\int_{\eta_0}^{\eta_{rec}} (1 - \cos\eta)\alpha(\eta)\, d\eta,$$

which can be solved numerically (only if $\chi_0 = \pi/2$ one can obtain a simple analytic formula for $\eta_0(\eta_{rec})$). Inserting $\eta_0(\eta_{rec})$ into (38), together with

$$t_{today} = \frac{a_M}{2}\left(\eta_{rec} - \sin\eta_{rec}\right), \quad z = \frac{1 - \cos\eta_0}{1 - \cos[\eta_0 + \chi_0 - \chi_*]} - 1,$$

one gets today's dependence $\bar{\Omega}^m_{star}/j_\Sigma$ on the redshift. Fig. 2 shows several typical curves parametrized by χ_0.

5.3 Apparent motion of the shell

Equation (34) allows one to determine the angular velocity of the shell as measured by an inertial observer at the centre of the void. If one replaces Ω_* by Ω_Σ, and η_* by η_0, one gets the apparent angular velocity of the shell as

$$\bar{\Omega}^m_\Sigma = \frac{1}{[\alpha + l_0\dot{a}]_0}[\Omega_\Sigma - \omega_0]_0 = \frac{1}{[\alpha + l_0\dot{a}]_0}[\bar{\Omega}_\tau]_0. \tag{39}$$

The last term, $\bar{\Omega}_\tau$, is its proper angular velocity defined in subsection 5.1.

- *Case of open universes*

Combining the above expression with the proper angular velocity of the shell (32) we find

$$\bar{\Omega}_\Sigma^m = \frac{192 G j_\Sigma}{a_M^3 \chi_0^3 \eta_0^6} \frac{\sqrt{1+4\chi_0^2/\eta_0^2}}{[2\chi_0/\eta_0 + \sqrt{1+4\chi_0^2/\eta_0^2}][1+6\chi_0^2/\eta_0^2 - \sqrt{1+4\chi_0^2/\eta_0^2}]}. \tag{40}$$

This expression is never negative. Hence, for any given finite χ_0, the shell is seen to be rotating *forwards*. The rotational rate decreases as χ_0 increases and vanishes as $\chi_0 \to \infty$. A similar behaviour is obtained for χ_0 fixed in the limit $\eta_0 \to \infty$ (fig. 3).

Open $k = -1$ universes yield qualitatively the same results.

- *Case of closed universes*

The same steps as in the previous case give

$$\bar{\Omega}_\Sigma^m = \frac{24 G j_\Sigma}{a_M^3 \sin^3 \chi_0 [1-\cos\eta_0]^3} \frac{\alpha(\eta_0)}{\left[\frac{\sin\chi_0 \sin\eta_0}{1-\cos\eta_0} + \alpha(\eta_0)\right]}$$
$$\times \frac{1}{\left[\alpha^2(\eta_0) - \cos\chi_0 \alpha(\eta_0) + \frac{\sin^2\chi_0}{1-\cos\eta_0}\right]}. \tag{41}$$

Just like in the open case, $\bar{\Omega}_\Sigma^m$ is never negative. The shell is hence always seen to be rotating *forwards*. However, in contrast to the open case $\bar{\Omega}_\Sigma^m$, as a function of η_0, shows a minimum that never falls down to zero. One gets a similar behaviour for η_0 fixed and varying χ_0 (fig. 3). We shall return to this point in the next section.

It is clear that the above results depend quantitatively on our choice of the function $f(r)$ expressing the initial condition on the motion of cosmic matter outside the shell. Another choice in (22) and (26) would have led to a different redshift dependence for the angular velocity of the star $\bar{\Omega}_{star}^m$. Hence, the actual measurement of the redshift dependence of $\bar{\Omega}_{star}^m$ would in principle determine $f(r)$. On the other hand, the angular velocity of the shell $\bar{\Omega}_\Sigma^m$ does not depend on the choice of $f(r)$, see eq. (39). Note that since $a^2(\Omega_\Sigma - \omega_0)$ depends on time we cannot require $\lim_{\chi \to \chi_0} f = a^2(\Omega_\Sigma - \omega_0)$. Consequently, a discontinuity in the angular velocities $\bar{\Omega}_\Sigma^m$, $\bar{\Omega}_{star}^m$ across the shell occurs for any choice of $f(r)$, see fig.2. This leaves the observers inside the void a possibility to determine observationally the two parameters of the problem, j_Σ and l_0 [20].

6 Dragging inside the void and Mach's principle

As we have seen, inertial frame S_{in} attached to the centre of the void as well as other inertial frames in the void all rotate with the same angular velocity with respect to the rotating frame S_{rot} or equivalently with respect to observers inside the shell fixed at constant $\bar{r}, \theta, \phi$. The two coordinate grids are related by $d\bar{\phi} = d\phi - \omega_0(t)dt$. From $\bar{\phi} = constant$ it follows that

$$\frac{d\phi}{dt} = \omega_0(t), \tag{42}$$

where the explicit expression for $\omega_0(t)$ is (21) or (25). The rotation of spatial axes of the frame $\mathcal{S}_{in}$ with respect to those of $\mathcal{S}_{rot}$ demonstrates ***dragging of inertial frames*** inside the void. It will be useful to define the *preferred* observers outside the shell as those who are at rest with respect to the outside coordinate grid in which the line element is given by (6), i.e. at constant r, θ, ϕ. The preferred observers are not freely falling in general. In expanding and collapsing stages when $\dot{a}/a \neq 0$ they fall freely at the radii where the rotational perturbation vanishes. The choice of $g(t)$ in (20) and (24) implies that the preferred observers fall freely at infinity in the open cases and at a certain $\tilde{\chi}$ in the closed cases (see discussion below eq. (30)).

In open universes, the rotating axes were fixed uniquely by the condition $r\,\omega(t,r) \ll 1$. In this frame, physically realizable preferred observers cover the whole outer universe and can be defined as *globally non-rotating.* Frame dragging inside a void in open universes can thus be regarded as ***absolute*** with respect to these observers.

In a closed universe on the other hand, the value of the rotational perturbation at the shell $\omega_0(t)$ is a gauge dependent quantity. The function $g(t)$ can be used to choose any arbitrary path $\tilde{\chi}(t)$ as the locus of freely falling preferred observers or to choose any value for $\omega_0(t)$ (cf. fig. 1). In particular, by choosing

$$g(t) = constant \equiv -\beta(l_0)$$

we obtain $\omega_0(t)$ vanishing at all times. In this case the preferred observers fall freely at the shell and the spatial axes of the two frames $\mathcal{S}_{in}$ and $\mathcal{S}_{rot}$ coincide. Hence dragging of inertial frames inside a void in closed universes is ***relative***, i.e. depending on the choice of gauge.

Now, since we have expressed all rotations in terms of the measurable quantities $\bar{\Omega}^m_{star}$ and $\bar{\Omega}^m_\Sigma$ (their gauge invariance being evident since $g(t)$ cancels out in both $(\Omega - \omega)$ and $(\Omega_\Sigma - \omega_0)$) we shall confront the observation of the angular velocity distribution of matter in the universe carried out by inertial observers inside the void directly with Mach's ideas. By ***perfect dragging*** we shall mean the following: An inertial observer at the centre of the void is ***dragged perfectly by the shell*** if the angular velocity of the shell $\bar{\Omega}^m_\Sigma$ ***measured*** in his frame $\mathcal{S}_{in}$ vanishes.

- ***Case of open universes***

In the case $k = 0$, $\bar{\Omega}^m_\Sigma$ decreases as the radius of the shell χ_0 or the expansion factor a increase, and dies away as (i) $\chi_0 \to \infty$ for a fixed (fig. 3) or (ii) $\eta \to \infty$ for χ_0 fixed. This behaviour is in agreement with Klein's result [16] and can be interpreted from a Machian point of view as follows:

(i) When the radius of the shell is getting large almost all matter of the universe can be regarded as redistributed on the shell. In this case, the angular velocity of the shell fully determines the angular velocity distribution of matter in the universe

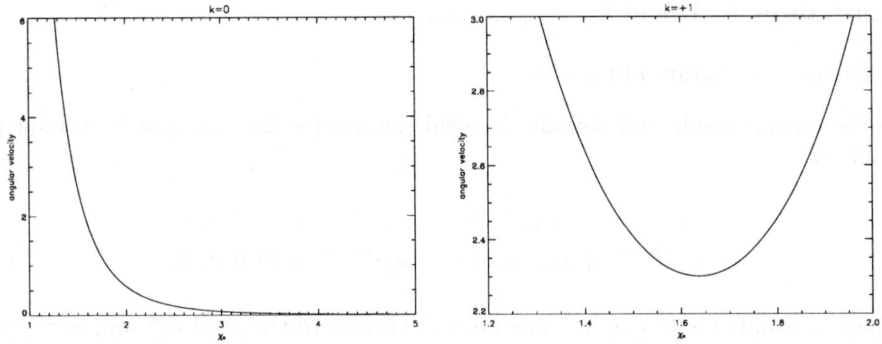

Figure 3. *Observed angular velocity of the shell* $\bar{\Omega}^m_\Sigma a^3_M/Gj_\Sigma$ *as a function comoving radius* χ_0*;* $\eta_0 = 1$ *arbitrarily. When* $k = 0$*,* $\bar{\Omega}^m_\Sigma$ *vanishes as* $\chi_0 \to \infty$*; for* $k = +1$ $\bar{\Omega}^m_\Sigma$ *shows a positive minimum. A similar behaviour is obtained for* χ_0 *fixed and* η_0 *varying.*

and frame dragging becomes perfect. This is in full conformity with Mach's ideas since 'the universe' does not rotate relative to local inertial frames inside the void.

(ii) In an expanding universe, the matter density as well as the rotational perturbation decrease as a function of time. Hence the influence of matter on inertial frames diminishes so that in this limiting case dragging inside the void becomes perfect.

For $k = -1$ we get similar results.

• *Case of closed universes*

In the closed case, $\bar{\Omega}^m_\Sigma$ is positive for all times η_0 and for all comoving positions of the shell χ_0. The rotational rate never vanishes (fig. 3) — there is no perfect dragging in closed universes. Since $\bar{\Omega}^m_\Sigma$ does not depend on the choice of function $f(r)$ this result holds for any initial angular momentum distribution of cosmic dust outside the shell. As a consequence of the zero angular momentum conservation in closed universes, cosmic dust (or, depending on $f(r)$, its part) is always observed to be in counter-rotation with respect to the shell.

In both cases, the fact that the totality of matter in the universe (the shell and cosmic dust) is not *observed* inside the void at any finite time to (slowly) rotate in a preferred direction is in full agreement with Mach's observation. Inertial frames inside the void are in this sense determined by the 'average motion of distant galaxies'.

Appendices A, B and C

A Extrinsic curvature of the shell

In the regions inside and outside the void the independent tangent vectors at the shell are

$$e_t^{\mu\ in} = (1, \dot{a}l_0, 0, 0) \quad , \quad e_t^{\mu\ out} = (1, 0, 0, 0),$$
$$e_\theta^{\mu\ in,out} = (0, 0, 1, 0) \quad , \quad e_\phi^{\mu\ in,out} = (0, 0, 0, 1), \tag{43}$$

where μ stands for (t, r, θ, ϕ). The normal vectors to the shell n_μ^{in} and n_μ^{out}, both pointing inside the shell, are determined by $(e^\mu n_\mu)^{in,out} = 0$ and normalized to unity

$$n_\mu^{in} = \alpha(t)(\dot{a}l_0, -1, 0, 0) \quad , \quad n_\mu^{out} = \frac{a(t)}{\sqrt{1 - kl_0^2}}(0, -1, 0, 0).$$

The extrinsic curvature is defined by

$$K_{ab}^{in,out} \equiv -e_a^\mu e_b^\nu \nabla_\mu n_\nu |^{in,out},$$

where the indices (a, b) stand for (t, θ, ϕ) and are raised and lowered by the induced metric on the shell (7). ∇_μ is the covariant derivative associated with the metric (4) or (6).

The metric being given by (4) the non-zero components of the extrinsic curvature of Σ calculated at first order in $\omega_0(t)$ are equal to

$$K_{tt}^{in} = -\frac{l_0 \ddot{a}}{\sqrt{1 + l_0^2 \dot{a}^2}} \quad , \quad K_{\theta\theta}^{in} = l_0 a \sqrt{1 + l_0^2 \dot{a}^2} \quad , \quad K_{\phi\phi}^{in} = K_{\theta\theta}^{in} \sin^2\theta,$$
$$K_{t\phi}^{in} = -l_0 a \omega_0 \sqrt{1 + l_0^2 \dot{a}^2} \sin^2\theta. \tag{44}$$

Similarly, the non-zero components of the extrinsic curvature of Σ as computed with the metric (6) are

$$K_{tt}^{out} = 0 \quad , \quad K_{\theta\theta}^{out} = l_0 a \sqrt{1 - kl_0^2} \quad , \quad K_{\phi\phi}^{out} = K_{\theta\theta}^{out} \sin^2\theta,$$
$$K_{t\phi}^{out} = -l_0 a \sqrt{1 - kl_0^2} \left(\frac{1}{2} l_0 \frac{\partial\omega}{\partial r}|_0 + \omega_0 \right) \sin^2\theta, \tag{45}$$

where $\partial\omega/\partial r|_0 \equiv \partial\omega(t, r)/\partial r|_{r=l_0}$.

An integration of the Einstein equations across the shell gives the stress-energy tensor $\mathcal{T}_{ab}$ of the shell as [22]

$$-8\pi G \mathcal{T}_{ab} = \hat{K}_{ab} - \gamma_{ab} \hat{K}_c^c, \tag{46}$$

$\hat{K}_{ab} \equiv K_{ab}^{out} - K_{ab}^{in}$ and γ_{ab} are the coefficients of the induced metric (7).

B Metric perturbation $\omega(t,r)$

The Einstein equations relate the rotational perturbation $\omega(t,r)$ and the dust angular velocity $\Omega(t,r)$. They can be easily found in the literature (see e.g. [25], [26]) recalling that $\omega(t,r)$ is Bardeen's [24] vector metric perturbation in the Poisson gauge. In our notations, they read (with $p=0$)

$$aD_{(k}\dot{\bar{B}}_{l)} + 2\dot{a}D_{(k}\bar{B}_{l)} = 0 \quad , \quad \frac{1}{2}D_iD^i\bar{B}_k + k\bar{B}_k = 8\pi Ga^2\rho\,(\bar{B}_k + \bar{v}_k), \tag{47}$$

where (k,l) stand for (r,θ,ϕ), D_k is the covariant derivative associated with the spatial part of the metric (6), $\bar{B}_k = (0,0,-a\omega r^2\sin^2\theta)$ and $\bar{v}_k = (0,0,a\Omega r^2\sin^2\theta)$. The first equation (47) implies

$$3\dot{a}\frac{\partial\omega}{\partial r} + a\frac{\partial\dot{\omega}}{\partial r} = 0, \tag{48}$$

the general solution of which is

$$\omega(t,r) = -\frac{1}{a^3}(\beta(r) + g(t)). \tag{49}$$

The second equation (47) translates as

$$(1-kr^2)\frac{\partial^2\omega}{\partial r^2} + (4-5kr^2)\frac{1}{r}\frac{\partial\omega}{\partial r} = -16\pi Ga^2\rho\,(\Omega-\omega). \tag{50}$$

This can be integrated and, using (15) and (49), written in terms of the constant of motion j_Σ as

$$\frac{d\beta}{dr} = \frac{6G}{r^4\sqrt{1-kr^2}}(j_\Sigma + j(r)), \tag{51}$$

where $j(r)$ is the angular momentum of cosmic dust defined by (19). Further integration of the previous equation finally gives

$$\beta(r) = -\frac{2G\sqrt{1-kr^2}}{r^3}(1+2kr^2)j_\Sigma + 6G\int_{l_0}^{r}\frac{j(r')}{r'^4\sqrt{1-kr'^2}}\,dr'. \tag{52}$$

This result corresponds to Klein's solution [17] for all $k=-1,0,+1$.

C Angular velocity of a distant star measured by inertial observers inside the void

In order to find $\bar{\Omega}^m_{star}$, we need to compute the trajectories of two photons emitted by the star and propagating first in the perturbed FLRW universe outside the shell, then in the void inside the shell and arriving at the centre of the void.

Consider a star comoving with cosmic dust at some radius χ_* emitting photons radially inwards and, for the sake of simplicity, in the equatorial plane. The trajectory of a photon inside the shell as described in the inertial frame $\mathcal{S}_{in}$ is given by the null geodesic

$$\bar{r} = \bar{t}^P_{rec} - \bar{t} \quad ; \quad \bar{\phi} = \bar{\phi}^P_{rec} \quad ; \quad \theta = \pi/2 \quad \text{for} \quad \bar{r} \le al_0, \tag{53}$$

where $\bar{t}^P_{rec}$ and $\bar{\phi}^P_{rec}$ are constants and where P stands for the photon 1 or 2. These photons cross the shell at times $\bar{t}^P_0$ defined by $a(\bar{t}^P_0)l_0 = \bar{t}^P_{rec} - \bar{t}^P_0$. Recalling that the coordinate $\bar{\phi}$ is related to the coordinate ϕ outside the void by $d\bar{\phi} = d\phi - \omega_0(t)dt$ and the proper time in $\mathcal{S}_{in}$ is related to the time coordinate of the outside grid by $d\bar{t} = \alpha(t)dt$, and introducing also the conformal time η ($dt = a(\eta)d\eta$), we have

$$\delta\bar{t}_{rec} = [a(\alpha + l_0\dot{\alpha})]_0 \delta\eta_0 \quad \text{where} \quad \delta\bar{t}_{rec} \equiv \bar{t}^2_{rec} - \bar{t}^1_{rec},$$
$$\delta\bar{\phi}_{rec} = \delta\phi_0 - [\omega a]_0 \delta\eta_0 \quad \text{where} \quad \delta\bar{\phi}_{rec} \equiv \bar{\phi}^2_{rec} - \bar{\phi}^1_{rec}, \tag{54}$$

which relate $\delta\bar{t}_{rec}$ to $\delta\eta_0$ (FLRW conformal time interval between the arrivals of the two photons on the shell) and $\delta\bar{\phi}_{rec}$ to $\delta\phi_0$ (angular deviation on the shell of the two photons read off with respect to the outside coordinate grid). The subscript '0' means that the quantities are evaluated at the time the photons reach the shell.

Outside the shell, the ingoing photons propagate to the zeroth order along the radial null geodesics $d(\bar{k}^0 a^2)/d\lambda = 0$, $d\bar{k}^r/d\lambda + 2a^{-1}da/d\eta\, \bar{k}^0\bar{k}^r + (kr)/(1 - kr^2)\,(\bar{k}^r)^2 = 0$, $\bar{k}^\theta = \bar{k}^\phi = 0$, where λ is an affine parameter and a bar denotes geodesics in the FLRW unperturbed universe. The non-zero components of the null wave vector are $\bar{k}^0 \equiv d\eta/d\lambda = (1/a^2)$ and $\bar{k}^r \equiv dr/d\lambda = -(1/a^2)\sqrt{1-kr^2}$. The coordinates $\chi(\eta)$ and $\theta(\eta)$ along the photons are given by

$$\chi - \chi_0 = \eta^P_0 - \eta \quad , \quad \theta = \pi/2, \tag{55}$$

η^P_0 is the conformal time at which the photon P reaches the shell at χ_0, $\delta\eta_0 = \eta^2_0 - \eta^1_0$. The photons are emitted at $\eta = \eta_*$ and $\eta = \eta_* + \delta\eta_*$ by a star placed at $\chi = \chi_*$ so that we see from (55) that simply

$$\delta\eta_0 = \delta\eta_*, \tag{56}$$

which relates $\delta\eta_0$ to $\delta\eta_*$.

Now, to the first order, the wave vector of the light propagating in the *perturbed* spacetime outside the shell is $k^\mu = (\bar{k}^0, \bar{k}^r, 0, k^\phi)$, $k^\phi \equiv d\phi/d\lambda$ solves the null geodesic equation up to the first order in perturbations

$$\frac{dk^\phi}{d\lambda} - \left\{\dot{a}a\omega + \frac{\partial}{\partial\eta}(a\omega)\right\}(\bar{k}^0)^2 + 2\dot{a}\bar{k}^0 k^\phi - \frac{1}{r^2}\frac{\partial}{\partial r}\{r^2 a\omega\}\bar{k}^0\bar{k}^r +$$
$$+\frac{2}{r}\bar{k}^r k^\phi + \frac{1}{1-kr^2}\{\dot{a}a\omega\}(\bar{k}^r)^2 = 0, \tag{57}$$

where $\omega(t,r)$ is given by (23) for $k = 0, -1$ and by (28) for $k = +1$. Equation (57) has a simple solution $k^\phi = \omega/a$, where ω can now be expressed only as a function of conformal time since the trajectory of the photon is given at leading order by (55): $\omega(\eta, \chi(\eta)) = \omega(\eta, \chi_0 + \eta^P_0 - \eta)$. Hence, using $d\eta/d\lambda = \bar{k}^0 = 1/a^2$, the $\phi(\eta)$ coordinates of the photon trajectories can be expressed as

$$\phi - \phi^P_0 = \int_{\eta^P_0}^{\eta} a(\eta')\omega(\eta', \chi_0 + \eta^P_0 - \eta')d\eta', \tag{58}$$

where ϕ^P_0 are their angles at which they reach the shell, $\delta\phi_0 = \phi^2_0 - \phi^1_0$. From (58), taking also into account that $\omega = -a^{-3}(\eta)(\beta(\chi) + g(\eta))$ (eq. (17)), we obtain

$$\delta\phi_* - \delta\phi_0 = \delta\eta_0 \int_{\eta^1_0}^{\eta_*}\left[\beta(\chi_0 + \eta^1_0 - \eta) \times \left(\frac{2}{a^3}\frac{da}{d\eta}\right) - \frac{d}{d\eta}\left(\frac{g}{a^2}\right)\right]d\eta, \tag{59}$$

which relates $\delta\phi_0$ to $\delta\eta_*$ and $\delta\phi_*$ (angular deviation of the two photons at the time of emission with respect to the outside coordinate grid). Again the subscript '*' means that the quantities are evaluated at the time of emission. Integrating by parts and changing the integration variable η to χ, (59) can be written as

$$\delta\phi_* - \delta\phi_0 = \delta\eta_* \left([a\omega]_* - [a\omega]_0 - \int_{\chi_0}^{\chi_*} \frac{d(a^3\omega)}{d\chi} \frac{1}{a^2} d\chi \right), \tag{60}$$

where a is now a function of $(\chi_0 + \eta_0 - \chi)$ and η_0 stands for η_0^1. The index 1 denoting the first ray is dropped out here and in the following.

In order to relate $\delta\phi_*$ to $\delta\eta_*$ we use the fact that the star is comoving with cosmic dust so that

$$\delta\phi_* = [a\Omega]_* \delta\eta_*, \tag{61}$$

where Ω is the angular velocity of cosmic matter introduced in (16) and related to the rotational perturbation ω by (22) for $k = -1, 0$ or by (27) for $k = +1$.

Putting the results (33), (54), (56), (60) and (61) together, one finally arrives at

$$\bar{\Omega}^m_{star} = \frac{1}{[a(\alpha + l_0 \dot{a})]_0} \left([a\Omega]_* - [a\omega]_* + \int_{\chi_0}^{\chi_*} \frac{d(a^3\omega)}{d\chi} \frac{1}{a^2} d\chi \right), \tag{62}$$

where we recall that $\alpha = \sqrt{1 + l_0^2 \dot{a}^2}$.

References

1. J. Barbour, H. Pfister, eds., *Mach's Principle: from Newton's Bucket to Quantum Gravity*, Birkhäuser, Basel (1995)
2. E. Mach, *The Economical Nature of Physical Inquiry* (1882), in Popular Scientific Lectures, ed. T. J. McCormach, Illinois, Open Court, p. 186 (1943)
3. E. Mach, *Die Mechanik in ihrer Entwicklung. Historisch-kritisch dargestellt.* Leipzig: F. A. Brockhaus (1883)
4. A. Einstein, *Vierteljahrsschrift für gerichtliche Medizin und öffentliches Sanitätswesen* **44**, 37 (1912)
5. A. Einstein, *Preussische Akademie der Wissenschaften, Sitzungsberichte* 142 (1917)
6. A. Einstein, *Annalen der Physik* **55**, 240 (1918)
7. H. Bondi, J. Samuel, *Phys. Lett. A* **228**, 121 (1997)
8. H. Thirring, *Phys. Z.* **19**, 33 (1918); *Phys. Z.* **22**, 29 (1921)
9. J. Lense, H. Thirring, *Phys. Z.* **19**, 156 (1918)
10. D. Brill, J. M. Cohen, *Phys. Rev.* **143**, 1011 (1966)
11. J. M. Cohen, *Phys. Rev.* **173**, 1258 (1968)
12. L. Lindblom, D. R. Brill, *Phys. Rev.* D **10**, 3151 (1974)
13. J. Katz, D. Lynden-Bell, J. Bičák, *Class. Quantum Grav.* **15**, 1 (1998)
14. A. Lausberg, *Astron. Astrophys.* **3**, 150 (1969)
15. S. M. Lewis, *Gen. Rel. Grav.* **12**, 917 (1980)
16. C. Klein, *Class. Quantum Grav.* **10**, 1619 (1993)

17. C. Klein, *Rotative Störungen und Mitführungseffekte in einem Friedmann-Universum.*, Dissertation, University of Tübingen (1993)
18. C. Klein, *Class. Quantum Grav.* **11**, 1539 (1994)
19. D. Lynden-Bell, J. Katz, J. Bičák, *Mon. Not. R. Astron. Soc.* **272**, 150; errata: **277**, 1600 (1995)
20. T. Doležel, J. Bičák, N. Deruelle, *Class. Quantum Grav.* **17**, 2719 (2000)
21. G. F. Smoot et al., *Astron. J.* **371**, L1 (1991)
22. W. Israel, *Nuovo Cim. B* **44**, 1 (1966); errata: **48**, 463 (1967)
23. C. Barrabès, W. Israel, *Phys. Rev.* D **43**, 1129 (1991)
24. J. M. Bardeen, *Phys. Rev.* D **22**, 1882 (1980)
25. J. M. Stewart, *Class. Quantum Grav.* **7**, 1169 (1990)
26. L. Bombelli, W. E. Couch, R. J. Torrence, *Class. Quantum Grav.* **11**, 139 (1994)

THE ROLE OF A REPULSIVE COSMOLOGICAL CONSTANT IN ASTROPHYSICAL PROCESSES

ZDENĚK STUCHLÍK
Faculty of Philosophy and Science, Silesian University at Opava, Bezručovo nám. 13, 746 01 Opava, Czech Republic
E-mail: Zdenek.Stuchlik@fpf.slu.cz

The role of a nonzero cosmological constant is discussed for black-hole or naked-singularity spacetimes, Einstein–Strauss vacuola spacetimes, and general relativistic polytropic fluid spheres. In case of the black-hole spacetimes, motion of test particles and photons and equilibrium configurations of test perfect fluid are considered. Their relevance to astrophysically very important accretion processes in disk regime is established.

1 Introduction

Recent observations give strong evidence that the energy sources of quasars and active galactic nuclei are accretion disks around central massive black holes [1–4]. The quasars and active galactic nuclei contain massive (10^6–$10^{11}\,M_\odot$) and very compact ($< 10^{-3}$ pc) gravitating center radiating during the active phase the power 10^{45}–10^{49} erg/s. The accretion disks around central massive black holes represent an efficient mechanism for extracting gravitational potential energy and converting it into radiation or gigantic outflows of matter, and give the most probable explanation of the main features of active galactic nuclei, i.e., high luminosities, compactness, jet formation, rapid time variability, and profiled emission lines. In the accretion disks, the potential gravitational energy of matter orbiting the central black hole is liberated and transformed into heat, due to viscous stresses acting against shearing motion, and radiated (at least partly) away. During the process, angular momentum of the accreting matter has to be transported outwards.

It is well known that at low accretion rates the pressure is negligible and the accretion disk is geometrically thin. Its basic properties are determined by the circular geodesic motion in the black-hole background. The radius $r_{\rm ms}$ of the marginally stable circular orbit represents the inner edge of the Keplerian disks since matter, following quasikeplerian orbits down to $r_{\rm ms}$, falls freely into the black hole from this radius [5,6].

At high accretion rates, the pressure is relevant and the accretion disk must be geometrically thick. Its basic properties are determined by equipotential surfaces of test perfect fluid (i.e., perfect fluid that does not alter the black-hole geometry) rotating in the black-hole background. The accretion is possible if a toroidal equilibrium configuration of the test fluid containing a critical, self-crossing equipotential surface can exist in the background. The cusp of this equipotential surface corresponds to the inner edge of the disk and the accretion inflow of matter into the black hole is possible due to a mechanical nonequilibrium process, i.e., because of matter slightly crossing over the critical equipotential surface. The pressure gradients push the inner edge of the thick disks under the radius $r_{\rm ms}$ [7,8]. The simplest, but quite illustrative case of the equipotential surfaces of the test fluid can be constructed for the configurations with uniform distribution of the angular momentum density. This case is fully governed by the geometry of the spacetime, however, it contains all the characteristic features of more complex cases of rotating fluid [9]. Moreover, this case is also very important physically since it corresponds to marginally stable equilibrium configurations [10].

The equipotential surfaces were analyzed for both Schwarzschild and Kerr black-hole spacetimes. The critical closed surfaces with a cusp can exist for angular momentum density higher (lower) than the one corresponding to the marginally stable (bound) circular geodesic and the location of the cusp shifts from $r_{\rm ms}$ to the radius of the marginally bound geodesic orbit $r_{\rm mb}$. The cusp close to the horizon enables the inflow of matter into the black hole. However, the character of the equilibrium configurations does not allow outflow of matter and transfer of the angular momentum for disks around isolated black holes. Similarly, the character of the circular geodesic motion does not allow outflow of matter from thin accretion disks around isolated black holes. In binary systems, the outflow is possible through the Lagrange point L2 (see, e.g., Ref. [5]).

Recently available data from a wide variety of cosmological observations (measurements of the present value of the Hubble parameter and dynamical estimates of the present energy density of the Universe, details of the anisotropy of the cosmic relict radiation, statistics of gravitational lensing of quasars, galaxy number counts, and high-redshift supernovae) indicate convincingly that in the framework of the inflationary paradigm [11], a nonzero, although very small, relict repulsive cosmological constant $\Lambda > 0$ or an analogous concept of the quintessence, has to be invoked in order to explain the dynamics of the recent Universe [12–18]. The presence of a repulsive cosmological constant changes dramatically the asymptotic structure of black-hole backgrounds as such backgrounds become asymptotically de Sitter spacetimes, not flat spacetimes. In such spacetimes, an event horizon always exists, behind which the geometry is dynamic; we call it a cosmological horizon. Therefore, it is interesting to clarify the influence of the repulsive cosmological constant on astrophysically relevant properties of black-hole spacetimes.

For these purposes, analysis of geodesic motion of test particles and photons and studies of equilibrium configurations of test perfect fluid orbiting the black holes are among the most fundamental techniques. Further, the curvature of the spacetime under consideration can be conveniently demonstrated by embedding 2-dimensional, appropriately chosen, spacelike surfaces into 3-dimensional Euclidean space [19]. The 3-dimensional optical reference geometry [20], associated with the spacetime under consideration, enables a natural 'Newtonian' concept of gravitational and inertial forces and reflects some hidden properties of the test particle motion [21, 22]. Properties of the inertial forces can be reflected by the embedding diagrams of appropriately chosen 2-dimensional sections of the optical geometry, as discussed in a parallel contribution in this volume [23]. (Note that in accord with the spirit of general relativity, alternative approaches to the concept of inertial forces are possible, e.g., the 'special relativistic' one [24].)

The most general, Kerr–Newman spacetimes with a nonzero cosmological constant and their associated electromagnetic field are given and the pseudosingularities of the geometry are discussed in Sec. 2. The motion of test particles and photons is determined by the Carter equations [25] presented and discussed in Sec. 3. From the astrophysical point of view, the most interesting classes of the test-particle motion are circular orbits that are directly related to the properties of the thin accretion disks and the radial geodesics that can be relevant in estimating properties of spherical accretion onto a black hole [5]. Due to the presence of a repulsive cosmological constant, an outer marginally stable circular orbit enabling the outflow of matter from thin accretion disks can exist. Moreover, the spherical geodesics of the Schwarzschild–de Sitter spacetimes can be used to construct the model of Einstein–Strauss–de Sitter vacuola [26,27] presented in Sec. 4. Refraction of light at the matching surface of the dynamic Friedman and static Schwarzschild–-de Sitter regions of the Einstein–Strauss–de Sitter vacuola is considered and its influence on the Rees–Sciama effect explaining anisotropies of the relict microwave background is estimated.

Considering equilibrium configurations of test perfect fluid orbiting a black-hole spacetime with a repulsive cosmological constant, it can be shown that the outflow of matter from the thick accretion disk is possible because equipotential surfaces with an outer cusp in vicinity of the so called static radius can exist (beside the critical surfaces with the inner cusp nearby the horizon) if the mass of the black hole is small enough to admit the existence of stable circular geodesics. Moreover, if the uniform angular momentum density of the equilibrium configurations corresponds to the marginally bound orbit of the background, the critical equipotential surface has both the inner and outer cusps. In this situation, any mechanical nonequilibrium in a thick accretion disk leads to both an inflow into the black hole, and an outflow from the disk near the static radius [28]. All of these results are presented in Sec. 5.

It is also useful to have an idea of the influence of a nonzero cosmological constant on selfgravitating mass configurations. In Sec. 6, structure equations governing spherically symmetric general relativistic polytropes are given. Generally, these equations can be solved by numerical methods. However, there is a special case of polytropes of index $n = 0$, corresponding to the configurations of constant energy density, for which the structure equations can be integrated in terms of elementary functions [29]. It can be shown that extremely compact configurations can exist, allowing the existence of bound null geodesics [30].

There is another very important family of solutions with a cosmological constant, namely the Robinson–Trautman–de Sitter 'radiating' solutions of the Einstein equations. These were extensively studied by Jiří Bičák and Jiří Podolský and many interesting results were obtained, say, e.g., the cosmic 'no hair' theorem important in the context of inflationary cosmology. However, we shall not discuss the 'radiating' solutions here because of their cosmological, rather than astrophysical, relevance. An interested reader is referred to the original papers [31,32] and [33] for other references.

Since anti-de Sitter spacetimes play a crucial role in the framework of superstring theories [34, 35], we shall also consider, in some illustrative cases, the black-hole spacetimes with an attractive cosmological constant; nevertheless, we shall concentrate on the discussion of the influence of a repulsive cosmological constant.

Similarly, in some cases, we shall consider naked singularity spacetimes related to the black-hole spacetimes as they still could be considered conceivable models of quasars and active galactic nuclei. Of particular interest should be those effects that could distinguish a naked singularity from black holes.

For simplicity, we shall use the geometric system of units ($c = G = 1$) if not stated otherwise.

2 Kerr–Newman–de Sitter geometry and its pseudosingularities

In the standard Boyer–Lindquist coordinates (t, r, θ, ϕ), the most general Kerr–Newman–(anti-)de Sitter dyon geometry is determined by the line element

$$\mathrm{d}s^2 = -\frac{\Delta_r}{I^2\rho^2}(\mathrm{d}t - a\sin^2\theta\,\mathrm{d}\phi)^2 + \frac{\Delta_\theta \sin^2\theta}{I^2\rho^2}\left[a\,\mathrm{d}t - \left(r^2+a^2\right)\mathrm{d}\phi\right]^2 + \frac{\rho^2}{\Delta_r}\mathrm{d}r^2 + \frac{\rho^2}{\Delta_\theta}\mathrm{d}\theta^2 \qquad (1)$$

where

$$\Delta_r = -\tfrac{1}{3}\Lambda r^2(r^2+a^2) + r^2 - 2Mr + a^2 + e^2 + p^2, \qquad (2)$$

$$\Delta_\theta = 1 + \tfrac{1}{3}\Lambda a^2\cos^2\theta, \qquad (3)$$

$$I = 1 + \tfrac{1}{3}\Lambda a^2, \qquad (4)$$

$$\rho^2 = r^2 + a^2\cos^2\theta. \qquad (5)$$

The electromagnetic field connected with the geometry is given by the electromagnetic potential

$$A_\mu = \frac{1}{I\varrho^2}\left\{(er + ap\cos\theta)\delta^t_\mu - \left[aer\sin^2\theta + (r^2+a^2)p\cos\theta\right]\delta^\phi_\mu\right\}. \qquad (6)$$

The parameters of the spacetime and its electromagnetic field are: mass (M), specific angular momentum (a), electric charge (e), magnetic charge (p), and cosmological constant (Λ). It is convenient to define a dimensionless cosmological parameter

$$y = \tfrac{1}{3}\Lambda M^2. \qquad (7)$$

For simplicity, we put $M = 1$ hereafter. Equivalently, also the coordinates t, r, and the spacetime parameters a, e, p are expressed in units of M and become dimensionless.

The event horizons of the spacetime are given by the pseudosingularities of the radial coordinate determined by the condition $\Delta_r = 0$. In the asymptotically anti-de Sitter spacetimes with $\Lambda < 0$, two black-hole horizons can exist but no cosmological horizon behind which the spacetime is dynamic—contrary to the case of asymptotically de Sitter spacetimes with $\Lambda > 0$. The loci of the event horizons are determined by the condition

$$y = y_{\mathrm{h}}(r; a, e) \equiv \frac{r^2 - 2r + a^2 + e^2}{r^2\left(r^2+a^2\right)}. \qquad (8)$$

(The magnetic charge p enters the Kerr–Newman–(anti-)de Sitter metric in the same way as the electric charge e; therefore, a re-definition $e^2 + p^2 \to e^2$ can be introduced here.) The function $y_{\rm h}(r; a, e)$ diverges at $r = 0$, while it goes to zero from above for $r \to \infty$. If $a^2 > 0$ and/or $e^2 > 0$, $y_{\rm h} \to \infty$ for $r \to 0$. However, in the special case $a^2 = e^2 = 0$ there is $y_{\rm h} \to -\infty$ for $r \to 0$.

The local extrema of the function $y_{\rm h}(r; a, e)$, giving the boundary between black-hole and naked-singularity spacetimes in the space of parameters y-a^2-e^2, are determined (due to the condition $\partial y_{\rm h}/\partial r = 0$) by the relation

$$\begin{aligned} a^2 &= a^2_{\rm ex(h)\pm}(r; e) \\ &\equiv \frac{1}{2}\left\{-2r^2 + r - e^2 \pm \left[r^2(8r+1) - e^2\left(4r^2 + 2r - e^2\right)\right]^{1/2}\right\}. \end{aligned} \tag{9}$$

The relevant extrema of the '+' branch of the function (9) exist for $e^2 < 9/8$. In the limiting case of $e = 0$, the '+' branch of the function (9) has its maximum at $r_{\rm crit} = 1.61603$, with the corresponding critical value of the rotation parameter $a^2_{\rm crit} = 1.21202$ and the critical value of the cosmological parameter $y_{\rm crit} = 0.0592$. If $0 < e^2 < 9/8$, the critical value $a^2_{\rm crit}(e)$ governs an inflex point of $y_{\rm h}(r; a, e)$. For $a^2 < a^2_{\rm crit}(e)$, the function $y_{\rm h}(r; a, e)$ has two local extrema $y_{\rm h(min)}(a, e)$ and $y_{\rm h(max)}(a, e)$. For $a^2 = a^2_{\rm crit}(e)$, these extrema coincide at $y_{\rm crit}(e)$, which is the limiting value for black-hole spacetimes with a fixed parameter e. The black-hole spacetimes exist for $y_{\rm h(min)}(a, e) < y < y_{\rm h(max)}(a, e)$ (for details, see Ref. [36]).

Distribution of black-hole and naked-singularity spacetimes in the parameter space is given by the functions $y_{\rm h(min)}(a, e)$, $y_{\rm h(max)}(a, e)$ and can be determined by a numerical code. The results are given in Fig. 1. We can see that black-hole

Figure 1. Parameter space (y-a^2-e^2) of the Kerr–Newman spacetimes with a nonzero cosmological constant separated into parts corresponding to the black-hole and naked-singularity spacetimes, respectively. The boundary between these states is given by the functions $y_{\rm h(max)}(a, e)$ and $y_{\rm h(min)}(a, e)$ determined by a numerical code. The black-hole spacetimes can exist for any asymptotically anti-de Sitter spacetime ($y < 0$), while for de Sitter spacetimes ($y > 0$), they are restricted by the upper limit $y_{\max} = 2/27$. In the limiting case, $y = y_{\max}$, we obtain a Reissner–Nordström–de Sitter extreme black hole with $e^2 = 9/8$ (and $a^2 = 0$). Taken from Ref. [36].

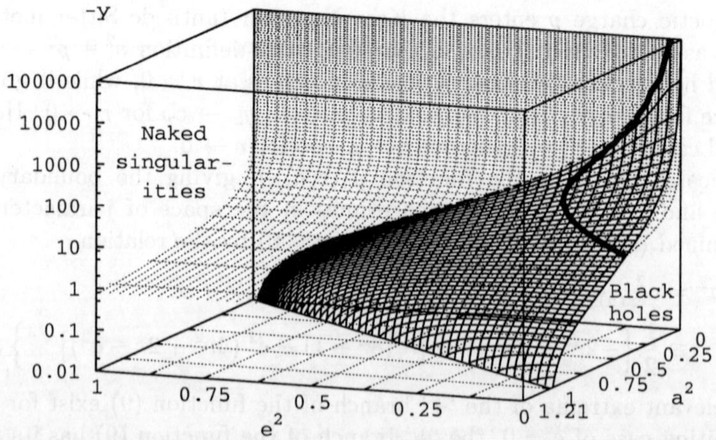

Figure 2. Parameter space of the Kerr–Newman–anti-de Sitter spacetimes. The black-hole spacetimes can exist for any $y < 0$; therefore, we give the $(-y)$ axis in the logarithmic scale. The boundary between the black-hole and naked-singularity states is given by the function $y_{\mathrm{h(min)}}(a, e)$, determined by a numerical code. The light surface corresponds to the boundary $(y_{\mathrm{d}}(a))$ between the spacetimes with and without the surface of degeneracy. Taken from Ref. [37].

spacetimes can exist for all values of attractive cosmological constant $y < 0$, contrary to the case of a repulsive cosmological constant $y > 0$ when black-hole spacetimes must have $y \leq 2/27$, and the extremal value $y_{\mathrm{cr}} = 2/27$ corresponds to the extreme Reissner–Nordström–de Sitter geometry with $e^2 = 9/8$ (see Ref. [36]).

The Kerr–Newman–de Sitter spacetimes can also have pseudosingularities of the latitudinal coordinate. In spacetimes with $ya^2 < -1$, a surface of degeneracy, determined by the condition $\Delta_\theta = 0$, is located at a constant latitude

$$\theta_{\mathrm{d}} = \arccos \frac{1}{\sqrt{-ya^2}}. \tag{10}$$

The combinations of the parameters $ya^2 = -1$ must be excluded due to the fact that for $I = 0$ the geometry is not well defined. Regions of the parameter space corresponding to the spacetimes with a surface of degeneracy are given by the condition

$$y < y_{\mathrm{d}}(a) \equiv -\frac{1}{a^2}. \tag{11}$$

The function $y_{\mathrm{d}}(a)$ is illustrated in Fig. 2.

It can be shown that the surfaces of degeneracy separate the spacetimes into two causally disconnected regions [37]. The geometry (1) with $ya^2 < -1$ has the usual form with the signature $+2$ in the region of $\theta > \theta_{\mathrm{d}} = \arccos(-ya^2)^{-1/2}$. In the black-hole spacetimes, the geometry is static outside the horizons and it is dynamic between the horizons (see Fig. 3a). The geometry is static at all radii in the naked-singularity spacetimes (see Fig. 3b). On the other hand, in the region of $\theta < \theta_{\mathrm{d}}$, the geometry (1) has the opposite signature -2 and the situation is exactly inverse to the situation in the region of $\theta > \theta_{\mathrm{d}}$. The black-hole spacetimes

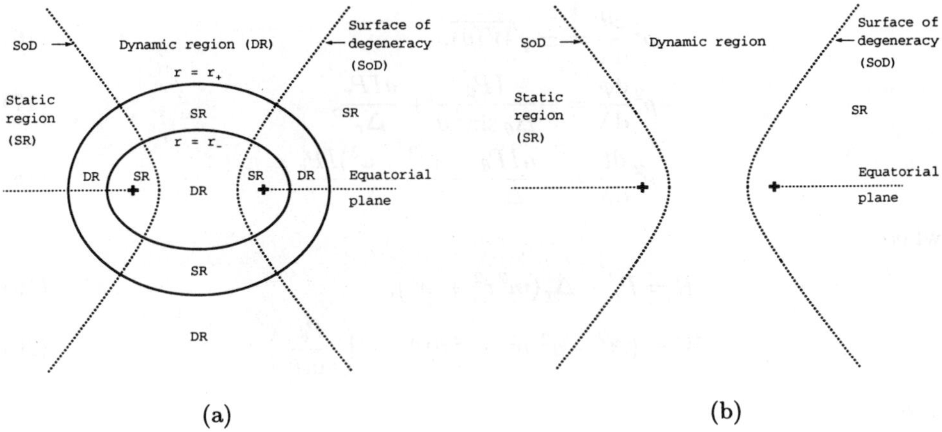

Figure 3. The spacetimes with a surface of degeneracy. The sections of $t = \text{const}$ and $\phi = \text{const}$ of the spacetimes are represented in (a) a black-hole case and (b) a naked-singularity case. Taken from Ref. [37].

are dynamic outside the horizon while they are static between the horizons; the naked-singularity spacetimes are dynamic at all radii.

3 Motion of test particles and photons

The motion of charged test particles is determined by the Lorentz equation

$$m\frac{\mathrm{D}u^\mu}{\mathrm{d}\tau} = (\epsilon F^\mu{}_\nu + g\,{}^*F^\mu{}_\nu)u^\nu \tag{12}$$

and the normalization condition

$$g_{\mu\nu}\frac{\mathrm{d}x^\mu}{\mathrm{d}\lambda}\frac{\mathrm{d}x^\nu}{\mathrm{d}\lambda} = -m^2. \tag{13}$$

The test particle has mass m and carries electric charge ϵ and magnetic monopole charge g; $u^\mu = \mathrm{d}x^\mu/\mathrm{d}\tau$ is its 4-velocity and τ is its proper time, related to the affine parameter λ by the relation $\tau = m\lambda$. For photons, $m = 0$. The tensor of the electromagnetic field $F_{\mu\nu} = A_{\nu;\mu} - A_{\mu;\nu}$; the tensor ${}^*F_{\mu\nu}$ is dual to the tensor $F_{\mu\nu}$ and it is determined by the relation

$${}^*F_{\mu\nu} = \epsilon_{\mu\nu\alpha\beta}F^{\alpha\beta}, \tag{14}$$

where $\epsilon_{\mu\nu\alpha\beta}$ is the completely antisymmetric Levi-Civita symbol. The dual tensor is determined by the vector potential (6) with interchanges $e \to p$, $p \to -e$. The motion of uncharged particles and photons is determined by the geodesic equations.

3.1 The Carter equations

In the integrated and separated form, the motion of test particles and photons is determined by the Carter equations [25]:

$$\rho^2\frac{\mathrm{d}r}{\mathrm{d}\lambda} = \pm\sqrt{R(r)}, \tag{15}$$

$$\rho^2 \frac{\mathrm{d}\theta}{\mathrm{d}\lambda} = \pm\sqrt{W(\theta)}, \tag{16}$$

$$\rho^2 \frac{\mathrm{d}\phi}{\mathrm{d}\lambda} = -\frac{IP_\theta}{\Delta_\theta \sin^2\theta} + \frac{aIP_r}{\Delta_r}, \tag{17}$$

$$\rho^2 \frac{\mathrm{d}t}{\mathrm{d}\lambda} = -\frac{aIP_\theta}{\Delta_\theta} + \frac{(r^2+a^2)IP_r}{\Delta_r}, \tag{18}$$

where

$$R = P_r^2 - \Delta_r(m^2r^2 + \mathscr{K}), \tag{19}$$

$$W = (\mathscr{K} - a^2m^2\cos^2\theta)\Delta_\theta - \left(\frac{P_\theta}{\sin\theta}\right)^2, \tag{20}$$

and

$$P_r = IE(r^2+a^2) - Ia\Phi - (\epsilon e - gp)r, \tag{21}$$

$$P_\theta = IaE\sin^2\theta - I\Phi + (\epsilon p - ge)\cos\theta. \tag{22}$$

The constants of motion are: energy (E), connected with the stationarity of the geometry, the axial angular momentum (Φ), connected with the axial symmetry of the geometry, and the 'total' angular momentum ($\mathscr{K}$), connected with the hidden symmetries of the geometry. It should be stressed that the radial motion is influenced by interactions of electric (or magnetic) charges of the source and the particle while the latitudinal motion is influenced by their mixed electro-magnetic interactions.

In general, investigation of the motion in the rotating and charged backgrounds is rather complex. Therefore, we concentrate our attention on characteristics of the latitudinal motion and then we discuss in detail the most simple case of motion in spherically symmetric, Schwarzschild–(anti-)de Sitter spacetimes. For comparison, we include at this point a detailed discussion of the influence of an attractive cosmological constant on the character of the motion. Most important from the astrophysical point of view are the properties of the circular orbits of test particles, due to their relevance in modeling thin accretion disks. We focus on the behavior of equatorial circular orbits in the Kerr–de Sitter spacetimes; in this case, properties of circular orbits around naked singularities will be mentioned.

3.2 Latitudinal motion

We shortly summarize properties of the latitudinal motion, which is influenced by the 'electro-magnetic' interaction of the source and the test particle, characterized by the parameter

$$\beta = \frac{\epsilon p - ge}{IaE}. \tag{23}$$

The motion can be determined by the 'effective potential' with respect to the constant of motion $K = \mathscr{K}/(aE)^2$, given by the relation

$$K_\mathrm{t} = \frac{(1+ya^2)^2(\sin^2\theta - l - \beta\cos\theta)^2}{(1+ya^2\cos^2\theta)\sin^2\theta} + \frac{\cos^2\theta}{\gamma^2}, \tag{24}$$

where the impact parameter $l = \Phi/(aE)$ and the specific energy $\gamma = E/m$ have been introduced.

If there is no interaction between the electric and magnetic charges of the source and the particle ($\beta = 0$), the motion is symmetric relative to the equatorial plane ($\theta = \pi/2$) of the spacetime and is allowed in this 2-dimensional plane. For the 'electro-magnetic' interaction switched on ($\beta \neq 0$), the motion is asymmetric relative to the equatorial plane and is not allowed there.

If $y > 0$, the latitudinal motion is of the same character as in the well known case of Kerr–Newman spacetimes (see Refs [38, 39]). However, in the case of Kerr–Newman–anti-de Sitter spacetimes ($y < 0$), a qualitative difference appears: beside

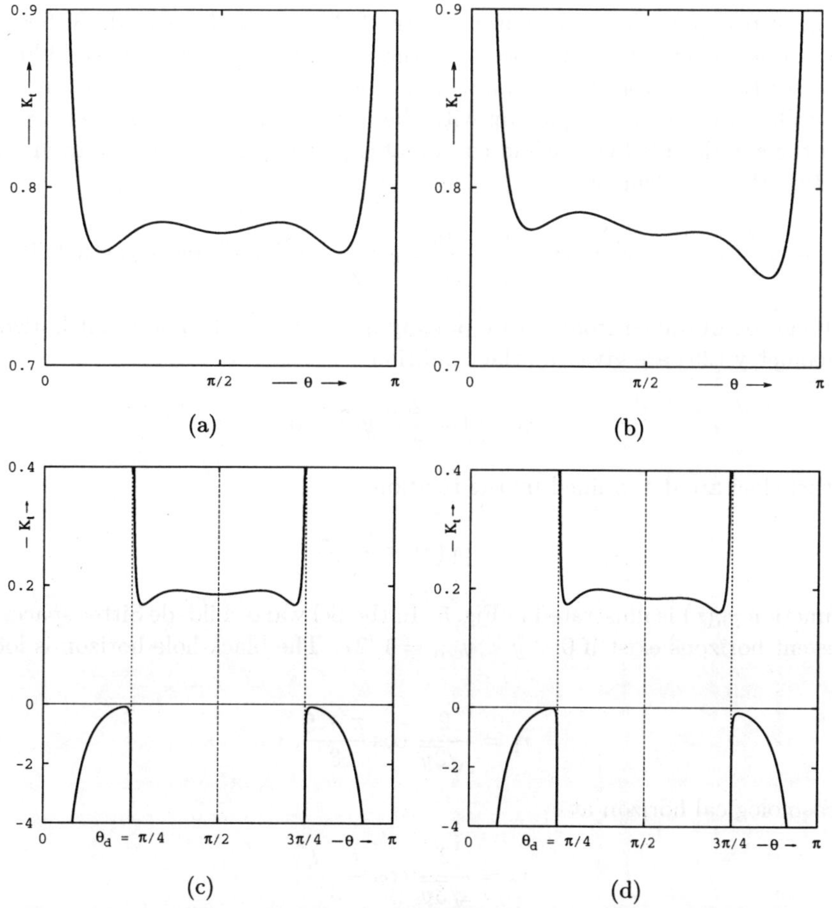

Figure 4. The effective potential of the latitudinal motion in the Kerr–Newman–anti-de Sitter dyon spacetimes. The typical situations are illustrated for the spacetimes with or without the surface of degeneracy and for uncharged or charged particles. For spacetimes without a surface of degeneracy, the effective potentials are given for $ya^2 = -0.2$, $\gamma^2 = 1.9$, and $l = -0.1$ with (a) $\beta = 0$ and (b) $\beta = -0.007$. For spacetimes with a surface of degeneracy, the effective potentials are given for $ya^2 = -2$, $\gamma^2 = 2.5$, and $l = 0.57$ with (c) $\beta = 0$ and (d) $\beta = -0.02$. Taken from Ref. [37].

the stable $\theta = \text{const}$ orbits located outside the equatorial plane, there exist unstable $\theta = \text{const}$ orbits outside the plane. The discussion has to be separated into two parts: if $ya^2 \geq -1$, the motion is allowed in regions where $K \geq K_\mathrm{t}$; if $ya^2 < -1$, the motion is allowed in regions where $K \geq K_\mathrm{t}$ for $\theta > \theta_\mathrm{d}$ and $K \leq K_\mathrm{t}$ for $\theta < \theta_\mathrm{d}$; of course, due to the plane symmetry of the spacetime, we assume $0 \leq \theta \leq \pi/2$. The typical behavior of the effective potential is in all of the characteristic cases illustrated in Fig. 4. For $\beta = 0$, a detailed analysis of the latitudinal motion is given in [26]. It can be checked easily from the behavior of the effective potential that the surface of degeneracy cannot be penetrated by test particles and photons [37].

3.3 Geodesic motion in the Schwarzschild–(anti-)de Sitter spacetimes

Motion of test particles and photons in the Schwarzschild–(anti-)de Sitter spacetimes is fully governed by the geodesic structure of these spacetimes and illustrates thus directly their properties. The role of a nonzero cosmological constant can be clearly illustrated by a comparison with the pure Schwarzschild case with $y = 0$.

In the standard Schwarzschild coordinates (t, r, θ, ϕ), the spacetimes are determined by the line element

$$\mathrm{d}s^2 = -\left(1 - \frac{2}{r} - yr^2\right)\mathrm{d}t^2 + \left(1 - \frac{2}{r} - yr^2\right)^{-1}\mathrm{d}r^2 + r^2(\mathrm{d}\theta^2 + \sin^2\theta\,\mathrm{d}\phi^2), \quad (25)$$

which can be obtained from (1) by putting $a = e = p = 0$. The event horizons of the geometry (25) are given by the condition

$$g_{tt} \equiv 1 - \frac{2}{r} - yr^2 = 0 \quad (26)$$

and their loci are determined by the relation

$$y = y_\mathrm{h}(r) \equiv \frac{r-2}{r^3}; \quad (27)$$

the function $y_\mathrm{h}(r)$ is illustrated in Fig. 5. In the Schwarzschild–de Sitter spacetimes, two event horizons exist if $0 < y < y_\mathrm{crit} = 1/27$. The black-hole horizon is located at

$$r_\mathrm{h} = \frac{2}{\sqrt{3y}}\cos\frac{\pi + \xi}{3}, \quad (28)$$

the cosmological horizon at

$$r_\mathrm{c} = \frac{2}{\sqrt{3y}}\cos\frac{\pi - \xi}{3}, \quad (29)$$

where

$$\xi = \cos^{-1}\left(3\sqrt{3y}\right). \quad (30)$$

The spacetime is dynamic at $r < r_\mathrm{h}$ and $r > r_\mathrm{c}$. If $y = y_\mathrm{crit} = 1/27$, the horizons coincide at $r_\mathrm{h} = 3$. If $y > 1/27$, the spacetime is dynamic for all $r > 0$. In the

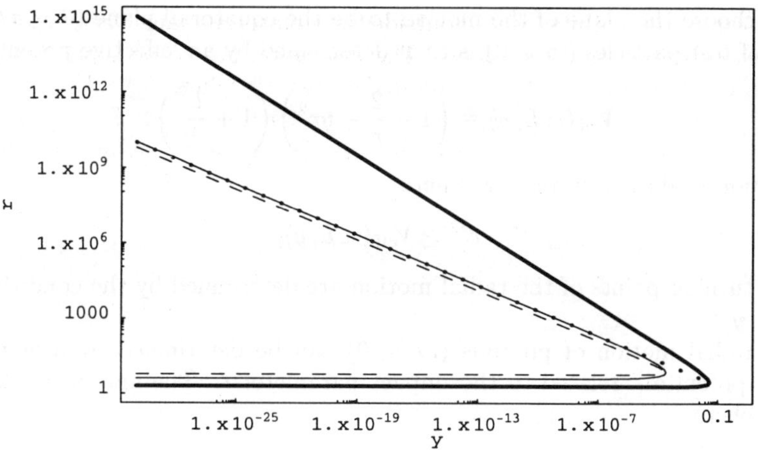

Figure 5. Characteristic radii of the Schwarzschild–de Sitter spacetimes as functions of the parameter y. The black hole (r_{h}) and cosmological (r_{c}) horizons are given by bold solid lines, the static radius (r_{s}) by bold dotted line, the radii of marginally stable orbits ($r_{\mathrm{ms(i)}}$ and $r_{\mathrm{ms(o)}}$) by thin dashed lines, and marginally bound orbits ($r_{\mathrm{mb(i)}}$ and $r_{\mathrm{mb(o)}}$) by thin solid lines. Taken from Ref. [28].

Schwarzschild–anti-de Sitter spacetimes ($y < 0$), a black-hole horizon always exist and its location is determined by the relation

$$r_{\mathrm{h}} = \left(-\frac{1}{y}\right)^{1/3} \left\{ \left[1 + \left(1 - \frac{1}{27y} \right)^{1/2} \right]^{1/3} + \left[1 - \left(1 - \frac{1}{27y} \right)^{1/2} \right]^{1/3} \right\}. \tag{31}$$

The geodesic equation reads

$$\frac{\mathrm{D}p^{\mu}}{\mathrm{d}\lambda} = 0, \tag{32}$$

where $p^{\mu} \equiv \mathrm{d}x^{\mu}/\mathrm{d}\lambda$ is the 4-momentum and λ is the affine parameter. It follows from the central symmetry of the geometry (25) that the motion of test particles and photons is allowed in the central planes only. Due to existence of the time Killing vector $\vec{\xi}_{(t)} = \partial/\partial t$ and the axial one $\vec{\xi}_{(\phi)} = \partial/\partial\phi$, two constants of motion, that are the projections of the 4-momentum onto the Killing vectors, must exist:

$$p_t = g_{t\mu}\, p^{\mu} = -E, \quad p_{\phi} = g_{\phi\mu}\, p^{\mu} = \Phi. \tag{33}$$

In the spacetimes with a nonzero cosmological constant, the constants of motion E and Φ cannot be interpreted as energy and axial component of the angular momentum at infinity, since their geometry is not asymptotically flat. It should be, therefore, interesting to discuss a possibility to find regions of these spacetimes that have a character 'close' to the Schwarzschild spacetime.

It is useful to introduce specific energy $\mathscr{E}$, specific angular momentum L, and impact parameter ℓ by the relations

$$\mathscr{E} = \frac{E}{m}, \quad L = \frac{\Phi}{m}, \quad \ell = \frac{\Phi}{E}. \tag{34}$$

We can choose the plane of the motion to be the equatorial plane ($\theta = \pi/2$). The motion of test particles ($m \neq 0$) is then determined by an 'effective potential' [27]

$$V_{\text{eff}}(r; L, y) \equiv \left(1 - \frac{2}{r} - yr^2\right)\left(1 + \frac{L^2}{r^2}\right). \tag{35}$$

The motion is allowed in regions where

$$\mathcal{E}^2 \geq V_{\text{eff}}(r; L, y), \tag{36}$$

and the turning points of the radial motion are determined by the condition $\mathcal{E}^2 = V_{\text{eff}}(r; L, y)$.

The radial motion of photons ($m = 0$) can be determined by a 'generalized effective potential,' related to the impact parameter ℓ. The motion is allowed in regions where

$$\ell^2 \leq \ell^2_{\text{R}}(r; y) \equiv \frac{r^3}{r - 2 - yr^3}, \tag{37}$$

the condition $\ell^2 = \ell^2_{\text{R}}(r; y)$ gives the turning points of the radial motion.

The functions $V_{\text{eff}}(r; L, y)$ and $\ell^2_{\text{R}}(r; y)$ are defined between the black-hole and cosmological horizons in the case of the Schwarzschild–de Sitter spacetimes; V_{eff} is zero at the horizons, while ℓ^2_{R} diverges there. The functions are defined at all radii above the black-hole horizons in the case of the Schwarzschild–anti-de Sitter spacetimes; V_{eff} is zero at the horizon, while ℓ^2_{R} diverges there.

Circular orbits of test particles correspond to local extrema ($\partial V_{\text{eff}}/\partial r = 0$) of the effective potential. Maxima ($\partial^2 V_{\text{eff}}/\partial r^2 < 0$) determine unstable circular orbits, minima ($\partial^2 V_{\text{eff}}/\partial r^2 > 0$) determine stable circular orbits. The specific energy and specific angular momentum of particles on circular orbits are determined by the relations

$$\mathcal{E}_{\text{c}}(r; y) = \left(1 - \frac{2}{r} - yr^2\right)\left(1 - \frac{3}{r}\right)^{-1/2}, \tag{38}$$

$$L_{\text{c}}(r; y) = \left[r\left(1 - yr^3\right)\right]^{1/2}\left(1 - \frac{3}{r}\right)^{-1/2}. \tag{39}$$

The circular orbits can exist at radii limited by

$$3 < r \leq r_{\text{s}} \equiv y^{-1/3}. \tag{40}$$

At $r = 3$, both $\mathcal{E}_{\text{c}}$ and L_{c} diverge; a photon circular orbit exists there, determined by a local minimum of the function $\ell^2_{\text{R}}(r; y)$, which is located at $r = 3$ independently of the value of the cosmological parameter y. Nevertheless, the impact parameter of the photon circular orbit depends on y:

$$\ell^2_{\text{c}}(y) = \frac{27}{1 - 27y}. \tag{41}$$

Thus, similarly to the 'pure' Schwarzschild case, the radii of circular orbits are limited from below by the photon circular orbit at $r = 3$. In the case of the Schwarzschild–anti-de Sitter spacetimes ($y < 0$), this is the only restriction.

If $y > 0$, radii of circular orbits are limited from above by the so called static radius, where the gravitational attraction invoked by the mass of the source is just balanced by the cosmological repulsion. The static radius is given by the condition

$$y = y_{\mathrm{s}}(r) \equiv \frac{1}{r^3}; \tag{42}$$

$y_{\mathrm{s}}(r)$ is illustrated in Fig. 5. At the static radius, particles with the specific energy

$$\mathscr{E}_{\mathrm{s}}(y) = (1 - 3y^{1/3})^{1/2} \tag{43}$$

(and $L = 0$) are in an unstable equilibrium.

The stable circular orbits exist at radii limited by the relation

$$4yr^4 - 15yr^3 - r + 6 \leq 0. \tag{44}$$

The marginally stable orbits are given by

$$y = y_{\mathrm{ms}}(r) \equiv \frac{r-6}{r^3(4r-15)}. \tag{45}$$

The function $y_{\mathrm{ms}}(r)$ is illustrated in Fig. 5. Its zero is at $r = 6$, corresponding to the marginally stable circular orbit in the Schwarzschild spacetime. The physically

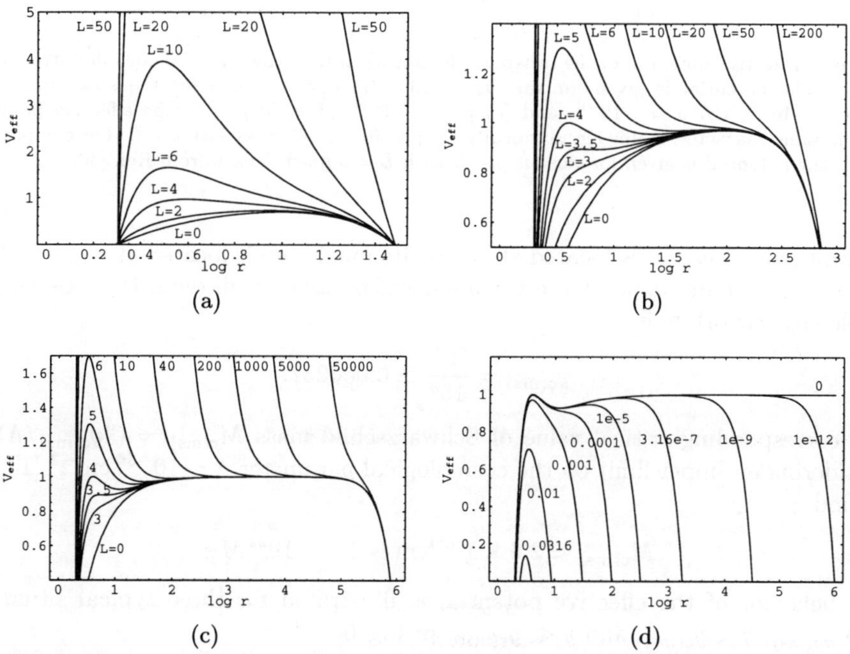

Figure 6. Effective potential of the test-particle motion in the Schwarzschild–de Sitter spacetimes. The potential is given for various specific axial momenta for the spacetimes with (a) $y = 10^{-3} > y_{\mathrm{c(ms)}} \approx 0.000237$, when stable circular orbits cannot exist, (b) $y = 10^{-6}$ and (c) $y = 10^{-12}$, when stable circular orbit corresponding to minima of the effective potential are possible. For comparison, the effective potential is given in (d) for various values of y, with $L = 4$ fixed. Taken from Ref. [40].

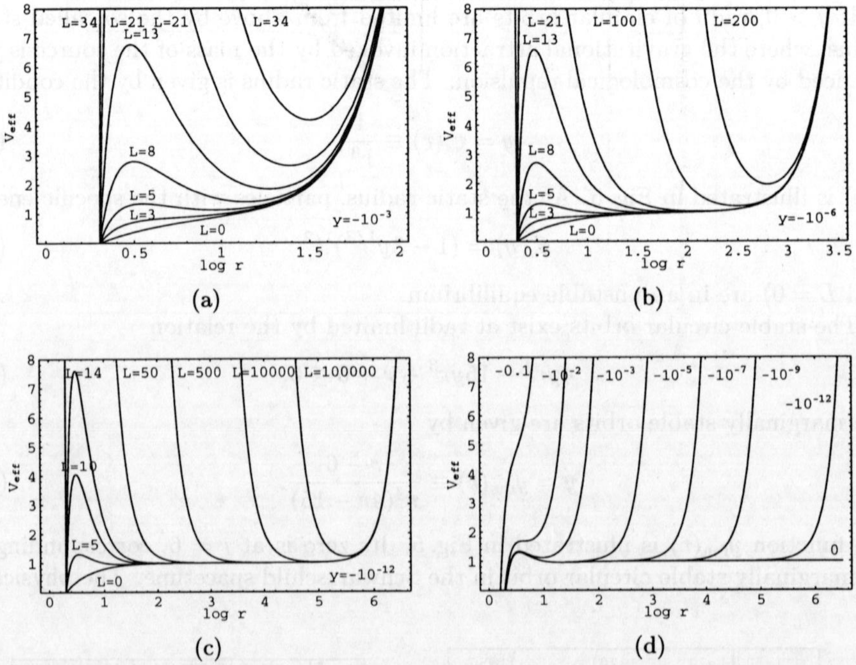

Figure 7. Effective potential of the test-particle motion in the Schwarzschild–anti-de Sitter spacetimes. The potential is given for various specific angular momenta for the spacetimes with (a) $y = -10^{-3}$, (b) $y = -10^{-6}$, and (c) $y = -10^{-12}$. For all $y < 0$, the effective potential has the same character; stable circular orbits are possible in all cases with $y < 0$. For comparison, in (d), the potential is given for various $y < 0$ with $L = 5$ fixed. Taken from Ref. [40].

relevant part of $y_{\rm ms}(r)$ is located at $r \geq 6$. Its maximum is located at $r_{\rm c(ms)} = 15/2$ and the maximum value of the cosmological parameter allowing the existence of stable circular orbits is

$$y_{\rm c(ms)} = \frac{12}{15^4} \simeq 0.000237. \tag{46}$$

The corresponding critical value of Schwarzschild mass $M_{\rm c(ms)} = (3y_{\rm c(ms)}/\Lambda)^{1/2}$; considering an upper limit on the cosmological parameter $\Lambda \sim 10^{-56}\,{\rm cm}^{-2}$ [14,28], we find

$$M_{\rm c(ms)} \approx 2.6 \times 10^{26}{\rm cm} \sim 1.8 \times 10^{21} M_{\odot}. \tag{47}$$

The behavior of the effective potential is illustrated for three typical situations $(y > y_{\rm c(ms)},\ y < y_{\rm c(ms)},\ \text{and}\ y \ll y_{\rm c(ms)})$ in Fig. 6.

If $y < 0$, the asymptotical behavior of the functions determining the motion of test particles and photons has a radically different character than for $y > 0$:

$$V_{\rm eff}(r \to \infty; y) \sim -yr^2, \quad \ell^2_{\rm R}(r \to \infty, y) \sim -y^{-1}. \tag{48}$$

The effective potential is illustrated again for three typical situations (with the same magnitude of y as in the Schwarzschild–de Sitter case) in Fig. 7. Now, the stable

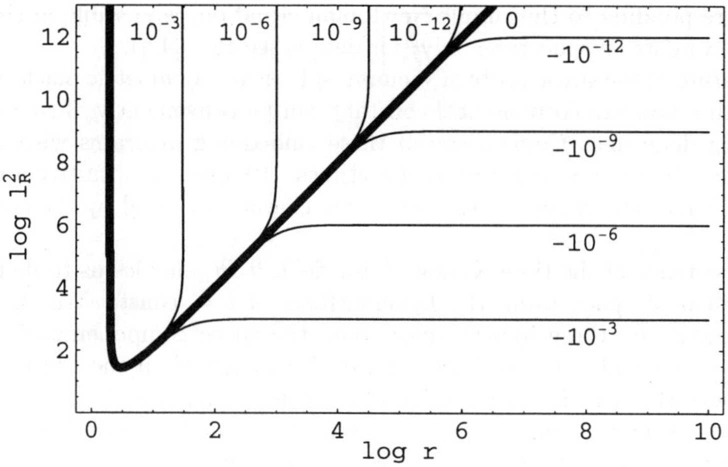

Figure 8. The generalized 'effective potential' of the photon motion in both the Schwarzschild–de Sitter and Schwarzschild–anti-de Sitter spacetimes. The generalized effective potential is related to the impact parameter. Taken from Ref. [40].

circular orbits can exist for all values of the attractive cosmological parameter. The marginally stable orbits are again determined by the relation (45); their radii shift from $r_{\rm ms} \to 6$ (for $y \to 0$) to $r_{\rm ms} \to 15/4$ (for $y \to -\infty$).

The 'generalized effective potential' for the photon motion $\ell^2_{\rm R}(r;y)$ is illustrated in Fig. 8 for the spacetimes with both $y > 0$ and $y < 0$.

The purely radial motion (with $L = 0$ or $\ell = 0$) is determined by the equation

$$\left(\frac{\mathrm{d}r}{\mathrm{d}\tau}\right)^2 + \left(1 - \frac{2}{r} - yr^2\right) = \mathscr{E}^2. \tag{49}$$

The radial motion in terms of both the proper time τ and the coordinate time t can be given by the elliptic integrals (see Ref. [26]).

Note that geodesic motion above the black-hole horizon of charged, Reissner–Nordström–(anti-)de Sitter black-hole spacetimes has the same character as in the case of Schwarzschild–(anti-)de Sitter black holes. On the other hand, in the naked-singularity charged backgrounds, two circular photon orbits exist, if $1 < e^2 < 9/8$, and even two separated regions of stable circular orbits exist in the naked-singularity backgrounds with $e^2 < 5/4$ (see Ref. [41] for details). Further, it can be shown that the equilibrium of spinning charged test particles in the Reissner–Nordström–(anti-)de Sitter spacetimes is independent of the spin due to the symmetry properties of the spherically symmetric spacetimes. Any charged particle can be in an unstable equilibrium state between the black-hole and cosmological horizons if $y > 0$ while sufficiently repulsed particles can be in an inner unstable equilibrium state and an outer stable equilibrium state above the horizon if $y < 0$. In the asymptotically de Sitter naked-singularity spacetimes, even two equilibrium positions are possible for uncharged particles; they correspond to the static radii of the spacetimes. The neutral, or separation-independent equilibrium, corresponding to the simple Newtonian situation, is possible in the extreme Reissner–Nordström spacetime only (for details see Ref. [42]).

Curvature of the static parts of vacuum, spherically symmetric black-hole spacetimes with a nonzero cosmological constant can be conveniently illustrated by the embedding diagrams. Comparison of these embedding diagrams with those constructed for the pure Schwarzschild spacetimes [19] gives an intuitive and illustrative insight into the change of the spacetime character caused by the cosmological constant.

The existence of the time Killing vector field, $\partial/\partial t$, enables us to define a privileged notion of space using the hypersurfaces of t = const. We shall use the induced metric on these hypersurfaces (i.e., the space components of the metric tensor g_{ik})—we call it the ordinary space. Geometry of all the central planes of these hypersurfaces is the same as that of the equatorial plane ($\theta = \pi/2$) as a consequence of the central symmetry of the spherically symmetric spacetimes. Therefore, we shall embed the surface described by the line element

$$\mathrm{d}l^2_{(\mathrm{S-dS})} = \left(1 - \frac{2}{r} - yr^2\right)^{-1} \mathrm{d}r^2 + r^2\,\mathrm{d}\phi^2 \tag{50}$$

into a flat Euclidean 3-dimensional space the line element of which is in the standard cylindrical coordinates given by

$$\mathrm{d}\sigma^2 = \mathrm{d}\rho^2 + \rho^2\,\mathrm{d}\phi^2 + \mathrm{d}z^2. \tag{51}$$

The embedding is given by the surface $z = z(\rho)$, which has to be isometric with the equatorial plane of the t = const hypersurface of the spacetime. Thus, we have to identify the line element given by

$$\mathrm{d}l^2_{(\mathrm{E})} = \left[1 + \left(\frac{\mathrm{d}z}{\mathrm{d}\rho}\right)^2\right] \mathrm{d}\rho^2 + \rho^2\,\mathrm{d}\phi^2 \tag{52}$$

with the line element (50).

We can identify both the azimuthal coordinates and the radial coordinates r and ρ . The embedding diagram can then be given by the formula $z = z(r)$, which can be obtained by integrating the relation

$$\frac{\mathrm{d}z}{\mathrm{d}r} = \pm\left(\frac{2 + yr^3}{r - 2 - yr^3}\right)^{1/2}. \tag{53}$$

The choice of the positive or negative sign in Eq. (53) leads to isometric surfaces and is therefore irrelevant.

If $y > 0$, the embedding can be constructed for complete static regions between the black-hole (r_{h}) and cosmological (r_{c}) horizons. Recall that the static region exists for $y \leq 1/27$ only. The presence of a repulsive cosmological constant alters the character of the embedding diagram substantially in comparison with the case of $y = 0$. Due to the asymptotical behavior of the Schwarzschild–de Sitter spacetimes, the embedding diagrams have a shape corresponding to a funnel—the throat corresponds to both the black-hole and cosmological horizons. With y growing, the distance between the horizons decreases and the height of the diagram decreases too. The shape of the embedding diagram becomes closer and closer to a cylindrical surface with more and more limited height. With $y \to 1/27$, the diagram is shrinking to the circle of radius $r = 3$.

If $y < 0$, the static region extends from the black-hole horizon to infinity. However, we can see directly from Eq. (53) that the embedding diagrams of the ordinary space can be constructed within a limited part of the static region, located between the black-hole horizon r_{h} and $r_{\mathrm{e(ord)}} = (-2/y)^{1/3}$. We shall characterize the limit of embeddability of the ordinary space by the condition

$$y \geq y_{\mathrm{e(ord)}}(r) \equiv -\frac{2}{r^3}. \tag{54}$$

If the embeddability condition is satisfied, the rate of change of the circumferential length per unit increase of the proper radial distance can be 'fit' into the Euclidean space. In the opposite case, the space is still well defined but the circumferential size grows faster than it does on a Euclidean plane and no embedding in the Euclidean space is then possible [43].

Qualitative features of the modifications in the ordinary geometry embedding diagrams are illustrated by Fig. 9 (for both $y > 0$ and $y < 0$). The behavior of the embedding diagrams is illustrative and we can see directly that the embedding diagrams of spacetimes with a repulsive cosmological constant low enough peel off the embedding diagram corresponding to the pure Schwarzschild spacetime just

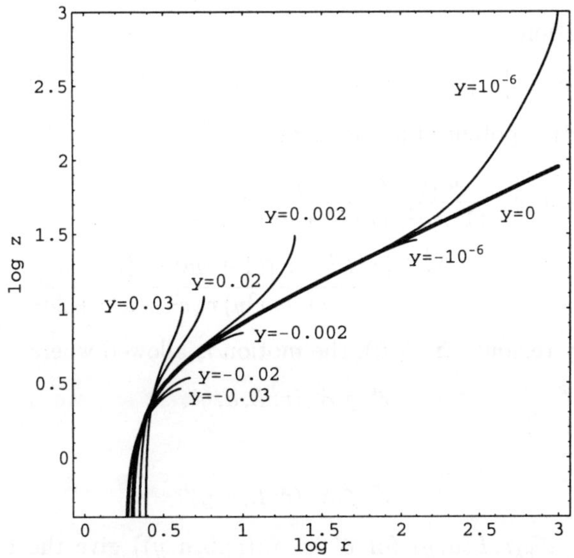

Figure 9. Qualitative features of the embedding diagrams of the ordinary geometry of the Schwarzschild–de Sitter and Schwarzschild–anti-de Sitter spacetimes in a log-log diagram. One can immediately see how the diagrams with $y \neq 0$ 'peel off' the pure Schwarzschild diagram ($y = 0$, bold curve). All sections with $y \neq 0$ are complete (i.e., the maximum that can be embedded into Euclidean space is shown), except for uninteresting lower parts of the throats. The diagrams clearly indicate modifications of the spacetime structure caused by the presence of a cosmological constant. Taken from Ref. [40].

near the static radius. This kind of behavior indicates that the region near the static radius could be considered to have properties that are close to the asymptotic region of the asymptotically flat spacetimes.

3.4 Equatorial circular orbits in the Kerr–de Sitter spacetimes

Studies of rotating, Kerr black-hole backgrounds are very important in order to understand astrophysically realistic situations in active galactic nuclei because their central black holes are assumed to be rotating very fast due to the matter accreted from the disk orbiting the black hole. It is crucial to understand the combined influence of the rotation and the repulsive cosmological constant on the equatorial circular motion due to its direct relevance in determining the structure of a thin accretion disk [44] and their relation to the structure of thick accretion disks [45]. In this case, it seems useful also to mention the properties of circular geodesics in the Kerr–de Sitter naked-singularity backgrounds.

The equatorial geodetical motion is, via the Carter equations, governed by the constants of motion E, Φ. (The equation of the latitudinal motion implies that the constant of motion $K = I^2(aE - \Phi)^2$.) The equatorial motion can be conveniently determined by an 'effective potential' given by the condition $R(r) = 0$. It is useful to define specific energy and specific angular momentum by the relations

$$\mathscr{E} = \frac{IE}{m}, \quad L = \frac{I\Phi}{m}. \tag{55}$$

Solving the equation

$$R(r) \equiv \left[\mathscr{E}\left(r^2 + a^2\right) - aL\right]^2 - \Delta_r\left[r^2 + (a\mathscr{E} - L)^2\right] = 0, \tag{56}$$

we find the effective potential in the form

$$\mathscr{E}_{\pm}(r; L, a, y) \equiv \frac{a\left[yr\left(r^2 + a^2\right) + 2\right]L}{\left(1 + a^2 y\right) r\left(r^2 + a^2\right) + 2a^2} \pm \frac{\Delta_r^{1/2}\left\{r^2 L^2 + r\left[\left(1 + ya^2\right) r\left(r^2 + a^2\right) + 2a^2\right]\right\}^{1/2}}{\left(1 + a^2 y\right) r\left(r^2 + a^2\right) + 2a^2}. \tag{57}$$

In the stationary regions ($\Delta_r \geq 0$), the motion is allowed where

$$\mathscr{E} \geq \mathscr{E}_{+}(r; L, a, y), \tag{58}$$

or

$$\mathscr{E} \leq \mathscr{E}_{-}(r; L, a, y). \tag{59}$$

Conditions $\mathscr{E} = \mathscr{E}_{+}(r, L, a, y)$ (or $\mathscr{E} = \mathscr{E}_{-}(r; L, a, y)$) give the turning points of the radial motion; in the dynamic regions ($\Delta_r < 0$), the turning points are not allowed. In the region outside the outer black-hole horizon, the motion of particles in the positive-root states, i.e., particles with positive energy as measured by local observers, and future-directed ($\mathrm{d}t/\mathrm{d}\lambda > 0$), having a direct physical meaning, is determined by the effective potential $\mathscr{E}_{+}(r; L, a, y)$. (For the character of motion in the entire background and the relevance of $\mathscr{E}_{-}(r; L, a, y)$, the results obtained in [46, 47] hold.)

Trajectories of the equatorial motion are determined by the equation

$$\frac{\mathrm{d}\phi}{\mathrm{d}r} = \pm\frac{r}{\Delta_r}\frac{a\mathscr{E}\left(r^2+a^2-\Delta_r\right)+\left(\Delta_r-a^2\right)L}{\sqrt{\left[\mathscr{E}\left(r^2+a^2\right)-aL\right]^2-\Delta_r\left[r^2+\left(a\mathscr{E}-L\right)^2\right]}}. \tag{60}$$

The circular orbits can most easily be determined by solving simultaneously the equations $R=0$ and $\mathrm{d}R/\mathrm{d}r=0$. After some algebraic manipulations (for details see Ref. [44]) we arrive at formulae determining the energy and angular momentum of a circular orbit in terms of its radius and the spacetime parameters:

$$\mathscr{E}_{\mathrm{c}\pm}(r;a,y) = \frac{r^{3/2}-2r^{1/2}\pm a\left(1-yr^3\right)^{1/2}-r^{3/2}\left(r^2+a^2\right)y}{r^{3/4}\left[r^{3/2}\left(1-ya^2\right)-3r^{1/2}\pm 2a\left(1-yr^3\right)^{1/2}\right]^{1/2}}, \tag{61}$$

and

$$L_{\mathrm{c}\pm}(r;a,y) = \pm\frac{\left(r^2+a^2\right)\left(1-yr^3\right)^{1/2}\mp\left[2ar^{1/2}+ar^{3/2}\left(r^2+a^2\right)y\right]}{r^{3/4}\left[r^{3/2}\left(1-a^2y\right)-3r^{1/2}\pm 2a\left(1-yr^3\right)^{1/2}\right]^{1/2}}, \tag{62}$$

that reduce to the well known formulae found by Bardeen, Press and Teukolsky [48]:

$$\mathscr{E}_{\mathrm{c}\pm}(r;a) = \frac{r^{3/2}-2r^{1/2}\pm a}{r^{3/4}\left[r^{3/2}-3r^{1/2}\pm 2a\right]^{1/2}} \tag{63}$$

$$L_{\mathrm{c}\pm}(r;a) = \pm\frac{r^2+a^2\mp 2ar^{1/2}}{r^{3/4}\left[r^{3/2}-3r^{1/2}\pm 2a\right]^{1/2}}. \tag{64}$$

In the limit of $a\to 0$, we arrive at formulae determined in the case of Schwarzschild–de Sitter black holes—see the preceding section.

The orbits corresponding to the '+' sign in the relations (61), (62) are called orbits of the first family by the authors, those corresponding to the '−' sign are called orbits of the second family. The orbits of the second family are all counterrotating relative to the family of locally nonrotating observers introduced by Bardeen [47]. In the field of Kerr black holes, orbits of the first family are all corotating relative to the locally nonrotating observers. However, in the field of a Kerr naked singularity with a sufficiently low parameter a, the orbits of the first family become counterrotating when approaching the ergosphere, and their energy relative to infinity becomes negative when entering the ergosphere allowing thus for extraction of the rotational energy from the Kerr naked singularity by a thin accretion disk (see Ref. [49]). This kind of separation between the orbits is useful even in the case of black-hole spacetimes if a repulsive cosmological constant is taken into consideration.

The formulae for the energy and angular momentum of circular orbits hold equally for both Kerr–de Sitter ($y>0$) and Kerr–anti-de Sitter ($y<0$) spacetimes. Here, we shall concentrate our discussion on the circular motion in the Kerr–de Sitter spacetimes.

Inspecting expressions (61) and (62), we find two reality conditions on the circular orbits:

$$y \leq y_{\rm s} \equiv \frac{1}{r^3}, \tag{65}$$

which gives the 'static radius' (compare with a formally identical result in the Schwarzschild–de Sitter spacetimes discussed in the preceding section), and

$$1 - \frac{3}{r} - a^2 y \pm 2a\left(\frac{1}{r^3} - y\right)^{1/2} \geq 0, \tag{66}$$

where the equality sign determines photon circular orbits with both $\mathscr{E}_{\rm c} \to \infty$ and $L \to \pm\infty$.

The photon circular orbits can be given by the formula

$$a^2_{\rm ph\pm}(r;y) = \frac{(1-yr^3) + (1-3yr^2) \pm 2\sqrt{(1-yr^3)(1-3yr^2)}}{y^2 r^3}. \tag{67}$$

The photon equatorial motion and the photon circular orbits were discussed in [36]. Here, we focus our attention on the astrophysically most important stable circular orbits of test particles because of their relevance to accretion disks.

The stable circular orbits are given by the condition

$$\frac{\mathrm{d}^2 R}{\mathrm{d}r^2} \geq 0, \tag{68}$$

which has to be satisfied simultaneously with $R(r) = 0$ and $\mathrm{d}R/\mathrm{d}r = 0$; we find

$$r\left[6 - r + r^3(4r - 15)y\right] \mp 8a\left[r\left(1 - yr^3\right)\right]^{1/2} + a^2\left[3 + r^2 y\left(1 - 4yr^3\right)\right] \leq 0. \tag{69}$$

The marginally stable orbits are determined by the relation

$$\begin{aligned} a^2_{\rm ms\pm}(r;y) = \left[3 + r^2 y\left(1 - 4yr^3\right)\right]^{-2} r\Big\{&\left[r - 6 - r^3(4r - 15)y\right] \\ &\times \left[3 + r^2 y\left(1 - 4yr^3\right)\right] + 32\left(1 - yr^3\right)^3 \\ &\pm 8\left(1 - yr^3\right)^{3/2}\left(1 - 4yr^3\right)^{1/2} \\ &\times \left\{r\left[3 - ry\left(6 + 10r - 15yr^3\right)\right] - 2\right\}^{1/2}\Big\}. \end{aligned} \tag{70}$$

We directly see that there are two reality conditions for $a^2_{\rm ms\pm}(r;y)$. It is the standard condition $y < y_{\rm s} \leq 1/r^3$ that is automatically satisfied for the circular orbits and the relevant conditions

$$y < y_{\rm ms} \equiv \frac{1}{4r^3}, \tag{71}$$

and

$$y < y_{\rm ms-}(r), \quad y > y_{\rm ms+}(r), \tag{72}$$

where

$$y_{\rm ms\pm}(r) = \frac{3 + 5r \pm \left(30r - 20r^2 + 9\right)^{1/2}}{15r^3}. \tag{73}$$

A detailed analysis of these functions allows for the following, briefly presented, conclusions. In the presence of a repulsive cosmological constant, circular orbits of the first family become counterrotating even in all black-hole backgrounds with $y > 0$. However, this happens near the static radius and for unstable circular orbits only. The stable orbits of the first family are always corotating in the black-hole backgrounds. In the naked singularity backgrounds, circular orbits of the first family become counterrotating near the static radius and, as in the case of Kerr naked singularities, they become counterrotating near the ergosphere if the rotation parameter a of the background is adjusted appropriately. Further, it can be shown that both black-hole and naked-singularity spacetimes exist in three types: the first one contains both the corotating and counterrotating stable orbits, the second type contains the corotating stable orbits only, and the third type contains no stable orbits. For details, see Ref. [44].

4 Einstein–Strauss–de Sitter vacuola

The Einstein–Strauss–de Sitter vacuola is a simple, fully nonlinear model representing gravitational field of a spherically symmetric mass condensation, or a black hole, immersed in the expanding ('dusty') Friedman universe with a nonzero cosmological constant [26,27]. In this model, the vacuum Schwarzschild–de Sitter spacetime characterized by mass parameter M and described by the line element

$$ds^2 = -\mathscr{A}^2(r)\,\mathrm{d}t^2 + \mathscr{A}^{-2}(r)\,\mathrm{d}r^2 + r^2\,\mathrm{d}\Omega, \tag{74}$$

$$\mathscr{A}^2 = 1 - \frac{2M}{r} - \frac{\Lambda}{3}r^2, \tag{75}$$

is smoothly matched to the (external) Friedman universe described in the standard comoving coordinates (T, χ, θ, ϕ) by the Robertson–Walker line element

$$ds^2 = -\mathrm{d}T^2 + a^2(T)\left[\mathrm{d}\chi^2 + \Sigma_k^2(\chi)\,\mathrm{d}\Omega\right], \tag{76}$$

where

$$\Sigma_k(\chi) = \begin{cases} \sin\chi & \text{if} \quad k = +1, \\ \chi & \text{if} \quad k = 0, \\ \sinh\chi & \text{if} \quad k = -1. \end{cases} \tag{77}$$

Evolution of the Friedman universe is determined by the dependence of the scale factor a and energy density ϵ on the cosmic time T through the equation

$$\left(\frac{\mathrm{d}a}{\mathrm{d}T}\right)^2 = \frac{8}{3}\pi\epsilon a^2 + \frac{1}{3}\Lambda a^2 - k, \tag{78}$$

and the equation expressing the conservation of energy

$$\tfrac{8}{3}\pi\epsilon a^3 = \text{const} = a_0. \tag{79}$$

The Schwarzschild–de Sitter spacetime can be a completely vacuum-spacetime, describing a black hole, or it can be joined to a part of an internal Friedman universe representing a mass condensation with parameters, in general, being different from those of the external Friedman universe.

4.1 Construction of the vacuola model

In this model, the interior of a spherical shell S_F in the dusty Friedman universe is replaced by the interior of the corresponding spherical shell S_S in the Schwarzschild–de Sitter spacetime. On the hypersurface S_F, chosen in such a way that $\chi = \chi_b =$ const, the length of main circles is determined by

$$2\pi r_b = 2\pi\, a(T)\, \Sigma_k(\chi_b). \tag{80}$$

Test particles at rest on the matching hypersurface from the point of view of the Robertson–Walker geometry follow radial geodesics in the Schwarzschild–de Sitter geometry. The proper time, as measured on the matching hypersurface in both geometries, must be synchronized in order to identify S_F and S_S. On $S_S = (t_b(\tau), r_b(\tau))$, the proper time is given by

$$\mathrm{d}\tau = \pm\sqrt{\frac{\tilde{R}}{2M}}\left(\frac{\Lambda \tilde{R} r_b^2}{6M} - k + \frac{\tilde{R}}{r_b}\right)^{-1/2} \mathrm{d}r, \tag{81}$$

with the parameter $\tilde{R}$ related to the specific energy of the radial geodesic $\mathscr{E}_b$ by

$$\mathscr{E}_b = 1 - \frac{2kM}{\tilde{R}}. \tag{82}$$

The matching conditions in the form

$$r_b = a(T)\, \Sigma_k(\chi_b), \tag{83}$$

$$\tilde{R} = a_0\, \Sigma_k(\chi_b), \tag{84}$$

$$\tilde{R}\sqrt{\tilde{R}/2M} = a_0 \tag{85}$$

then guarantee the identification of the proper time τ as measured along the radial geodesics with the cosmic time T.

The internal geometry of the matching hypersurface as measured in the Friedman universe is given by

$$\mathrm{d}s_+^2 = -\mathrm{d}T^2 + a^2(T)\, \Sigma_k^2(\chi_b)(\mathrm{d}\theta^2 + \sin^2\theta\, \mathrm{d}\phi^2), \tag{86}$$

while it is given by

$$\mathrm{d}s_-^2 = -\mathrm{d}T^2 + r_b^2(T)\, (\mathrm{d}\theta^2 + \sin^2\theta\, \mathrm{d}\phi^2), \tag{87}$$

if measured from the Schwarzschild–de Sitter spacetime. These geometries are identical due to the matching conditions. We can also show that the components of the matching hypersurface external curvature are identical. The external curvature of a 3-geometry representing a spacetime section determines the change of the unit vector $\vec{n}$ normal to the section when parallely transported in the 3-geometry of the section:

$$K_{ij} = -\vec{e}_i \cdot \nabla_j \vec{n} = \tfrac{1}{2}(g_{in,j} + g_{ij,n} - g_{nj,i}), \tag{88}$$

with $\vec{e}_i$ denoting the basis vectors of the 3-geometry; here, these are $\partial/\partial T$, $\partial/\partial\theta$, and $\partial/\partial\phi$. The normal vector to the hypersurface $\chi = \chi_b =$ const is

$$\vec{n} = \frac{1}{a}\frac{\partial}{\partial \chi}. \tag{89}$$

As measured in the Robertson–Walker geometry, the nonzero components of the external curvature are

$$K_{\theta\theta}^{(+)} = \frac{K_{\phi\phi}^{(+)}}{\sin^2\theta} = -a(T)\,\Sigma_k(\chi_{\rm b})\,\frac{{\rm d}\Sigma_k}{{\rm d}\chi}(\chi_{\rm b}), \tag{90}$$

while in the case of the Schwarzschild–de Sitter geometry the nonzero components are given by

$$K_{\theta\theta}^{(-)} = \frac{K_{\phi\phi}^{(-)}}{\sin^2\theta} = -r_{\rm b}(T)\left(1 - \frac{2kM}{\tilde{R}}\right)^{1/2}. \tag{91}$$

Again, these components can be identified due to the matching conditions and the construction of the Einstein–Strauss–de Sitter vacuola model is correct. The matching conditions also hold for the gravitational collapse dynamics of a homogeneous dust sphere with a cosmological constant. This situation is discussed in detail in [50].

4.2 Geodesics matching

A geodesic crossing both the Friedman universe and the Schwarzschild–de Sitter spacetime has to be smoothly matched on the matching hypersurface, i.e., the tangent vector to the geodesic must be the same on both sides of $S_{\rm F} = S_{\rm S}$. Therefore, we have to find the Lorentz transformation relating the comoving Friedman observers and the static Schwarzschild–de Sitter observers on the matching hypersurface.

In the Friedman universe, the 4-momentum components of the test-particle geodetical motion are given by [27]

$$p^T = \frac{{\rm d}T}{{\rm d}\lambda} = \left(m^2 + \frac{p^2}{a^2}\right)^{1/2}, \tag{92}$$

$$p^\chi = \frac{{\rm d}\chi}{{\rm d}\lambda} = \pm\frac{1}{a^2}\left(p^2 - \frac{L^2}{\Sigma_k^2}\right)^{1/2}, \tag{93}$$

$$p^\theta = \frac{{\rm d}\theta}{{\rm d}\lambda} = \pm\frac{1}{a^2\Sigma_k^2}\left(L^2 - \frac{\ell^2}{\sin^2\theta}\right)^{1/2}, \tag{94}$$

$$p^\phi = \frac{{\rm d}\phi}{{\rm d}\lambda} = \frac{\ell}{a^2\Sigma_k^2\sin^2\theta}; \tag{95}$$

the proper time of the particle τ, the affine parameter λ, and its mass m are related by $\tau = m\lambda$ and the three constants of motion are

$$\ell = p_\phi, \tag{96}$$

$$L^2 = p_\theta^2 + p_\phi^2\sin^{-2}\theta, \tag{97}$$

$$p^2 = p_\chi^2 + L^2\Sigma_k^{-2}; \tag{98}$$

$\ell(L)$ represents the azimuthal (total) angular momentum of the particle. In the Schwarzschild–de Sitter spacetime there is

$$p^t = \frac{{\rm d}t}{{\rm d}\lambda} = E\mathscr{A}^{-2}(r), \tag{99}$$

$$p^r = \frac{\mathrm{d}r}{\mathrm{d}\lambda} = \pm(E^2 - V_{\text{eff}}^2)^{1/2}, \tag{100}$$

$$p^\theta = \frac{\mathrm{d}\theta}{\mathrm{d}\lambda} = \pm\frac{1}{r^2}\left(L^2 - \frac{\ell^2}{\sin^2\theta}\right)^{1/2}, \tag{101}$$

$$p^\phi = \frac{\mathrm{d}\phi}{\mathrm{d}\lambda} = \frac{\ell}{r^2\sin^2\theta}, \tag{102}$$

with the effective potential of the motion determined by

$$V_{\text{eff}}^2 = \mathscr{A}^2(r)\left(m^2 + \frac{L^2}{r^2}\right). \tag{103}$$

The constants of motion ℓ a L have the same meaning as in the Friedman case, while the covariant energy

$$E = -p_t. \tag{104}$$

In the case of null geodesics (motion of photons) we have to put $m = 0$.

Considering Schwarzschild and Friedman coordinate systems moving relative to each other in the radial direction, their orthonormal basis vectors are related by the Lorentz transformation

$$\vec{e}_{(\mu')} = \Lambda_{\mu'}{}^{\nu}\,\vec{e}_{(\nu)} \tag{105}$$

with the Lorentz matrix

$$\Lambda_{\mu'}{}^{\nu} = \begin{pmatrix} \cosh\alpha & \sinh\alpha & 0 & 0 \\ \sinh\alpha & \cosh\alpha & 0 & 0 \\ 0 & 0 & 1 & 0 \\ 0 & 0 & 0 & 1 \end{pmatrix}. \tag{106}$$

The orthonormal basis of the static Schwarzschild–de Sitter observers is given by

$$\vec{e}_{(t)} = \mathscr{A}^{-1}(r)\,\partial/\partial t, \tag{107}$$

$$\vec{e}_{(r)} = \mathscr{A}(r)\,\partial/\partial r, \tag{108}$$

$$\vec{e}_{(\theta)} = r^{-1}\,\partial/\partial\theta, \tag{109}$$

$$\vec{e}_{(\phi)} = (r\sin\theta)^{-1}\,\partial/\partial\phi, \tag{110}$$

and the orthonormal basis of the Friedman comoving observers is given by

$$\vec{e}_{(T)} = \partial/\partial T, \tag{111}$$

$$\vec{e}_{(\chi)} = a^{-1}\,\partial/\partial\chi, \tag{112}$$

$$\vec{e}_{(\theta)} = (a\Sigma_k)^{-1}\,\partial/\partial\theta, \tag{113}$$

$$\vec{e}_{(\phi)} = (a\Sigma_k\sin\theta)^{-1}\,\partial/\partial\phi. \tag{114}$$

Due to the matching condition, the angular basis vectors $\vec{e}_\theta$, $\vec{e}_\phi$ are the same on both sides of the matching hypersurface. Since the 4-velocity of a test particle confined to the hypersurface $S_{\mathrm{F}} = S_{\mathrm{S}}$ is

$$\vec{u}_{(\mathrm{b})} = \partial/\partial T = \vec{e}_T = \mathscr{A}^{-1}(r_{\mathrm{b}})\,\mathscr{E}_{\mathrm{b}}\,\vec{e}_t + \left[\mathscr{E}_{\mathrm{b}}^2 - \mathscr{A}^2(r_{\mathrm{b}})\right]^{1/2}\mathscr{A}^{-1}(r_{\mathrm{b}})\,\vec{e}_r, \tag{115}$$

we directly obtain the parameter of the Lorentz transformation in the form

$$\cosh\alpha = \Lambda_T{}^t = \Lambda_\chi{}^r = \mathscr{A}^{-1}(r_\mathrm{b}) = \sqrt{1-\frac{2kM}{\tilde{R}}}\left(1-\frac{2M}{r_\mathrm{b}}-\frac{1}{3}\Lambda r_\mathrm{b}^2\right)^{-1/2}. \tag{116}$$

The parameter of the Lorentz shift

$$V_\mathrm{b}(r_\mathrm{b}) = \tanh\alpha = \left[1-\mathscr{E}_\mathrm{b}^{-2}\mathscr{A}^2(r_\mathrm{b})\right]^{1/2} \tag{117}$$

determines the expansion velocity of the matching hypersurface relative to the static Schwarzschild observers; with the corresponding Lorentz factor given by

$$\gamma = (1-V_\mathrm{b}^2)^{-1/2} = \cosh\alpha. \tag{118}$$

The Lorentz transformation enables an effective discussion of phenomena related to the vacuola model. For example, it can be shown how a spherical distribution of momentum of test particles or photons in the Friedman universe is transformed into a prolate-ellipsoid distribution relative to the static Schwarzschild observers [27]. This effect is of crucial importance in models where bound mass condensations of massive weakly interacting particles (e.g., neutrinos) are created in the expanding universe [51, 52].

Another phenomenon, namely the refraction of light, could be important when modeling anisotropies of the relict microwave background in the framework of the Rees–Sciama effect explaining the background temperature fluctuations as a result of mass distribution inhomogeneities in the expanding universe. The Einstein–Strauss model has been used in [53], however, the effect of refraction has not been considered there.

The refraction of light on the matching hypersurface can be determined by using the Lorentz transformation. The effect of refraction is that the directional angle as measured by a static Schwarzschild observer differs from the angle as measured by a comoving Friedman observer. The equations of the geodetical motion have to be taken with $m = 0$.

The directional angle ψ of a photon, assigned by an observer to its outward directed radial direction, is given by

$$\sin\psi = \frac{p^{(\phi)}}{p^{(t)}}, \quad \cos\psi = \pm\frac{p^{(r)}}{p^{(t)}}, \tag{119}$$

where $p^{(\phi)}$, $p^{(r)}$, and $p^{(t)}$ are nonzero components of the photon 4-momentum, as measured by the observer. The upper (lower) sign corresponds to the photons entering (leaving) the Schwarzschild–de Sitter spacetime. The components of the 4-momentum, as measured by local observers, are

$$p_{(\mu)} = p_\nu\, e^\nu_{(\mu)}. \tag{120}$$

Denoting the angle on the Friedmanian side of the matching hypersurface ψ_F and, correspondingly, ψ_S on the Schwarzschild side, the Lorentz transformation

$p^{(\mu_{\rm F})} = \Lambda^{\mu_{\rm F}}{}_{\nu_{\rm S}} p^{(\nu_{\rm S})}$ with the Lorentz matrix parameters $\cosh\alpha = \gamma$, $\sinh\alpha = -\gamma V(r_{\rm b})$ yields

$$\cos\psi_{\rm F}^{(\mp)} = \frac{\cos\psi_{\rm S}^{(\mp)} \pm V(r_{\rm b})}{1 \pm V(r_{\rm b})\cos\psi_{\rm S}^{(\mp)}}, \tag{121}$$

$$\sin\psi_{\rm F}^{(\mp)} = \frac{\sin\psi_{\rm S}^{(\mp)}}{\gamma\left[1 \pm V(r_{\rm b})\cos\psi_{\rm S}^{(\mp)}\right]}. \tag{122}$$

The magnitude of refraction is determined by the expansion velocity of the matching hypersurface relative to the static Schwarzschild observers but it depends on the angle too and for $V(r_{\rm b}) \sim c$ the refraction law is rather complex. In the case of the most simple, flat universe with $k = 0$ the velocity is determined by the relation

$$V(r_{\rm b}) = \left(\frac{2M}{r_{\rm b}} + \frac{\Lambda}{3} r_{\rm b}^2\right)^{1/2}. \tag{123}$$

If $\Lambda = 0$, the expansion velocity falls with $r_{\rm b}$ growing as the expansion continues and the refraction effects vanish at late expansion times. On the other hand, for $\Lambda > 0$, the expansion velocity grows up to the velocity of light as the matching hypersurface approaches the cosmological horizon. In this case, the refraction effect could be very important. It is under study now, whether recent accelerated expansion of the universe could give observable effects in the character of the microwave background anisotropies.

5 Perfect fluid orbiting black holes

The hydrodynamic structure of test perfect fluid orbiting in the field of black holes is crucial for understanding the behavior of thick accretion disks. We shall demonstrate an astrophysically very important result that due to the presence of a repulsive cosmological constant the structure of equilibrium configurations of perfect fluid allows an outflow of matter from thick accretion disks. This fact is possible only in the black-hole spacetimes allowing the existence of outer marginally stable circular geodesics, discussed before.

5.1 The Boyer equilibrium condition

We briefly summarize the results of a general theory of equipotential surfaces inside any relativistic, differentially rotating, perfect fluid body [54, 55], applied to test configurations of perfect fluid orbiting in the stationary and axially symmetric spacetimes [7]. In the standard coordinate system, the spacetimes are described by the line element

$$\mathrm{d}s^2 = g_{tt}\,\mathrm{d}t^2 + 2g_{t\phi}\,\mathrm{d}t\mathrm{d}\phi + g_{\phi\phi}\,\mathrm{d}\phi^2 + g_{rr}\,\mathrm{d}r^2 + g_{\theta\theta}\,\mathrm{d}\theta^2, \tag{124}$$

where the metric coefficients depend neither on the time coordinate, t, nor the azimuthal coordinate, ϕ, i.e., the spacetimes contain timelike and azimuthal Killing vector fields $\partial/\partial t$ and $\partial/\partial\phi$.

We shall consider test perfect fluid orbiting in the ϕ direction with 4-velocity vector field U^μ of the form

$$U^\mu = (U^t, U^\phi, 0, 0), \tag{125}$$

where U^t and U^ϕ can be functions of the coordinates r, θ. The stress-energy tensor of the perfect fluid is

$$T^\mu{}_\nu = (p+\epsilon)U^\mu U_\nu + p\,\delta^\mu_\nu, \tag{126}$$

where ϵ and p denote the total energy density and the pressure of the fluid. The rotating fluid can be characterized by the vector fields of the angular velocity Ω and the angular momentum per unit mass (angular momentum density) ℓ, defined by

$$\Omega = \frac{U^\phi}{U^t}, \quad \ell = -\frac{U_\phi}{U_t}. \tag{127}$$

These vector fields are related by

$$\Omega = -\frac{g_{t\phi} + \ell g_{tt}}{g_{\phi\phi} + \ell g_{t\phi}}. \tag{128}$$

In static spacetimes ($g_{t\phi} = 0$), the relation (128) reduces to a simple formula

$$\frac{\Omega}{\ell} = -\frac{g_{tt}}{g_{\phi\phi}}. \tag{129}$$

The surfaces of constant ℓ and Ω are called von Zeipel cylinders. The family of von Zeipel cylinders does not depend on the assumed fluid orbit characteristics, $\ell = \ell(\Omega)$, in the static spacetimes but it will depend on it in the stationary spacetimes (with $g_{t\phi} \neq 0$).

Projecting the energy conservation law $T^{\mu\nu}{}_{;\nu} = 0$ onto the hypersurface orthogonal to the 4-velocity U^μ by the projection tensor $h_{\mu\nu} = g_{\mu\nu} + U_\mu U_\nu$, we obtain the relativistic Euler equation in the form

$$\frac{\partial_\mu p}{p+\epsilon} = -\partial_\mu(\ln U_t) + \frac{\Omega\,\partial_\mu \ell}{1-\Omega\ell}, \tag{130}$$

where

$$(U_t)^2 = \frac{g_{t\phi}^2 - g_{tt}\,g_{\phi\phi}}{g_{\phi\phi} + 2\ell g_{t\phi} + \ell^2 g_{tt}}. \tag{131}$$

The solution of the relativistic Euler equation can be given by the Boyer condition determining the surfaces of constant pressure through the 'equipotential surfaces' of the potential $W(r,\theta)$ by the relations

$$\int_0^p \frac{\mathrm{d}p}{p+\epsilon} = W_{\text{in}} - W, \tag{132}$$

$$W_{\text{in}} - W = \ln(U_t)_{\text{in}} - \ln(U_t) + \int_{\ell_{\text{in}}}^{\ell} \frac{\Omega\,\mathrm{d}\ell}{1-\Omega\ell}; \tag{133}$$

the subscript 'in' refers to the inner edge of the disk. For an alternative definition of the Boyer condition, see Ref. [8]. The equipotential surfaces are determined by the condition

$$W(r,\theta) = \text{const}, \tag{134}$$

and in a given spacetime can be found from Eq. (133) if a fluid orbit law $\Omega = \Omega(\ell)$ is given. The surfaces of constant pressure $p(r,\theta) = \text{const}$ are given by Eq. (132). The structure of thick accretion disks can be also obtained in the framework of a very practical and accurate Newtonian model for the gravitational field of a nonrotating black hole, known as the Paczyński–Wiita potential [56,57].

5.2 Marginally stable configurations orbiting Schwarzschild–de Sitter black holes

Equilibrium configurations of test perfect fluid rotating around the axis of rotation in a given spacetime are determined by the equipotential surfaces where the gravitational and inertial forces are just compensated by the pressure gradient. (In an axially symmetric spacetime, the axis of rotation coincides with the axis of symmetry of the spacetime while in a spherically symmetric spacetime the axis of rotation can be any radial line; usually, the coordinate system is chosen so that the rotation axis corresponds to $\theta = 0$.)

The equipotential surfaces can be closed or open. Moreover, there is a special class of critical, self-crossing surfaces (with a cusp) that can be either closed or open. The closed equipotential surfaces determine stationary equilibrium configurations. The fluid can fill any closed surface—on the surface of the equilibrium configuration, the pressure vanishes but its gradient is nonzero. On the other hand, the open equipotential surfaces are important in dynamical situations, e.g., in jet models [58, 59]. The critical, self-crossing closed equipotential surfaces W_{cusp} are important in the theory of thick accretion disks because accretion onto the black hole through the equipotential surface cusp located in the equatorial plane is possible due to the Paczyński mechanism, i.e., due to the disk surface slightly exceeding the critical equipotential surface, $W = W_{\text{cusp}}$. The accretion is thus driven by a violation of the hydrostatic equilibrium rather than by viscosity of the accreting matter.

It is well known that all characteristic properties of the equipotential surfaces for a general fluid orbit law are reflected by the equipotential surfaces of the simplest configurations with a uniform distribution of the angular momentum density ℓ (see Ref. [9]). Moreover, these configurations are very important astrophysically, being marginally stable [10]. Under the condition

$$\ell(r,\theta) = \text{const}, \tag{135}$$

holding in the rotating fluid, a simple relation for the equipotential surfaces follows from Eq. (133):

$$W(r,\theta) = \ln U_t(r,\theta), \tag{136}$$

with $U_t(r,\theta)$ being determined by $\ell = \text{const}$ and the metric coefficients only.

The equipotential surfaces are described by the formula $\theta = \theta(r)$, given by the differential equation

$$\frac{\mathrm{d}\theta}{\mathrm{d}r} = -\frac{\partial p/\partial r}{\partial p/\partial \theta}, \tag{137}$$

which for the configurations with $\ell = \text{const}$ reduces to

$$\frac{\mathrm{d}\theta}{\mathrm{d}r} = -\frac{\partial U_t/\partial r}{\partial U_t/\partial \theta}. \tag{138}$$

The influence of a nonzero cosmological constant on the character of the equipotential surfaces of the marginally stable configurations rotating around a black hole will be examined in the simplest case of Schwarzschild–de Sitter spacetimes corresponding to a repulsive cosmological constant, $y > 0$. (For completeness, we briefly discuss the case of Schwarzschild–anti-de Sitter spacetimes corresponding to an attractive cosmological constant, $y < 0$.)

In the standard Schwarzschild coordinates, the equipotential surfaces are given by the formulae

$$W(r,\theta) = \ln \frac{(1 - 2r^{-1} - yr^2)^{1/2} r \sin\theta}{\left[r^2 \sin^2\theta - (1 - 2r^{-1} - yr^2)\ell^2\right]^{1/2}} \tag{139}$$

and

$$\frac{\mathrm{d}\theta}{\mathrm{d}r} = \tan\theta \, \frac{\left[r(1 - yr^3)\sin^2\theta - (1 - 2r^{-1} - yr^2)^2\ell^2\right] r}{(r - 2 - yr^3)^2\ell^2}; \tag{140}$$

for $y = 0$, these relations reduce to the well known Schwarzschild formulae [9].

The best insight into the nature of the $\ell = \text{const}$ configurations can be gained by examining the behavior of the potential $W(r,\theta)$ in the equatorial plane ($\theta = \pi/2$). There are two reality conditions of $W(r, \theta = \pi/2)$:

$$1 - 2r^{-1} - yr^2 \geq 0, \tag{141}$$

$$r^2 - (1 - 2r^{-1} - yr^2)\ell^2 \geq 0. \tag{142}$$

The first condition is identical to the condition for the static regions (located between the black-hole and cosmological horizons); the second condition can be expressed in the form

$$\ell^2 \leq \ell^2_{\mathrm{ph}}(r; y) \equiv \frac{r^3}{r - 2 - yr^3}. \tag{143}$$

The function $\ell^2_{\mathrm{ph}}(r; y)$ is the effective potential of the photon geodetical motion; recall that $\ell \equiv U_\phi/U_t$ corresponds to the definition of the impact parameter for the photon geodetical motion (see Ref. [40]). Further, the condition of the local extrema of the potential $W(r, \theta = \pi/2)$ is identical to the condition of a vanishing pressure gradient ($\partial U_t/\partial r = 0$, $\partial U_t/\partial \theta = 0$). Since in the equatorial plane there is $\partial U_t/\partial \theta = 0$ independently of the $\ell = \text{const}$ and

$$\frac{\partial U_t}{\partial r}(r, \theta = \pi/2) = \frac{\left[r(1 - yr^3) - (1 - 2r^{-1} - yr^2)^2\ell^2\right]}{(1 - 2r^{-1} - yr^2)^{1/2}\left[r^2 - (1 - 2r^{-1} - yr^2)\ell^2\right]^{3/2}}, \tag{144}$$

we arrive at the condition

$$\ell^2 = \ell_{\rm K}^2(r;y) \equiv \frac{r^3(1-yr^3)}{(r-2-yr^3)^2}. \tag{145}$$

The extrema of $W(r,\theta=\pi/2)$ correspond to the points where the fluid moves along a circular geodesic, since $\ell_{\rm K}^2(r;y)$ corresponds to the angular momentum density distribution of the circular geodetical orbits. Clearly,

$$W_{\rm extr}(r,\theta=\pi/2;y) = \ln E_{\rm c}(r,y), \tag{146}$$

where

$$E_{\rm c}(r,y) = \left(1-\frac{2}{r}-yr^2\right)\left(1-\frac{3}{r}\right)^{-1/2} \tag{147}$$

is the specific energy of the circular geodesics. The most important properties of the potential $W(r,\theta)$ are determined by its behavior in the equatorial plane and, especially, by the properties of the functions $\ell_{\rm ph}^2(r;y)$ and $\ell_{\rm K}^2(r,y)$.

If $y>0$, the function $\ell_{\rm ph}^2(r,y)$ diverges at the black-hole horizon, $r_{\rm h}$, and the cosmological horizon, $r_{\rm c}$, determined by the equality sign in the condition (141). The local minimum of $\ell_{\rm ph}^2(r,y)$ is located at $r_{\rm ph}=3$, independently of y, and determines the unstable photon circular geodesic with the impact parameter

$$\ell_{\rm ph(c)}^2 = \ell_{\rm ph(min)}^2(y) \equiv \frac{27}{1-27y}. \tag{148}$$

The function $\ell_{\rm K}^2(r;y)$, determining the Keplerian (geodetical) circular orbits, has a zero point at the static radius $r_{\rm s}(y)$ given by $r_{\rm s}=y^{-1/3}$ and it is not well defined for $r>r_{\rm s}$, as it is negative there.

The function $\ell_{\rm K}^2(r;y)$ diverges at the black-hole horizon: $\ell_{\rm K}^2(r\to r_{\rm h},y)\to+\infty$; at the cosmological horizon, there is $\ell_{\rm K}^2(r\to r_{\rm c},y)\to-\infty$. Since

$$\frac{\partial\ell_{\rm K}^2}{\partial r} = \frac{r^2[r-6+yr^3(15-4r)]}{(r-2-yr^3)^3}, \tag{149}$$

the local extrema of $\ell_{\rm K}^2(r,y)$ are given by the condition

$$y = y_{\rm ms}(r) \equiv \frac{r-6}{r^3(4r-15)}, \tag{150}$$

determining the marginally stable circular geodesics. The local maximum of $y_{\rm ms}(r)$ gives the critical value of the parameter y admitting stable circular orbits $y_{\rm ms} = 12/15^4 \sim 0.000237$. If $y<y_{\rm ms}$, there exists an inner (outer) marginally stable circular geodesic at $r_{\rm ms(i)}$ ($r_{\rm ms(o)}$), see Fig. 5. The angular momentum density of the marginally stable orbits $\ell_{\rm ms(i)}(y)$ and $\ell_{\rm ms(o)}(y)$ is simultaneously determined by Eqs (145) and (150)—see Fig. 10. The specific energy of these orbits $E_{\rm ms(i)}(y)$ and $E_{\rm ms(o)}(y)$ is simultaneously determined by Eqs (147) and (150)—see Fig. 11. There is another special value of y, corresponding to the situation where the value of the minimum of $\ell_{\rm ph}^2(r;y)$ equals the maximum of $\ell_{\rm K}^2(r;y)$. We denote this value $y_{\rm e}$. It can be found that

$$y_{\rm e} = 1/118125 = 1/(3^3 5^4 7) \sim 0.00000846. \tag{151}$$

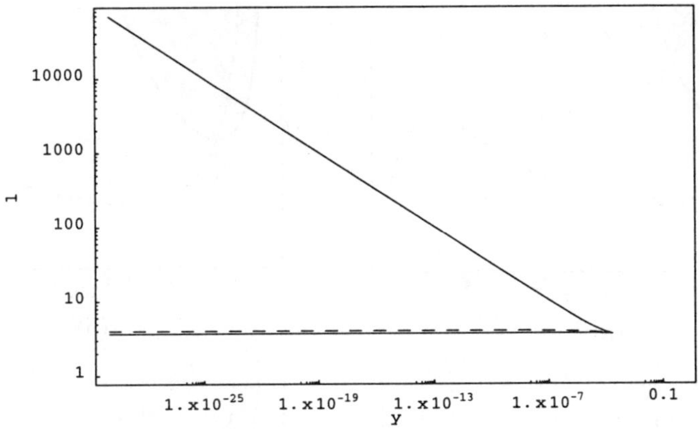

Figure 10. The angular momentum density of the marginally stable ($\ell_{\rm ms(i)}$ and $\ell_{\rm ms(o)}$, solid line) and marginally bound ($\ell_{\rm mb}$, dashed line) orbits as functions of the parameter y of the Schwarzschild–de Sitter spacetimes. Note that $\lim_{y\to 0}\ell_{\rm mb} = 4 \neq \lim_{y\to 0}\ell_{\rm ms(i)} = 3\sqrt{3}/2$, $\lim_{y\to 0}\ell_{\rm ms(o)} = +\infty$. Taken from Ref. [28].

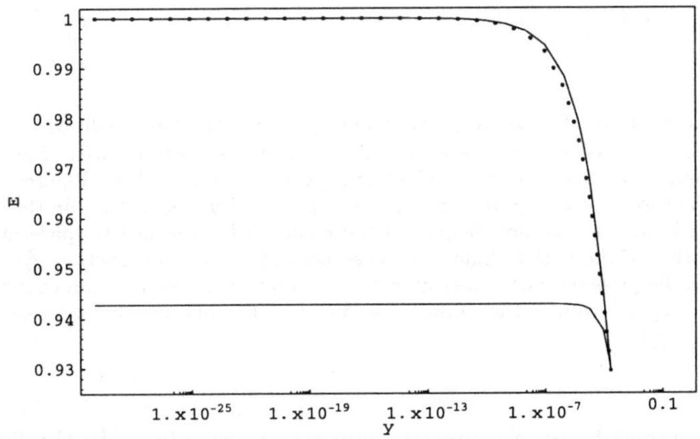

Figure 11. The specific energy of the marginally stable ($E_{\rm ms(i)}$ and $E_{\rm ms(o)}$, solid line) and marginally bound ($E_{\rm mb}$, bold dotted line) orbits as functions of the parameter y of the Schwarzschild–de Sitter spacetimes. Note that $\lim_{y\to 0} E_{\rm ms(o)} = \lim_{y\to 0} E_{\rm mb} = 1$, $\lim_{y\to 0} E_{\rm ms(i)} = 2\sqrt{2}/3$. Taken from Ref. [28].

In the Schwarzschild–de Sitter spacetimes, there is another important class of circular geodesics—namely the marginally bound orbits. These orbits exist in the Schwarzschild–de Sitter spacetimes admitting the existence of stable circular orbits, i.e., spacetimes with $y < y_{\rm ms}$. In these spacetimes, there exists an inner, $r_{\rm mb(i)}$ (outer, $r_{\rm mb(o)}$), marginally bound orbit close to the black-hole horizon (static radius). These orbits are defined by the condition

$$E_{\rm mb}(r_{\rm mb(i)}, \ell_{\rm mb}) = E_{\rm mb}(r_{\rm mb(o)}, \ell_{\rm mb}), \tag{152}$$

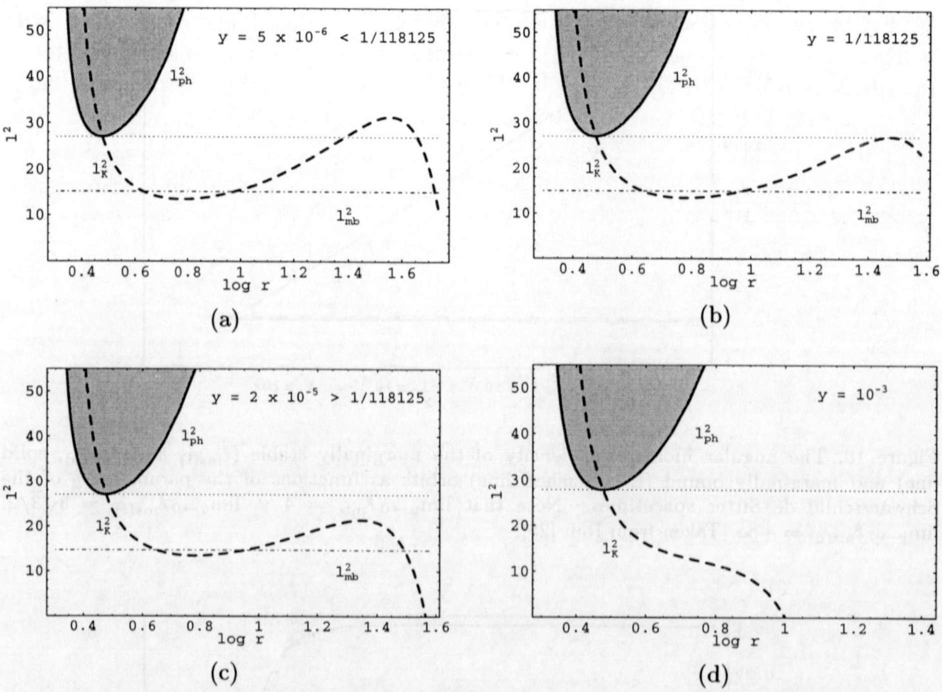

Figure 12. Behavior of the functions $\ell^2_{\rm ph}(r;y)$ and $\ell^2_{\rm K}(r;y)$ in the four qualitatively different cases determining the four classes of the Schwarzschild–de Sitter spacetimes with different properties of the equipotential surfaces (both r and ℓ^2 are given in units of M). Figures (a)–(d) reflect subsequently the cases $0 < y < y_{\rm e}$, $y = y_{\rm e}$, $y_{\rm e} < y < y_{\rm ms}$, and $y_{\rm ms} < y < y_{\rm c}$. In the shaded region, the equipotential surfaces are not defined in the equatorial plane of the spacetime, defined by the axis of rotation of the perfect fluid. The descending parts of the function $\ell^2_{\rm K}(r;y)$ determine the cusps, while the growing parts determine central rings of the equilibrium configurations. The dotted line ($\ell^2_{\rm ph(c)}$) determines the impact parameter of the photon circular geodesic at $r = 3$. Taken from Ref. [28].

and are determined by an appropriate numerical procedure. In the Schwarzschild spacetime ($y = 0$) the marginally bound orbit is located at $r_{\rm mb} = 4$ and $E_{\rm mb} = 1$—this is because the effective potential of the geodetical motion $V_{\rm eff} \to 1$ at $r \to \infty$ independently of the angular momentum density in the Schwarzschild spacetime. In the Schwarzschild–de Sitter spacetimes with $y_{\rm ms} < y < y_{\rm c}$, the marginally bound circular orbits are not defined because only unstable circular orbits exist in these spacetimes; particles can always escape to infinity from these orbits.

We can distinguish four qualitatively different cases of behavior of the functions $\ell^2_{\rm ph}(r;y)$, $\ell^2_{\rm K}(r;y)$ that give four classes of the Schwarzschild–de Sitter black holes with a different character of the rotating perfect fluid equipotential surfaces:

(A) $0 < y < y_{\rm e}$,

(B) $y = y_{\rm e}$,

(C) $y_{\rm e} < y < y_{\rm ms}$,

(D) $y_{\rm ms} \le y < y_{\rm c}$.

For these classes, the typical behavior of the functions $\ell^2_{\rm ph}(r;y)$, $\ell^2_{\rm K}(r;y)$, with y fixed, is given in Figs 12(a)–(d). For completeness, the corresponding value of $\ell_{\rm mb}(y)$ is exhibited in these figures. Note that the descending parts of the curve $\ell^2_{\rm K}(r;y)$ (with y fixed) correspond to the unstable circular geodesics, while its growing part (if it exists) corresponds to the stable circular geodesics. The extrema of $\ell^2_{\rm K}(r;y)$, if they exist, have an important role: the minimum $\ell_{\rm ms(i)}$, at $r_{\rm ms(i)}$, determines the inner marginally stable circular geodesic, while the maximum $\ell_{\rm ms(o)}$, at $r_{\rm ms(o)}$, determines the outer marginally stable circular geodesic.

The properties of the equipotential surfaces can be established easily, using the behavior of the potential $W(r,\theta)$ in the equatorial plane. The properties of the potential $W(r,\theta=\pi/2;y)$ are closely related to the properties of the effective potential of the geodetical motion and, at their local extrema located at the same radii, the condition (146) is satisfied. Further, $W(r;\theta=\pi/2,y)\to-\infty$ if $r\to r_{\rm h}$ or $r\to r_{\rm c}$. The topological properties of the equipotential surfaces can be directly inferred from the properties of the potential $W(r,\theta=\pi/2;y)$. The local extrema of the potential $W(r,\theta=\pi/2;y)$ are determined by the condition

$$\ell^2=\ell^2_{\rm K}(r;y); \tag{153}$$

therefore, at the radii determined by the local extrema of $W(r,\theta=\pi/2;y)$, perfect fluid follows free, geodetical circular orbits. The maxima of the potential are determined by the descending part of $\ell^2_{\rm K}(r;y)$; they correspond to the cusps of the equipotential surfaces and the matter moves along unstable geodetical orbits at the corresponding radii. The minima of the potential are determined by the rising part of $\ell^2_{\rm K}(r;y)$; they correspond to the central rings of the equilibrium configurations and the matter moves along stable geodetical orbits at the corresponding radii.

Now, we give a complete survey of the behavior of the equipotential surfaces and the related potential $W(r,\theta=\pi/2;y)$.

(A) $\mathbf{0<y<y_e}$. From Fig. 12a, we obtain nine qualitatively different cases of the behavior of the potential $W(r,\theta=\pi/2)$ and the corresponding nine

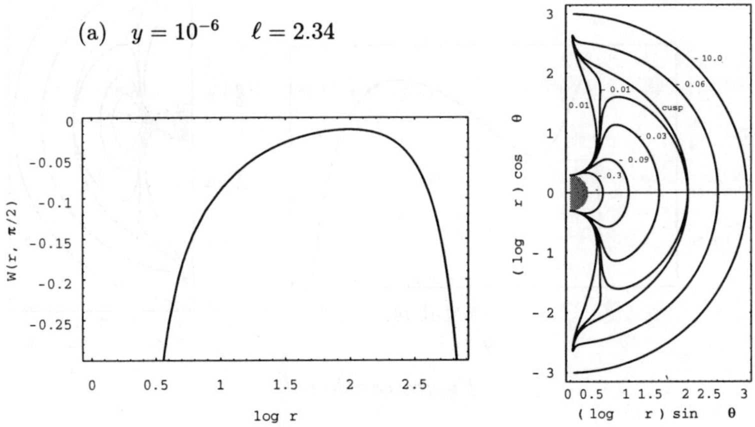

(Figure continued)

(b) $y = 10^{-6}$ $\ell = 3.67403603$

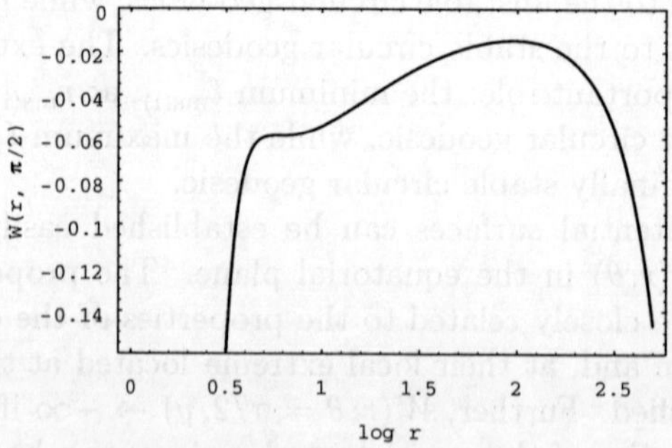

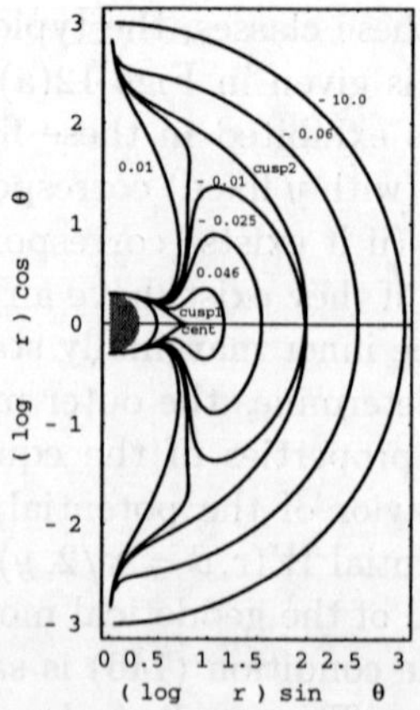

(c) $y = 10^{-6}$ $\ell = 3.84$

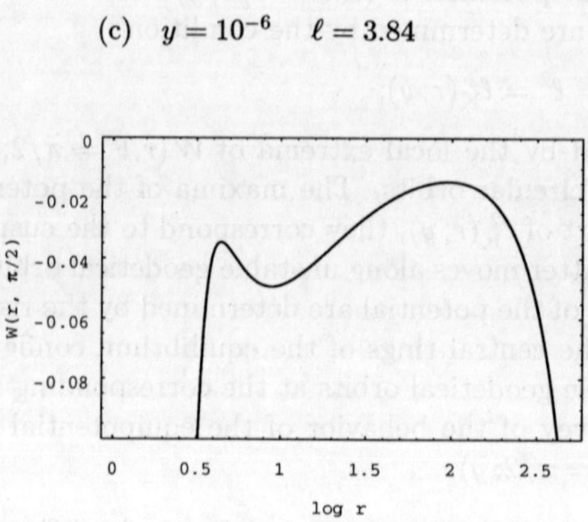

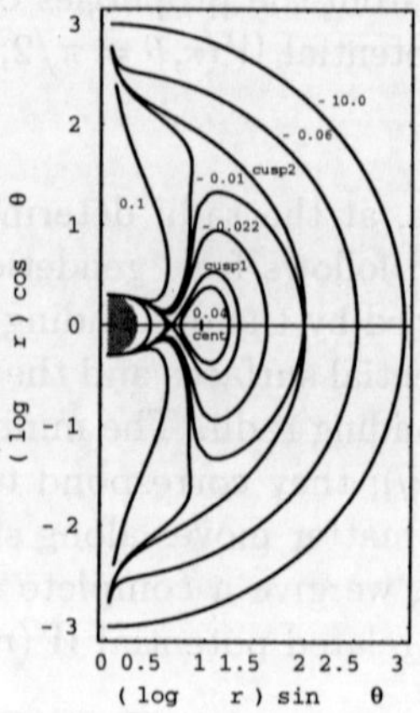

(d) $y = 10^{-6}$ $\ell = 3.93920702$

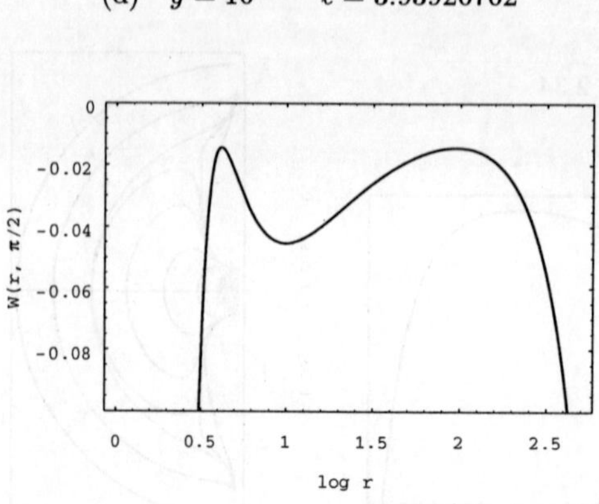

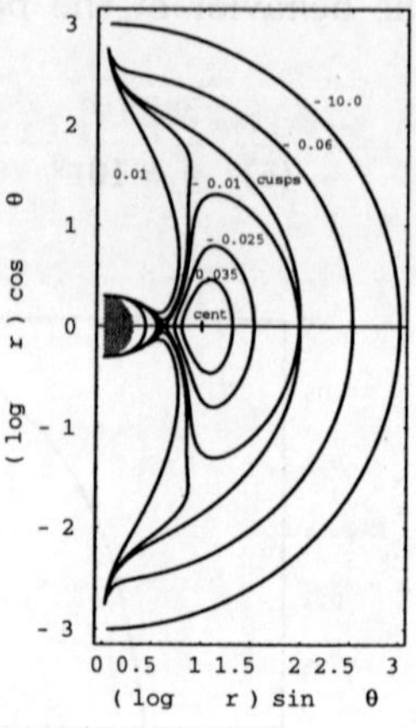

(Figure continued)

(e) $y = 10^{-6}$ $\ell = 4.18$

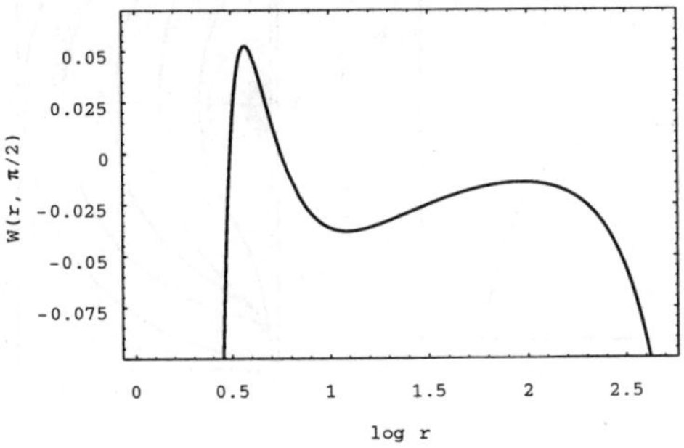

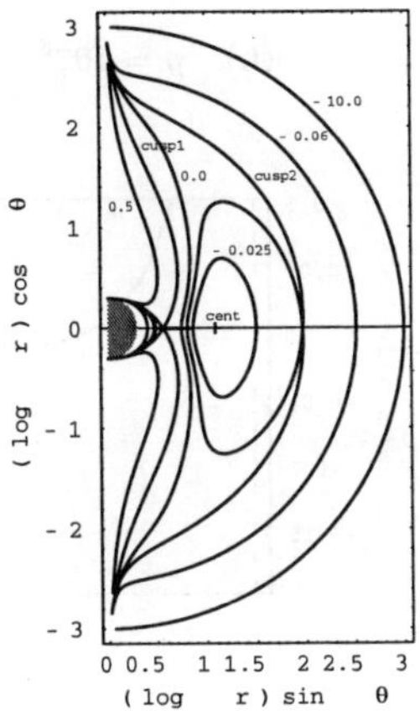

(f) $y = 10^{-6}$ $\ell = 5.19622257$

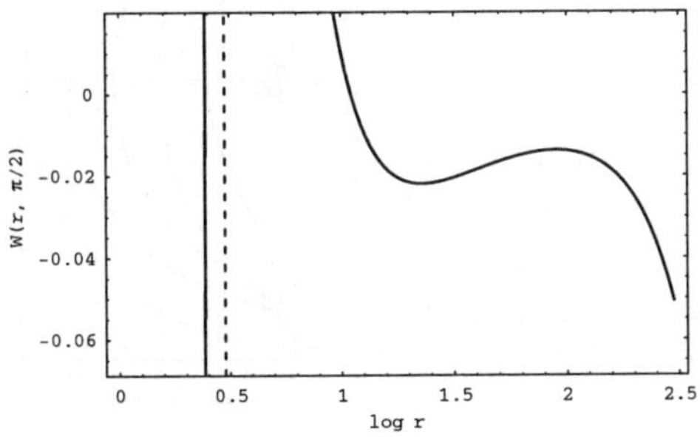

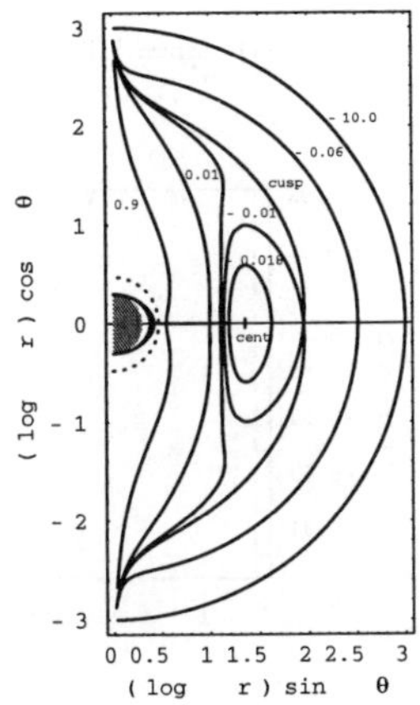

(g) $y = 10^{-6}$ $\ell = 5.9$

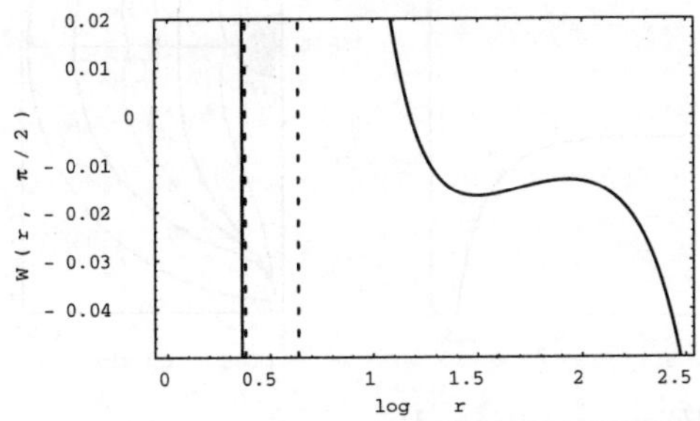

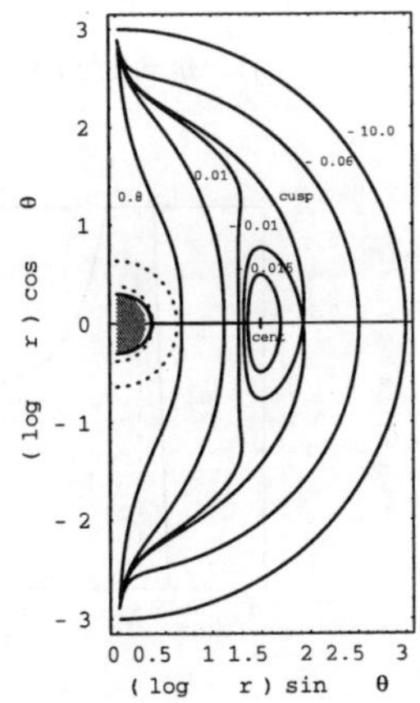

(Figure continued)

(h) $y = 10^{-6}$ $\quad \ell = 7.12938367$

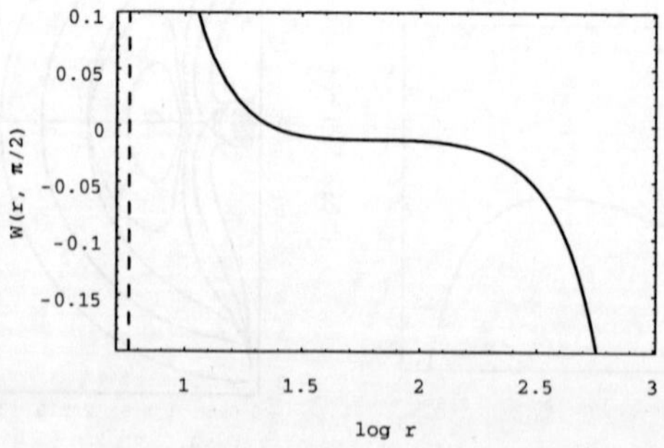

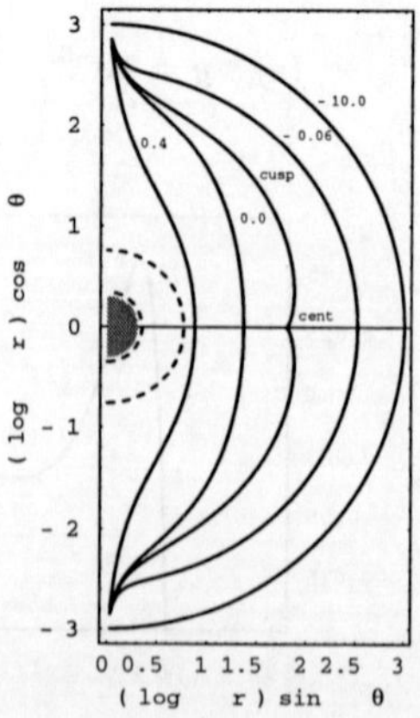

(i) $y = 10^{-6}$ $\quad \ell = 8.11$

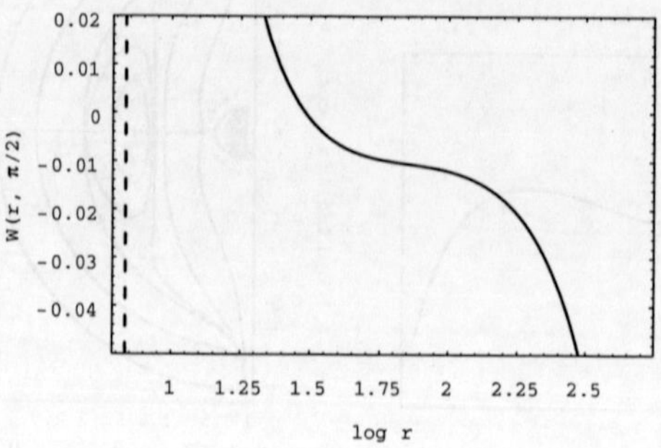

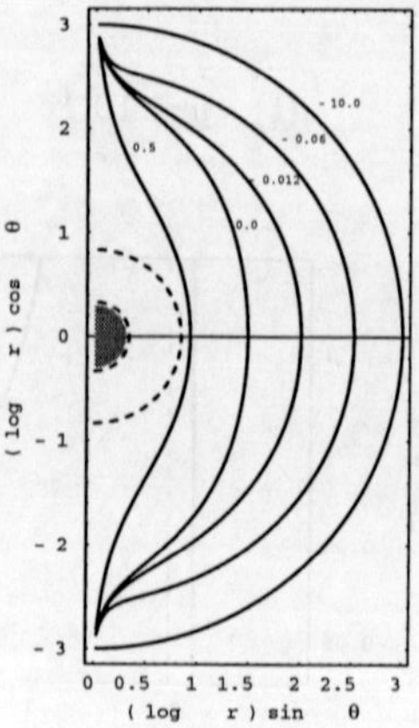

(j) $y = 1/118125$ $\quad \ell = 5.19674637$

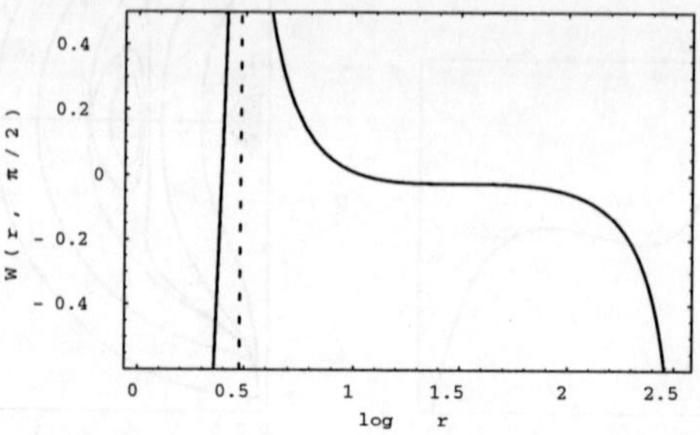

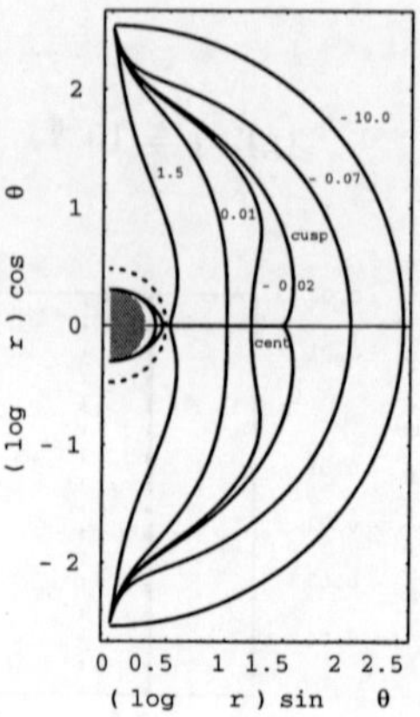

(Figure continued)

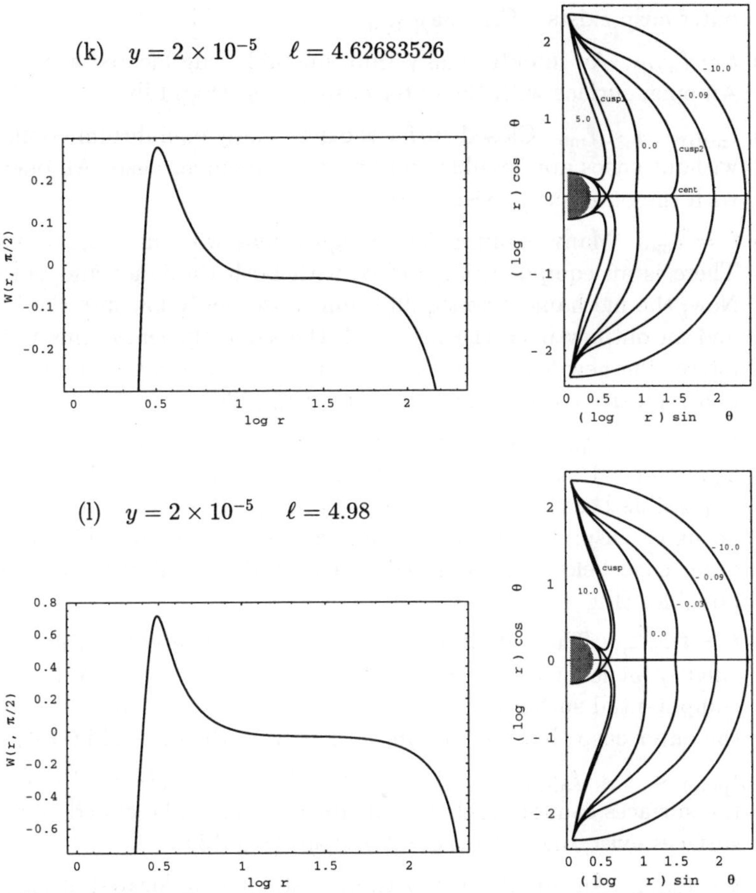

Figure 13. Equipotential surfaces (meridional sections) for the marginally stable (ℓ = const) configurations of test perfect fluid orbiting the Schwarzschild–de Sitter black-holes and the related potential $W(r, \theta = \pi/2; y)$. The radial coordinate is expressed in units of M; the logarithmic scale is used in order to cover the whole range between the inner and outer cusps. The central black hole is shaded. The sequence of figures (a)–(l) covers all the possibilities of the behavior of the equipotential surfaces for black holes in spacetimes with a repulsive cosmological constant. The sequence (a)–(i) gives successively all the possibilities for the behavior of the equipotential surfaces in the spacetimes of class A, with $0 < y < y_{\mathrm{e}}$, which is the astrophysically most plausible class. For the spacetimes of the classes B–D, the relevant sequences of the equipotential surfaces are determined in the text. The cusps of the toroidal disks correspond to the local maxima of $W(r, \theta = \pi/2)$, the central rings correspond to their local minima. The dashed lines give the asymptotics of $W(r, \theta = \pi/2)$ and determine the interval of radii where the equipotential surfaces cannot reach the equatorial plane. Taken from Ref. [28].

qualitatively different families of the equipotential surfaces, according to the values of ℓ = const. (In the following, we consider $\ell > 0$ only. This can be done due to the symmetry of the spacetimes under consideration.)

(I) $\ell < \ell_{\mathrm{ms(i)}}$. Open surfaces only; no disks are possible. Surface with the outer cusp exists. (Fig. 13a)

(II) $\ell = \ell_{ms(i)}$. An infinitesimally thin, unstable ring located at $r_{ms(i)}$ exists. An open surface with the outer cusp exists. (Fig. 13b)

(III) $\ell_{ms(i)} < \ell < \ell_{mb}$. Closed surfaces exist. Many equilibrium configurations without cusps are possible and one with the inner cusp. An open surface with the outer cusp exists. (Fig. 13c)

(IV) $\ell = \ell_{mb}$. Many equilibrium configurations without cusps are possible. There is an equipotential surface with both the inner and outer cusps. Now, the mechanical nonequilibrium causes an inflow into the black hole and an outflow from the disk with the same efficiency; this is the most interesting new feature of the accretion processes caused by the presence of a repulsive cosmological constant. (Fig. 13d)

(V) $\ell_{mb} < \ell < \ell_{ph(c)}$. Equilibrium configurations are possible because closed equipotential surfaces exist. However, accretion into the black hole is impossible because the equilibrium configurations (closed surfaces) have no inner cusp; the inner cusp has an open equipotential surface. The outer cusp belongs to a closed surface and the outflow from the disk is possible. (Fig. 13e)

(VI) $\ell = \ell_{(ph(c))}$. The potential $W(r, \theta = \pi/2; y)$ diverges at the photon circular orbit located at $r = 3$ and the inner cusp disappears. The closed equipotential surfaces still exist, with the most extended one containing the outer cusp that enables an outflow from the disk. (Fig. 13f)

(VII) $\ell_{ph(c)} < \ell < \ell_{ms(o)}$. In the region defined by $\ell^2_{ph}(r; y)$, the equipotential surfaces cannot reach the equatorial plane. The closed equipotential surfaces exist, one with the outer cusp. (Fig. 13g)

(VIII) $\ell = \ell_{ms(o)}$. An infinitesimally thin, unstable ring located at $r_{ms(o)}$ exists (the center and the outer cusp coalesce). (Fig. 13h)

(IX) $\ell > \ell_{ms(o)}$. Open equipotential surfaces exist only. There is no cusp in this case. (Fig. 13i).

(B) $\boldsymbol{y = y_e}$. For this special value of y (Fig. 12b), we still obtain the families of equipotential surfaces given by (A-I)–(A-V) and (A-IX). However, the case (A-VII) disappears and the cases (A-VI) and (A-VIII) coalesce, giving the case

(X) $\ell = \ell_{ph(c)} = \ell_{ms(o)}$. The inner cusp just disappears while the outer cusp coalesces with the center. (Fig. 13j)

(C) $\boldsymbol{y_e < y < y_{ms}}$. It follows from Fig. 12c that the intervals of ℓ and the families of equipotential surfaces (A-I)–(A-IV) remain. The following new intervals of the angular momentum density must be introduced.

(XI) $\ell_{mb} < \ell < \ell_{ms(o)}$. This case is equivalent to the case (A-V).

(XII) $\ell = \ell_{ms(o)}$. There is the inner cusp of an open equipotential surface but the center and the outer cusp coalesce – this corresponds to an infinitesimally thin unstable ring, located at $r_{ms(o)}$. (Fig. 13k)

(XIII) $\ell_{\rm ms(o)} < \ell < \ell_{\rm ph(c)}$. There are open surfaces only, one with the inner cusp. (Fig. 13l)

(XIV) $\ell \geq \ell_{\rm ph(c)}$. This case corresponds to the case (A-IX).

(D) $y_{\rm ms} \leq y < y_{\rm c}$. For this interval of y, the function $\ell_{\rm K}^2(r;y)$ is descending everywhere (see Fig. 12d). Only maxima of the potential $W(r, \theta = \pi/2; y)$ are possible (if $\ell^2 < \ell_{\rm ph(c)}^2$) and only open equipotential surfaces can exist. Equilibrium configurations corresponding to toroidal disks are not possible. This is a quite natural result since in the spacetimes under consideration stable circular geodesics cannot exist. Now, there are only two different intervals of the parameter ℓ.

(XV) $\ell < \ell_{\rm ph(c)}$. This family of equipotential surfaces corresponds to the case (A-I).

(XVI) $\ell \geq \ell_{\rm ph(c)}$. This family of equipotential surfaces corresponds to the case (A-IX).

Values of the potential at the central ring and the cusps (provided that they exist) are given in [28]; we can conclude that the presence of a repulsive cosmological constant makes the structure of the disk 'smoother.' The critical self-crossing surface for the marginally bound configurations $W_{\rm cusp}(\ell = \ell_{\rm mb}; y = 0) = 0$ while $W_{\rm cusp}(\ell = \ell_{\rm mb}, y > 0) < 0$.

If $y < 0$, the function $\ell_{\rm ph}^2(r, y)$ diverges at infinity and at the black-hole horizon. There is no zero point of $\ell_{\rm K}^2(r, y)$ and $\ell_{\rm K}^2(r \to r_{\rm h}, y) \to +\infty$, $\ell_{\rm K}^2(r \to \infty, y) \to +\infty$. Now, Eq. (150) determines only one marginally stable circular geodesic close to the horizon. On the other hand, in the Schwarzschild–anti-de Sitter spacetimes the notion of a marginally bound circular geodesic looses any meaning because particles from the unstable circular orbits never escape to infinity since the effective

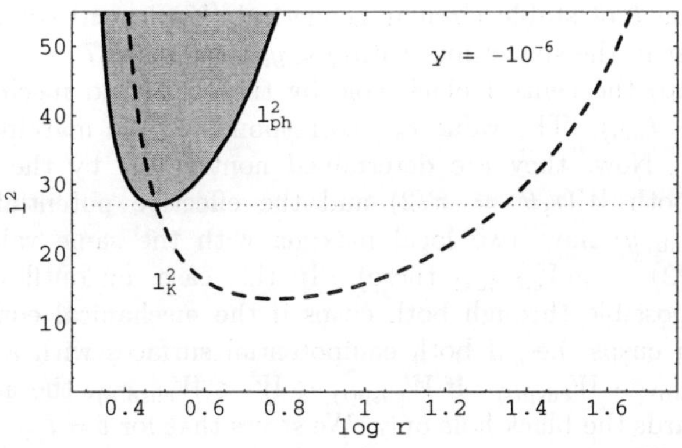

Figure 14. Behavior of the functions $\ell_{\rm ph}^2(r;y)$ and $\ell_{\rm K}^2(r;y)$ for the Schwarzschild–anti-de Sitter spacetimes, given for $y = -10^{-6}$ (both r and ℓ^2 are given in units of M). The dotted line determines $\ell_{\rm ph(c)}^2$, as in Fig. 12. It is qualitatively similar to the pure Schwarzschild case ($y = 0$) and it has the same character for all $y < 0$. In the shaded region, the equipotential surfaces are not defined in the equatorial plane. Taken from Ref. [28].

potential diverges at infinity for each value of the angular momentum density, as shown in the preceding section. The behavior of the functions $\ell^2_{\rm ph}(r;y)$ and $\ell^2_{\rm K}(r;y)$ is qualitatively the same as in the Schwarzschild case. This is illustrated in Fig. 14. The function $\ell^2_{\rm K}(r;y)$ has a minimum $\ell_{\rm ms}$ at $r_{\rm ms}$ corresponding to the marginally stable circular geodesic. The unstable geodesics are given by the descending part of $\ell^2_{\rm K}(r;y)$, while the stable ones are given by the rising part. It is immediately clear that for all the Schwarzschild–anti-de Sitter spacetimes we always obtain four possible cases of the behavior of the potential $W(r,\theta=\pi/2;y)$ and four corresponding families of the equipotential surfaces; notice that $W(r,\theta=\pi/2,y)\to\infty$ as $r\to\infty$. These cases are given by the following intervals of ℓ:

(I) $\ell<\ell_{\rm ms}$. There are open equipotential surfaces only. (Fig. 15a)

(II) $\ell=\ell_{\rm ms}$. An infinitesimally thin unstable ring is located at $r_{\rm ms}$. (Fig. 15b)

(III) $\ell_{\rm ms}<\ell\leq\ell_{\rm ph(c)}$. Closed equipotential surfaces exist, one with the cusp that enables accretion from the toroidal disk into the black hole. (Fig. 15c)

(IV) $\ell>\ell_{\rm ph(c)}$. Closed equipotential surfaces exist, but none with a cusp in the equatorial plane. In the vicinity of the horizon (in the region limited by radii determined by the equation $\ell^2_{\rm ph}(r;y)=\ell^2$), the equipotential surfaces cannot cross the equatorial plane. (Fig. 15d)

5.3 *Astrophysical implications*

The accretion disks around black holes can exist if an inner cusp appears near the black-hole horizon in addition to the outer cusp, located nearby the static radius $r_{\rm s}=\sqrt[3]{1/y}$. Closed equipotential surfaces, necessary for the existence of toroidal accretion disks, can exist for $\ell\in(\ell_{\rm ms(i)},\ell_{\rm ms(o)})$. Here $\ell_{\rm ms(i)}$ ($\ell_{\rm ms(o)}$) corresponds to the local minimum (maximum) of the function $\ell^2_{\rm K}(r;y)$, giving the minimum (maximum) value ℓ of stable circular geodetical (Keplerian) orbits. The closed surfaces can exist in the spacetimes with $y<y_{\rm ms}\sim 0.000237$.

Accretion onto the central black hole by the Paczyński mechanism is possible if $\ell\in(\ell_{\rm ms(i)},\ell_{\rm mb})$. The value $\ell_{\rm mb}$ corresponds to the marginally bound circular geodesics. Now, they are determined nontrivially by the condition that for $\ell=\ell_{\rm mb}$ both $W(r,\theta=\pi/2)$ and the effective potential of geodetical motion $V_{\rm eff}(r;\ell_{\rm mb},y)$ have two local maxima with the same value (recall that $W_{\rm max}(r,\theta=\pi/2)=\ln V_{\rm eff(max)}$ there). In this case, an outflow from the accretion disk is possible through both cusps if the mechanical equilibrium is destroyed for both cusps, i.e., if both equipotential surfaces with a cusp are filled up: $W>W_{\rm cusp(o)}>W_{\rm cusp(i)}$. If $W_{\rm cusp(i)}<W<W_{\rm cusp(o)}$, the accretion flow is directed downwards the black hole only. We stress that for $\ell=\ell_{\rm mb}$, the equipotential surface with $W=W_{\rm max}(r_{\rm mb(i)},y)=W_{\rm max}(r_{\rm mb(o)},y)$ has two cusps. The mass outflow due to the mechanical nonequilibrium, i.e., overfilling of the (both-sided) marginally closed equipotential surface, is equally efficient for the inflow down the black hole and the outflow near the static radius. Of course, we could expect significant differences in details of the accretion inflow near the black-hole horizon and the outflow near the static radius.

(a) $y = -10^{-6} \quad \ell = 2.18$

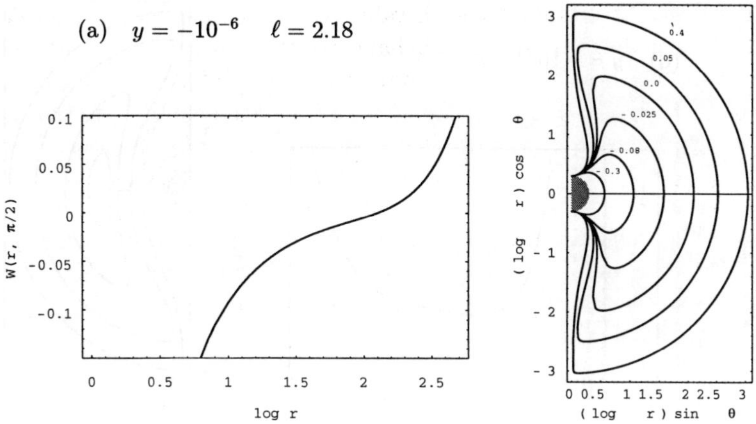

(b) $y = -10^{-6} \quad \ell = 3.67443285$

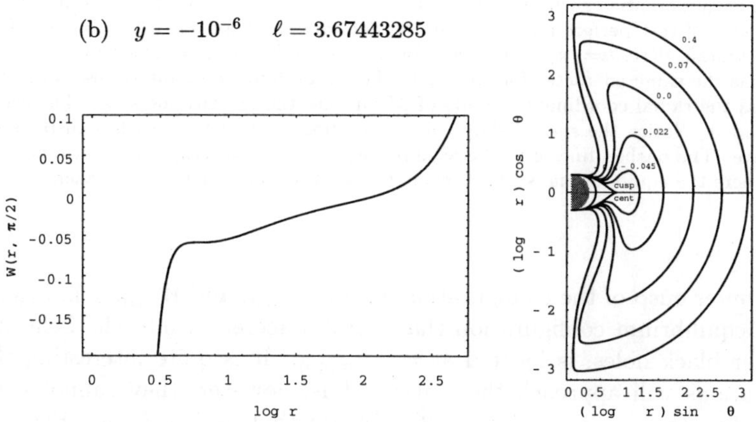

(c) $y = -10^{-6} \quad \ell = 4.12$

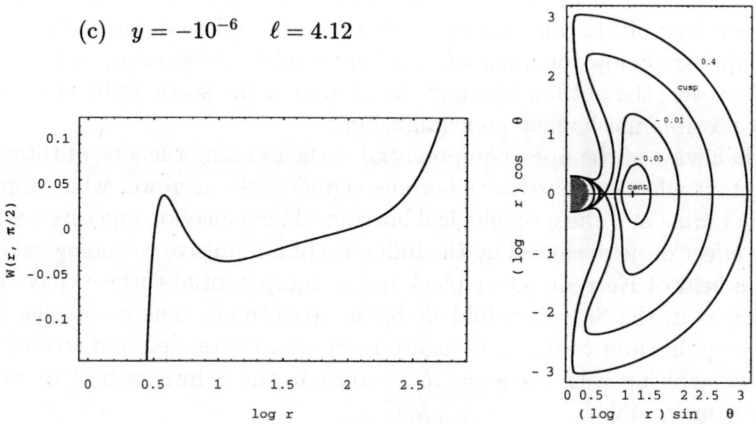

(Figure continued)

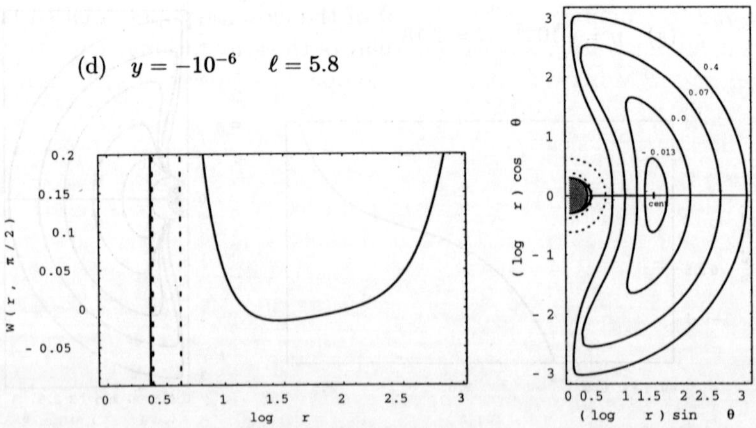

Figure 15. Equipotential surfaces (meridional sections) for the marginally stable (ℓ = const) configurations of test perfect fluid orbiting the Schwarzschild–anti-de Sitter black holes and the related potential $W(r, \theta = \pi/2, y)$, given for $y = -10^{-6}$. The behavior of the equipotential surfaces has the same character for all $y < 0$. There are four possibilities described in the text. We express the radial coordinate in units of M and use the logarithmic scale. The central black hole is shaded. Notice the special shape of the equipotential surfaces with a cusp, resembling a falling wave. The dashed lines give the asymptotics of $W(r, \theta = \pi/2)$ and determine the interval of radii where the equipotential surfaces cannot reach the equatorial plane. Taken from Ref. [28].

The outer cusp of the configuration with $\ell = \ell_{\rm mb}$, and $W_{\rm cusp} = \ln E_{\rm mb}$, i.e., the limiting equilibrium configuration that enables accretion into the Schwarzschild–de Sitter black holes, is located at $r = r_{\rm mb(o)}$. It is quite interesting that such configurations will approach the static radius, however, they cannot exceed the static radius ($r_{\rm mb(o)} \to r_{\rm s}$ if $y \to 0$). Notice that $\ell_{\rm mb} \sim \ell_{\rm ms(i)}$ (Fig. 10), while $E_{\rm mb} \sim E_{\rm ms(o)}$ (Fig. 11). For $\ell \in (\ell_{\rm mb}, \ell_{\rm ms(o)})$, the accretion flow downwards the hole is 'switched-off' because an open self-crossing equipotential surface with $W = W_{\rm cusp(i)}$ appears below the inner edge of the toroidal configuration in the equatorial plane. However, the outflow through the cusp near the static radius can still occur due to a possible mechanical nonequilibrium.

The behavior of the open equipotential surfaces along the axis of rotation gives an important effect—the surfaces become significantly narrower when approaching the static radius and the cosmological horizon. This behavior suggests a strong collimation effect on jets, caused by the influence of a repulsive cosmological constant.

In the field of Kerr–de Sitter black holes, equipotential surfaces have the same properties as in the Schwarzschild–de Sitter spacetimes. The corotating (counter-rotating) equilibrium configurations are more (less) extended and steeper (flatter) than those orbiting with the same $|\ell|$ = const in the Schwarzschild–de Sitter field with the same y [45].

In the case of Schwarzschild–anti-de Sitter spacetimes, the situation is different. The presence of an attractive cosmological constant brings no qualitatively new phenomena in comparison with the Schwarzschild case, concerning the character of the equilibrium configurations related to accretion disks. Notice, however, the special shape (resembling a falling wave) of the closed equipotential surfaces that manifests in an illustrative way the interplay of the gravitational, cosmological, and centrifugal forces. Moreover, there exist no open equipotential surfaces around the rotation axis in these spacetimes.

From the astrophysical point of view, the most important phenomena were discovered in spacetimes with a repulsive cosmological constant if they admit stable circular geodetical orbits. The first is the presence of an outer cusp of toroidal disks nearby the static radius that enables an outflow of mass and angular momentum from the accretion disks by the Paczyński mechanism, i.e., due to a violation of the hydrostatic equilibrium. This is the same mechanism that drives the accretion into the black hole through the inner cusp. (Recall that outflow from toroidal disks around a Schwarzschild or Kerr black hole by the Paczyński mechanism is impossible because no outer cusp of toroidal disks exists in the asymptotically flat black-hole spacetimes [7–9].) The second is the possibility of strong collimation effects on jets escaping along the rotation axis of toroidal disks following the open equipotential surfaces that are narrowing strongly when approaching the static radius (and the cosmological horizon). We give an explicit illustration of these two principally new phenomena caused by the repulsive cosmological constant in Fig. 16. Of course, both of those very interesting phenomena deserve further, more detailed studies. Further, the runaway instability of the toroidal disks with respect to the outflow through the outer cusp and the influence of self–gravitation on their structure have to be examined. We plan these studies in the near future. Some approaches to the problem of self-gravitating disks around black holes can be found in [60].

It is interesting to find astrophysically plausible situations in which these two phenomena could be relevant. We should consider their role in quasars and active galactic nuclei during the present period of the expansion of Universe and in the accretion processes onto primordial black holes during the very early stages of the expansion of Universe when phase transitions connected to symmetry breaking of physical interactions due to Higgs mechanism (e.g., the breaking of electroweak interactions) could take place and the effective cosmological constant could have values exceeding its present value by many orders [61].

We can clearly see that the relict cosmological constant $\Lambda_0 \sim 1.1 \times 10^{-56}\,\mathrm{cm}^{-2}$ puts a natural limit on the size of equilibrium configurations rotating around black holes [28]. In fact, the outer edge of the accretion disks, where the outflow goes through the outer cusp of the toroidal structure, is located near the static radius. It is quite interesting that for black holes of masses $\sim 10^8\, M_\odot$–$10^9\, M_\odot$, corresponding to black holes located in the central parts of quasars and active galactic nuclei, the outer edge of the largest accretion disks is located at $r_{\mathrm{mb(o)}} \sim$ 50–100 kpc and is comparable to the maximum extension of large galaxies. Note that the extension of quasikeplerian, thin accretion disks is limited by the outer marginally stable circular orbit; if y is small enough ($y \leq 10^{-8}$), it can be shown that

$$r_{\mathrm{ms(o)}} \sim 0.63 r_{\mathrm{s}}, \tag{154}$$

and dimensions of these disks are comparable to the static radius, too. Therefore, the relict repulsive cosmological constant can radically influence the behavior of accretion disks in large galaxies with active nuclei and can even be connected to

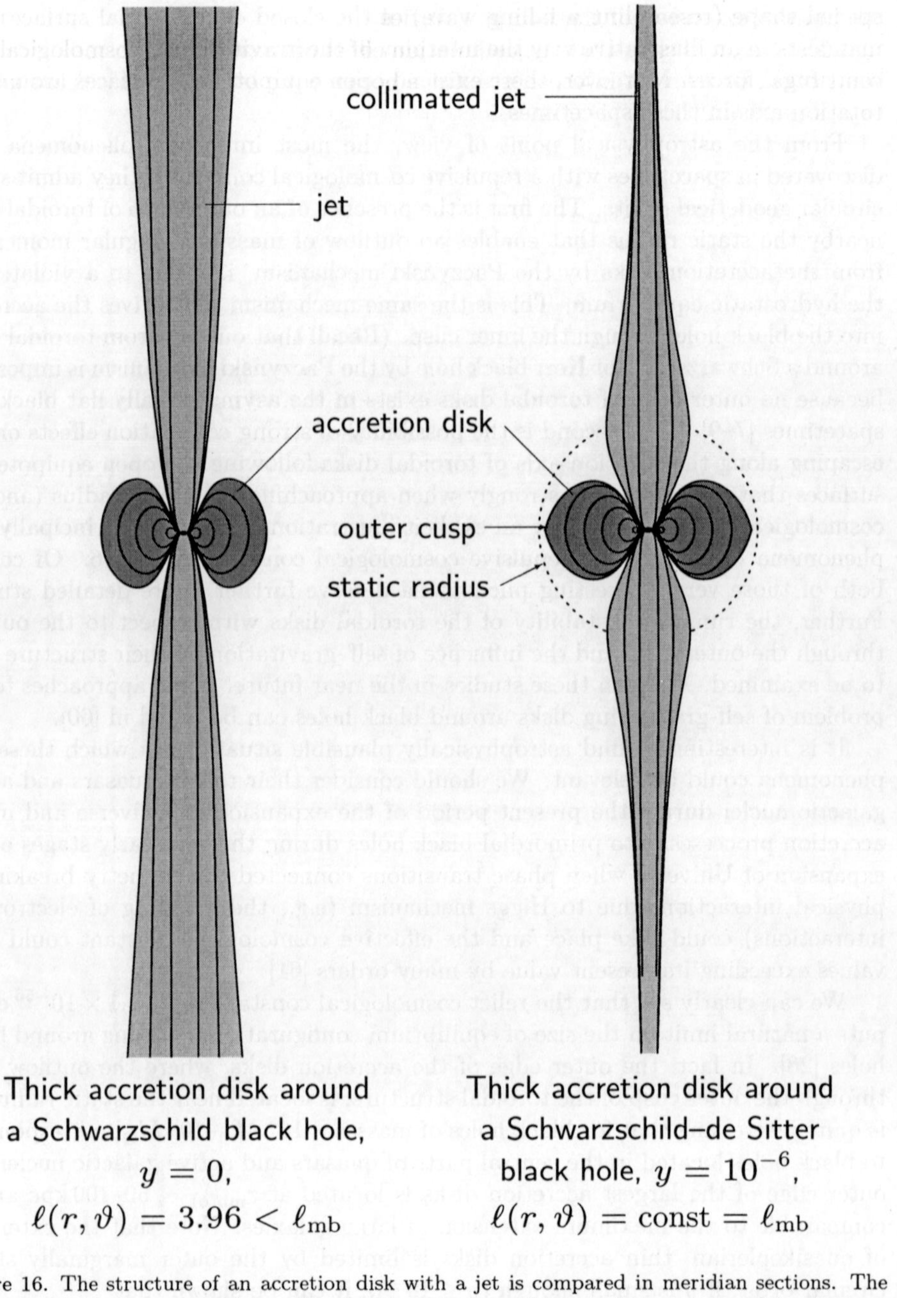

Figure 16. The structure of an accretion disk with a jet is compared in meridian sections. The radial coordinate is expressed in units of M but the logarithmic scale is not used here since we are interested in the regions near the static radius where both the outer cusp and the collimation effect are evident. Taken from Ref. [28].

the limit of extension of these large galaxies.

Moreover, it is clear that the collimation effect of the repulsive cosmological constant could be relevant in these situations because the largest observed jets extend to distances $\sim 200\,\mathrm{kpc}$ [1], exceeding dimensions of the 'seed' galaxy (comparable to the static radius).

6 General relativistic polytropes

Besides the case of vacuum black-hole (naked-singularity) spacetimes, we can expect unusual phenomena also in nonvacuum spacetimes with a nonzero cosmological constant, representing static mass configurations. Here, we shall discuss spherically symmetric configurations of perfect fluid with a polytropic equation of state.

In terms of the standard Schwarzschild coordinates, the line element of a spherically symmetric, static spacetime is given in the form

$$\mathrm{d}s^2 = -\mathrm{e}^{2\Phi}\,\mathrm{d}t^2 + \mathrm{e}^{2\Psi}\,\mathrm{d}r^2 + r^2(\mathrm{d}\theta^2 + \sin^2\theta\,\mathrm{d}\phi^2) \tag{155}$$

with just two unknown functions, $\Phi(r)$ and $\Psi(r)$. Matter inside the configuration is assumed to be a perfect fluid with $\rho = \rho(r)$ being the density of mass-energy in the rest-frame of the fluid and $p = p(r)$ being the isotropic pressure.

In a static configuration, each element of the fluid must remain at rest in the static coordinate system. Therefore, the 4-velocity field is given by

$$u^r = \frac{\mathrm{d}r}{\mathrm{d}\tau} = 0, \quad u^\theta = \frac{\mathrm{d}\theta}{\mathrm{d}\tau} = 0, \quad u^\phi = \frac{\mathrm{d}\phi}{\mathrm{d}\tau} = 0 \tag{156}$$

and

$$u^t = \frac{\mathrm{d}t}{\mathrm{d}\tau} = \mathrm{e}^{-\Phi}. \tag{157}$$

The structure of a relativistic star is determined by the Einstein field equations, $G_{\mu\nu} \equiv R_{\mu\nu} - \frac{1}{2}Rg_{\mu\nu} + \Lambda g_{\mu\nu} = 8\pi T_{\mu\nu}$, and by the law of local energy-momentum conservation, $T^{\mu\nu}{}_{;\nu} = 0$. It is convenient to express the equations in terms of the orthonormal tetrad components using 4-vectors carried by the fluid elements:

$$\vec{e}_{(t)} = \frac{1}{\mathrm{e}^\Phi}\frac{\partial}{\partial t}, \tag{158}$$

$$\vec{e}_{(r)} = \frac{1}{\mathrm{e}^\Psi}\frac{\partial}{\partial r}, \tag{159}$$

$$\vec{e}_{(\theta)} = \frac{1}{r}\frac{\partial}{\partial\theta}, \tag{160}$$

$$\vec{e}_{(\phi)} = \frac{1}{r\sin\theta}\frac{\partial}{\partial\phi}. \tag{161}$$

Projection of $T^{\mu\nu}{}_{;\nu} = 0$ orthogonal to u^μ (by the projection tensor $P^{\mu\nu} = g^{\mu\nu} + u^\mu u^\nu$) gives the relevant equation

$$(\rho + p)\frac{\mathrm{d}\Phi}{\mathrm{d}r} = -\frac{\mathrm{d}p}{\mathrm{d}r}, \tag{162}$$

which is the equation of hydrostatic equilibrium describing the balance between the gravitational force and pressure gradient.

6.1 Equations of structure

There are two relevant structure equations following from the Einstein equations. These are determined by the $(t)(t)$ and $(r)(r)$ tetrad components of the field equations (the $(\theta)(\theta)$ and $(\phi)(\phi)$ components give dependent equations). First we shall discuss the $(t)(t)$ component:

$$G_{(t)(t)} = \frac{1}{r^2} - \frac{e^{-2\Psi}}{r^2} - \frac{1}{r}\frac{d}{dr}e^{-2\Psi} - \Lambda = 8\pi\rho. \tag{163}$$

We can transfer it into the form

$$\frac{d}{dr}\left[r\left(1 - e^{-2\Psi}\right) - \frac{1}{3}\Lambda r^3\right] = \frac{d}{dr}2m(r) \tag{164}$$

where

$$m(r) = \int_0^r 4\pi r^2 \rho\, dr. \tag{165}$$

The integration constant in (165) is chosen to be $m(0) = 0$ because then the spacetime geometry is smooth at the origin (see Ref. [19]). We find the relation

$$e^{2\Psi} = \left[1 - \frac{2m(r)}{r} - \frac{1}{3}\Lambda r^2\right]^{-1}. \tag{166}$$

The $(r)(r)$ component of the field equations reads

$$G_{(r)(r)} = -\frac{1}{r^2} + \frac{e^{-2\Psi}}{r^2} + \frac{2e^{-2\Psi}}{r}\frac{d\Phi}{dr} + \Lambda = 8\pi p. \tag{167}$$

Using (166), we obtain the relation

$$\frac{d\Phi}{dr} = \frac{m(r) - \frac{1}{3}\Lambda r^3 + 4\pi p r^3}{r\left[r - 2m(r) - \frac{1}{3}\Lambda r^3\right]}, \tag{168}$$

which enables us to put the equation of hydrostatic equilibrium (162) into the Tolman–Oppenheimer–Volkoff (TOV) form modified by the presence of a nonzero cosmological constant:

$$\frac{dp}{dr} = -(\rho + p)\frac{m(r) - \frac{1}{3}\Lambda r^3 + 4\pi p r^3}{r\left[r - 2m(r) - \frac{1}{3}\Lambda r^3\right]}. \tag{169}$$

Introducing a dimensionless mass parameter $u(r)$ by the relation

$$u(r) \equiv \frac{m(r)}{M} = \frac{r}{2M}\left(1 - e^{-2\Psi} - \frac{1}{3}\Lambda r^2\right), \tag{170}$$

we can express the $(t)(t)$ component of the Einstein equations in the form

$$M\frac{du(r)}{dr} = 4\pi\rho(r)r^2. \tag{171}$$

Now, it is necessary to introduce an explicit form of the equation of state. We shall assume the polytropic equation of state

$$p = K\rho^{1+\frac{1}{n}}, \tag{172}$$

where n is the 'polytropic index' assumed to be a given constant (not necessarily an integer) and K is a constant that has to be determined by the thermal characteristics of a given fluid sphere, by specifying the density ρ_c and pressure p_c at the center. Since, for a given pressure, the density is a function of temperature, K contains the temperature implicitly. It can be shown that K is determined by the total mass, radius, and p_c/ρ_c ratio [62].

Note that the polytropic equation is the limiting form of the parametric equations of state for a completely degenerate gas at zero temperature, relevant, e.g., for neutron stars. Then both n and K are universal physical constants.

When integrating the equations of structure, it is convenient to introduce a new variable θ related to the density $\rho(r)$ and the central density ρ_c, by

$$\rho = \rho_c \theta^n, \tag{173}$$

with n being the polytropic index. The boundary condition on $\theta(r)$ is $\theta(r=0) = 1$. The pressure dependence is then given by the relation

$$p = K \rho_c^{1+\frac{1}{n}} \theta^{n+1}. \tag{174}$$

The conservation law (170) can then be expressed in the form

$$\sigma(n+1)\,\mathrm{d}\theta + (\sigma\theta + 1)\,\mathrm{d}\Phi = 0, \tag{175}$$

where the parameter σ is given by

$$\sigma = K \rho_c^{1/n} = \frac{p_c}{\rho_c}. \tag{176}$$

At the edge of the configuration, $r = R$, there is $\rho(R) = p(R) = 0$. Outside the mass configuration, the spacetime is described by the vacuum Schwarzschild–(anti-)de Sitter metric. Solving Eq. (175) and using the boundary condition that the internal and external metric coefficients are smoothly matched at $r = R$, we obtain

$$\mathrm{e}^{2\Phi} = (1+\sigma\theta)^{-2(n+1)} \left(1 - \frac{2M}{R} - \frac{1}{3}\Lambda R^2\right); \tag{177}$$

thus, the metric coefficient g_{tt} is determined by θ and σ. Function $\mathrm{e}^{2\Psi}$, or equivalently $u(r)$, remains to be expressed in terms of θ, and we need to find $\theta = \theta(r)$. First, we rewrite Eq. (175) in the form

$$\frac{\mathrm{d}\Phi}{\mathrm{d}r} = \frac{\sigma(n+1)}{1+\sigma\theta}\frac{\mathrm{d}\theta}{\mathrm{d}r}. \tag{178}$$

Then we can express the $(r)(r)$ component of the Einstein equations and Eq. (170), in the form

$$\frac{\sigma(n+1)}{1+\sigma\theta}\, r \frac{\mathrm{d}\theta}{\mathrm{d}r}\left(1 - \frac{2Mu}{r} - \frac{1}{3}\Lambda r^2\right) + \frac{Mu}{r} - \frac{1}{3}\Lambda r^2 = -M\sigma\theta\frac{\mathrm{d}u}{\mathrm{d}r}, \tag{179}$$

$$M\frac{\mathrm{d}u}{\mathrm{d}r} = 4\pi r^2 \rho_c \theta^n. \tag{180}$$

Introducing the quantity L with the dimension of length by

$$L = \left[\frac{(n+1)K\rho_c^{1/n}}{4\pi\rho_c}\right]^{1/2} = \left[\frac{\sigma(n+1)}{4\pi\rho_c}\right]^{1/2}, \tag{181}$$

Eqs (179) and (180) can then be transformed into dimensionless form by introducing a dimensionless radial coordinate

$$\xi = \frac{r}{L}, \tag{182}$$

and quantities

$$v(\xi) = \frac{M}{4\pi L^3 \rho_c} u(r), \tag{183}$$

$$\lambda = \frac{\rho_{\text{vac}}}{\rho_c}, \tag{184}$$

where the vacuum energy density is related to the cosmological constant by

$$\rho_{\text{vac}} = \frac{\Lambda}{8\pi}. \tag{185}$$

The dimensionless form of Eqs (179) and (180) becomes

$$\xi^2 \frac{\mathrm{d}\theta}{\mathrm{d}\xi} \frac{1 - 2\sigma(n+1)\left(\frac{v}{\xi} + \frac{1}{3}\lambda\xi^2\right)}{1+\sigma\theta} + v(\xi) - \frac{2}{3}\lambda\xi^3 = -\sigma\xi\theta\frac{\mathrm{d}v}{\mathrm{d}\xi}, \tag{186}$$

$$\frac{\mathrm{d}v}{\mathrm{d}\xi} = \xi^2\theta^n. \tag{187}$$

Equations (186) and (187) have to be solved for given n and σ under boundary conditions

$$\theta(0) = 1, \quad v(0) = 0. \tag{188}$$

It follows from (187) and (188) that $v(\xi) \sim \xi^3$ for $\xi \to 0$ and, according to Eq. (186),

$$\lim_{\xi \to 0_+} \frac{\mathrm{d}\theta}{\mathrm{d}\xi} = 0. \tag{189}$$

The boundary of the fluid sphere ($r = R$) is represented by the first zero point of $\theta(\xi)$, say at ξ_1:

$$\theta(\xi_1) = 0. \tag{190}$$

6.2 Properties of the polytropes

One of the basic characteristics of the polytropes is the mass-radius (M-R) relation. Using Eq. (183), we obtain

$$v(\xi) = \frac{M}{4\pi\rho_c L^2}\frac{\xi}{r}u(r) = \frac{M}{\sigma(n+1)}\xi\frac{u(r)}{r}, \tag{191}$$

and the M-R relation can be given in the form

$$\frac{M}{R} \equiv \frac{1}{2}\frac{r_g}{R} = \frac{\sigma(n+1)v(\xi_1)}{\xi_1}, \tag{192}$$

where $r_g = 2M$ is the gravitational radius of the configuration. The radial metric coefficient in terms of the dimensionless quantities ξ and λ reads

$$\mathrm{e}^{-2\Psi} = 1 - 2\sigma(n+1)\left[\frac{v(\xi)}{\xi} + \frac{1}{3}\lambda\xi^2\right]. \tag{193}$$

Assuming Λ, n, σ, and ρ_c are given, then mass M, radius R, and the internal structure of the polytropes can be easily determined. First, the length scale L given by Eq. (181) has to be found. By numerical integration of Eqs (186) and (187), functions $\theta(\xi)$ and $v(\xi)$ are found and ξ_1, where $\theta(\xi_1) = 0$, is determined together with $v(\xi_1)$. The radius of the sphere is

$$R = L\xi_1, \tag{194}$$

and the mass of the sphere is then given by

$$M = L\sigma(n+1)v(\xi_1). \tag{195}$$

Then the density, pressure, and mass-distribution profiles are determined by the relations

$$\rho(\xi) = \rho_c\theta^n(\xi), \tag{196}$$

$$p(\xi) = \sigma\rho_c\theta^{n+1}(\xi), \tag{197}$$

$$M(\xi) = M\frac{v(\xi)}{v(\xi_1)}. \tag{198}$$

The temporal metric coefficient can be expressed in the form

$$\mathrm{e}^{2\Phi} = (1+\sigma\theta)^{-2(n+1)}\left\{1 - 2\sigma(n+1)\left[\frac{v(\xi_1)}{\xi_1} + \frac{1}{3}\lambda\xi_1^2\right]\right\}. \tag{199}$$

In accordance with the equivalence of matter and energy, the total energy E of the mass configuration, including the internal energy and gravitational potential energy, is given by the mass M producing the external gravitational field:

$$E = M = 4\pi L^3\rho_c v(\xi_1) = 4\pi\int_0^R \rho r^2\,\mathrm{d}r. \tag{200}$$

The proper energy E_0 is defined as the integral of the energy density over the proper volume of the fluid sphere

$$E_0 = 4\pi L^3\rho_c\int_0^{\xi_1}\frac{\theta^n\xi^2\,\mathrm{d}\xi}{\left\{1 - 2\sigma(n+1)\left[\frac{v(\xi)}{\xi} + \frac{1}{3}\lambda\xi^2\right]\right\}^{1/2}}. \tag{201}$$

The gravitational potential energy is then given by

$$\mathscr{G} = E - E_0. \tag{202}$$

Since $\mathrm{e}^{\Psi} \geq 1$, in general, $E_0 \geq E$ and $\mathscr{G} \leq 0$—the gravitational potential energy is negative. The gravitational binding energy $(-\mathscr{G})$ is the energy representing the work that has to be applied to the system to disperse the matter against the gravitational forces. The intensity of the gravitational bound can be represented by the ratio

$$g = \frac{\mathscr{G}}{E} = 1 - \frac{1}{v(\xi_1)} \int_0^{\xi_1} \frac{\theta^n \xi^2 \, d\xi}{\left\{1 - 2\sigma(n+1)\left[\frac{v(\xi)}{\xi} + \frac{1}{3}\lambda\xi^2\right]\right\}^{1/2}}. \tag{203}$$

Of course, the proper energy consists of the rest energy of gas, the kinetic energy of microscopic motion of the gas, and the radiation energy. In the relativistic polytropes, the special case of adiabatic processes implies a unique relation between the gas density ρ_g and the total mass density ρ, or between ρ_g and θ. Then, the total rest energy of gas can be integrated over the proper volume of the sphere, giving the binding energy E_b that can be compared with the gravitational potential energy (see Ref. [62] for $\lambda = 0$ and [63] for $\lambda \neq 0$).

The numerical analysis shows that the cosmological constant has a relevant influence on the structure of general relativistic polytropes when the length scale L becomes comparable with the cosmological length scale $\sim \Lambda^{-1/2}$; it is clear that L grows with σ increasing and ρ_c decreasing. The influence is amplified for polytropes with the polytropic index high enough. For example, in the case of $n > 3.5$, the influence of $\lambda > 0$ can lead to an instability of static polytropic configurations found in the case of $\lambda = 0$. For details, see Ref. [63].

6.3 Configurations of uniform density

It is interesting that there is a special class of general relativistic polytropes of the index $n = 0$ where the structure equations can be integrated in terms of elementary functions. We shall discuss these polytropes in detail because they can give an intuitive insight into the role of the cosmological constant.

The $n = 0$ polytropes correspond to the special class of the internal Schwarzschild–(anti-)de Sitter spacetimes [29] where the distribution of density ρ is uniform although the pressure qrows monotonically from its zero value on the surface of the configuration to a maximum value at its center. Recall that in the configurations with $\rho = \text{const}$, it is not necessary to use the unrealistic notion of an incompressible fluid—one can consider fluids with pressure growing as radius decreases, being 'hand tailored' [19]. Assuming $n = 0$, Eq. (187) can be integrated to give

$$v(\xi) = \tfrac{1}{3}\xi^3, \tag{204}$$

and the Eq. (186) takes the form

$$\left[3 - 2\sigma(1+\lambda)\xi^2\right] \frac{d\theta}{d\xi} + (1 - 2\lambda + 3\sigma\theta)(1 + \sigma\theta) = 0, \tag{205}$$

and can be integrated directly after separation of variables. Using the boundary condition $\theta(0) = 1$, we obtain

$$\sigma\theta = \frac{(1 - 2\lambda + 3\sigma)\left[1 - \frac{2}{3}\sigma(1+\lambda)\xi^2\right]^{1/2} - (1 - 2\lambda)(1 + \sigma)}{3(1 + \sigma) - (1 - 2\lambda + 3\sigma)\left[1 - \frac{2}{3}\sigma(1+\lambda)\xi^2\right]^{1/2}}. \tag{206}$$

This solution determines the dependence of pressure on the radial coordinate since for $n = 0$, there is $\theta = p(r)/p_c$. The dependence is given in units of energy density since $\sigma = p_c/\rho_c$. From the condition $\theta(\xi_1) = 0$, we find the radius of the configuration to be determined by

$$\xi_1^2 = \frac{6[1 + 2\sigma - \lambda(2 + \sigma)]}{(1 - 2\lambda + 3\sigma)^2}. \tag{207}$$

It is convenient to express the pressure profile in terms of r, R, instead of ξ, ξ_1, and σ. Introducing a new parameter a with a dimension of length by the relation

$$\frac{1}{a^2} = \frac{8\pi}{3}(\rho_c + \rho_{vac}), \tag{208}$$

we find the relation of R, σ, and λ to be given by

$$\left(1 - \frac{R^2}{a^2}\right)^{1/2} = \frac{(1 - 2\lambda)(1 + \sigma)}{1 - 2\lambda + 3\sigma}. \tag{209}$$

Then the central pressure p_c and the pressure profile $p(r)$ can be expressed in the known form [29]

$$p_c = \rho_c \frac{(1 - 2\lambda)\left[1 - \left(1 - R^2/a^2\right)^{1/2}\right]}{3\left(1 - R^2/a^2\right)^{1/2} - (1 - 2\lambda)}, \tag{210}$$

$$p(r) = \rho_c \frac{(1 - 2\lambda)\left[\left(1 - r^2/a^2\right)^{1/2} - \left(1 - R^2/a^2\right)^{1/2}\right]}{3\left(1 - R^2/a^2\right)^{1/2} - (1 - 2\lambda)\left(1 - r^2/a^2\right)^{1/2}}. \tag{211}$$

The radial metric coefficient is given by

$$e^{2\Psi(r)} = \left(1 - \frac{r^2}{a^2}\right)^{-1}, \tag{212}$$

and the total mass is

$$M = \frac{1}{3} L\sigma\xi_1^3 = \frac{4\pi}{3}\rho_c R^3. \tag{213}$$

The temporal metric coefficient is determined by formulae

$$e^{\Phi(r)} = \frac{[1 - \frac{2}{3}\sigma(1 + \lambda)\xi^2]^{1/2}}{1 + \sigma\theta}, \tag{214}$$

and

$$e^{\Phi(r)} = \frac{3\left(1 - R^2/a^2\right)^{1/2} - (1 - 2\lambda)\left(1 - r^2/a^2\right)^{1/2}}{2(1 + \lambda)}. \tag{215}$$

The special case of $\lambda = -1$ has to be treated separately as $1/a^2 = 0$. Now, Eq. (205) reduces to

$$\frac{d\theta}{(1 + \sigma\theta)^2} = -\xi\, d\xi, \tag{216}$$

which leads, after integration with the boundary condition $\theta(0)=1$, to the formula

$$\theta = \frac{1-\frac{1}{2}(1+\sigma)\xi^2}{1+\frac{1}{2}(1+\sigma)\xi^2}. \tag{217}$$

The boundary of the configuration is given by

$$\xi_1^2 = \frac{2}{1+\sigma}. \tag{218}$$

The central pressure and the pressure profile can be expressed in terms of r, R in the form

$$p_c = \frac{\rho_c}{\left(\frac{2R}{3M}-1\right)}, \tag{219}$$

$$p(r) = p_c \frac{1-\frac{r^2}{R^2}}{1+\frac{1}{\frac{2R}{3M}-1}\frac{r^2}{R^2}}. \tag{220}$$

The metric coefficients have a special form now, too. The radial component corresponds to flat $t = \text{const}$ sections

$$e^{2\Psi(r)} = 1, \tag{221}$$

while the temporal component has the form [29]

$$e^{\Phi(r)} = 1 + \frac{3M}{2R}\left(\frac{r^2}{R^2}-1\right). \tag{222}$$

The reality conditions on the general solution given by Eqs (211), (212), and (215) must guarantee that the pressure is positive and nondivergent and the metric coefficients must have regular behavior for $r \leq R$. Therefore, two conditions must be satisfied:

$$1-2\lambda > 0, \tag{223}$$

and

$$3\left(1-\frac{R^2}{a^2}\right)^{1/2} - (1-2\lambda) > 0. \tag{224}$$

Considering the case of $\lambda = 1/2$, we find

$$\theta(\xi) = \frac{(1-\sigma\xi^2)^{1/2}}{1+\sigma+\sigma(1-\sigma\xi)^{1/2}}. \tag{225}$$

The boundary of such a configuration is at

$$\xi_1^2 = \frac{1}{\sigma}, \tag{226}$$

and it can be shown that

$$R^2 = a^2. \tag{227}$$

Therefore, the metric coefficients are singular because there is

$$e^{-2\Psi(R)} = e^{2\Phi(R)} = e^{2\Phi(r)} = 1 - \frac{R^2}{a^2}. \tag{228}$$

For configurations of given M and Λ, condition (223) gives an upper limit on admissible values of the external radius R. Configurations with $\lambda \sim 1/2$ can be considered as nearly 'geodetical' since the pressure gradient almost vanishes on the surface, which is close to the static radius of the external geometry. (For $\lambda = 1/2$, the surface of the static configuration has to be located at the horizon of the external configuration with $R = 3M$; however, no static configuration can have its boundary at a black-hole horizon.)

The lower limit on the external radius of the $n = 0$ polytropes is determined by the condition (224) that can be transformed into the relations

$$\frac{R^2}{a^2} < \frac{4(1+\lambda)(2-\lambda)}{9}, \tag{229}$$

and

$$R > 2M\frac{9}{4(2-\lambda)}. \tag{230}$$

For $\lambda = 0$, we obtain the well known limit $R > \frac{9}{4}M$. The restrictions on physically realistic $n = 0$ polytropes can also be transformed into a form containing dimensionless quantities $x \equiv R/M$, $y \equiv \Lambda M^2/3$ (see Refs [29, 30]).

Finally, we determine the gravitational energy of the $n = 0$ polytropes and the parameter g. The total energy

$$E = M = \frac{4\pi}{3}\rho_c R^3. \tag{231}$$

The formula for the total proper energy reads

$$E_0 = 4\pi\rho_c \int_0^R \left(1 - \frac{r^2}{a^2}\right)^{-1/2} r^2 \,\mathrm{d}r, \tag{232}$$

and after integration

$$E_0 = \frac{3}{2}M\left(\frac{a}{R}\right)^3 \left[\arcsin\left(\frac{R}{a}\right) - \frac{R}{a}\left(1 - \frac{R^2}{a^2}\right)^{1/2}\right], \tag{233}$$

where

$$\frac{R}{a} = \frac{2(1+\lambda)^{1/2}}{1 - 2\lambda + 3\sigma}\left\{\sigma[1 + 2\sigma - \lambda(2+\sigma)]\right\}^{1/2}. \tag{234}$$

The gravitational potential energy

$$\mathscr{G} = M\left\{1 - \frac{3}{2}\left(\frac{a}{R}\right)^3 \left[\arcsin\left(\frac{R}{a}\right) - \frac{R}{a}\left(1 - \frac{R^2}{a^2}\right)^{1/2}\right]\right\}; \tag{235}$$

is naturally negative and the gravitational binding factor $g = \mathscr{G}/E$ is given by

$$g = 1 - \frac{3}{2}\left(\frac{a}{R}\right)^3 \left[\arcsin\left(\frac{R}{a}\right) - \frac{R}{a}\left(1 - \frac{R^2}{a^2}\right)^{1/2}\right]. \tag{236}$$

It expresses the gravitational binding in a 'pure' form as there in no internal energy in 'incompressible' configurations. On the other hand, we can directly conclude that the relation of the binding energy of gas $E_{\rm b}$, and the gravitational energy is given by

$$\frac{E_{\rm b}}{\mathscr{G}} = -1. \tag{237}$$

In polytropes with $n > 0$, the binding energy $E_{\rm b}$ will differ from the gravitational binding energy $(-\mathscr{G})$ as some of the work of the gravitational field is converted into the kinetic energy of microscopic motion in these polytropes. For the role of $\lambda \neq 0$ in the effect of gravitational binding, see Ref. [63].

It can be shown [30] that in extremely compact configurations (with $R < 3M$), a stable circular null geodesic exists around which null geodesics captured by the strong gravitational field are concentrated. Neutrinos, moving along these bound null geodesics, can influence cooling of extremely compact neutron stars. The potential well of the captured geodesics becomes deeper with the repulsive cosmological constant growing while it gets flatter with the attractive cosmological constant decreasing.

Acknowledgments

The present work was supported by the grant GAČR No. 202/99/0261, by the Committee for collaboration of the Czech Republic with CERN, and Bergen Computational Physics Laboratory project, an EU Research Infrastructure at the University of Bergen, Norway, supported by the European Community Access to Research Infrastructure Action of the Improving Human Potential Programme. The author would like to express his gratitude to Prof. Jiří Bičák for introducing into the fascinating field of relativistic astrophysics and cosmology, for helpful advice, and numerous stimulating discussions on many of the problems of black-hole physics. The present work results from a series of papers written in collaboration with Jiří Bičák, Stanislav Hledík, Petr Slaný, Marek Abramowicz, Massimo Calvani, John Miller, and Erlend Østgaard. Contributions of all these collegues is acknowledged.

References

1. R. D. Blandford. Physical processes in active galactic nuclei. In T. J.-L. Courvoisier and M. Mayor, editors, *Active Galactic Nuclei. Saas-Fee Advanced Course 20, Lectures Notes 1990*, page 161, Berlin, 1990. Swiss Society for Astrophysics and Astronomy, Springer-Verlag.
2. M. A. Abramowicz and M. J. Percival. *Classical Quantum Gravity*, 14:2003, 1997.
3. V. Karas. Active galactic nuclei. In J. Dušek and M. Zejda, editors, *Proceedings of the Conference on Variable Stars Research*, pages 49–62, 1998.
4. V. Karas, L. Šubr, and M. Šlechta. Stellar dynamics in a galactic centre surrounded by an accretion disc. This volume, 2001.
5. I. D. Novikov and K. S. Thorne. Black hole astrophysics. In C. De Witt and B. S. De Witt, editors, *Black Holes*, page 343, New York–London–Paris, 1973. Gordon and Breach.

6. W. R. Stoeger. *Astronomy and Astrophysics*, 53:267, 1976.
7. M. Kozłowski, M. Jaroszyński, and M. A. Abramowicz. The analytic theory of fluid disks orbiting the Kerr black hole. *Astronomy and Astrophysics*, 63:209–220, 1978.
8. M. A. Abramowicz, M. Jaroszyński, and M. Sikora. *Astronomy and Astrophysics*, 63:221, 1978.
9. M. Jaroszyński, M. A. Abramowicz, and B. Paczyński. *Acta Astronom.*, 30:1, 1980.
10. F. H. Seguin. *Astrophys. J.*, 197:745, 1975.
11. A. D. Linde. *Particle Physics and Inflationary Cosmology.* Gordon and Breach, New York, 1990.
12. L. M. Krauss and M. S. Turner. *Gen. Relativity Gravitation*, 27:1137, 1995.
13. J. P. Ostriker and P. J. Steinhardt. *Nature*, 377:600, 1995.
14. L. M. Krauss. The end of the age problem, and the case for a cosmological constant revisited. *Astrophys. J.*, 501(2):461–466, 1998.
15. N. Bahcall, J. P. Ostriker, S. Perlmutter, and P. J. Steinhardt. The cosmic triangle: Revealing the state of the universe. *Science*, 284:1481–1488, 1999.
16. R. R. Caldwell, R. Dave, and P. J. Steinhardt. Cosmological imprint of an energy component with general equation of state. *Phys. Rev. Lett.*, 80(8):1582, 1998.
17. C. Armendariz-Picon, V. Mukhanov, and P. J. Steinhardt. Dynamical solution to the problem of a small cosmological constant and late-time cosmic acceleration. *Phys. Rev. Lett.*, 85(21):4438, 2000.
18. L. Wang, R. R. Caldwell, J. P. Ostriker, and P. J. Steinhardt. Cosmic concordance and quintessence. *Astrophys. J.*, 530(1):17–35, 2000.
19. C. W. Misner, K. S. Thorne, and J. A. Wheeler. *Gravitation.* Freeman, San Francisco, 1973.
20. M. A. Abramowicz, B. Carter, and J.-P. Lasota. *Gen. Relativity Gravitation*, 20:1173, 1988.
21. M. A. Abramowicz. Centrifugal force – a few surprises. *Monthly Notices Roy. Astronom. Soc.*, 245(4):733–746, 1990.
22. M. A. Abramowicz, J. Miller, and Z. Stuchlík. Concept of radius of gyration in general relativity. *Phys. Rev. D*, 47(4):1440–1447, 1993.
23. S. Hledík. Optical geometry, inertial forces, and embedding diagrams. This volume, 2001.
24. O. Semerák. What forces drive the relativistic motion? *Nuovo Cimento B*, 110:973, 1995.
25. B. Carter. Black hole equilibrium states. In C. De Witt and B. S. De Witt, editors, *Black Holes*, page 57, New York–London–Paris, 1973. Gordon and Breach.
26. Z. Stuchlík. The motion of test particles in black-hole backgrounds with non-zero cosmological constant. *Bull. Astronom. Inst. Czechoslovakia*, 34(3):129–149, 1983.
27. Z. Stuchlík. An Einstein–Strauss–de Sitter model of the universe. *Bull. Astronom. Inst. Czechoslovakia*, 35(4):205–215, 1984.
28. Z. Stuchlík, P. Slaný, and S. Hledík. Equilibrium configurations of perfect fluid

orbiting Schwarzschild–de Sitter black holes. *Astronomy and Astrophysics*, 363(2):425–439, 2000.

29. Z. Stuchlík. Spherically symmetric static configurations of uniform density in spacetimes with a non-zero cosmological constant. *Acta Phys. Slovaca*, 50(2):219–228, 2000.
30. Z. Stuchlík, S. Hledík, J. Šoltés, and E. Østgaard. Null geodesics and embedding diagrams of the interior Schwarzschild–de Sitter spacetimes with uniform density. *Phys. Rev. D*, 64(4):044004 (17 pages), 2001.
31. J. Bičák and J. Podolský. Cosmic no-hair conjecture and black-hole formation: An exact model with gravitational radiation. *Phys. Rev. D*, 52(2):887–895, 1995.
32. J. Bičák and J. Podolský. Global structure of robinson–trautman radiative space-times with cosmological constant. *Phys. Rev. D*, 55(4):1985–1993, 1997.
33. J. Podolský. Exact impulsive gravitational waves in spacetimes of constant curvature. This volume, 2001.
34. A. Sen. Developments in superstring theory. In A. Astbury, D. Axen, and J. Robinson, editors, *29th International Conference on High Energy Physics, Vancouver, 23–29 July 1998*, Singapore, 1999. World Scientific.
35. C. Schmidhuber. AdS_5 and the 4d Cosmological Constant. *Nuclear Phys. B*, 580:140, 2000.
36. Z. Stuchlík and S. Hledík. Equatorial photon motion in the Kerr–Newman spacetimes with a non-zero cosmological constant. *Classical Quantum Gravity*, 17(21):4541–4576, 2000.
37. Z. Stuchlík. Kerr–Newman–anti-de Sitter black-hole spacetimes with a surface of degeneracy. *Acta Phys. Slovaca*, 48(5):535–548, 1998.
38. J. Bičák and Z. Stuchlík. On the latitudinal and radial motion in the field of a rotating black hole. *Bull. Astronom. Inst. Czechoslovakia*, 27(3):129–133, 1976.
39. M. Calvani and Z. Stuchlík. The latitudinal motion of test particles in the Kerr–Newman dyon space-time. *Nuovo Cimento B (11)*, 70(1):128–140, 1982.
40. Z. Stuchlík and S. Hledík. Some properties of the Schwarzschild–de Sitter and Schwarzschild–anti-de Sitter spacetimes. *Phys. Rev. D*, 60(4):044006 (15 pages), 1999.
41. Z. Stuchlík and S. Hledík. Properties of the Reissner–Nordström spacetimes with a nonzero cosmological constant. (unpublished, preprint TPA 003/Vol. 2, 2001), 2001.
42. Z. Stuchlík and S. Hledík. Equilibrium of a charged spinning test particle in Reissner–Nordström backgrounds with a nonzero cosmological constant. *Phys. Rev. D*, 64(10):104016 (12 pages), 2001.
43. S. Kristiansson, S. Sonego, and M. A. Abramowicz. Optical space of the Reissner–Nordström solutions. *Gen. Relativity Gravitation*, 30(2):275–288, 1998.
44. Z. Stuchlík and P. Slaný. Equatorial circular orbits in the Kerr–de Sitter spacetimes. (unpublished, preprint TPA 005/Vol. 2, 2001), 2001.
45. Z. Stuchlík and P. Slaný. Perfect fluid orbiting the Kerr–de Sitter black holes and naked singularities. (In preparation), 2001.

46. J. Bičák, Z. Stuchlík, and V. Balek. The motion of charged particles in the field of rotating charged black holes and naked singularities – I. The general features of the radial motion and the motion along the axis of symmetry. *Bull. Astronom. Inst. Czechoslovakia*, 40(2):65–92, 1989.
47. J. M. Bardeen. Timelike and null geodesics in the Kerr metric. In C. De Witt and B. S. De Witt, editors, *Black Holes*, page 215, New York–London–Paris, 1973. Gordon and Breach.
48. J. M. Bardeen, W. H. Press, and S. A. Teukolsky. Rotating black holes: locally nonrotating frames, energy extraction, and scalar synchrotron radiation. *Astrophys. J.*, 145:505–513, 1973.
49. Z. Stuchlík. Equatorial circular orbits and the motion of the shell of dust in the field of a rotating naked singularity. *Bull. Astronom. Inst. Czechoslovakia*, 31(3):129–144, 1980.
50. D. Markovic and S. L. Shapiro. Gravitational collapse with a cosmological constant. *Phys. Rev. D*, 61(8):084029 (7 pages), 2000.
51. R. Fabbri, R. Jantzen, and R. Ruffini. *Astronomy and Astrophysics*, 114:219, 1982.
52. Z. Stuchlík and M. Calvani. On the role of the cosmological constant in the formation of neutrino halos. *Astrophys. Space Sci.*, 98:405–409, 1984. Letter to the Editor.
53. A. Mészáros and Z. Molnár. On the alternative origin of the dipole anisotropy of microwave background due to the Rees–Sciama effect. *Astrophys. J.*, 470:49, 1996.
54. R. H. Boyer. *Proc. Cambridge Phil. Soc.*, 61:527, 1965.
55. M. A. Abramowicz. *Acta Astronom.*, 24:45, 1974.
56. B. Paczyński and P. Wiita. *Astronomy and Astrophysics*, 88:23, 1980.
57. M. A. Abramowicz, M. Calvani, and L. Nobili. *Astrophys. J.*, 242:772, 1980.
58. D. Lynden-Bell. *Nature*, 223:690, 1969.
59. R. D. Blandford. Astrophysical black holes. In S. W. Hawking and W. Israel, editors, *Three hundred years of gravitation*, page 277, Cambridge, 1987. Cambridge University Press.
60. O. Semerák. Gravitating discs around stationary black holes. This volume, 2001.
61. E. W. Kolb and M. S. Turner. *The Early Universe.* Addison-Wesley, Redwood City, California, 1990. The Advanced Book Program.
62. R. F. Tooper. General relativistic polytropic fluid spheres. *Astrophys. J.*, 140(2):434–459, 1964.
63. Z. Stuchlík and S. Hledík. General relativistic polytropes with a nonzero cosmological constant. (unpublished, preprint TPA 004/Vol. 2, 2001), 2001.

STELLAR DYNAMICS IN A GALACTIC CENTRE SURROUNDED BY AN ACCRETION DISC

VLADIMÍR KARAS§ AND LADISLAV ŠUBR
Astronomical Institute, Charles University Prague, V Holešovičkách 2, CZ-180 00 Praha, Czech Republic

MIROSLAV ŠLECHTA
Astronomical Institute, Academy of Sciences, CZ-251 65 Ondřejov, Czech Republic

Structure of a stellar cluster is examined assuming that it is embedded in the gravitational field of a supermassive black hole. Members of the cluster experience orbital decay via collisions with an accretion disc and several other dissipative processes. The attention is concentrated on the region of gravitational dominance of the central body. It is shown how mutual interaction between stars and the surrounding environment establishes a non-spherical shape and anisotropy of the cluster. In some cases, the stellar sub-system acquires ring-type geometry.

1 Introduction

This paper is focused on the region of gravitational dominance of massive dark objects residing in cores of numerous galaxies where the central compact body is surrounded by a rather dense population of stars and gaseous environment of an accretion flow. The dissipative interaction between stellar-mass satellites and gas results in an orbital decay that proceeds on the time-scale given by the properties of both components of the system. In the inner regions, the disc exerts hydrodynamical friction that acts on passing through satellites. Beyond certain radius, the medium becomes self-gravitating and the disc structure is changed. Gravitational relaxation takes over a relaxed cluster in still more remote regions. The inner cluster is thus dynamically distinct from the rest of the system, which, however, acts as its outer boundary. We want to discuss how an initially spherical outer cluster becomes gradually flattened in its inner region towards the disc plane. We examine growing anisotropy of the cluster and segregation of its members, changes in their mass function, and we calculate the expected velocity dispersions and profiles of averaged spectral lines originating from the nuclear cluster.

Galactic nuclei are firmly believed to host highly massive black holes ($M_{\bullet} \approx 10^8 M_{\odot}$) and we adopt this scenario here, too [13, 22]. The energy of infalling matter is liberated via an accretion flow surrounding the core in the form of a gas disc or a torus. It follows from the observations of nearby galaxies that a remarkable concentration of stars can be expected around the nucleus, although sub-parsec scales remain mostly unresolved. Three ingredients will be thus essential in following considerations: a compact massive core, the diluted medium of an accretion disc, and stellar-mass satellites forming a central cluster. Current evidence for the central black hole and for strong gravitational fields is provided mainly by

§ This contribution was prepared while VK was a visiting astronomer in DAEC, Observatoire de Paris-Meudon.

spectroscopy of accretion flows in X-rays; these are able to probe the innermost regions of just a few or a few tens of gravitational radii from the centre [6], however, understanding of local physics and of turbulent motions of the glowing gas at those distances is still insufficient. Optical and radio observations provide complementary information but these methods typically resolve regions where general-relativity effects due to the central black hole are already unimportant (however, cf. [10]). Sub-parsec scales (≈ 0.1 pc) can provide a link between the two mentioned domains. Let us emphasize that we focus in this contribution on such systems where the density of gas and number density of stars are rather high in the core (we do not discuss the case of Milky Way here).

We start with a brief account of simple scenarios that have been explored to great detail by various people: a geometrically thin disc perturbs motion of stellar bodies on long time-scales, responds to repetitive passages of the satellites, becomes modified by them, and, possibly, even disrupted. A possibility was proposed by Syer, Clarke & Rees [28] that a solar-mass body is captured in a quasi-stable bound orbit around the central black hole. Zurek, Siemiginowska & Colgate [33], Armitage, Zurek & Davies [1], and Ivanov, Igumenshchev & Novikov [9] examined various effects of repetitive passages of individual stars through dilute gaseous environment, assuming that it forms an accretion disc. There are, however, additional more subtle effects calling for thorough investigations; e.g., periodicities and quasi-periodicities that are introduced by general-relativity effects [12], influence of the disc self-gravity [25, 30] and of various other perturbations (gravitational radiation decay of satellites' orbits, mutual close encounters among the satellites, etc. [11, 16]), stellar collisions [21], and gradual changes of the physical and chemical state of the satellites and of the disc itself [3, 29]. These effects must be known with sufficient accuracy if one intends to study the state of stellar population in systems that are not sufficiently clean (a situation typical in those considerations that aspire to be astrophysically realistic), have rich stellar content, and therefore a relatively short relaxation time.

This paper is organized as follows. In the next section we describe the details of the model. Then, in Sec. 3, we give the results of computations showing modified quasi-stationary structure of the inner cluster (further details can be found in Šubr [27]). We consider a continuous inflow of the satellites from the outer parts of the cluster. The adopted scheme brings new time-scales and length-scales into the system and introduces interesting effects, such as the creation of transition regions where the satellites become accumulated in a ring-type structure. Finally, in Sec. 4, we briefly discuss the predicted velocity dispersions, averaged spectral-line profiles, and the modified mass function in terms of the model parameters.

2 Description of the model

The three constituents of our model system have been treated in the following manner. A central compact body has the mass $10^{-2} \lesssim M_8 \lesssim 10^2$ ($M_8 \equiv M_\bullet/10^8 M_\odot$). It introduces $r_{\rm g} \equiv GM_\bullet/c^2 \doteq 1.5 \times 10^{13} M_8$ cm as the smallest natural length-scale relevant for our considerations.

An accretion flow is characterized by its surface density $\Sigma_{\rm d}$, geometrical semi-thickness h, radial transport velocity v_r, and the speed of sound within the medium $v_{\rm s}$ (all quantities are defined as functions of r in the disc plane). Medium of the disc influences passing satellites either by direct hydrodynamical collisions (the case of inclined trajectories), or density waves and gap formation (embedded trajectories) [32]. Gravitational radiation is also taken into account in a simple approximation [18] (apart from that, all computations in this paper are performed in the purely Newtonian regime).

A cluster of satellites obeys, initially, spherical symmetry that is determined by a stationary state in the potential well of the central body. In its inner regions the cluster becomes gradually flattened and anisotropic as a result of its dissipative interaction with the disc material.

Rigorous treatment of non-spherical and anisotropic stellar systems is a difficult task that remains beyond the scope of this contribution. Instead, we adopt a simplified approach and divide the cluster into two subsystems: Long-term dynamics of the inner cluster (in terms of evolving orbital semi-axes of its members) is dominated by interaction with the disc; the exact prescription how this interaction proceeds depends on the parameters of individual satellites and hence it is sensitive to their mass and size, orbital inclination, and also to the properties of the dissipative environment at the corresponding radius. The outer cluster, on the other hand, is governed by the gravitational relaxation, and it acts as a reservoir from which fresh satellites are injected inwards, maintaining the overall steady state.

The evolution due to the gravitational relaxation takes place on the time-scale $t_{\rm r} = \sigma^3/(C \ln\Lambda\, G^2 M_*^2 n)$ where σ represents the velocity dispersion, n stands for the number density of the stellar system, and Λ is usual ratio of separations for the most distant and the closest encounters. As explained in the next section, we focus our attention on central regions where the gas environment is supposed to be relatively dense, and hence $t_{\rm r}$ is an order of magnitude longer than time $t_{\rm incl}$ required for the satellites to align with the disc. Relaxation plays a secondary role in our present discussion, nevertheless, we do incorporate the relaxed cluster as an initial and boundary state of the system. Geometry and the main ingredients of the model are illustrated in Figure 1, including the assumed distinction between the outer and the inner regions of the system. A computational procedure accounting for the mentioned effects is described hereafter.

In order that the computations be well defined and easier to compare, we considered the same disc models as in ref. [11]: (i) standard ("SS") radiation pressure dominated disc; (ii)–(iii) gas-pressure dominated SS disc with electron scattering and with free-free opacity, respectively; and (iv)–(v) marginally unstable self-gravitating disc with solar metallicity and with zero metallicity, respectively [5]. These models capture qualitatively different situations (because of their different profiles of density and thickness), and one can expect that more realistic and complicated situations will exhibit similar behaviour. Radial dependencies of the disc variables are in all these cases expressed by power-laws. For instance, the disc surface density can be written in the form $\Sigma_{\rm d}/\Sigma_* \propto (r/r_{\rm g})^s$, where s is a constant slope and $\Sigma_* = M_*/\,(\pi R_*)^2$ stands for the column density of the star; see Table 1 in Karas & Šubr [11] for a list of adopted parameter values that are relevant to different regimes of the star–disc interaction. The interaction between individual satellites and the

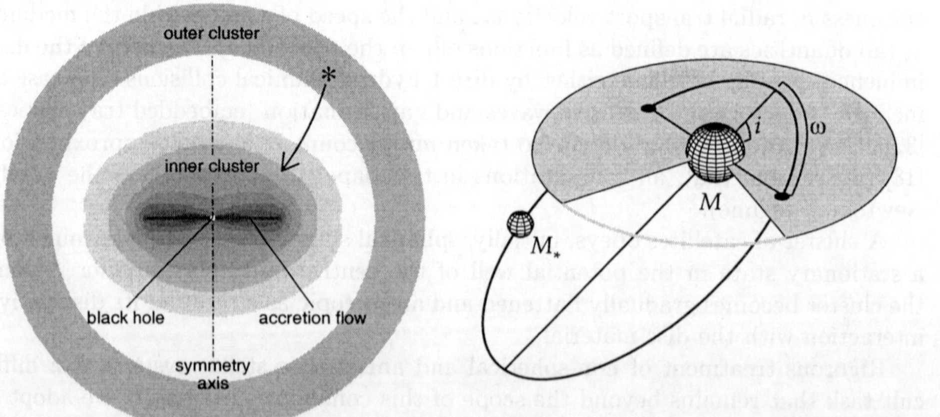

Figure 1. Left: A schematic drawing of the model adopted in computations. Its main ingredients are a massive compact body in the centre, an accretion disc or a torus, and a stellar cluster. The latter one has been divided here into two components: an inner cluster is formed by stars on orbits bound gravitationally to the centre and evolving due to interactions with the disc, while an outer cluster serves as a reservoir of satellites injected into the inner region. Gray levels used within the inner cluster indicate the number density of the satellites, as it follows from computations. Right: Geometrical situation of an orbit followed by a satellite in the inner cluster.

disc has been treated in the same manner as in the quoted paper but, instead of a finite sample of stars, here we consider a steady-state system with continuous replenishment. We also explore the case of more complicated composite disc models, such as (i)+(iv), constructed by concatenating simple models and reflecting the fact that each component can be relevant only within a limited range of radii. Transitions between different regions of an accretion disc leave their imprint in the modified cluster.

2.1 Star–disc interaction

We assume that the accretion rate is sufficiently large and, at the same time, the cluster is not excessively dense so that the disc is not completely removed by the passing satellites. In such circumstances, the ratio $\tau_{\rm incl}(r) \equiv \Sigma_*/\Sigma_{\rm d}$ gives the characteristic time-scale of the orbital evolution when a single satellite follows an inclined trajectory and hits the disc repetitively (units of the dynamical time at the corresponding radius). It is evident that different members of the cluster, ranging from compact stars to giants, populate an enormous interval of $\tau_{\rm incl}$ (orders of magnitude at a given radius), suggesting that different bodies must follow very different routes in the phase space of the evolving stellar sample. In order to simplify comparisons of the time scales among mutually competing effects, we estimate the mean grinding time as the period required for a circular orbit of a given radius a_0 to decline from a perpendicular inclination $i_0 = \pi/2$ down to $i \approx 10^{-3}$ (for more

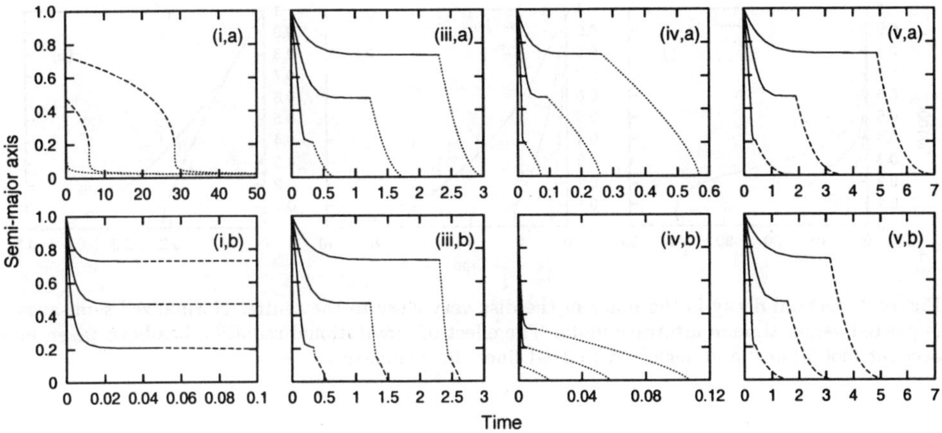

Figure 2. Evolution of a single orbiter semi-axis expressed in units of its initial value, $a_0 = 10^3 r_g$. Initially, the orbiter is put on an inclined trajectory and it experiences a prolonged period of repetitive collisions with the disc (indicated by solid lines). Then, being aligned with the plane of the disc, the evolution proceeds in one of two possible regimes: gap formation (dashed) or density waves (dotted). Three curves in each panel represent different initial inclinations: $i_0 = \pi/6$, $\pi/3$, and $\pi/2$ (the last one decays most quickly). Time is given in arbitrary units, common in all panels.

details and further references, see Appendix and ref. [27]):

$$t_{\rm incl} = t_0 M_8 \left(\frac{\Sigma_*}{\Sigma_\odot}\right)^{-1} \left(\frac{a_0}{r_{\rm g}}\right)^{3/2-s} {\rm yr}\,, \tag{1}$$

where t_0 includes model dependent terms and varies only slightly with a_0; e.g., in the cases (ii), ..., (v), t_0 acquires the values of 0.92, 0.22, 1.89×10^{-6} and 9.40×10^{-4}, respectively.

Repetitive collisions bring the satellites into the plane of the disc and onto almost circular orbits [11,20,27]. Subsequent orbit evolution is then reduced to the situation that was addressed by several people in connection with the formation and migration of bodies inside the disc [32]. Two basic modes can be distinguished: First, a gap is cleared in the disc if the satellite's Roche radius exceeds the disc thickness, $r_{\rm L} \approx (M_*/M_\bullet)^{1/3} r \gtrsim h$, and simultaneously $M_* \gtrsim M_{\rm gap} \approx \sqrt{40\alpha}\,(h/r)^{5/2} M_\bullet$ [15]. Motion of the satellite is then coupled with the disc inflow. On the other hand, if the satellite is unable to create the gap, it migrates inwards mainly due to the effect of density waves excitation on the time-scale

$$t_{\rm dw} = (C M_* \Omega)^{-1} \left(\frac{r}{h}\right)^2 \left(\frac{M_\bullet^2}{\pi r^2 \Sigma_{\rm d}}\right), \tag{2}$$

where C is a dimensionless constant of the order of unity and Ω stands for the disc rotation angular frequency.

With the help of Figure 2 we can briefly summarize how the interaction of individual satellites with the disc medium is treated. Examples of orbital evolution are shown in which the initial period of direct collisions with the disc is followed by

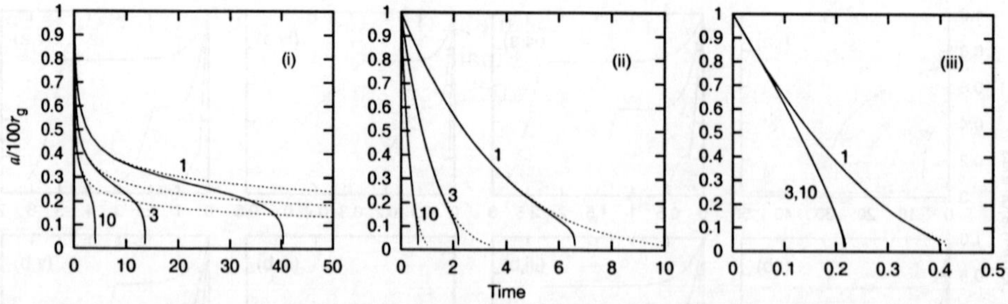

Figure 3. Orbital decay in the plane of the disc very close to the centre. Normalized semi-axes a is plotted versus time in arbitrary units. The effect of gravitational radiation has been taken into account (solid lines) and neglected (dotted lines) for comparison.

an appropriate regime of radial migration in the disc plane. Panels are arranged in columns corresponding to each selected disc model; the cases (ii) and (iii) are very similar and hence we show only the latter one. Common parameters of the plots are: the central mass $M_\bullet = M_8$; accretion rate $\dot{M}_\bullet = 0.1\dot{M}_{\rm Edd}$; and $\alpha = 0.1$ for the cases (i), ..., (iii), while $\zeta = 5$ for the cases (iv) and (v). The satellite in this example has one solar mass, $M_* = M_\odot$, and its column density is $\Sigma_* = \Sigma_\odot$. Furthermore, in all panels of the first row, initial values of the orbital elements are $a_0 = 10^3 r_{\rm g}$, $e_0 = 0.4$, and $i_0 = \pi/6$, $\pi/3$, and $\pi/2$ (corresponding to three different evolutionary tracks; orbits with the lowest i_0 evolve most slowly). We assume that the regime of direct collisions goes over to the regime of an embedded satellite when inclination passes through a critical value given, roughly, by $\tan i_{\rm cr} = h(r)/r$. Abrupt changes are clearly noticeable in the plots at moments when the evolution experiences transition between different regimes.

One notices in Fig. 2 that the orbital decay via collisions with the disc is decelerated as the orbit becomes aligned with the disc. Later on, with the single exception of disc (i), the decay gets accelerated again as the orbit becomes embedded in the disc. Also, by comparing panels (iii,a) and (iii,b), where $M_* = 3M_\odot$ was considered, we see that the decay is faster for more massive stars moving under the regime of density waves. The initial phase proceeds faster when it starts from higher eccentricities, which is demonstrated in panel (iv,b) with $e_0 = 0.8$. In the case (v) a gap develops when the orbiter enters the disc (for a solar type star this happens below $r = 5 \times 10^4 r_{\rm g}$, and the radius of the gap opening is even larger for more massive satellites). Comparison of the two panels in the last column illustrates that the variation of $i_{\rm cr}$ (its value is ten times larger in (v,b) than in (v,a)) does not change evolutionary tracks in any qualitative manner, although it somewhat speeds up the evolution of the satellites near co-rotation.

In the following computations we considered also the orbital decay caused by the gravitational radiation but the effect has a visible impact only very near the centre. This can be seen in Figure 3 where the evolution in the plane of the disc is captured for cases (i), (ii), and (iii), and for three values of the satellite mass, $M_* = 1M_\odot$, $3M_\odot$, and $10M_\odot$ (indicated in the plots). Sinking satellites follow almost identical trajectories if their mass is large enough and gap opening takes place ($M_* \gtrsim 3M_\odot$, the case (iii)), while in all other cases the curves proceed under the regime of density waves induction. Finally, one needs to consider the effect of the dynamical friction acting on each satellite by surrounding gas [17, 24], taking into account slab geometry of the disc [26].

2.2 The cluster

We employ the Bahcall & Wolf [4] solution, denoted as B&W hereafter, to define the asymptotic structure of the cluster at large distances from the centre, i.e. outside the region affected by direct interaction with the disc medium (but still influenced by the black hole). In contrast to ref. [4], we relax the assumption of spherical symmetry in the inner parts of the cluster, even if it were at the expense of some other simplifications (these are specified in the following paragraphs). Nuclear clusters were examined by Rauch [21] who concentrated on the effects of collisions and relativistic effects. He employed a picture of a cluster divided into two intervals of radius, the inner part and the outer part, that are treated in different ways and matched on the boundary. We adopt a similar approach using a simpler computational scheme and we examine situations when the gas content exerts a significant drag.

The relaxation time-scale can be estimated by substituting σ with the Keplerian velocity and setting $C \ln \Lambda \approx 10^2$ in $t_{\rm r}$:

$$t_{\rm r} = 1.3 \times 10^{19} \left(\frac{M_*}{M_\odot}\right)^{-2} \left(\frac{n(r)}{10^6\,{\rm pc}^{-3}}\right)^{-1} \left(\frac{r}{r_{\rm g}}\right)^{-\frac{3}{2}} {\rm yr}\,. \tag{3}$$

In case of the B&W distribution, the stellar number density follows a power-law radial profile, $n(r) = n_0 (r/r_{\rm h})^{-7/4}$. In the velocity space we assume that the distribution function of the boundary cluster corresponds to the Maxwellian velocity dispersion, and that it is constant on length-scale of $r_{\rm h}$. Density of the boundary cluster is linked to the gravitationally bound region via a proportionality factor, n_0. Hence, the relaxation time in the B&W domain is

$$t_{\rm r} = 4 \times 10^8 n_6 \left(\frac{M_*}{M_\odot}\right)^{-2} \left(\frac{r_{\rm h}}{10^6 r_{\rm g}}\right)^{-\frac{7}{4}} \left(\frac{r}{r_{\rm g}}\right)^{\frac{1}{4}} {\rm yr}\,, \tag{4}$$

with $n_6 = (n_0/10^6)\,{\rm pc}^{-3}$. Let us recall that we already have met, in eq. (1), another characteristic time, $t_{\rm incl}$, for the orbiter being aligned with the disc. The ratio of the two characteristic times is essential here. The relaxation time increases with r at a lower rate than $t_{\rm incl}$ and, therefore, we may introduce radius $r_{\rm i}$ where both times are comparable:

$$\left(\frac{r_{\rm i}}{r_{\rm i_0}}\right)^{\frac{5}{4}-s} = (n_6 M_8)^{-1} \left(\frac{\Sigma_*}{\Sigma_\odot}\right) \left(\frac{M_*}{M_\odot}\right)^{-2} \left(\frac{r_{\rm h}}{10^6 r_{\rm g}}\right)^{-\frac{7}{4}}, \tag{5}$$

where $r_{\rm i_0} = 4.73 \times 10^4 r_{\rm g}$, $4.36 \times 10^4 r_{\rm g}$, $1.68 \times 10^4 r_{\rm g}$, and $9.80 \times 10^3 r_{\rm g}$ for the cases (ii), ..., (v), respectively. In the model case (i), the presented estimate of drag time results in a more complicated expression, nevertheless, we may safely consider the relaxation time to be longer than $t_{\rm incl}$ for any reasonable choice of model (i)

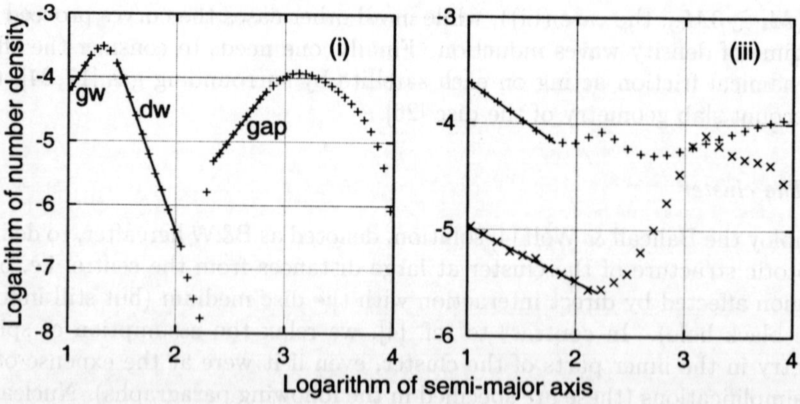

Figure 4. Left panel: a quasi-stationary $n_t(a)$ profile is plotted (crosses) for the case (i), $t = 10^{13}$. A power-law approximation of $n_t(a) \propto a^q$ is constructed in the appropriate ranges of radii (solid lines). Suppression of the profile occurs for a small a as a consequence of quite efficient gravitational radiation losses near the center, and for large a where the sample is depleted. Right panel: Two snapshots are shown for the model (iii): $t = 10^{10}$ (upper profile) and $t = 10^{12}$ (lower profile) in arbitrary time units; explicit time dependence of density profiles will be eliminated hereafter in the models with continuous replenishment.

parameters [27], which, in any case, is only relevant in the innermost region and must be truncated and matched to a different solution farther away.

In our present model, we want to incorporate the effects of orbital decay in the inner cluster via dissipative interaction with gas (and gravitational radiation [11]), and we account for a steady inflow of stars via relaxation across the outer boundary. To this aim, we developed two schemes how the satellites are injected from the outer region (models M1 and M2 hereafter). We compared the results obtained using both models (and found them mutually compatible). The adopted scheme is described in the next section.

2.3 Cluster–disc interaction

Before we embark on the description of inner cluster replenishment, it would be instructive to recall the situation that develops when there are no fresh satellites injected from the outer region. In that case, a quasi-stationary state lasts for a limited period until the inner cluster is depleted. But even during that period one can recognize characteristic density profiles being established in the regions where one of the orbital decay mechanisms dominates. This is clearly seen in Figure 4 and further discussed in ref. [11]. For example, in the case (i) and for solar-type satellites (see left panel), we find that the initial slope ($n_0(a) \propto a^{1/4}$) reaches a quasi-stationary state with $q \to -4.9$ in the region $30 \lesssim a \lesssim 120$; here the evolution is driven by density waves, denoted as dw in Fig. 4. Farther away, $q \to 2.3$ for $200 \lesssim a \lesssim 700$ (i.e. beyond the transition point where gap formation takes over and acts as the main channel of orbital decay). On the other hand, the slope is influenced by gravitational waves (gw) emission on very small radii: $q \to 2.8$ for $r \lesssim 20$.

Substantial span of values acquired by the slope q reflects the presence of ripples in $n(a)$ that develop in certain areas of the disc. The right panel of Fig. 4 shows how the $n(a)$ profile is gradually adjusted as time proceeds. In both panels, one notices a gradual depletion of the sample that starts at large a's and gradually depresses the whole profile. Such depletion is prevented in models with continuous replenishment, as described hereafter.

An outer relaxed cluster provides a continuous supply of fresh stars in the sample. Hence, a stationary state is ensured in the inner region where each member of the sample is replaced by another one, newly generated, once it has been removed from the inner cluster via orbital decay and subsequent disruption or capture. New bodies are injected at a rate equal to $1/t_{\mathrm{r}}$, where t_{r} is now the relaxation time in the outer cluster. They are generated with parameters complying with the outer cluster distribution function, $f_0(\boldsymbol{r},\boldsymbol{v})$. The probability to find a star evolving from state $(\boldsymbol{r},\boldsymbol{v})$ and occupying a phase space element $\mathrm{d}\Omega$, with semi-major axis within interval $\langle a, a+\mathrm{d}a\rangle$ then reads:

$$n(a;\boldsymbol{r},\boldsymbol{v})\mathrm{d}a = \mathrm{d}\Omega\, f_0(\boldsymbol{r},\boldsymbol{v})\, t_{\mathrm{r}}^{-1} J^{-1}\,, \tag{6}$$

$J \equiv |\mathrm{d}a(a;\boldsymbol{r},\boldsymbol{v})/\mathrm{d}t|$. The meaning of this relation can be also clarified by integrating eq. (6) over a. The integration gives the total number of stars inserted from $\mathrm{d}\Omega$ in the inner cluster, and this fraction is equal to the number of stars in $\mathrm{d}\Omega$ multiplied by the ratio of drag to relaxation periods. At the same time, an integration over the relevant domain Ω in the phase space of the initial parameters gives the desired probability density, $n(a)$. The way how the transition zone is introduced in the scheme represents the main simplification (and limitation) of the described approach: It is assumed that Ω is emptied *only* due to dissipative effects of interaction with the disc, and it is replenished, at a constant rate, *only* via relaxation processes. This approach helps us to focus on the main effects in which we are interested.

We used two different regimes of satellite insertion in the inner cluster, i.e. two different choices of Ω and $f_0(\boldsymbol{r},\boldsymbol{v})$. Within the first model, denoted as M1 hereafter, we define Ω as a spherical layer in configuration space between radii $r = \chi r_{\mathrm{h}}$ and $r = r_{\mathrm{h}}$ from where the stars are injected into the inner system (χ is an order-of-unity parameter; we set $\chi = 0.5$ for definiteness of examples). In the velocity space, the boundaries are fixed by the requirement that the orbits intersect the disc and have $a < r_{\mathrm{h}}$ (otherwise, the orbits traveling for extensive periods of time in the outer cluster are, predominantly, subject to relaxation and are not really influenced much by the centre). In line with these considerations is also our choice of the starting distribution, f_0, which we assumed to be homogeneous in the configuration space and isotropic in velocities,

$$f_0(\boldsymbol{r},\boldsymbol{v}) = \frac{n_0}{(2\pi\sigma^2)^{3/2}} \exp\left[\frac{v^2}{2\sigma^2}\right]\,, \tag{7}$$

with a number density n_0 and a velocity dispersion $\sigma \equiv \sqrt{\langle v^2\rangle}$. The probability density of the semi-axis values in the inner cluster then reads

$$n(a)\mathrm{d}a = \int_\Omega \mathrm{d}\boldsymbol{r}\mathrm{d}\boldsymbol{v}\, \frac{n_0}{(2\pi\sigma^2)^{3/2}} \exp\left[\frac{v^2}{2\sigma^2}\right] t_{\mathrm{r}}^{-1} J^{-1}\,. \tag{8}$$

We used a Monte-Carlo scheme to evaluate the integral on the right-hand-side of eq. (8) and we followed $N \approx 10^6$ orbits with randomly generated starting velocities and positions in the region $\chi r_{\rm h} < r < r_{\rm h}$. Each time the orbit reaches a predetermined value of a_i (these are distributed linearly over a logarithmical grid) the appropriate bin of the probability density is increased by a factor $|\dot{a}|^{-1}$:

$$n_i \Delta a_i = \frac{1}{t_{\rm r}} \frac{N_0}{N} \sum_1^{N_{\rm c}} \left| \frac{{\rm d}a}{{\rm d}t}(a; r, v) \right|^{-1} . \tag{9}$$

Here, N_0 is the number of stars within the specified range of radii, $N_{\rm c}$ is the number of the generated orbits intersecting the disc and satisfying the condition $a_0 < r_{\rm h}$. For example, in case of a disc with the outer radius $10^4 r_{\rm g}$ we adopted $r_{\rm h} = 10^6 r_{\rm g}$ and velocity dispersion $\sigma = 300\,{\rm km/s}$. Hence the resulting $N_{\rm c}$ comes out roughly $(1/200)\,N$ for $\chi = 0.5$. Obtained profiles of $n(a)$ remain rather insensitive to the exact value of the ad-hoc parameter χ, although their absolute value decreases as $\chi \to 1$.

The second scheme (M2) employs Bahcall-Wolf distribution function to describe the reservoir of stars. In this approach it is possible to eliminate parameter χ by generating semi-axes of replacement satellites with a probability density $n(a) \propto a^{1/4}$ over the interval $\langle a_{\rm min}, r_{\rm h} \rangle$. We set $a_{\rm min} = 10 r_{\rm g}$ in a typical simulation. The remaining orbital parameters, $x\ (\equiv \cos i)$ and e, are distributed randomly; this ensures spherical symmetry of the outer cluster. We have also tested other distributions, such as $n(e) \propto e$ for eccentricity, but without any substantial change in the results; cf. ref. [19]. Finally, in M2, an additional condition was imposed requiring that the total time $t_{\rm d}$ of orbital decay be shorter than the relaxation time (4) in the outer cluster, i.e. $t_{\rm d} \equiv a_0/|\dot{a}_0| \le t_{\rm r}$.

3 Results

We first carried out computations for the model M1 with simple density profiles $\Sigma_{\rm d}(r)$ of the disc (single power-law) and varying free parameters (such as the prescribed outer edge, $R_{\rm d}$). The typical results are summarized in Figure 5. From top to bottom, each row corresponds to one of the disc models, (i), ..., (v) extending to $R_{\rm d} = 10^4 r_{\rm g}$. The first column: single-mass satellites with $M_* = 0.1 M_\odot$ and $\Sigma_* = 4\Sigma_\odot$ (solid lines), $M_* = M_\odot$ and $\Sigma_* = \Sigma_\odot$ (dashed lines), and $M_* = 10 M_\odot$ and $\Sigma_* = 0.25\Sigma_\odot$ (dotted lines). The second column: fraction of $n(a)$ formed only by satellites in the disc plane. Formation of ring-type structures can be traced in these graphs. The third column: an example of $n(a)$ formed by solar type stars with special initial parameters ($a_0 = 10^6 r_{\rm g}$, $e_0 = 0.995$, and $i_0 = \pi/2$). The fourth column: a cluster replenished with stars of masses varying over the range $0.1 M_\odot \le M_* \le 50 M_\odot$.

In the left column of Fig. 5, each subplot shows three curves for different masses of the cluster members. The innermost region of the cluster is, similarly to the above discussed quasi-stationary profiles, dominated by stars embedded in the disc – satellites number density, $n(a)$, is well described by a broken power-law the slope of which depends on the regime of star–disc interaction at corresponding distance. Transitions occur where the regime is changed. However, a power-law fit is now

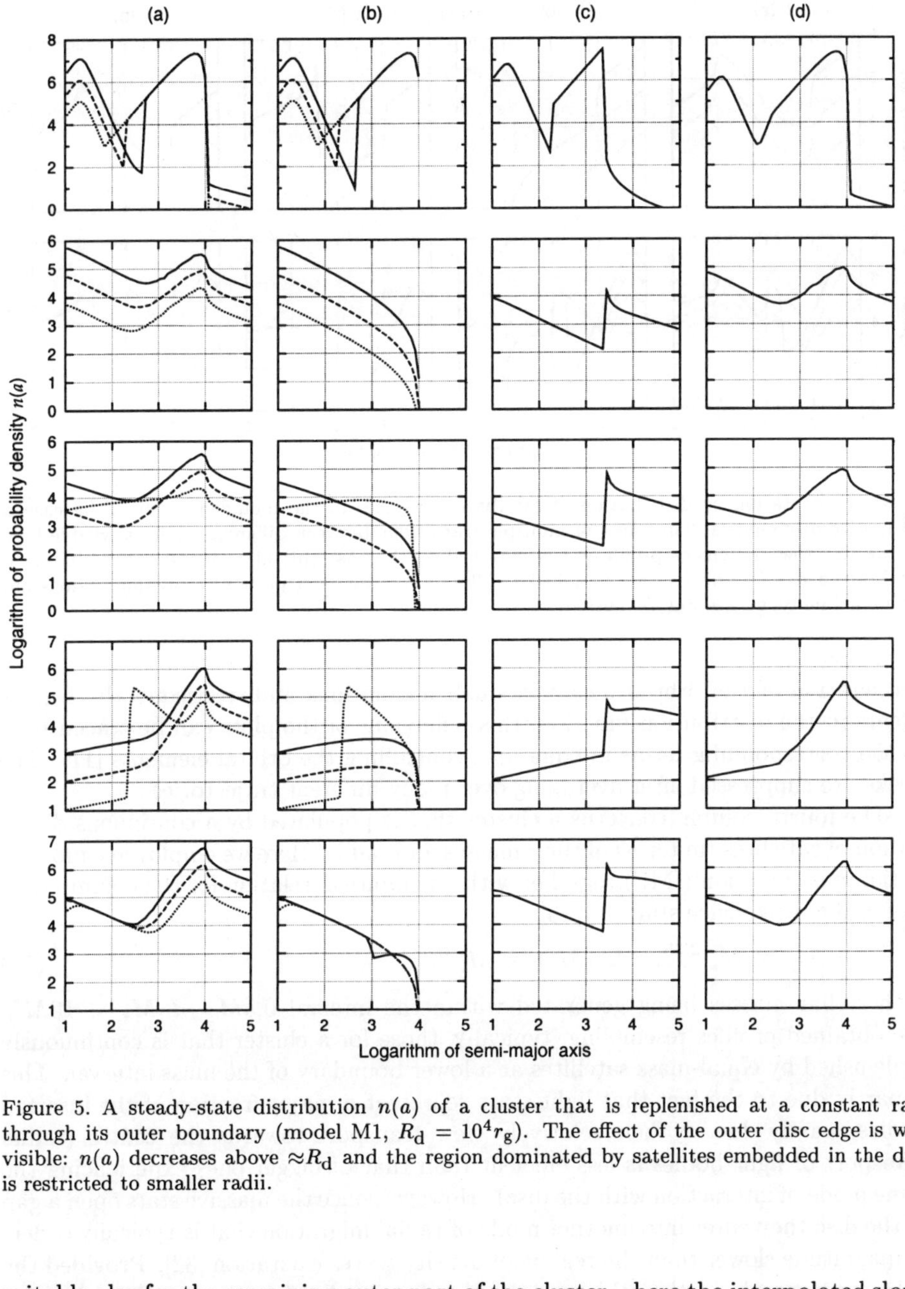

Figure 5. A steady-state distribution $n(a)$ of a cluster that is replenished at a constant rate through its outer boundary (model M1, $R_{\rm d} = 10^4 r_{\rm g}$). The effect of the outer disc edge is well visible: $n(a)$ decreases above $\approx R_{\rm d}$ and the region dominated by satellites embedded in the disc is restricted to smaller radii.

suitable also for the remaining outer part of the cluster – here the interpolated slope is $q = 1.4$, 1.6, 3.0, and 2.4 for disc models (ii), (iii), (iv), and (v) respectively.

A qualitatively different plot is obtained when $n(a)$ is calculated with regard to only one particular set of initial parameters, a_0, e_0, and i_0. Such an example is presented in the third column of Fig. 5. In this case, the inclination and eccentricity dependent parts of $({\rm d}a/{\rm d}t)_{\rm col}$ play an important role. They cause a significant

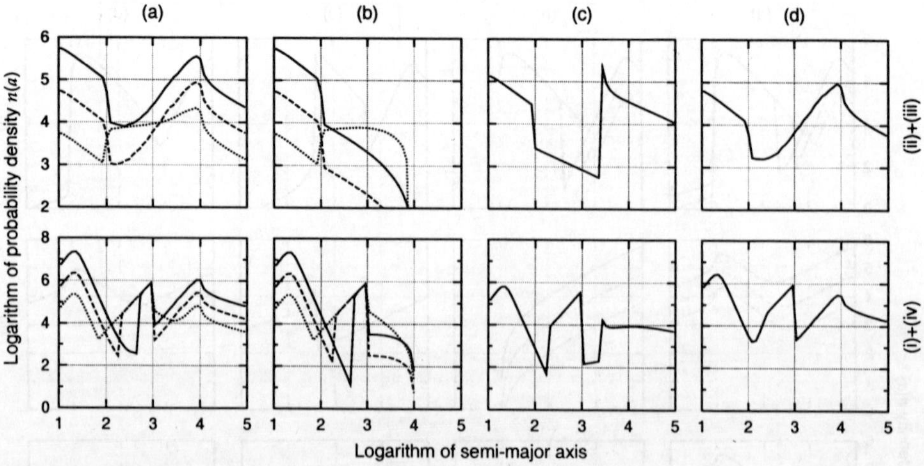

Figure 6. A stationary distribution of satellites semi-axes in the modified cluster: the case of composite disc models. The effect of multiple transitions between different modes is visible here. In the top row, a density profile (ii) is matched to the case (iii) at a distance $r = 10^2 r_g$; in the bottom row (i) and (iv) are matched at $10^3 r_g$. For the description of individual panels and different line types, see Fig. 5 caption.

deceleration of the orbital decay at small inclinations and, consequently, a peak occurs in the distribution function; this behaviour of the plots can be traced back to the corresponding terms in equations controlling the orbital elements [11]. The peaks are suppressed after averaging over many different trajectories.

The fourth column concerns a cluster that is populated by a continuous distribution of satellites (in terms of their masses and radii). Here we employed canonical initial mass function (IMF) together with an empirical relation for the column density of main sequence stars [14,23]:

$$\mathcal{N}_0(M_*) \propto M_*^{-2.35}, \quad \Sigma_*(M_*) \propto M_*^{-0.6} \tag{10}$$

with stellar masses being generated within the interval $0.5M_\odot < M_* < 10M_\odot$. We obtained profiles resembling, typically, those for a cluster that is continuously replenished by equal-mass satellites at a lower boundary of the mass interval. This is partly due to the fact that light stars represent a major fraction of the inserted bodies, partly due to their relatively slow transport towards the centre (radial transport of light bodies is less efficient than that of bigger ones experiencing the same mode of interaction with the disc). However, once the massive stars open a gap in the disc they enter into another mode of radial migration that is typically orders of magnitude slower than the regime of density waves excitation [32]. Provided the light stars remain in the other regime, the distribution function is affected – this is a consequence of different migration modes for different stars is observed in the panel (d,iii), where the slope of the $n(a)$ curve is modified with respect to an equal-mass cluster below $\approx 200r_g$. For $M_* = 0.1M_\odot$ the slope attains the theoretically predicted value for the density waves regime, i.e. $q = -0.5$ [27], while, in case of the IMF, the slope is shifted up towards a positive value that corresponds to massive

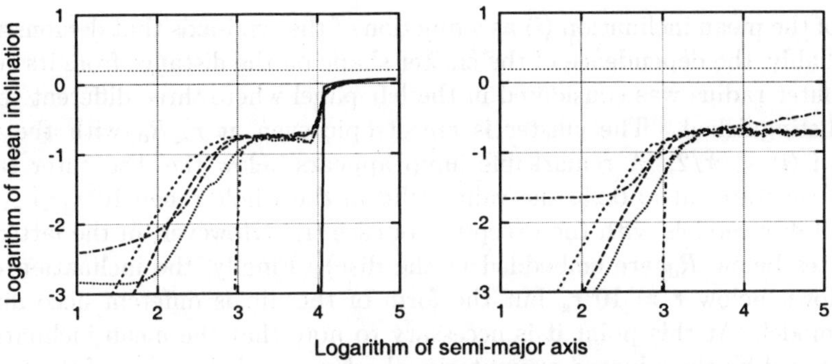

Figure 7. Mean orbital inclination as a function of the semi-axis for a cluster replenished at a constant rate with solar-type stars. The left panel corresponds to the case of a disc with an outer boundary at $R_{\rm d} = 10^4 r_{\rm g}$, while, in the right panel, an infinite radius of the disc is assumed. Disc models are indicated by line-types: Solid, long-dashed, short-dashed, dotted, long-dash-dotted, and short-dash-dotted lines correspond to the discs (i), (ii), (iii), (iv), (v), and (i)+(iv), respectively. The horizontal dotted line indicates the value of $i = \pi/2$.

stars migrating in the gap opening mode (in particular, $q \approx -0.41$ for the discussed panel).

The overall picture is rich in possible situations. Given the stellar mass function of the cluster, there is one more difference from the case of equal-mass satellites distributions that have not been mentioned yet: it is the smoothing of abrupt steps in the distribution around the transition radii, the exact location of which depends on each satellite's mass and compactness. To see this clearly, we compare the panel (a,i) in Fig. 5 (equal-mass bodies; area around $r = 10^2 r_{\rm g}$) with the corresponding graph (d,i) (universal $\mathcal{N}_0$, eq. (10)).

An additional jump in the profile occurs when the disc is treated as a composite conjunction of different power-law profiles. This is clearly visible in the first row of Figure 6 where two gas-pressure dominated disc models (ii) and (iii) are matched at $r = 10^2 r_{\rm g}$. In the second row, another case is shown with a radiation-pressure dominated SS disc switched to a marginally unstable disc (iv) at $10^3 r_{\rm g}$. Let us point out that a modification of $n(a)$ is only observed near to the characteristic radius (just like the case of the disc outer edge); the rest of the profile remains unchanged. We recall that similar profiles of $n(a)$ were obtained in a quasi-stationary phase of an isolated cluster [11], and we found very similar results also for the model M2 (discussed below). Hence we can conclude that local properties of the disc do influence the cluster profile in the presented manner.

In the computational procedure that refills the cluster and maintains its steady state, the satellites are injected in the ensemble with randomly oriented trajectories; only then does the phase of the orbital evolution take place in the regime of star–disc collisions and the satellites are dragged towards the centre and to the plane of the disc with a rate that increases with the initial inclination. Hence, it appears natural that the initial steady cluster is quickly changed during the phase of continuous replenishment. We demonstrate this distortion in two ways: First, in Figure 7, there is a plot of the mean inclination $\langle i \rangle$ as a function of the semi-axis that demonstrates, rather reliably, the dependence of the cluster shape on the distance from its centre. A finite outer radius was considered in the left panel where three different regions can be distinguished. The cluster is anisotropic even at $r \gtrsim R_{\rm d}$ with the mean inclination $\langle i \rangle < \pi/2$. A remarkable jump appears below the disc outer radius, where $\langle i \rangle$ saturates at ≈ 0.25, the value valid in the whole range $10^3 r_{\rm g} \lesssim a \lesssim 10^4 r_{\rm g}$ and for all disc models with the exception of case (i). (However, in the latter case all satellites below $R_{\rm d}$ are embedded in the disc). Finally, the inclination drops further down below $r \approx 10^3 r_{\rm g}$ but the form of the dip is different according to the disc model. At this point it is necessary to note that the mean inclination is also influenced by the adopted computational scheme and the value of the limiting inclination $i_{\rm cr}$ (the inclination is frozen as the orbit aligns with the disc plane). An abrupt fall off is observed in the case of the disc model (i)+(iv) (similar also to the case (i) at $r = R_{\rm d}$), where the disc (i) extends below $10^3 r_{\rm g}$ and drags the stars into its plane there on very short time-scales.

As shown in the right panel of Figure 7, the picture appears less complicated when the disc extends to infinity and the transition area is missing at the disc outer radius. Let us point out a rather outstanding profile of the cluster if the disc (v) is involved: below $r = 10^3 r_{\rm g}$, $\langle i \rangle$ remains several orders of magnitude larger than in other cases. Even above $10^3 r_{\rm g}$, the mean inclination increases outwards much faster in this case, which means that the cluster stays less flattened than the others.

A complementary discussion is provided in Figure 8, which gives a two-dimensional representation of the cluster spatial distribution. We realized that this image contains a wealth of information about our model and we thus embark on its more detailed description at this point. The number density is shown within a meridional section across the cluster (the first column on the left shows the range of $4R_{\rm d}$ while the second one captures a small detail around the centre). Further, two projections are shown along the line of sight perpendicular to and colinear with the symmetry axis (the third and the fourth columns). Gray levels express the number and column densities in the logarithmical scale. Five disc models are presented, (ii), ..., (v) and (i)+(iv), each of them occupying one row. In all these cases the accretion flow is characterized by the $\dot{M}$, α, and ζ parameters and by the disc outer radius, $R_{\rm d} = 10^4 r_{\rm g}$. The cluster is formed by satellites supplied continuously from a layer $0.5 r_{\rm h} \leq r \leq r_{\rm h} = 10^6 r_{\rm g}$. Furthermore, the outer cluster is characterized by a number density $n_0 = 10^6 {\rm pc}^{-3}$ and a velocity dispersion $\sigma = 300 {\rm km\,s}^{-1}$.

In all these cases, the inner cluster is clearly flattened and its dimension is closely related to the size of the disc. Qualitative differences are clearly visible between the models: First, notice that the growing thickness of the disc (as a function of radius) leaves its imprints on isocontours of the number density, as can be seen in the first and the second columns. While, in case of models (ii) and (iii), the isocontours depart from the plane of the disc with gradually decreasing slope (they are bent towards the equator), their shape is distorted in the opposite direction for the marginally unstable disc models (iv) and (v). Hence, the characteristic thickness of the latter is much larger and the satellites are removed from the region above and below the centre. Still another picture would be obtained for the radiation pressure dominated SS disc. In this case, the inner cluster takes the form of a thin

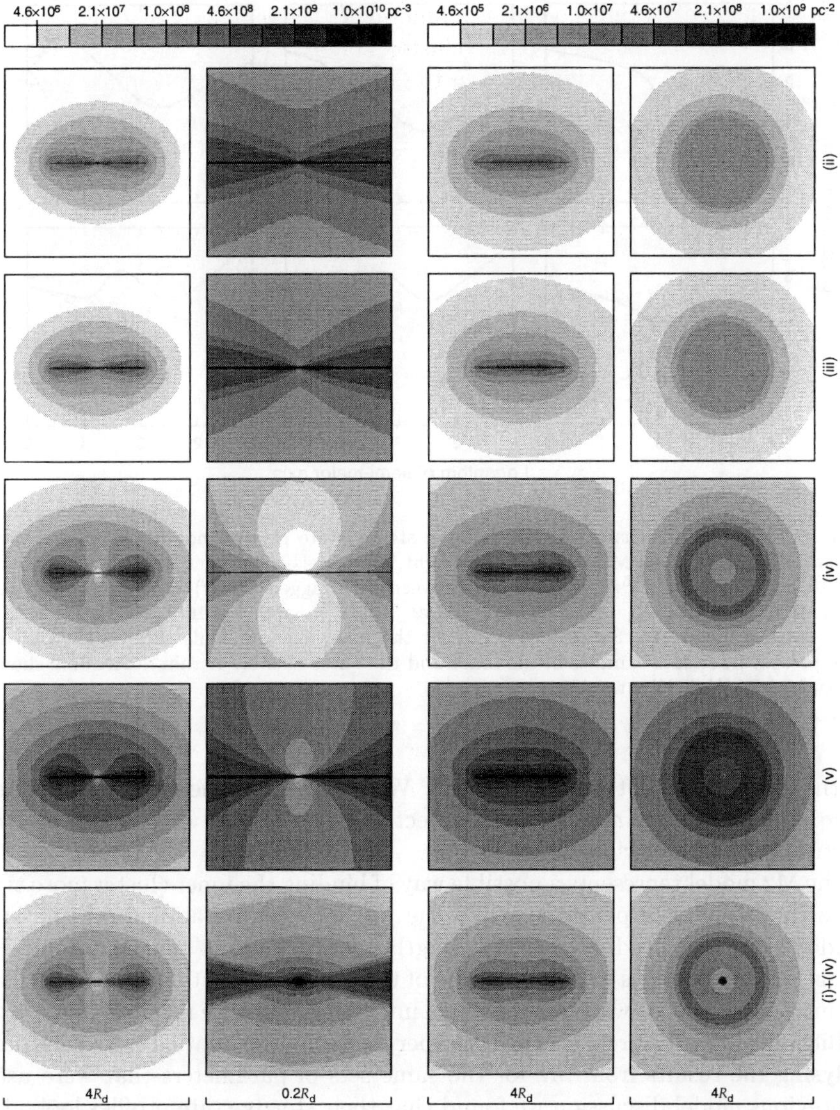

Figure 8. The number density $n(a)$ of the cluster is plotted in different representations. Model M1 and different disc profiles have been used. See the text for details.

disc without any vertical structure; this is simply due to the fact that it drags the satellites into the disc on much shorter time-scales than the time they subsequently spend in the disc. The radial structure of the inner cluster is then given directly by the probability density $n(a)$ presented in Fig. 5.

Within the scope of the model M1, the inner cluster is coupled to the outer one through a (somewhat artificial) parameter χ, which defines their mutual boundary

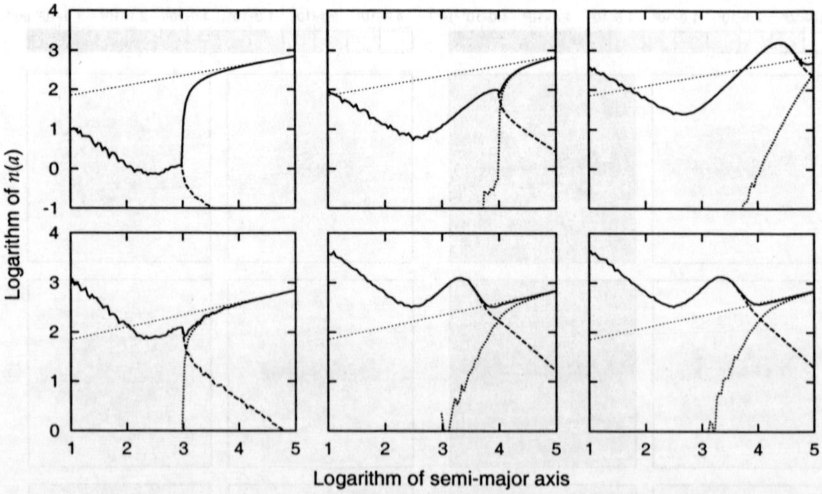

Figure 9. The number density vs. radius for a steady-state cluster modified by a gas pressure dominated SS disc (model M2). From left to right, the panels correspond to the disc outer radius $R_{\rm d} = 10^3 r_{\rm g}$, $10^4 r_{\rm g}$, and $10^5 r_{\rm g}$. Transitions between the dragged (inner) and the reservoir (outer) regions occur either at $R_{\rm d}$ or at $r_{\rm i}$, which is $\approx 5 \times 10^3 r_{\rm g}$ and $5 \times 10^4 r_{\rm g}$ in the first and the second rows, respectively. Solid lines stand for the total density, while the dashed and dotted lines represent fractional samples of the inner and the outer cluster. A thin dotted line shows the unperturbed B&W distribution for reference.

(i.e. the lower radius of the outer cluster). We verified that the distribution function of the inner cluster is not markedly affected either by the dimension, or by the structure of the outer cluster.

The M2 model represents a possible way of binding the inner cluster more tightly to the reservoir. The parameter χ of the previous computational scheme is now abandoned and replaced by a natural length-scale, defined by the condition $t_{\rm d} \leq t_{\rm r}$. The size of the cluster is calculated to be of the order of $r_{\rm i}$, but clearly the transition area introduced in this way spans a whole interval of radii (we recall the estimate (5) in which only strictly circular orbits and perpendicular inclination were considered). Analysing the results from M2 for the same sets of parameters that were used in the previous model discussion we found that their steady-state profiles look almost identical. Again, the inner cluster becomes progressively flattened towards the centre; furthermore, $n(a)$ is suitable for a power-law fit in those disjoint regions where one of the modes of radial migration dominates over the others. For this reason, we do not repeat the whole discussion and, instead, we concentrate on additional effects related to the newly introduced length-scale $r_{\rm i}$.

In Figure 9, we explore the stationary density profile that is established when the disc (ii) is present. Here, the panels show $n(a)$ as the outer radius gradually increases (plotted with a solid line). The profile can be divided into two contributions that are also plotted: A part of $n(a)$ formed by the satellites dragged by the disc and the remaining fraction of those initial distribution members that have not yet come into contact with the disc. These are represented by dashed and dotted

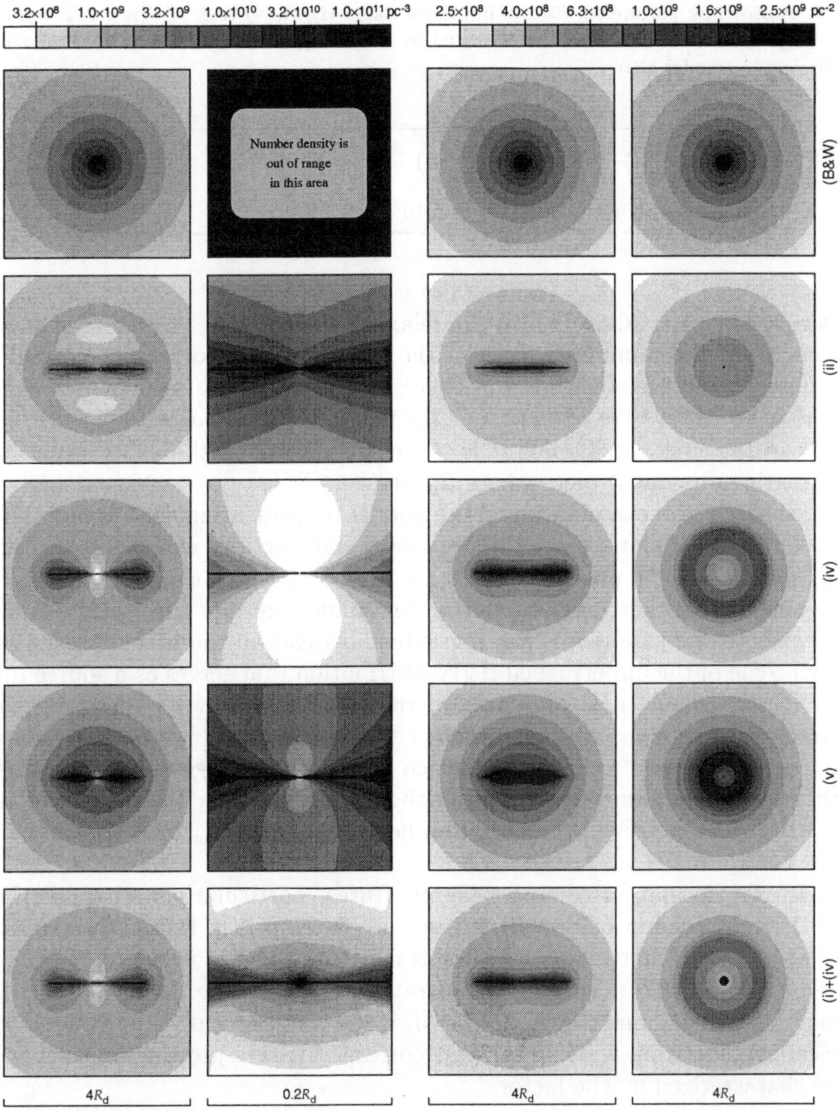

Figure 10. As in Fig. 8 but for M2 model. The first row corresponds to the reference B&W cluster, then modified clusters are shown as indicated on the right side.

lines, respectively. In the first row of graphs, $n_0 = 10^6$ and $r_{\rm h} = 10^6 r_{\rm g}$ was used (implying $t_{\rm r} \approx 10^9$yr in the observed region), while a much higher cluster density, $n_0 = 10^8 {\rm pc}^{-3}$, was adopted in the second row.

A sharp boundary exists between the inner (dragged) and the outer cluster for the case of $R_{\rm d} = 10^3 r_{\rm g}$ and $10^4 r_{\rm g}$. This transition almost coincides with the outer disc edge, but it is rather smoothed for $R_{\rm d} = 10^5 r_{\rm g}$ – evidently due to the limiting

Table 1. Anisotropy parameter β_a computed for cases (i), ..., (v) and (i)+(iv), $M_\bullet = M_8$ and $R_d = 10^4 r_g$. The outer cluster is characterized by radius $r_h = 10^6 r_g$ and number density $n(r) = 10^6 (r/r_h)\,\mathrm{pc}^{-3}$ (model M2).

disc	(i)	(ii)	(iii)	(iv)	(v)	(i)+(iv)
β_a	<-40	-1.54	-1.45	-1.52	-1.72	-4.57

radius $r_i \approx 5 \times 10^4 r_g$ above which the relaxation overwhelms the drag of the disc. An increase in the number density of the outer cluster shortens relaxation time and, consequently, r_i is decreased. This is exhibited in the second row of Fig. 9 where $r_i \approx 5 \times 10^3 r_g$. We observe that the inner cluster scales with the disc radius R_d only in the first panel while it is rather insensitive to the exact value of this parameter in subsequent plots where $R_d \gg r_i$.

We conclude the examination of our model by presenting spatial distributions in two-dimensional plots of azimuthal (meridional) sections across the cluster and its edge-on and face-on projections. Figure 10 is arranged in the same way as Fig. 8 with a single exception that the cluster containing the disc (iii), which would be very similar to the case (ii), is not presented. Instead, in the first row we plot the density profile of the unperturbed B&W distribution that serves as a source (outer) cluster for the simulations presented in the subsequent rows. The outer cluster is characterized by $r_h = 10^6 r_g$ and $n(r_h) \equiv n_0 = 10^6\mathrm{pc}^{-3}$. (Its density formally exceeds $1.7 \times 10^{11}\mathrm{pc}^{-3}$ below $0.1 R_d$ which is out of the range used in the Figure.) Another relevant parameters that enter the simulations are: the disc outer radius $R_d = 10^4 r_g$, parameters of the accretion flow $\dot{M}_\bullet = 0.1 \dot{M}_{\mathrm{Edd}}$, $\alpha = 0.1$, and $\zeta = 5$, and the central mass $M_\bullet = M_8$.

Notice the formation of ring-type structures, particularly well visible in the case (iv) (right column of Fig. 10). Rings are developed in those situations when star–disc collisions first align the satellites with the plane, where their transport proceeds further with a varying radial rate. Although the detailed structure of the modified cluster may be complicated, a tendency to flattening of an initially spherical configuration is obvious. This corresponds to a growing anisotropy that can be characterized by the factor

$$\beta_a \equiv 1 - \frac{\langle v_T^2 \rangle}{2\langle v_r^2 \rangle}, \tag{11}$$

which is defined in terms of the transverse and radial velocity components averaged over the system. For spherical self-gravitating systems, β_a varies only within several tenths around zero. In case of an embedded system in the central field, the anisotropy of the cluster is increased due to the loss cone effect acting on low angular momentum orbits; B&W distribution used for the outer cluster in our simulations has $\beta_a \approx -0.87$. The computed values of β_a for the present system are given in Table 1. The case (i) is somewhat exceptional because in this model all satellites are rapidly aligned with the disc, which leads to an exceedingly nega-

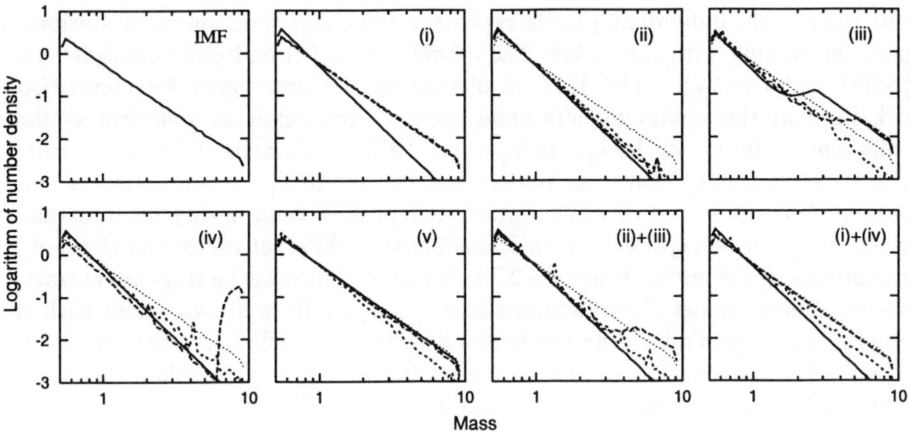

Figure 11. The modified mass function in the inner cluster. Curves represent the mass function of a sample restricted below a given radius; In particular, solid, dashed and dotted lines correspond to the limiting radii of $10r_{\rm g}$, $10^2r_{\rm g}$, and $10^3r_{\rm g}$, respectively. The thin dotted line stands for the IMF (slope -2.35).

tive anisotropy factor. That is of course unrealistic in view of the fact that the radiation-pressure dominated disc is actually relevant only in a very limited region of the system. Other cases are better suited for the entire inner cluster, but even here an increased anisotropy is evident.

4 Discussion

One can immediately recognize numerous consequences connected to the non-sphericity, anisotropy, and segregation in the cluster. Here we mention several examples where these effects are visible: the modified mass function, the expected spectral-line profiles, and the velocity dispersions.

The disc-driven evolution of the cluster results in a gradual modification of the stellar mass function $\mathcal{N}(M_*)$. The effect of the satellites' orbital decay changes its initial form and contributes to the gradual separation of the satellites according to their physical properties. We have analyzed the changing mass function with the adopted toy model in which three of the four considered dissipative interactions act selectively on bodies with different properties. (Even the fourth one, i.e. the decay of a satellite orbit coupled to the radial migration of the disc medium, is implicitly selective as it preferentially affects massive stars that have a better chance to form a gap in the disc.) The effects depend on satellite sizes and masses that, for the purpose of our discussion, are conveniently expressed in terms of Σ_* (in the case of star–disc collisions) and M_* (in the case of density waves governing the migration in the disc and for the gravitational radiation orbital decay). Hence, different stellar types are expected to concentrate in certain segments of the cluster or to diffuse with varying rates in the radial direction. Here we discuss a modification of the Salpeter's IMF, eq. (10).

In Figure 11, individual panels represent the mass function in a stationary cluster interacting with power-law disc models (i), ..., (v) and two composite discs (ii)+(iii) and (i)+(iv). The IMF is plotted in the first panel for comparison. Thick lines in the remaining sub-plots show $\mathcal{N}$ restricted to spherical sections of different radii ($r < 10^2 r_g$, $10^3 r_g$, and $10^4 r_g$), normalized to unity within $M_* \in (0.5M_\odot, 10M_\odot)$; One can notice that the modified mass function is again represented by power-law or broken power-law profiles valid in disjoint intervals of masses. With the exception of transitions between different areas, the slope of $\mathcal{N}$ does not exceed the initial value of -2.35. It means that massive stars are extracted from the cluster sooner than low-mass stars. This result is in agreement with the view of massive stars migrating towards the centre at a higher rate than light stars (assuming the same regime of interaction for both); Fig. 5 (first column) is a good aid for qualitative interpretations of mass function profiles.

Let us consider the case (i). It is only in the innermost considered sector of $r < 100r_g$ that nearly all stars enter the mode of mass dependent radial migration with density waves excitation, which makes the mass function modified in favor of light stars. Such a trend is not observed when a larger region with $r < 10^3 r_g$ (or even $r < 10^4 r_g$) is considered because in that case a major fraction of stars opens a gap and, hence, they move with the same radial velocity regardless of their masses.

A different picture arises in the case (ii) where the orbital decay depends on Σ_* everywhere and, therefore, $\mathcal{N}$ is decreased from the initial profile in all three considered intervals. Closer to the IMF is the mass function in the largest region, where a major fraction of stars follows inclined trajectories. In the latter case, the speed of radial drift scales as $\Sigma_*^{-1} \propto M_*^{0.6}$ in contrast to $\dot{a} \propto M_*$ for a star that is embedded without opening a gap.

A rather complicated pattern is shown in the graph devoted to a case (iii) disc, concerning the different profiles for light and massive stars in the two observed regions. Again, Fig. 5 helps us to understand the reason for such profiles: The closer to the centre, the less massive stars succeed in opening gaps in the disc. At $r = 100r_g$, all stellar bodies with $M_* > 2.6M_\odot$ migrate inwards in the same way, while at the radius of $10^3 r_g$ this only happens for $M_* > 5.4M_\odot$.

In the scheme described above, it is straightforward to interpret other profiles of $\mathcal{N}(M_*)$ that are obtained with the remaining models. We note that an extension of the considered domain further out to larger radii ($\gtrsim 10^4 r_g$) does not change much the profiles with respect to those obtained for $r < 10^4 r_g$, as the stars at the outskirts remain on inclined trajectories or, in case of the disc (i), as there are no additional stars involved when the observed region is extended above R_d.

Now we turn to integrated line profiles expected from the central cluster, unresolved to individual stars. Imprints of the central black hole on the line profiles were examined already in the paper [4], which we can compare with the present model to reveal the effect of the disc. We assume an infinitesimally narrow line profile in each satellite local frame and we neglect all other effects influencing the final spectrum. Left panel in Figure 12 shows the rms velocity profile integrated over a column of radius $r_s = 10^4 r_g$ centered on the cluster. In this example we assumed the case (iii), $R_d = 10^4 r_g$, and the model M2 (cp. ref. [27] to see analogous profiles obtained with the M1 model). Another view of the same system is shown with a

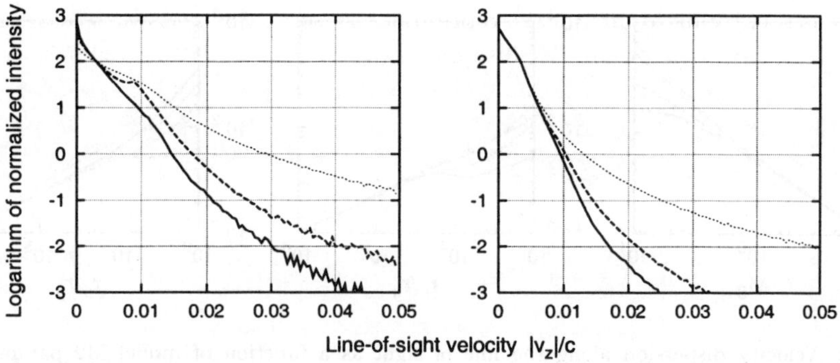

Figure 12. The inner cluster velocity distribution along the line of sight, integrated across a column of radius $r_s = 10^4 r_g$ (left panel) and $10^5 r_g$ (right panel). Solid and dashed lines represent different inclinations towards the observer: $\xi = 0°$ and $60°$, respectively. The thin dotted line stands for a reference B&W distribution.

lower resolution in the right panel.

Also, one can recover in these profiles a characteristic tendency of system flattening and ring formation. An increased abundance of slowly moving stars develops near the disc outer edge if the radial density of the cluster increases outwards and reaches a maximum at $r \approx R_d$. Consequently, a local maximum occurs in the line profile around $v \approx \sin(\xi) v_K(R_d)$; in some cases, this secondary peak may even overwhelm the central maximum. Another effect influencing the line profiles is the fast fall-off of their high-velocity tails when the observed region extends to large distances. This can be clearly demonstrated by comparing the two panels of Fig. 12. The majority of newly inserted satellites contributes to the low-velocity part of the profiles and hence this part is brought closer to the reference case of the B&W cluster. Different disc models and sight parameters – the view angle and spatial resolution, create a rather rich variety of synthetic lines.

Finally, we computed velocity dispersions whose dependence on the central mass is similar to an analogous empirical relation between σ_{rms} and the bulge mass [7,8], although it is obvious that we now explore a relatively small volume of direct dynamical influence of the central mass. Hence, the obtained $\sigma_{rms}(M_\bullet)$ relation is determined by the presence of the black-hole and the gas environment. We present the calculated relation in Figure 13 (left) by plotting σ_{rms} vs. $M_\bullet$ in a section of radius $r_s = 1\,\text{pc}$ of the cluster. Here, the M2 model and the disc (iii) were used with $R_d = 10^4 r_g \approx 0.1 M_8\,\text{pc}$. The outer cluster is characterized by a radius $r_h = 10^6 r_g \approx 10 M_8\,\text{pc}$ and a number density $n(r) = 10^6 (r/r_h)^{-7/8}\,\text{pc}^{-3}$. Hence, the two length-scales vary with the central mass, while the parameter r_s is fixed according to the presumed resolution. The dispersion increases with $M_\bullet$ almost independently of the inclination (but only below a certain threshold mass). As $M_\bullet$ increases, the lines separate from each other with the lowest values of σ_{rms} corresponding to the face-on view. This is another consequence of the system flattening in its central part, dimension of which increases (proportionally to $M_\bullet$) with respect to the observed region. A systematical shift of σ_{rms} towards lower values with respect

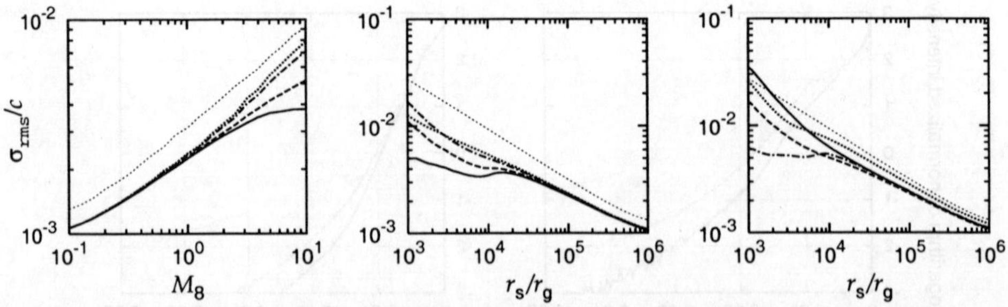

Figure 13. Velocity dispersion along the line of sight as a function of model M2 parameters. Left: Central mass is taken as independent variable (in units of $10^8 M_\odot$) while the line style indicates different values of inclination; $\xi = 0°$ (solid), 30° (dashed), 60° (dash-dotted), and 90° (dotted). Thin dotted line corresponds to an unperturbed outer cluster with B&W profile. Middle: $M_\bullet = 1 M_8$ is fixed while r_s is varied for the case (iii). Line types indicate inclination as in the left panel. Right: inclination is now fixed at 60°. Cases (ii), (iii), (iv) and (v) are represented by solid, dashed, dash-dotted and dotted lines, respectively.

to the reference cluster is caused by relatively small total number of satellites in the modified sample below R_d. Indeed, a simulation with a smaller disc shifts the σ_{rms} curves towards the B&W profile. A complementary representation of the velocity dispersion is shown in the remaining two panels of Fig. 13: σ_{rms} is plotted as a function of the column cross-sectional radius r_s for different view angles (middle) and different disc models (right), respectively.

5 Conclusions

We explored several aspects of dissipative stellar dynamics in the gravitational field of a central compact body. The motion of a stellar-mass orbiter can serve as an accurate tool to probe the physical properties of galactic cores provided that various intervening influences are well understood. However, even if the stellar motion is less sensitive to non-gravitational forces than the gas swirling in an accretion flow, mutual interaction between stars and the interstellar environment still needs to be determined with sufficient accuracy. Assuming that hydrodynamical drag in the disc is the main perturbing agent, we showed how an initially spherical system attains a flattened configuration and how a part of the system forms a ring, typically on sub-parsec scales. There are interesting possibilities for future research in this direction, especially in connection with new observational techniques that should reach a better resolution or even implement novel approaches. For example, a question arises to what degree a ring-shaped stellar sub-system that we notice in our simulations can contribute to the background gravitational waves signal. Furthermore, the enhanced rate of star–disc collisions and mutual stellar encounters near the black hole may contribute to the variability that many galactic nuclei exhibit and it may partly explain the origin of material swept out of the disc.

Acknowledgments

We acknowledge support from grants GACR 205/00/1685 and GAUK 188/2001.

APPENDIX

One can define a critical value of inclination by comparing the relative velocity of an orbiter on a circular orbit (with respect to material in a Keplerian disc), $v_{\rm rel} = 2\sin(i/2)v_{\rm k} \approx 4.2 \times 10^{10}\sin(i/2)(r/r_{\rm g})^{-1/2}{\rm cm\,s}^{-1}$, to the speed of sound. For example, for the middle region of the SS disc with $\dot{M} = 0.1\dot{M}_{\rm Edd}$, $\alpha = 0.1$, the condition $v_{\rm rel} \gg v_{\rm s}$ implies $\sin(i/2) \gg 10^{-4}M_8^{-1/10}(r/r_{\rm g})^{1/20}$. The corresponding value of i is actually an overestimate of the critical inclination. Notice that, for elliptical orbits, $v_{\rm rel}$ is larger than in the case of a circular orbit, which implies smaller values of the limiting inclination.

Another limitation comes from the geometrical thickness $2h$ of the disc. Instead of impulsive collisions, the interaction would be dominated by other effects if a non-negligible fraction of the trajectory lay too close to the disc plane or even inside it. Hence, we write $\tan i > i_{\rm min} \equiv \eta\, h/r$, where $\eta \gtrsim 1$ is a factor determining what fraction of its orbit a satellite is allowed to spend in the disc. Again, we take the case (ii) to illustrate the explicit form of this condition: $\tan i > 1.3 \times 10^{-3}M_8^{-1/10}\eta(r/r_g)^{1/20}$. It follows from the disc structural equations that this inequality is essentially equivalent to the preceding one, reflecting the requirement of supersonicity of the star–disc collision.

The influence of satellite impacts and transitions across the disc depends on many factors. The interaction is particularly complex at a relatively low speed (less than that of sound), especially if the satellite looses part of its atmosphere. Now we ignore such effects. Provided the orbiter is compact enough, its mass and size determine the magnitude of local collisional interaction by means of the column density Σ_*; the compactness ensures that no substantial stripping of the stellar envelope occurs [1]. We recall that, for a solar type star, this condition is fulfilled in all considered models of thin discs.

Let us denote $\boldsymbol{v} = (v_r, v_\vartheta, v_\varphi)$ and $\boldsymbol{v}_{\rm d} = v_{\rm d}\boldsymbol{e}_\varphi$ the satellite and the disc velocities before the interaction takes place. According to this scheme, the velocity of the star and the expelled material of the disc after the interaction will be both equal to $\boldsymbol{v}'$. The momentum conservation law then gives

$$M_*\boldsymbol{v} + \pi R_*^2\Sigma_{\rm d}\frac{v_{\rm rel}}{|v_\perp|}\boldsymbol{v}_{\rm d} = \left(M_* + \pi R_*^2\Sigma_{\rm d}\frac{v_{\rm rel}}{|v_\perp|}\right)\boldsymbol{v}', \tag{A.1}$$

where $v_{\rm rel}$ is the magnitude of the star relative velocity with respect to the disc, $\boldsymbol{v}_{\rm rel} \equiv \boldsymbol{v} - \boldsymbol{v}_{\rm d}$, and $v_\perp = v_\vartheta$ is its component perpendicular to the plane of the disc. Setting

$$A \equiv \frac{\Sigma_{\rm d}}{\Sigma_*}\frac{v_{\rm rel}}{|v_\vartheta|} = \frac{\pi R_*^2\Sigma_{\rm d}}{M_*}\frac{v_{\rm rel}}{|v_\vartheta|}, \tag{A.2}$$

eq. (A.1) can be rewritten in the form

$$\delta\boldsymbol{v} \equiv \boldsymbol{v}' - \boldsymbol{v} = -\frac{A}{A+1}\left(\boldsymbol{v} - \boldsymbol{v}_{\rm d}\right). \tag{A.3}$$

Let us remark that even without an explicit knowledge of the star–disc interaction (in terms of the function A) eq. (A.3) is valid whenever the momentum change is collinear with the relative velocity of the star and the disc material. For a solar type star, $\Sigma_* \approx 10^9$–$10^{12}\,\mathrm{g\,cm^{-2}}$, while for the disc $\Sigma_{\mathrm{d}} \approx 10^4$–$10^5\,\mathrm{g\,cm^{-2}}$; therefore, A is typically orders of magnitude less than unity. The parameter A can reach values order of unity only if $v_{\mathrm{rel}} \gg |v_\vartheta|$, which, however, is the case of orbits embedded in the disc whose interaction is treated in a different way.

Using formula (A.3), we derive changes of orbital parameters caused by a single star–disc interaction. It is useful to introduce variables a, $x = \cos i$, $y = 1 - e^2$, and $z = 1 + e\cos\omega$. The radius of the ascending node and the satellite velocity components are:

$$r = \frac{ay}{z}, \quad v_r = \pm\frac{\sqrt{2z - z^2 - y}}{\sqrt{ay}}, \quad v_\vartheta = -\frac{z\sqrt{1-x^2}}{\sqrt{ay}}, \quad v_\varphi = \frac{zx}{\sqrt{ay}}. \tag{A.4}$$

The upper sign in the formula for v_r holds when $\omega \in \langle 0, \pi\rangle$, while the lower sign applies otherwise (the same notation is used in subsequent equations). Next, changes of the satellite binding energy per unit mass E, angular momentum L, its z-component L_z, and the radial component of velocity are calculated from eq. (A.3) and expressed by means of the orbital parameters:

$$\delta E = A\frac{1}{ay}\left(2z - y - \sqrt{ay}\,xzv_{\mathrm{d}}\right), \qquad \delta L = A\frac{ay}{z}\left(xv_{\mathrm{d}} - \frac{z}{\sqrt{ay}}\right), \tag{A.5}$$

$$\delta L_z = A\frac{ay}{z}\left(v_{\mathrm{d}} - \frac{xz}{\sqrt{ay}}\right), \qquad \delta v_r = \mp A\left(\frac{2z - z^2 - y}{ay}\right)^{1/2}. \tag{A.6}$$

Here, an expansion to $\mathcal{O}(A)$ was used assuming $A \ll 1$. Since $E = 1/(2a)$, $L = \sqrt{ay}$, $L_z = \sqrt{ay}\,x$, and $v_r = \mp\sqrt{(2z - z^2 - y)/ay}$, we may write

$$\delta E = -\frac{1}{2a^2}\,\delta a, \tag{A.7}$$

$$\delta L = \frac{1}{2}\sqrt{\frac{y}{a}}\,\delta a + \frac{1}{2}\sqrt{\frac{a}{y}}\,\delta y, \tag{A.8}$$

$$\delta L_z = x\,\delta L + \sqrt{ay}\,\delta x, \tag{A.9}$$

$$\delta v_r = \pm\frac{(1-z)\delta z - \frac{1}{2}\delta y}{\sqrt{(2z - z^2 - y)ay}} \mp \frac{\sqrt{2z - z^2 - y}}{ay}\,\delta L. \tag{A.10}$$

Combining these equations we obtain at the ascending node

$$\delta a = 2Aay^{-1}\left(y - 2z + \sqrt{ay}\,xzv_{\mathrm{d}}\right), \quad \delta x = A\sqrt{ay}\,z^{-1}\left(1 - x^2\right)v_{\mathrm{d}}, \tag{A.11}$$

$$\delta y = 2A\left(2z - 2y + \frac{xy}{z}\sqrt{ay}\,v_{\mathrm{d}} - xz\sqrt{ay}\,v_{\mathrm{d}}\right), \; \delta z = 2A\left(x\sqrt{ay}\,v_{\mathrm{d}} - z\right). \tag{A.12}$$

Analogous formulae can be derived for the corresponding changes of the orbital parameters at the descending node. We obtain them by replacing z with $(2 - z)$

and simultaneously changing the sign of v_r in eq. (A.4). As a result, a set of equations similar to eqs. (A.11)–(A.12) is obtained with $z \to (2-z)$ and $\delta z \to -\delta z$. Summing up two subsequent contributions, we obtain a set of equations determining the parametric evolution of the orbit:

$$\begin{aligned} \mathrm{d}a &= 2A_1 a y^{-1}\Big(y - 2z + \sqrt{ay}xzv_\mathrm{d}\Big) + \\ &\quad + 2A_2 a y^{-1}\Big[y - 2(2-z) + \sqrt{ay}x(2-z)v_\mathrm{d}\Big], \qquad (A.13) \\ \mathrm{d}x &= A_1\sqrt{ay}z^{-1}\left(1-x^2\right)v_\mathrm{d} + A_2\sqrt{ay}(2-z)^{-1}\left(1-x^2\right)v_\mathrm{d}, \qquad (A.14) \\ \mathrm{d}y &= 2A_1\left(2z - 2y + xyz^{-1}\sqrt{ay}v_\mathrm{d} - xz\sqrt{ay}v_\mathrm{d}\right) + \\ &\quad + 2A_2\Big[2(2-z) - 2y + xy(2-z)^{-1}\sqrt{ay}v_\mathrm{d} - x(2-z)\sqrt{ay}v_\mathrm{d}\Big], \qquad (A.15) \\ \mathrm{d}z &= 2A_1\left(x\sqrt{ay}v_\mathrm{d} - z\right) + 2A_2\Big(x\sqrt{ay}v_\mathrm{d} + 2 - z\Big). \qquad (A.16) \end{aligned}$$

The differences of orbital parameters summed over two subsequent intersections represent the changes per one entire revolution with period $T = 2\pi a^{3/2}$.

Functions $A_{1,2}$ and v_d describe the disc surface density and velocity profile, respectively. Setting $v_\mathrm{d} = v_\mathrm{K}$, equations (A.13)–(A.16) take the form

$$\begin{aligned} \mathrm{d}a &= 2A_1 a y^{-1}\Big[y - 2z + xz^{3/2}\Big] + 2A_2 a y^{-1}\Big[y + 2z - 4 + x(2-z)^{3/2}\Big], \qquad (A.17) \\ \mathrm{d}x &= A_1\left(1-x^2\right)z^{-1/2} + A_2\left(1-x^2\right)(2-z)^{-1/2}, \qquad (A.18) \\ \mathrm{d}y &= 2A_1\left(2z - 2y + xyz^{-1/2} - xz^{3/2}\right) \\ &\quad + 2A_2\left[2(2-z) - 2y + xy(2-z)^{-1/2} - x(2-z)^{3/2}\right], \qquad (A.19) \\ \mathrm{d}z &= 2A_1\left(xz^{1/2} - z\right) + 2A_2\left[2 - z - x(2-z)^{1/2}\right], \qquad (A.20) \end{aligned}$$

where

$$A_1^2 = \tau_\mathrm{incl}^{-2}(r)\,\frac{3z - y - 2xz^{3/2}}{\left(1-x^2\right)z^2}, \qquad (A.21)$$

$$A_2^2 = \tau_\mathrm{incl}^{-2}(r)\,\frac{6 - 3z - y - 2x\,(2-z)^{3/2}}{\left(1-x^2\right)(2-z)^2}. \qquad (A.22)$$

We concentrated in this paper on the discs with power-law surface density profiles, $\Sigma_\mathrm{d} \propto r^s$, which determines a particular profile of $A_{1,2}(r)$. Nevertheless, eqs. (A.17)–(A.20) can be solved for special configurations independent of any particular form of Σ_d [27]. Some general characteristics of the solution are also evident, e.g. the fact that the inclination decreases monotonically ($\mathrm{d}x \geq 0$). One can further show that $\mathrm{d}a \leq 0$, $\mathrm{d}x \geq 0$, and $\mathrm{d}y \geq 0$ for the surface density decreasing outwards. On the other hand, for Σ_d increasing outwards (the case of a radiation pressure dominated SS disc) collisions may cause an increase of the satellite orbital energy and angular momentum.

References

1. Armitage P. J., Zurek W. H., Davies M. B., 1996, ApJ, 470, 237
2. Artymowicz P., 1994, ApJ, 423, 581
3. Artymowicz P., Lin D. C. N., Wampler E. J., 1993, ApJ, 409, 592
4. Bahcall J. N., Wolf R. A., 1976, ApJ, 209, 214
5. Collin S., Huré J.-M., 1999, A&A, 341, 385
6. Fabian A. C., Iwasawa K., Reynolds C. S., Young A. J., 2000, PASP, 112, 775
7. Ferrarese L., Merritt D., 2000, ApJ, 539, L9
8. Gebhardt K., Bender R., Bower G. et al., 2000, ApJ, 539, L13
9. Ivanov P. B., Igumenshchev I. V., Novikov I. D., 1998, ApJ, 507, 131
10. Junor W., Biretta J. A., Livio M., 1999, Nature, 401, 891
11. Karas V., Šubr L., 2001, A&A, 376, 687
12. Karas V., Vokrouhlický D., 1994, ApJ, 422, 208
13. Kormendy J., Richstone D., 1995, AR&AA, 33, 581
14. Lang K. R., 1980, *Astrophysical formulae* (Springer Verlag, Berlin)
15. Lin D. N. C., Papaloizou J. C. B., 1986, ApJ, 309, 846
16. Narayan R., 2000, ApJ, 536, 663
17. Ostriker E. C., 1999, ApJ, 513, 252
18. Peters P. C., Mathews J., 1963, Phys. Rev., 131, 435
19. Pineault S., Landry S., 1994, MNRAS, 267, 557
20. Rauch K. P., 1995, MNRAS, 275, 628
21. Rauch K. P., 1999, ApJ, 514, 725
22. Rees M. J., 1998, in *Black Holes and Relativistic Stars*, ed. R. M. Wald (The University of Chicago Press, Chicago)
23. Salpeter E. E., 1964, ApJ, 121, 161
24. Sánchez-Salcedo F. J., Brandenburg A., 2001, MNRAS, 322, 67
25. Shlosman I., Begelman M. C., 1989, ApJ, 341, 685
26. Šlechta M., 2001, PhD Thesis (Charles University, Prague)
27. Šubr L., 2001, PhD Thesis (Charles University, Prague)
28. Syer D., Clarke C. J., Rees M. J., 1991, MNRAS, 250, 505
29. Vilkoviskij E. Y., Czerny B., 2002, A&A, submitted
30. Vokrouklický D., Karas V., 1998, MNRAS, 298, 53
31. Ward W. R., 1986, Icarus, 67, 164
32. Ward W. R., 1997, Icarus, 126, 261
33. Zurek W. H., Siemiginowska A., Colgate S. A., 1994, ApJ, 434, 46

TOWARDS GRAVITATING DISCS AROUND STATIONARY BLACK HOLES

OLDŘICH SEMERÁK
Institute of Theoretical Physics, Charles University, Prague
E-mail: oldrich.semerak@mff.cuni.cz

This article outlines the search for an exact general relativistic description of the exterior (vacuum) gravitational field of a rotating spheroidal black hole surrounded by a realistic axially symmetric disc of matter. The problem of multi-body stationary spacetimes is first exposed from the perspective of the relativity theory (section 1) and astrophysics (section 2), listing the basic methods employed and results obtained. Then (in section 3) basic formulas for stationary axisymmetric solutions are summarized. Sections 4 and 5 review what we have learnt with Miroslav Žáček and Tomáš Zellerin about certain static and stationary situations recently. Concluding remarks are given in section 6. Although the survey part is quite general, the list of references cannot be complete. Our main desideratum was the informative value rather than originality — novelties have been preferred, mainly reviews and those with detailed introductions.

1 The fields of multi-body systems involving black holes

The subject of self-gravitating sources around rotating black holes is interesting in several respects, relevant from the point of view of the relativity theory itself as well as in the astrophysical context.

First, due to the non-linearity of Einstein's equations, the field of a multi-body system is a traditional challenge where one typically does not manage with a simple superposition.[a] On the first post-Newtonian level, the "celestial mechanics" of gravitationally interacting bodies can be kept linear [89,90], but in the strong-field region the interaction may bring surprising features that have only been described in a very few cases yet (for a two-body problem — today at the centre of attention because of the expected gravitational waves from colliding compact binaries, "the current state of art is the third post-Newtonian approximation" [91]).

The second point in which the problem embodies the essence of general relativity is the effect of inertial frame dragging due to the rotation of the sources. Contrary to the Newtonian treatment which does not discriminate between static and stationary situation, the field is now determined not only by mass-energy configuration, but also by its motion within the bodies. The inertial space can be imagined as a viscous fluid mixed by the sources. In today's "gravito-electromagnetic" language, gravity has not only an electric component, generated by the mass, but also a magnetic one, generated by mass currents; see [79,103,147,163,201,202,215,245,298] and references therein. (The long-time effort to measure the gravitomagnetic effect of the Earth is just culminating [78]; cf. [238,239].)

The third point is that our setting involves a black hole, perhaps the most bizarre

[a]Even the problem of an "isolated" object is very difficult (mainly if its motion must be found as well, not mentioning the interior field) unless one makes some simple assumptions about its multipole structure; see e.g. [17,98,99] for a treatment of bodies with matter interior, [297] for that also valid for singular bodies such as black holes, and [40] for the case of point-like particles.

of the new predictions of general relativity.[b] Within and around it, the deviations from Newtonian theory become dominant, in particular the above mentioned implications of non-linearity and rotation reveal themselves prominently. Perhaps the most extreme example (of an exact spacetime) involving *all the three* aspects[c] is a "double Kerr(-Newman)" solution for an aligned pair of rotating ultracompact objects [37, 96, 210, 211] (cf. [176] where an approximate method for axially symmetric superpositions of rotating black holes was proposed as a starting point of computations of black hole collisions).

Relativistic spacetimes are found in three ways: by numerical solution of Einstein equations, by perturbation of previously known spacetimes, and by exact analytical solution of Einstein equations. Let us mention what the above approaches told us about stationary axisymmetric spacetimes describing rotating black holes with additional matter (ring, disc or torus).[d]

1.1 Numerical solution

Although computers have been naturally employed in relativity for decades, in algebra as well as in numerics, it is only quite recently that they embarked on numerical solution of Einstein's equations in the most interesting cases involving black holes. Teukolsky [295] lists today's peak parallel-type hardware and software and the appropriate, hyperbolic formulation of the field equations as "three reasons why we are on the verge of important advances in the computer solution of Einstein's equations". Whereas analytical prospects are restricted, the present-day computational facilities can handle almost any situation, at least for certain period of time. On the other hand, it is never sure whether a given numerical solution represents a typical or just a marginal case. Due to this lack of generality, it is difficult to discern, analyse and interpret different classes of solutions within boundless ranges of possibilities. However, numerical solutions can provide explicit *examples* of spacetimes and processes that might otherwise remain only conjectural [187].

Numerical spacetimes containing a rotating black hole surrounded by an additional stationary axisymmetric source were constructed by [181] (a hole with a thin finite equatorial disc) and by [237] (a hole with a thick toroid). To mention just one particular point, [181] ended with a prolate horizon (stretching along the axis) in certain cases (when the disc was strongly counter-rotating with respect to the hole), which was not observed in [237]. The horizons have been known to inflate towards the external sources, so it would be an interesting consequence of the interplay between dragging from the hole and from the disc if it were confirmed that they can indeed become prolate, even though under extreme circumstances only.

[b] The prologue to a major reference [69] (not speaking about its entire content) however convinces us that black holes are "the simplest objects" (also "the most perfect macroscopic objects") in the universe. Cf. also [115, 298].

[c] Within general relativity, there is only one more ingredient that could make the situation even more dramatic: non-stationarity. The result would be a gravitational collapse or a collision of already collapsed objects. These events are currently devoted an exceptional attention as the strongest sources of gravitational waves.

[d] In the following, we do not review superpositions, obtained by any of the methods, of black holes with external *electromagnetic* fields. We refer to [34–36, 196] for test EM fields and to [11, 12, 106] for exact solutions. A more recent account can be found in [39].

Note, however, that the main stream of numerical relativity is focused on *non-stationary* problems important in generation of gravitational waves, in particular on black hole inspiral collision (e.g. [9, 295, 296] and a thorough review [187]).

1.2 Perturbative solution

A great deal of literature was devoted to perturbations of black-hole spacetimes and several formalisms were developed, none of which can be reviewed here.[e] In [318], a fully explicit solution for the perturbation of the Schwarzschild metric by a (rotating) axisymmetric weakly gravitating thin equatorial ring was found by solving the perturbed Einstein equations. This direct approach is however not practicable for a rotating hole and/or for an extended external source with pressure (not mentioning more complicated situations).

The rotating case (specifically, the algebraically special vacuum case) was seized notably by Teukolsky [294] who succeeded in separating the decoupled equations for the first order perturbations of a Kerr black hole into the second-order ordinary differential equations for the scalar field (for scalar perturbations) or for the Newman-Penrose scalars constructed from the electromagnetic field tensor (for electromagnetic perturbations), from the neutrino spinor (for neutrino perturbations) or from the Weyl tensor (for gravitational perturbations). By solving the "gravitational" Teukolsky equation, the perturbative deformation of the Kerr horizon was calculated by [94]. The stationary axisymmetric Green's function of the Teukolsky equation was provided by [195]. The most comprehensive expositions of the first-order black hole perturbation theory were given by Chandrasekhar [68, 69] in Newman-Penrose formalism. More recently, [183, 213, 214] solved the Teukolsky equation in terms of Coulomb wave functions and hypergeometric functions.

In a related method of handling the first-order perturbations of rotating fields, first devised for electromagnetic perturbations by Cohen & Kegeles and then generalized to gravitational ones [76], the perturbation components are expressed in terms of a single (Debye) potential that obeys a wavelike equation. Wald showed that this equation is just the adjoint of the Teukolsky master equation, while [160] gave a covariant extension of the method of Debye superpotentials to neutrino, electromagnetic and gravitational perturbations of all algebraically special spacetimes (see [300, 303] for summary, references and generalization). In the meantime, [77] treated the problem of perturbation of the Kerr horizon after having learnt [76] how to fix the perturbed metric from the Newman-Penrose functions. General perturbations of Schwarzschild solution were treated in this manner by [301]. Latest advance of the approach consists mainly in formulating the conservation laws for perturbations [60, 303].

Let us mention yet another approach [269], also dealing with curvature components but not restricted to algebraically special backgrounds. It has been shown, in the static case and to the linear order so far, that it can be generalized naturally to self-gravitating matter fields; in a spherically symmetric case, the connection with

---

[e] In monograph [69], the author admits that "...the account, in large parts, is hardly more than an outline" at the beginning of chapter ***The gravitational perturbations of the Kerr black hole*** which has 101 pages.

older, metric formalisms of Regge & Wheeler and Zerilli was elucidated.

In the charged case, the perturbation problem introduces coupling between gravitational and electromagnetic quantities, implying e.g. conversion between gravitational and electromagnetic waves. Interacting perturbations of the Reissner-Nordström black hole were studied by [32] (using the earlier formalisms of Regge & Wheeler, Zerilli and Moncrief) and by [302] (using the method of scalar potentials). A gauge invariant derivation of the basic equations of different formalisms was provided in [110], while [111] related their rotating analog to Teukolsky equation.

The second-order perturbation theory has also been under development since 1970s. For Schwarzschild black holes it is surveyed in [125], while the rotating case is tackled in [57].

It should be remarked that solutions describing stationary sources around black holes were not the primary aim of the black-hole perturbation strategy. Historically, attention has rather been devoted to the stability of black-hole solutions [255], to the relaxation of black holes into a stationary "no hair" state [134], involving the issue of back reaction of a perturbation [2] propagating out of [20] or into the hole [240, 241] (cf. [144]), to non-stationary processes of astrophysical significance such as the behaviour of weakly gravitating particles or waves in black-hole backgrounds [148, 253], and, of course, to gravitational waves [231] — presently mainly to black-hole collisions as their most prominent source (e.g. [58, 59, 161]; for a survey, see [296]). The latter is exactly an example of a problem where numerical relativity and perturbative analysis may — and should — coexist while mutually checking each other [18, 199, 270].

1.3 Exact solution

Perturbative approach is adequate in situations when the external source has only a very small effect on the field of the main body. Whenever this is not the case, the superpositions have to be described by solutions of the full, non-linear theory. At the end of Jiří Bičák's recent survey [33], one is encouraged "not to cease in embarking upon journeys for finding them, and perhaps even more importantly, for revealing new roles of solutions already known", because "Is there another so explicit way of how to learn more about the rich possibilities embodied in Einstein's field equations?" For a more technical demonstration, see the canonical monograph [174] (an updated version is awaited); more recent reviews of vacuum fields can be found in [45, 48].

The only stationary (in fact static) equilibrium configuration containing more than one black hole is the Majumdar-Papapetrou case with extreme centres of the Reissner-Nordström type (e.g. [52, 75, 123, 250]), where the electrostatic repulsion exactly compensates the gravitational attraction. With rotating sources, gravitomagnetic interaction between spins (repulsive in the parallel case — see e.g. [252]) and magnetic interaction between magnetic dipole moments (repulsive in the antiparallel case) are also present. Analysis of the momentarily stationary and axisymmetric system of two identical sources was carried out by [96, 210] for the Kerr components (with mass M and specific rotational angular momentum a) and by [37, 211] for the Kerr-Newman components (with M, a and charge Q). Equilibrium was found to be only possible for the extreme values of the charges ($Q = M$). In all other cases, supporting singularities ("struts") occur indicating that a given system of sources cannot remain stationary (or static) according to the field equations. Hence, rotating centres can only remain in equilibrium in a super-extreme case of two naked singularities (that have $Q^2 + a^2 > M^2$) [50], both the magnetic dipole-dipole and the gravitational spin-spin interactions being too weak to keep apart black holes (having $Q^2 + a^2 \leq M^2$). (On the other hand, a non-singular equilibrium *is* possible with two spinning Curzon particles [95]. Most recent discussions of both cases, in particular of the character of supporting singularities, can be found in [46,47,191].)

Accretion discs of astrophysical interest are likely to lie in the equatorial plane of the centre and unlikely to bear a considerable charge, so that neither the spin-spin nor the electromagnetic repulsion can act within them. One must rather refer to a centrifugal force resulting from the orbital motion of the material or to hoop stresses when interpreting the situation.

It turned out to be difficult to superpose a Kerr centre with an additional axisymmetric ring, disc or torus and no explicit exact solutions describing the systems of this kind have been found until now. Nevertheless, many stationary axisymmetric solutions of the (electro)vacuum Einstein equations are known that do generalize those containing only isolated black holes. These were mostly obtained by indirect methods known as "generating techniques". Two major approaches, developed by the end of the 1970's — the group-theoretic techniques and the soliton-theoretic (or inverse-scattering) techniques, work for spacetimes with two commuting symmetries. (See [170] for a review and [83–85,135,189,206] for more detailed analysis and interrelations between different formulations, e.g. those of Kinnersley & Chitre; Maison; Belinskii & Zakharov; Harrison; Hoenselaers, Kinnersley & Xanthopoulos; Hauser & Ernst; Neugebauer; Kramer & Neugebauer; or Alekseev.) Other related methods have been proposed more recently, e.g. the simplification of the Hauser-Ernst integral equation by Sibgatullin [281], "monodromy data transform" by Alekseev [10] (also [165]), the static gravitational multipoles [128] (cf. [260]) and their superposition with stationary fields [129] by Gutsunaev & Manko, the linear transformation by Quevedo [260] or the "finite-gap" (algebraic-geometric) solutions by Korotkin & Matveev ([172] and references therein). Most of them are briefly compared in (appendix 6. of) [11]. (For other approaches, see e.g. [82,290] or the results of Nakamura and Kyriakopoulos, referred to and worked out by [293]; cf. also [308].)

The crucial point of the soliton generating techniques is a solution of two linear differential equations (Lax pair) the integrability conditions of which are exactly the Einstein's equations (namely the Ernst equation). The linear problem can be tackled in order to generate new solutions from the known ones: given some ("seed") metric with two Killing vectors, it yields a different metric of the given type (the procedure is often called Bäcklund transformation in analogy with the technique used for the KdV equation). In such a manner, many known spacetimes were reproduced, but also broad families of new solutions were provided characterized by arbitrarily large sets of free parameters. However, only a very restricted number of them have been given a clear physical interpretation. Though several of these results perhaps represent a rotating black hole in an "external" gravitational field (e.g. [51, 71, 72, 208, 209, 261]), none of the latter has yet been specified to the case generated by a concrete realistic body such as ring, disc or torus (cf. the last paragraph of section IV in [190]).

Making the "random" Bäcklund transformation with some metric, it is, however, difficult to require specific physical properties of the spacetime being constructed — one rather "takes pot luck" and looks what comes out (see [260] for a hint how to overcome this); section 5.1 will illustrate what problems may arise. It would be more appropriate to express the physical boundary conditions of the Ernst equation in terms of the quantities which appear in the linear problem and then solve the latter. This leads to Riemann-Hilbert problems known from the theory of completely integrable differential equations [15, 165, 234]. Tackling the stationary axisymmetric boundary value problem with the help of the linear system, the fields of two physically relevant types of sources have been discussed: that of black holes [219, 235] and that of finite thin discs ([14, 114, 219, 235] and references therein). There is some hope that the above methods could also be used to describe superpositions of both (e.g. [235]).

2 Astrophysical relevance

The system of a black hole with an accretion disc is very important in astrophysical considerations. In particular, it plays the key role in models of a whole range of active galactic nuclei and of some X-ray binaries.

2.1 Galactic nuclei and X-ray binaries

Extraordinary luminosity generated within a small volume (as manifested by its rapid variability), the presence of very hard (X and gamma) radiation, high rotation speeds and broad velocity distribution of the material around indicate that there is an interacting ultracompact object in the inner regions of these sources [63, 66, 182, 262].[f] The ambient gas, present in the galactic nucleus or overflowing from the double-star primary, typically has enough angular momentum not to fall directly (almost radially) onto the object; it rather forms a disc and only gradually spirals down. The flow is strongly sheared due to a considerable non-homogeneity of the field, which enables viscous torques to heat the gas to high temperatures. (See e.g. [4, 8, 65, 159, 264] for various aspects of the modern accretion disc theory. In particular, the most realistic grid of non-LTE disc models was constructed in the paper series [136–139].) In some cases, the excess of angular momentum drives outflows through the empty regions along the rotation axis of the system. Jets of matter-energy emanate from a number of active nuclei, often at relativistic speeds, feeding giant outer lobes distinguished in radio surveys. The formation mechanism of the jets is studied by several groups using relativistic-MHD numerical simulations (e.g. [43, 169]).[g] The high degree of jet collimation evidences an important role of

---

[f] In fact, the best evidence for such an object was provided by the nuclei of several "normal" galaxies [171] (led by our own [121]) which are not so over-exposed with matter and radiation and where it is easier to study the motion of individual sources.

[g] For the jet in M87, the first known and the best resolved one, simulations are already fitted to observations at distances from several dozens of Schwarzschild radii from the centre [109, 150]!

magnetic fields [44, 93, 154, 218] and poses questions concerning the interaction of jets with the ambient gas and radiation. (See chapter 9 of [315] and [197] for a review.) The entire topic of active galactic nuclei is covered in [177] (for earlier references, see [317]) while that of X-ray binaries in [193]. Recent status of both areas can be learned from [152].

In spite of the observational variety of active galaxies (blazars, quasars, Seyfert galaxies, radiogalaxies, ...), it now seems that we simply look at a similar type of source from different directions [102]. The discovery of jet structures accompanying several X-ray binaries in our Galaxy (the literature speaks of "microquasars" [228, 282]) has strengthened the conviction that the stellar-size active sources are also powered by the same accretion mechanism. Further evidence is provided by observations of compact X-ray sources in several nearby galaxies [207].

The suspicion of the presence of ultracompact objects has also been supported by more detailed considerations. Certain spectral features (the iron Kα line in particular) of some active galactic nuclei (a celebrated example is the Seyfert 1 galaxy MCG–6-30-15) have been interpreted as originating as close to the very centre as a few to a few dozens of Schwarzschild radii,[h] so they may well represent data from the strongest fields ever met (see [107] for a review, and e.g. [88, 155, 266, 267] for models and interpretation issues). The iron lines have already been discovered in the X-ray spectrum of several Galactic microquasars ([224] and references therein).

Also, the frequencies of quasi-periodic oscillations, found at a growing number of stellar-size sources, have been identified as the values characteristic for the accretion flow close to a black hole or a neutron star [86, 97, 151, 225, 230, 256, 286, 287]. This allows us to consider models involving orbital frequencies of "hot spots" in the inner parts of an accretion disc (e.g. [153]), beat of these frequencies with the centre's spin [257], blobs on tilted orbits affected by Lense-Thirring precession [16, 87, 221], action of a secondary, intersecting the disc repeatedly while orbiting the centre [289], oscillations of the (thin) disc itself, perturbed off its circular (equatorial) flow [157, 299, 310], or waves in tilted discs resulting from the Bardeen-Peterson effect [113]. (Clearly, the processes can proceed in a symbiosis, for instance the secondary would probably actuate hot spots as well as oscillations in the disc, pulling some of its material out into a tilted orbit.) See [306] for a survey and [86, 232, 285–287] for discussions of the individual models in the light of new observations.

Among the sources displaying the presence of an interacting ultracompact body there is a class whose luminosity is still very low and fluctuating within a wide range. This seems to point towards the existence of a horizon, because a quasi-stationary inflow onto a neutron star should yield a relatively strong and steady output, produced mainly by the gas streams colliding with the star's surface. With a horizon, the accretion may (perhaps occasionally) switch over to an "advection-dominated" regime when the material plunges in without having released any significant fraction of its binding energy [119, 233]. See [5] for a general model of advection-dominated discs, [141, 142, 179, 204, 212] for recent numerical simulations of their structure and spectrum, [25, 217] for a study of jet production in such discs, and [226] for a review.

The main goal of precise measurements of sources suspected of core collapse is

[h] Even a distance of only half the Schwarzschild radius has already been announced (!), giving support for a rotating black hole.

a determination of three independent parameters — mass of the compact centre M, its specific angular momentum a, and radius r where a given process happens. Describing the core field by the Kerr metric, the above three parameters appear in the formulae employed in models of the observed phenomena. These can be inverted if the necessary independent observables (at least three, in general) are really measured; see e.g. [276] for a simple example of such an inference, [120] for main techniques of fixing the central M in galaxies ([205] proposed a new one), and e.g. [7, 19, 200, 222] for other prospects.

The dividing line between neutron stars and black holes being given by the Landau-Oppenheimer-Volkoff mass limit of about $M = (2 \div 2.5)M_{\odot}$ ($M_{\odot}$ is the solar mass), attention has recently been focused on determination of the rotation rate of the more massive cases. There is some evidence that highly active sources (in "hard X-state", "radio-loud", with strong and fast jets) host rapidly rotating holes and vice versa [87, 124, 207, 217, 286, 310, 326]. In particular, the popular microquasars GRS 1915+105 and GRO J1655-40 favour rapid central rotation, perhaps even close to the extreme value $a = M$ which separates black holes from objects without a horizon (naked singularities).[i] Such a rapid rotation is expected due to theoretical reasons — calculations of the effect of disc accretion on the central body [8,314]. On the other hand, a hyper-extreme object would probably be slowed down to $a = M$ by accretion [56, 288].[j] The centre (and the disc) could also lose rotational energy to the benefit of the material outflowing from the system. The significance of the Blandford-Znajek mechanism of jet fuelling [41,263] for the active galactic nuclei and X-ray binaries is still under discussion [122, 185, 198, 258] (while it is rather generally accepted for gamma-ray bursts [186]). There also exist other extraction possibilities (see [309] for a review), mainly the Punsly-Coroniti variant of Penrose mechanism [259] (contrary to the Blandford-Znajek one, it does not require a magnetic field threading the horizon) which was indeed noticed in numerical simulations [169]. Another efficient scenario has been proposed recently [194].

Black holes are certainly favoured by the "cosmic censorship" hypothesis [246] which defends predictability by excluding the naked singularities as the outcome of a realistic gravitational collapse. However, this "question is still very much open" [248] (also [81, 133]), with some amount of evidence as well as counter-indications already given (e.g. [1, 70, 127, 131, 149, 265] and references therein) whose actual significance is debatable (e.g. [49, 313]).[k]

The experimental task of deciding between black holes and naked singularities thus brings us back to the fundamental problems of the theory. It should be em-

[i] Note that GRS 1915+105 provided the first direct evidence of the disc-jet interaction [100]; GRO J1655-40 is likely to have originated in a supernova explosion [145]. Their central spins however remain uncertain — cf. the disjoint ranges estimated in [124], [3] and [311]. The spin of the central black hole in our Galaxy is discussed in [220].

[j] Conversely, cf. a recently proposed gedanken experiment [92] which appears to show that it is possible, in principle, to turn an extreme (Reissner-Nordström) black hole into a (Kerr-Newman) naked singularity by a radial infall of a (strongly bound, extended and electrically neutral) spinning body. In a related example, [140] tried to overcharge a near-extremal Reissner-Nordström hole.

[k] Black holes are sometimes said to reside where God divided by zero. Regardless of whether he or she used Good coordinates or not, this is consistent with his/her being Almighty. Certain current readings of the cosmic censorship conjecture ("God abhors a naked singularity" [133]) sound somewhat disturbing, on the contrary.

phasized here that Jiří Bičák has always taught us to respect the decisive validity of observations, notwithstanding his obvious esteem for the penetrating "pure reasoning" (on the back of an envelope). Jiří often shows romantic affinity to astronomy and direct contact with nature, though majority of his results concern mathematical aspects of the theory. I heard him mentioning the need for a balance between the "right and left deviation" in one's work. As a star on the stage of our Institute, he also alternates a glowing guitar in "Rock around the clock" and a delicate violin in Vivaldi.

2.2 Gravity of accretion discs around black holes

It is the subject of a standard university course to discuss the features of the Schwarzschild, Reissner-Nordström, Kerr and Kerr-Newman metrics (see [33] for a review), exactly describing isolated stationary black holes (or naked singularities). Black holes expected in galactic nuclei and in X-ray binaries are not isolated, however: matter is concentrated rather than absent there. Most of the evidence for ultracompact bodies is in fact *based* on their interaction with the surroundings. In theoretical models, the external matter is supposed to form an accretion disc around the equatorial plane of the centre (this is justified by the approximate reflectional symmetry of the host systems and, at least in the discs' inner parts, by Bardeen-Peterson effect [23]; for later results, see [113]). Accretion is a problem governed by complicated magneto-hydrodynamics and radiative transfer which is usually tackled under the assumptions of smoothness, stationarity, axisymmetry and simple parametrization of the flow. Gravity of the disc itself is neglected, the field is thus fully determined by the centre, either described exactly by the Kerr metric, approximated by the Schwarzschild one, simulated by some pseudo-Newtonian potential, or quite reduced to a Newtonian $(-M/r)$.

The main topic of the present paper is an inclusion of the disc gravitational field in a fully general relativistic description. *Newtonian* treatments of the problem began with Ostriker's equilibria of uniformly rotating, polytropic, self-gravitating slender rings [242]. The paper opened with references to previous centuries and concluded with remarks on extragalactic radio sources and stability of rings, both of which were to boom soon. Later, polytropic self-gravitating configurations with a realistic equation of state and opacity were calculated by [173] (cold case) and [42, 116, 132] (hot case), already suggesting the possibility of a black hole present in galactic centres. [227] incorporated viscosity (in a self-similar case).[l] In the meantime, basic formalism for studying self-gravitating discs around black holes in a stationary axisymmetric *general relativistic* case was given by [22, 112].

Though the mass of real inner accretion discs is usually claimed not to exceed few percent of the central mass,[m] the role of their self-gravity is not fully understood

[l]The papers must be studied in detail in order to understand precisely what was done. The topic of accretion discs is very wide and conclusions found in the literature are often hardly comparable due to different kinds of simplifications made, relating not only to the gravitational field and symmetries, but mainly to the rotation law, accretion rate, equation of state, viscosity, heat production, radiation pressure, opacity, boundary conditions (material supply), etc.

[m]This only applies to the ultracompact centre (cf. e.g. [231, 316], however), not to protostellar discs [321] where even the opposite may well happen [320].

yet. Are any properties of accretion discs so sensitive to the details of the field that even a small effect of the disc itself can alter the flow significantly? It was shown by [280] that the very global structure of the disc is an example of such properties. This paper presented an extension of the disc "α-models" (both in their Newtonian version of Shakura & Sunyaev and in the relativistic version of Novikov & Thorne) which includes the effects of self-gravitation, radiation pressure and variable opacity source. The self-consistent configurations thus obtained were considerably thinner than their test counterparts, showing sharp radial decrease of the vertical dimension in the middle region. The implications of self-gravity concerning global properties of the disc (e.g. distribution of angular momentum and outer radius) were discussed and interpreted in a Newtonian analysis [6], the roles of the disc density and total mass were distinguished and specified in particular. With increasing gravity of the disc, regions occurred where free circular motion became unstable or even impossible, thus it was stated that "astrophysically relevant models of thick accretion discs must incorporate self-gravity effects in order to be self-consistent" (the same had already been indicated in [280, 316] before).

Changes in the properties of circular equatorial motion were later observed in the pseudo-Newtonian study [162] of a rotating black hole surrounded by a massive thin equatorial ring, in a relativistic analysis of the corresponding static case [64], and also in the perturbative solution [318] for a slowly rotating, weakly gravitating thin ring around a Schwarzschild black hole [319]. These changes were claimed to result in modifications of the observable characteristics, e.g. in higher temperature of the disc. The latter was confirmed by numerically constructing equilibrium structures of self-gravitating polytropic thick discs in a pseudo-Newtonian potential [42]. The accretion-disc "self-field" was also discussed in connection with the disc's vertical structure and stability. It turns out that certain modes of instability can be damped whereas other are amplified by the disc's gravity (see [13, 244]). The latest fully *relativistic* results — employing numerical spacetimes obtained in [181, 237] — include the computation of emission-line profiles from self-gravitating discs around rotating black holes; the case of light, thin finite discs was treated by [156] while that of heavy toroids by [305].

Self-gravitating accretion discs also appeared in other contexts, some of which are interesting from the point of view of general relativity (e.g. heavy discs around white dwarfs as a ponderable stage of a double-white-dwarf merger [130]), whereas other are not relevant for the present review (young stars [320, 321], outer parts of nuclear discs, galactic discs [279], effect on bipolar jets [117]).

Recently we have studied exact solutions for a static (Bach-Weyl) ring [64] and static thin annular disc [188] around a non-rotating black hole; some of the results are presented in section 4. We also checked the possibility to generalize them to a stationary situation [325] (section 5) when the hole and the disc were allowed to rotate and the self-gravity effects could be even more pronounced. Actually, in relativity, the kinetic (in fact *any*) energy also generates the field and mass-energy currents give rise to frame dragging. Whereas mass of the hole should be dominant in a real accretion system, the disc can bear much (even most) of the angular momentum, thus modifying the gravitomagnetic field of the centre significantly. This could be important for the mechanisms of rotational energy extraction.

We should not forget the accretion system where the mass of the disc or torus can actually be comparable to that of the central hole. It is considered in the literature as a transient state of gravitational collapse that could power the gamma-ray bursts.

In a present-day relativistic astrophysics, perhaps the most popular system is a close binary of ultracompact objects. The famous Hulse-Taylor binary pulsar (PSR 1913+16) turned out to spiral in at the rate predicted by general relativistic loss of angular momentum via gravitational radiation [291], thus providing the first, indirect confirmation of the existence of gravitational waves. Nowadays late stages of evolution of such systems (inspiral, merger and "ringdown") are studied numerically or perturbatively by a number of research groups, mainly as the most promising source of gravitational waves [296], and also as an engine suggested for the gamma-ray bursts [223]. In particular, at late stages of the neutron star–black hole binary inspiral, the neutron star tidally breaks up. Loosely speaking, if this happens outside the innermost stable orbit around the hole, the neutron debris spreads into a strongly magnetized torus (otherwise it plunges into the hole). Similar outcome is expected in case of a binary with two neutron stars (which is expected to be less frequent than the black hole–neutron star case [29]). It was pointed out in numerical analyses that the black hole–neutron torus system was liable to runaway instability (see [216] and references therein). Depending on the exact shape of the created magnetosphere, the energy release is mainly ensured by leptonic winds from pair-creation along the axis of rotation or/and axially collimated Poynting outflow resulting from the Blandford-Znajek mechanism [186,307]. In [307], "Long/short gamma-ray bursts are identified with suspended/hyper- accretion onto rapidly/slowly rotating black holes. ...In long bursts, the torus is expected to radiate most of the black hole luminosity in gravitational waves. This predicts that long gamma-ray bursts are potentially the most powerful burst-sources of gravitational waves in the Universe." (This thorough review also conjectures about connections between gamma-ray bursts and active galactic nuclei and micro-quasars.)

The black hole with a disc/torus can also arise as the outcome of a gravitational collapse of a massive rotating star. This was already mentioned in [231], while more recently in scenarios of "failed supernovae" from the cores of young stars, either isolated [203] or in a binary (this case is referred to as "hypernovae") [54, 243]; for a review, see discussions in [53, 307].

3 Stationary axisymmetric spacetimes

It is natural first to try to include the gravitational effect of the external source in the simplest case when the system is stationary and axisymmetric. Stationary axisymmetric spacetimes are of obvious astrophysical importance: they describe the exterior of bodies like stars, galaxies or accretion discs in equilibrium. The issue has been exposed in many places (e.g. [22, 55, 61, 67, 74, 103, 143, 192, 318] or chapters 17–19 of [174]); for more information about the most important particular solutions, see [33, 45].

In the Weyl-Lewis-Papapetrou coordinates (t, ρ, ϕ, z) of the cylindrical type, the metric can be written as[n]

$$ds^2 = -e^{2\nu}dt^2 + \rho^2 B^2 e^{-2\nu}(d\phi - \omega dt)^2 + e^{2\lambda - 2\nu}(d\rho^2 + dz^2), \tag{1}$$

where the unknown functions ν, B, ω and λ only depend on ρ and z; ω is interpreted as the angular velocity of inertial space rotation with respect to observers at rest in spatial infinity. The metric coefficients g_{ab} have invariant meaning, they can be expressed in terms of Killing fields $k^\mu = \partial x^\mu/\partial t$ and $m^\mu = \partial x^\mu/\partial\phi$:

$$g_{tt}\,(= -e^{2\nu} + \omega^2 g_{\phi\phi}) = k_\iota k^\iota, \tag{2}$$

$$g_{t\phi}\,(= -\omega g_{\phi\phi}) = k_\iota m^\iota, \tag{3}$$

$$g_{\phi\phi}\,(= \rho^2 B^2 e^{-2\nu}) = m_\iota m^\iota. \tag{4}$$

The simple result for the corresponding subdeterminant, $\det(g_{ab}) = -\rho^2 B^2$, implies

$$g^{tt} = -e^{-2\nu}, \qquad g^{t\phi} = -e^{-2\nu}\omega, \qquad g^{\phi\phi} = -e^{-2\nu}\omega^2 + \rho^{-2}B^{-2}e^{2\nu}. \tag{5}$$

3.1 Field equations

The Einstein equations read

$$\vec{\nabla}\cdot(\rho\vec{\nabla}B) = 8\pi\rho B(T_{\rho\rho} + T_{zz}), \tag{6}$$

$$\vec{\nabla}\cdot(\rho^2 B^3 e^{-4\nu}\vec{\nabla}\omega) = -16\pi B e^{2\lambda - 2\nu} T^t_\phi, \tag{7}$$

$$\vec{\nabla}\cdot(B\vec{\nabla}\nu) = \frac{1}{2}\rho^2 B^3 e^{-4\nu}(\vec{\nabla}\omega)^2 + 4\pi B e^{2\lambda}(2T^{tt} + e^{-2\nu}T), \tag{8}$$

where $T \equiv T^\sigma_\sigma$ and $\vec{\nabla}$ and $\vec{\nabla}\cdot$ stand for the gradient and divergence in a flat three-dimensional space with coordinates (ρ, ϕ, z); thus $\vec{\nabla}X = (X_{,\rho}, 0, X_{,z})$ and $\vec{\nabla}\cdot\vec{X} = \rho^{-1}[(\rho X^\rho)_{,\rho} + (\rho X^z)_{,z}]$ in the axially symmetric case. Once B, ω and ν are known, λ can be integrated from equations

$$B\lambda_{,z} - B_{,z} + \rho\,(B_{,\rho}\lambda_{,z} + B_{,z}\lambda_{,\rho} - B_{,\rho z} - 2B\nu_{,\rho}\nu_{,z}) + \frac{1}{2}\rho^3 B^3 e^{-4\nu}\omega_{,\rho}\omega_{,z} = 8\pi\rho B T_{\rho z}, \tag{9}$$

$$2B\lambda_{,\rho} - 2B_{,\rho} + \rho\left[2B_{,\rho}\lambda_{,\rho} - 2B_{,z}\lambda_{,z} - B_{,\rho\rho} + B_{,zz} - 2B(\nu^2_{,\rho} - \nu^2_{,z})\right] + \frac{1}{2}\rho^3 B^3 e^{-4\nu}(\omega^2_{,\rho} - \omega^2_{,z}) = 8\pi\rho B(T_{\rho\rho} - T_{zz}). \tag{10}$$

---

[n] The quantities will be given in geometrized units in which $c = G = 1$; the signature of the metric tensor $g_{\mu\nu}$ is $(-+++)$. Greek indices run from 0 to 3 and Latin indices $(i, j, \ldots)$ run from 1 to 3; indices from the beginning of the Latin alphabet $(a, b, \ldots)$ represent cyclic coordinates t and ϕ. Partial differentiation is denoted by ∂ or by a subscript comma, covariant derivative is denoted by ∇ or by a subscript semicolon.

Although a suitable combination of solenoidal motions can also satisfy the assumption of axial symmetry, it is natural to expect that the axially symmetric source just rotates in the azimuthal direction, with an angular velocity $\Omega = \mathrm{d}\phi/\mathrm{d}t$. The corresponding four-velocity is

$$u^\mu = u^t(k^\mu + \Omega m^\mu) = u^t(1,0,\Omega,0), \tag{11}$$

where $(u^t)^{-2} = -g_{tt} - 2\Omega g_{t\phi} - \Omega^2 g_{\phi\phi} = e^{2\nu}(1-\hat{v}^2)$ and $\hat{v} = \rho B e^{-2\nu}(\Omega - \omega)$ is the linear speed with respect to the local zero-angular-momentum observer.

For a source made of ideal fluid, having a total mass-energy density ϵ and pressure P as measured in a co-moving system, the energy-momentum tensor is $T_{\mu\nu} = (\epsilon + P)u_\mu u_\nu + P g_{\mu\nu}$ and the equations of motion read $(\epsilon + P)a_\mu = -P_{,\mu}$, where the four-acceleration $a_\mu = u_{\mu;\nu}u^\nu$ can, for instance, be written as

$$a_\mu = -\frac{1}{2}g_{\alpha\beta,\mu}u^\alpha u^\beta = -\frac{g_{tt,\mu} + 2\Omega g_{t\phi,\mu} + \Omega^2 g_{\phi\phi,\mu}}{2e^{2\nu}(1-\hat{v}^2)}$$

$$= \frac{2e^{2\nu}\nu_{,\mu} + 2(\Omega-\omega)g_{\phi\phi}\omega_{,\mu} - (\Omega-\omega)^2 g_{\phi\phi,\mu}}{2e^{2\nu}(1-\hat{v}^2)}. \tag{12}$$

From equations (6)–(8) one then obtains

$$\vec{\nabla}\cdot(\rho\vec{\nabla}B) = 16\pi\rho B e^{2\lambda-2\nu}P, \tag{13}$$

$$\vec{\nabla}\cdot(\rho^2 B^3 e^{-4\nu}\vec{\nabla}\omega) = -16\pi\rho B^2 e^{2\lambda-4\nu}(\epsilon+P)\frac{\hat{v}}{1-\hat{v}^2}, \tag{14}$$

$$\vec{\nabla}\cdot(B\vec{\nabla}\nu) = \frac{1}{2}\rho^2 B^3 e^{-4\nu}(\vec{\nabla}\omega)^2 + 4\pi B e^{2\lambda-2\nu}\left[(\epsilon+P)\frac{1+\hat{v}^2}{1-\hat{v}^2} + 2P\right], \tag{15}$$

and λ is given by

$$\lambda_{,\rho} = \frac{B}{(\rho B)^2_{,\rho} + \rho^2 B^2_{,z}}\Big\{B_{,\rho} + \rho B^{-1}(B^2_{,\rho} + B^2_{,z})$$
$$+\rho\left[(\rho B)_{,\rho}(\nu^2_{,\rho} - \nu^2_{,z}) + 2\rho B_{,z}\nu_{,z}\nu_{,\rho}\right]$$
$$-\frac{1}{4}\rho^3 B^2 e^{-4\nu}\left[(\rho B)_{,\rho}(\omega^2_{,\rho} - \omega^2_{,z}) + 2\rho B_{,z}\omega_{,z}\omega_{,\rho}\right]\Big\}, \tag{16}$$

$$\lambda_{,z} = \frac{B}{(\rho B)^2_{,\rho} + \rho^2 B^2_{,z}}\Big\{B_{,z} + \rho\left[\rho B_{,z}(\nu^2_{,z} - \nu^2_{,\rho}) + 2(\rho B)_{,\rho}\nu_{,\rho}\nu_{,z}\right]$$
$$-\frac{1}{4}\rho^3 B^2 e^{-4\nu}\left[\rho B_{,z}(\omega^2_{,z} - \omega^2_{,\rho}) + 2(\rho B)_{,\rho}\omega_{,\rho}\omega_{,z}\right]\Big\}. \tag{17}$$

The unknown metric functions are subject to boundary conditions on the horizon (if there is one), on the symmetry axis and at spatial infinity. The metric must be regular on the horizon and on the axis; in the case of an isolated source one requires asymptotic flatness. The conditions were discussed by [22, 61], for example.

3.2 *Vacuum case and the Ernst formulation*

In the following sections, we will be interested in vacuum solutions or in those with infinitely thin sources. In both cases $P=0$ and the equation (13) has only one solution with a satisfactory asymptotic behaviour [namely (36)], $B=1$. Only two of the field equations then remain, (7,8), in the form

$$\vec{\nabla}\cdot(\rho^2 e^{-4\nu}\vec{\nabla}\omega)=0, \tag{18}$$

$$\vec{\nabla}^2\nu=\frac{1}{2}\rho^2 e^{-4\nu}(\vec{\nabla}\omega)^2, \tag{19}$$

and the relations (16,17) reduce to

$$\lambda_{,\rho}=\rho(\nu_{,\rho}^2-\nu_{,z}^2)-\frac{1}{4}\rho^3 e^{-4\nu}(\omega_{,\rho}^2-\omega_{,z}^2), \tag{20}$$

$$\lambda_{,z}=2\rho\nu_{,\rho}\nu_{,z}-\frac{1}{2}\rho^3 e^{-4\nu}\omega_{,\rho}\omega_{,z}. \tag{21}$$

Many other formulations of the stationary axisymmetric problem are possible, starting from a different definition of the four (or three) unknown metric functions. In the vacuum case the formulation by Ernst [104, 105] is often considered, where the field equations are translated into the (Ernst) equation

$$-g_{tt}\vec{\nabla}^2\mathcal{E}=(\vec{\nabla}\mathcal{E})^2 \tag{22}$$

for a complex (Ernst) potential $\mathcal{E}\equiv -g_{tt}+\mathrm{i}\psi$ whose imaginary part ψ is given by

$$\rho\psi_{,z}=g_{tt}g_{t\phi,\rho}-g_{tt,\rho}g_{t\phi}, \qquad \rho\psi_{,\rho}=g_{tt,z}g_{t\phi}-g_{tt}g_{t\phi,z}; \tag{23}$$

λ is found by a line integral as above.

3.3 *Horizon*

(See e.g. [61, 62, 115] for thorough accounts.) A black-hole horizon is a null hypersurface below which the spacetime is dynamical. In Weyl-Lewis-Papapetrou coordinates, it is located where $\det(g_{ab})=-\rho^2B^2$ is zero. Also vanishing there is the *lapse* (or *redshift factor*) $\alpha\equiv e^\nu$ or the magnitude of the Killing bivector $k_{[\kappa}m_{\lambda]}$; this is clear from relations

$$-2k_{[\kappa}m_{\lambda]}k^{[\kappa}m^{\lambda]}=-\det(g_{ab})=\alpha^2 g_{\phi\phi}, \tag{24}$$

because $g_{\phi\phi}$ (as well as $g_{\rho\rho}$) must be regular (and positive) on the horizon. In the vacuum case $B=1$, so the horizon lies on the axis ($\rho=0$). The interior of the black hole thus has to be studied in different coordinates, e.g. in spheroidal coordinates of the Boyer-Lindquist type (t,r,θ,ϕ), introduced by the transformation

$$\rho=\sqrt{\Delta}\sin\theta, \qquad z=(r-M)y, \tag{25}$$

where $\Delta=(r-M)^2-k^2$ and $y=\cos\theta$, M being a scale parameter (it usually represents the total mass) and $(\pm)k$ determining where the horizon reaches up along the z axis. Sometimes an isotropic radial coordinate R is employed, given by

$$\Delta=R^2\left(1-\frac{k^2}{4R^2}\right)^2. \tag{26}$$

On the horizon, $\Delta = 0$ and both radial coordinates are constant, $r = M + k \equiv r_{\rm H}$ (hence, $z = ky$), $R = k/2 \equiv R_{\rm H}$ (there are more black-hole horizons in general, we mean the outermost one here).

There are three important parameters of the horizon: its surface area (proportional to the entropy of the black hole), the surface gravity (proportional to the horizon temperature) and the angular velocity relative to infinity,

$$A = 2\pi \int_0^\pi \sqrt{(g_{\theta\theta}g_{\phi\phi})_{\rm H}}\,{\rm d}\theta, \qquad \kappa = |\nabla e^\nu|_{\rm H}, \qquad \omega_{\rm H} = \omega(r = r_{\rm H}); \tag{27}$$

κ and $\omega_{\rm H}$ are constant all over the horizon.

For spacetimes with a symmetry lower than spherical, the question of the "true shape" of the black hole, not distorted by coordinates, is non-trivial. How is the hole influenced by rotation, electromagnetic field or an external source? The answer is obtained by representing the horizon as a two-dimensional surface in a three-dimensional Euclidean space. Following [283], the two-dimensional metric is first rewritten as

$${\rm d}s^2 = \frac{A}{4\pi}\,[h^{-1}(y){\rm d}y^2 + h(y){\rm d}\phi^2], \tag{28}$$

where $h(y) = (4\pi/A)(g_{\phi\phi})_{\rm H}$. An isometric embedding of the two-surface (y, ϕ) in $\mathbb{E}^3$ with coordinates (X, Y, Z) is then given by

$$X = \frac{A}{4\pi}\sqrt{h}\cos\phi, \qquad Y = \frac{A}{4\pi}\sqrt{h}\sin\phi, \qquad Z = \frac{A}{4\pi}\int_0^y \sqrt{\frac{1}{h}\left(1 - \frac{1}{4}h_{,y}^{\,2}\right)}\,{\rm d}y. \tag{29}$$

If its Gaussian curvature

$$\mathcal{C}_{\rm H} = -\frac{8\pi^2}{A^2}\,h_{,yy} \tag{30}$$

is negative anywhere, the horizon cannot be (globally) embedded in $\mathbb{E}^3$. It was shown in [283] that rotation can flatten the horizon considerably, the Gaussian curvature of a Kerr black hole finally turns negative at the axis for $a/M > \sqrt{3}/2$ (a denotes the angular momentum per unit M). Some of the black holes swirling in galactic nuclei may even be impossible to imagine!

3.4 Static case

In the static case $\omega = 0$ and (1) acquires the Weyl canonical form

$${\rm d}s^2 = -e^{2\nu}{\rm d}t^2 + \rho^2 e^{-2\nu}{\rm d}\phi^2 + e^{2\lambda-2\nu}({\rm d}\rho^2 + {\rm d}z^2). \tag{31}$$

More precisely, the metric can be written in this way provided that $T^\rho_\rho + T^z_z = 0$ (which is fulfilled by incoherent dust, certain electromagnetic fields or certain infinitely thin sources, but not by a fluid with non-zero pressure), otherwise there remains B^2 in $g_{\phi\phi}$.

Equations (6)–(10) reduce to the Poisson equation

$$\vec{\nabla}^2\nu \equiv \rho^{-1}\nu_{,\rho} + \nu_{,\rho\rho} + \nu_{,zz} = 4\pi e^{2\lambda-2\nu}(T^\phi_\phi - T^t_t) \tag{32}$$

for ν and to relations

$$\lambda_{,z} - 2\rho\nu_{,\rho}\nu_{,z} = 8\pi\rho T_{\rho z}, \qquad \lambda_{,\rho} - \rho(\nu_{,\rho}^2 - \nu_{,z}^2) = 4\pi\rho(T_{\rho\rho} - T_{zz}) \tag{33}$$

for λ.

3.5 *The axis and the equatorial plane*

On the symmetry axis ($\rho = 0$), the regularity of metric requires $e^\lambda = B$, so

$$ds^2 = -e^{2\nu}dt^2 + B^2 e^{-2\nu}dz^2. \tag{34}$$

The formulas can also be expected to simplify in the equatorial plane ($z = 0$) if the spacetime is reflectionally symmetric with respect to it. This is usually the case in astrophysically motivated considerations. As a special case, the solutions with stronger, cylindrical symmetry (where another Killing field, $\partial x^\mu/\partial z$, exists) can be mentioned, where *any* of the planes $z = \text{const}$ is equatorial in this sense; however, there is no place for black holes in these spacetimes.

3.6 *Singularities*

From astrophysical point of view, a given solution is problematic mainly if it contains physical singularities on or above the horizon, if it shows bad asymptotic properties or if it does not correspond to any realistic source. The existence of a singularity follows — according to its type (see [80, 101]) — e.g. from the divergence of scalars constructed from the metric tensor and its derivatives or from the divergence of physical (tetrad) components of the Riemann tensor. (Usually, the Kretschmann invariant $R_{\mu\nu\rho\sigma}R^{\mu\nu\rho\sigma}$ is checked first.) However, the spacetimes are known which do contain singularities, although all their curvature invariants vanish. It is thus difficult to verify the *non*-existence of singularities; the only proof of regularity at a given point is to find a (local) coordinate system in which the metric is smooth enough there.

3.7 *Analytic extension and global structure*

Astrophysically relevant is that part of spacetime which can communicate with spatially remote regions, i.e. the "region of outer communications" outside the horizon. General relativity is interested in the black hole interior as well — in fact even there an observer can perform physical observations (and perhaps even survive). Unfortunately, in Weyl-Lewis-Papapetrou coordinates the metric is not defined below the horizon (this would correspond to imaginary radius ρ) and one has to extend it there. Such a task may be solvable but numerically and hardly expectable regions may open. What is the type and shape of the singularity under the horizon? How does it change when an additional source is "switched on" above the horizon? The answer is unknown even for quite simple superpositions (on the other hand, in a highly symmetric case it may well be possible to solve the dynamical problem [178]).

Far away from the horizon, one inquires for the asymptotic behaviour of the field. That of an isolated stationary source falls to zero there in a specific manner

— the spacetime is said to be asymptotically flat (see e.g. [312], chapter 11, or [26], section 3). In case of the metric (1), it must hold, for $r \to \infty$,

$$\nu = -\frac{\text{total mass}}{r} + O(r^{-2}), \tag{35}$$

$$B = 1 + O(r^{-2}), \tag{36}$$

$$\omega = \frac{2 \cdot (\text{total angular momentum})}{r^3} + O(r^{-4}), \tag{37}$$

$$\lambda = O(r^{-2}). \tag{38}$$

The courses can be expressed, in the same manner, in terms of the isotropic radial coordinate R or in terms of the radius $\sqrt{\rho^2 + z^2}$, since $R = r - M + O(r^{-1})$[o] and $\sqrt{\rho^2 + z^2} = r - M + O(r^{-1})$ for $r \to \infty$.

3.8 Circular orbits and ergosphere

The simplest type of motion in stationary axisymmetric spacetimes is that along spatially circular orbits (ρ = const, z = const) with steady angular velocity Ω: such a motion takes place along the symmetries, namely tangent to each individual circular orbit is a Killing field (one speaks of "quasi-Killing" trajectories); physically speaking, an observer on a circular orbit sees time-independent field around and is thus called the stationary observer. In order that the corresponding four-velocity (11) lie inside the light cone, Ω must fall within the values

$$\Omega_{\substack{\max \\ \min}} = \omega \pm \frac{\alpha}{\sqrt{g_{\phi\phi}}}; \tag{39}$$

on the horizon (where $\alpha = 0$) the permitted range narrows down to a single value $\omega_{\rm H}$.

Calculations involving acceleration are more difficult as they also contain derivatives of the metric. The circular-orbit four-acceleration (12) has at most two non-zero components (a_ρ and a_z); μ-component is zero if

$$\Omega = \Omega_\pm = \frac{-g_{t\phi,\mu} \pm \sqrt{(g_{t\phi,\mu})^2 - g_{tt,\mu} g_{\phi\phi,\mu}}}{g_{\phi\phi,\mu}}, \tag{40}$$

where the upper/lower sign corresponds to a "co-rotating"/"counter-rotating"[p] orbit.

In the equatorial plane, $a_z = 0$ due to the symmetry. If $\Omega = \Omega_\pm$ (here with $\mu = \rho$), $a_\rho = 0$ as well and the worldline is a geodesic. Among equatorial geodesics, three particular cases are most important: the photon, the marginally bound and the marginally stable orbits. The photon orbits bound the regions where circular motion is time-like; they are given by equalities

$$\Omega_\pm = \Omega_{\substack{\max \\ \min}} \tag{41}$$

[o] This is clear from an explicit form of (26), $R = \frac{1}{2}(r - M + \sqrt{\Delta})$, or $r = R + M + \frac{k^2}{4R}$.

[p] These terms may be misleading in spacetimes with large angular momentum.

[$\mu = \rho$ being the only non-trivial index value in (40) now]. The marginally bound orbits limit the regions where particles on circular orbits have lower energy than is necessary for the existence at spatial infinity; the limiting case is given by $\gamma(\Omega = \Omega_\pm) = 1$, where

$$\gamma = -u_t = -u^t(g_{tt} + \Omega g_{t\phi}) = u^t e^{2\nu} + \omega\ell \quad \text{and} \quad \ell = u_\phi = u^t g_{\phi\phi}(\Omega - \omega) \quad (42)$$

are specific energy and specific azimuthal angular momentum at infinity. The marginally stable orbits bound the regions where circular motion is stable with respect to perturbations within the orbital plane; they satisfy $[\ell(\Omega = \Omega_\pm)]_{,\rho} = 0$. They play a key role in the theory of accretion discs: the matter is "swept away" from unstable sectors; in particular, the innermost marginally stable orbit should represent the inner disc rim. (We will see in section 4 that perturbations *perpendicular* to the orbital plane can also be important.)

A peculiar feature of rotating fields are the dragging effects. Their history starts from the famous Newton's bucket experiment (if not earlier, see Doležel in this volume) and continues notably by Mach's ideas of relativity of motion and inertia towards Einstein and Lense & Thirring who showed, in the early times of general relativity, that the effect is present in this theory (see the 1913's letter of Einstein to Mach [166]).[q] The effect has an obvious analogy in electromagnetism, where (electric) currents generate the magnetic component of the field. This relationship was already noticed by Einstein (see his Prague paper of 1912 [167]) and today the gravito-electro-magnetic approach to rotating fields prevails.

An extreme implication of dragging is the occurrence of the ergosphere in the vicinity of ultracompact rotating objects. In this region $\Omega_{\min} > 0$, thus the stationary observer cannot be static (at rest relative to infinity), albeit he still withstands the radial attraction. The static-limit surface, given by $\Omega_{\min} = 0$ and hence by $g_{tt} = 0$, also represents the set of points from where signals emitted by static observers ($\Omega = 0$) reach infinity with an infinite redshift. Ergospheres received considerable attention (and the name from $\epsilon\rho\gamma o\nu$) after processes had been suggested (working only there) [246] by means of which the rotational energy of the centre could be extracted (without violating the second law of black-hole dynamics that forbids the horizon area to decrease).

With the metric (1), $g_{tt} = 0$ corresponds to $\rho B\omega = \alpha^2$ which can only be satisfied by $\rho \geq 0$, so the static limit really lies outside the horizon;[r] in particular, $\rho = 0$ implies $\alpha^2 = 0$, so the static limit touches the horizon at the axis.

3.9 Sources

In order to complete the relativistic solution, one must also describe their interior (where $T_{\mu\nu} \neq 0$). Finding a realistic interior solution is a remarkable achievement in general, mainly if it should match smoothly the vacuum exterior (e.g. [73, 180, 249] and references therein); this is also confirmed by the history of searching for a source of the Kerr field (see [38, 236] and references therein).

[q] Machian inspiration keeps vivid within contemporary relativity — see, for example, the last-decade references [79, 147, 163, 201, 202] or the whole conference [21].

[r] In coordinates which also describe the black hole interior, *two* horizons and *two* static limits are found in general. We mean the outer horizon and the outer static limit everywhere.

The problem is much simpler if the source is infinitely thin: the solution is then vacuum-type everywhere, with the energy-momentum tensor $T_{\mu\nu} = g_{zz}^{-1/2} S_{\mu\nu}\delta(z)$ found from the jump of the normal field across the source like in electrodynamics; the appropriate covariant method is known as the Israel's formalism [24]. It starts from a projection of the exterior (here vacuum) metric $g_{\mu\nu}$ onto a three-dimensional hypersurface S, representing the history of the source: the projection from one (+) and the other (−) side of S yields the induced three-metrics

$$^{\pm}g_{\hat{A}\hat{B}} = g_{\mu\nu}\,{}^{\pm}e^{\mu}_{\hat{A}}\,{}^{\pm}e^{\nu}_{\hat{B}}, \tag{43}$$

where $\left\{{}^{\pm}e^{\mu}_{\hat{A}}\right\}_{A=0,1,2}$ are bases tangent to $^{\pm}S$. The hyperplanes $^{\pm}S$ are however the back and the face of the same surface, so we can choose $^{+}e^{\mu}_{\hat{A}} = {}^{-}e^{\mu}_{\hat{A}}$ and only the normals $^{+}n^{\mu}$ and $^{-}n^{\mu}$ differ. The surface density of the energy-momentum tensor $S_{\hat{A}\hat{B}}$ reads

$$S_{\hat{A}\hat{B}} = (8\pi)^{-1}\left([K_{\hat{A}\hat{B}}] - g_{\hat{A}\hat{B}}[K^{\hat{C}}_{\hat{C}}]\right), \tag{44}$$

where $[K_{\hat{A}\hat{B}}]$ denotes the jump of the exterior curvature across S, i.e. $[K_{\hat{A}\hat{B}}] = {}^{+}K_{\hat{A}\hat{B}} - {}^{-}K_{\hat{A}\hat{B}}$, with

$$^{\pm}K_{\hat{A}\hat{B}} = e^{\mu}_{\hat{A}}e^{\nu}_{\hat{B}}\nabla_{\mu}{}^{\pm}n_{\nu} = -{}^{\pm}n_{\nu}e^{\mu}_{\hat{A}}\nabla_{\mu}e^{\nu}_{\hat{B}}. \tag{45}$$

The case of a thin source (a disc) in a stationary axisymmetric spacetime has notably been studied by [126, 184]. In a disc lying at z = const the radial pressure is zero, $S_{\rho\rho} = 0$, and for a disc in the equatorial plane ($z = 0$) of the reflectionally symmetric spacetime $S_{zz} = 0$, too. Only $S_{t\phi}$ is non-zero off the diagonal. In the Weyl-Lewis-Papapetrou coordinates the formula (44) can be written as ([184], chapters 3 and 4)[s]

$$S_{ab} = -\frac{\sqrt{g_{\rho\rho}}}{8\pi}\left(\frac{g_{ab}}{g_{\rho\rho}}\right)_{,z}. \tag{46}$$

Using (5) and (21), this yields

$$S^{t}_{t} = -\frac{e^{\nu-\lambda}}{8\pi}\left[4\nu_{,z}(1-\rho\nu_{,\rho}) - \rho^2 e^{-4\nu}\omega_{,z}(\omega - \rho\omega_{,\rho})\right], \tag{47}$$

$$S^{t}_{\phi} = -\frac{e^{\nu-\lambda}}{8\pi}\rho^2 e^{-4\nu}\omega_{,z}, \tag{48}$$

$$S^{\phi}_{t} = -\frac{e^{\nu-\lambda}}{8\pi}\left[4\omega\nu_{,z} - (1+\rho^2 e^{-4\nu}\omega^2)\omega_{,z}\right], \tag{49}$$

$$S^{\phi}_{\phi} = \frac{e^{\nu-\lambda}}{8\pi}\rho\left[4\nu_{,\rho}\nu_{,z} - \rho e^{-4\nu}\omega_{,z}(\omega + \rho\omega_{,\rho})\right]. \tag{50}$$

[s]The indices $a, b, \ldots$ only run through t and ϕ as anywhere.

The total mass and angular momentum of the disc are fixed by the Komar integrals that lead to

$$\mathcal{M} = \frac{1}{2}\int_b^\infty g^{tc} g_{tc,z}\rho\,\mathrm{d}\rho = -\frac{1}{4}\int_b^\infty \frac{g_{\phi\phi}^2}{\rho}\left(\frac{g_{tt}}{g_{\phi\phi}}\right)_{,z}\mathrm{d}\rho, \tag{51}$$

$$\mathcal{J} = \frac{1}{4}\int_b^\infty g^{tc} g_{c\phi,z}\rho\,\mathrm{d}\rho = \frac{1}{4}\int_b^\infty \frac{g_{\phi\phi}^2}{\rho}\omega_{,z}\mathrm{d}\rho, \tag{52}$$

where $\rho = b > 0$ is the position of the disc inner rim. All the z-derivatives are understood to be calculated in the limit $z \to 0^+$.

In the event that there is a black hole present in the centre of the disc, its contributions M_H and J_H need also to be included; for a given type of solution, the aggregate parameters of the spacetime can be written as the sums $M_\mathrm{H} + \mathcal{M}$ and $J_\mathrm{H} + \mathcal{J}$ (see e.g. [62]). The black-hole mass is related through the Smarr formula ([62], equation (6.297))

$$M_\mathrm{H} = 2\omega_\mathrm{H} J_\mathrm{H} + \frac{\kappa A}{4\pi} = 2\omega_\mathrm{H} J_\mathrm{H} + k \tag{53}$$

with the angular momentum

$$J_\mathrm{H} = -\frac{1}{8}\int_0^\pi (\Delta^2 e^{-4\nu}\omega_{,r})_\mathrm{H} \sin^3\theta \mathrm{d}\theta. \tag{54}$$

The last step is a physical interpretation of the source. First, if

$$D \equiv (S_\phi^\phi - S_t^t)^2 + 4S_\phi^t S_t^\phi = \frac{e^{2\nu-2\lambda}}{16\pi^2}(4\nu_{,z}{}^2 - \rho^2 e^{-4\nu}\omega_{,z}{}^2) \tag{55}$$

is not negative, then there exists a stationary observer with the angular velocity

$$\Omega_\mathrm{iso} = (2S_\phi^t)^{-1}\left(S_\phi^\phi - S_t^t - \sqrt{D}\right) = \omega - \frac{2e^{4\nu}\nu_{,z}}{\rho^2\omega_{,z}} + \sqrt{\left(\frac{2e^{4\nu}\nu_{,z}}{\rho^2\omega_{,z}}\right)^2 - \frac{e^{4\nu}}{\rho^2}}, \tag{56}$$

with respect to whom the energy-momentum tensor assumes a diagonal ("isotropic") form $S^{\mu\nu} = \hat{w} u_\mathrm{iso}^\mu u_\mathrm{iso}^\nu + \hat{P} v_\mathrm{iso}^\mu v_\mathrm{iso}^\nu$, where $u_\mathrm{iso}^\mu = u_\mathrm{iso}^t(1, 0, \Omega_\mathrm{iso}, 0)$ and $v_\mathrm{iso}^\mu = -\rho^{-1}(\ell_\mathrm{iso}, 0, \gamma_\mathrm{iso}, 0)$ are the observer's four-velocity and unit base vector in the ϕ-direction, ℓ_iso and γ_iso being the corresponding specific angular momentum and energy at infinity (see section 3.8).[t] The observer measures the surface density

$$\hat{w} = \frac{\gamma_\mathrm{iso}^2 S^{tt} - \ell_\mathrm{iso}^2 S^{\phi\phi}}{u_\mathrm{iso}^t(\gamma_\mathrm{iso} + \Omega_\mathrm{iso}\ell_\mathrm{iso})} \tag{57}$$

[t]The case of $D < 0$ (not involved here) represents discs with non-zero heat flow. Then the energy-momentum tensor must be written in a more general form $S^{\mu\nu} = \hat{w} u_\mathrm{iso}^\mu u_\mathrm{iso}^\nu + \hat{P} v_\mathrm{iso}^\mu v_\mathrm{iso}^\nu + \hat{K}(u_\mathrm{iso}^\mu v_\mathrm{iso}^\nu + v_\mathrm{iso}^\mu u_\mathrm{iso}^\nu)$ with $\hat{K} = \sqrt{-D/2}$, and $\Omega_\mathrm{iso} = (2S_\phi^t)^{-1}(S_\phi^\phi - S_t^t) = \omega - \frac{2e^{4\nu}\nu_{,z}}{\rho^2\omega_{,z}}$. See [126].

and tangential pressure

$$\hat{P} = \frac{\rho^2 u^t_{\rm iso}(S^{\phi\phi} - \Omega^2_{\rm iso} S^{tt})}{\gamma_{\rm iso} + \Omega_{\rm iso}\ell_{\rm iso}}. \tag{58}$$

If $\hat{w} \geq \hat{P} > 0$, the energy-momentum tensor can represent two equal streams of particles moving on (accelerated) circular orbits in opposite directions at the same speed $\sqrt{\hat{P}/\hat{w}}$. This "speed of sound" was shown in [126] to be equal to the geometric mean of the local prograde and retrograde circular geodesic speeds $\hat{v}_\pm$ as measured by the observer $u^\mu_{\rm iso}$:

$$|\hat{v}_+\hat{v}_-| = \hat{P}/\hat{w}, \tag{59}$$

where

$$\hat{v}_\pm = \frac{(v_{\rm iso})_t + (v_{\rm iso})_\phi \Omega_\pm}{(u_{\rm iso})_t + (u_{\rm iso})_\phi \Omega_\pm} \tag{60}$$

and $\Omega_\pm$ are the angular velocities of the prograde and retrograde circular geodesics (40).

It was shown in [184] that in the absence of radial pressure the source disc can be composed of two counter-rotating streams of non-interacting particles on time-like circular *geodesics*, provided that the following conditions are satisfied: both geodesic frequencies (40) lie within the sub-luminal interval $(\Omega_{\rm min},\Omega_{\rm max})$, the tensor (46) yields $S_{ab}X^aX^b \geq 0$ for at least one two-vector X^a, and $\det(S_{ab}) > 0$. The last two assumptions ensure that the superposition

$$S^{ab} = w_+ u^a_+ u^b_+ + w_- u^a_- u^b_- \tag{61}$$

of the two dust components on circular geodesics $u^a_\pm = u^t_\pm(1, \Omega_\pm)$ comes out with positive proper surface densities of the streams

$$w_\pm = \pm \frac{S^{t\phi} - \Omega_\mp S^{tt}}{(u^t_\pm)^2(\Omega_+ - \Omega_-)}. \tag{62}$$

In a static case, the above formulae simplify to $\Omega_{\rm iso} = 0$,

$$\hat{w} = \frac{2w_\pm}{\sqrt{1-\hat{v}^2_\pm}} = -S^t_t = \frac{e^{\nu-\lambda}}{2\pi}(1-\rho\nu_{,\rho})\nu_{,z}, \qquad \hat{P} = S^\phi_\phi = \frac{e^{\nu-\lambda}}{2\pi}\rho\nu_{,\rho}\nu_{,z}, \tag{63}$$

$$\mathcal{M} = \int_b^\infty \nu_{,z}\rho {\rm d}\rho, \qquad \mathcal{J} = 0, \qquad M_{\rm H} = k = M, \qquad J_{\rm H} = 0, \tag{64}$$

$$\hat{v}_\pm = \rho e^{-2\nu}\Omega_\pm = \pm\sqrt{\frac{\rho\nu_{,\rho}}{1-\rho\nu_{,\rho}}} = \pm\sqrt{\hat{P}/\hat{w}}\,. \tag{65}$$

Hence, the disc particles circumscribe circular geodesics in the spacetime which they generate themselves.

Now further physical requirements can be raised, namely the weak energy condition ($\hat{w} \geq 0$, $\hat{P} \geq -\hat{w}$), the dominant energy condition ($\hat{w} \geq 0$, $|\hat{P}| \leq \hat{w}$), and the non-negativity of pressure ($\hat{P} \geq 0$). Their combination yields $\hat{w} \geq \hat{P} \geq 0$. The requirements are very simple in the static case: if $\nu_{,z}(z=0^+)>0$ (which is fulfilled for a realistic matter), then $\hat{P}\geq 0$ is equivalent to $\nu_{,\rho}\geq 0$ which is satisfied below the Lagrangian point of zero field (where $\nu_{,\rho}=0$ and $\hat{v}_\pm^2=0$). In regions with tension ($\hat{P}<0$), the particles are more attracted by the outer parts of the disc than by the central black hole, thus no circular geodesics exist (then hoop stresses have to be employed to interpret the disc matter); this typically happens in discs with too much matter on larger radii. The other condition $\hat{w}\geq\hat{P}$ is equivalent to $\hat{v}_\pm^2\leq 1$. Hence, the above constraints are in fact ensured by the obvious condition $0\leq\hat{v}_\pm^2\leq 1$ which can be written explicitly as $1\geq 1-\rho\nu_{,\rho}\geq 1/2$.

4 Static thin discs around non-rotating black holes

In a static spacetime rotation is excluded or it has to be compensated exactly as in the case of counter-rotating streams of matter. Within the vacuum outside of the sources, the static axisymmetric Einstein equations (32) and (33) yield the Laplace equation for ν and a simple quadrature for λ (calculated along a vacuum path going from the axis to the given point),

$$\vec{\nabla}^2\nu=0,\qquad \lambda=\int_{\text{axis}}^{\rho,z}\rho\,[(\nu_{,\rho}^2-\nu_{,z}^2)\,\mathrm{d}\rho+2\nu_{,\rho}\nu_{,z}\mathrm{d}z].\tag{66}$$

Thanks to the linearity of the Laplace equation, the solutions can simply be added to obtain fields of multiple sources. However, such a superposition may not be physically acceptable — supporting singularities ("struts") may occur in calculating λ or it may not be possible to interpret the resulting system of sources in a reasonable way (for instance, the matter works out with negative density or pressure or moving at a superluminal speed). Consequently, even the static axisymmetric case has only afforded a few realistic superpositions. Let us refer to two of them involving a Schwarzschild black hole: the superposition with an infinite annular thin disc, obtained by Lemos & Letelier [188] by inversion of the first counter-rotating finite disc of Morgan & Morgan [229], and the one with an inverted isochrone thin disc, obtained by Klein [164]. The Lemos-Letelier solution has turned out to allow for physically satisfactory situations. We briefly summarize its properties below.

4.1 Superposition with the inverted first Morgan-Morgan disc

The inverted first Morgan-Morgan counter-rotating disc has a surface density

$$w(\rho)=\frac{2\mathcal{M}b}{\pi^2\rho^4}\sqrt{\rho^2-b^2}\,,\tag{67}$$

$\mathcal{M}$ and b denoting the mass and the Weyl inner radius of the disc. The matter is thus concentrated near the rim (w is maximal at $\rho=2b/\sqrt{3}\doteq 1.115b$) which is supposed to be the case in real accretion discs. The resulting spacetime is given by a sum of potentials which in Weyl coordinates read

$$\nu_{\text{Schw}}=\frac{1}{2}\ln\frac{d_1+d_2-2M}{d_1+d_2+2M}\tag{68}$$

and

$$\nu_{\rm disc} = -\frac{\mathcal{M}}{\pi(\rho^2+z^2)^{3/2}}\left[\left(2\rho^2+2z^2-b^2\frac{\rho^2-2z^2}{\rho^2+z^2}\right)\operatorname{arccot}\sqrt{\frac{\sigma-(\rho^2-b^2+z^2)}{2(\rho^2+z^2)}}\right.$$

$$\left.-(3\sigma-3b^2+\rho^2+z^2)\sqrt{\frac{\sigma-(\rho^2-b^2+z^2)}{8(\rho^2+z^2)}}\right], \qquad (69)$$

where $d_{1,2} = \sqrt{\rho^2+(z\mp M)^2}$ and $\sigma = \sqrt{(\rho^2-b^2+z^2)^2+4b^2z^2}$; M is the Schwarzschild mass. On the axis ($\rho = 0$), $\nu_{\rm disc}$ reduces to

$$\nu_{\rm disc} = -\frac{2\mathcal{M}}{\pi|z|^3}\left[(z^2+b^2)\arctan\frac{|z|}{b} - b|z|\right], \qquad (70)$$

while in the equatorial plane ($z = 0$)

$$\nu_{\rm disc}(\rho > b) = -\frac{\mathcal{M}}{\rho}\left(1-\frac{b^2}{2\rho^2}\right), \qquad (71)$$

$$\nu_{\rm disc}(\rho < b) = -\frac{\mathcal{M}}{\pi\rho}\left[\left(2-\frac{b^2}{\rho^2}\right)\arcsin\frac{\rho}{b}+\sqrt{\frac{b^2}{\rho^2}-1}\right]. \qquad (72)$$

In [277] we illustrated the shape of the superposed field by drawing its field-lines, namely the integral curves of the four-acceleration $a^\mu = \nabla^\mu\nu$ of a static congruence. Then we plotted the flattening of the black hole with increasing relative mass of the disc or with decreasing inner disc radius. The explicit (static axisymmetric) form of the horizon Gaussian curvature (30),

$$\mathcal{C}_{\rm H} = \frac{1-4y\nu_{{\rm ext},y}-(1-y^2)(2\nu_{{\rm ext},y}^2-\nu_{{\rm ext},yy})}{4M^2e^{2\nu_{\rm ext}(y)-4\nu_{\rm ext}(1)}} \qquad (73)$$

(the exterior potential $\nu_{\rm ext}$ is represented by $\nu_{\rm disc}$ in our case), implies that starting from certain (though unrealistic) values, $\mathcal{C}_{\rm H}$ becomes negative at the axis.

The next paper [278] dealt with the motion of free test particles in the superposed fields, especially with the influence of the disc parameters on the positions of important equatorial circular geodesics (the photon, the marginally bound and the marginally stable one) which are critical for the disc accretion flows. The following questions arose: with increasing disc mass, how does the inner rim have to shift if the whole disc is to be stable permanently? Does it shift towards or away from the horizon starting from the Schwarzschild radius $r = 6M$ where the marginally stable orbit lies in the case of a *test* disc ($\mathcal{M} = 0$)? How do other properties of the resulting spacetime evolve? The answers were given in papers [274, 275, 322].

A sequence of superposed solutions was generated (figure 4 in [274]) by gradually increasing the relative disc mass $\mathcal{M}/M$, while demanding that (i) all the disc matter can be interpreted as two equal counter-rotating streams of particles on stable time-like equatorial circular geodesics and that (ii) the inner disc rim is fixed at the smallest possible radius. Keeping the rim at the marginally stable orbit of the complete spacetime turned out not to be the correct recipe for maintaining the *whole* disc stable while changing the parameters; namely an instability first appears *inside* the disc, not at the rim. With increasing disc mass, the rim may first shift towards the horizon:[u] the minimal possible Schwarzschild radius r of the rim goes from $6.00M$ to $3.60M$ with $\mathcal{M}$ growing from 0 to $0.42M$; the minimal possible circumferential radius R of the rim goes from $6.00M$ to $3.86M$ with $\mathcal{M}$ growing from 0 to $0.29M$; and the minimal possible proper distance from the horizon d_ρ of the rim goes from $7.19M$ to $4.26M$ with $\mathcal{M}$ growing from 0 to $0.33M$. Along the generated sequence of stable discs with minimal inner radii, the radius of the marginally bound orbit below the disc was also found to decrease, whereas the photon orbit went slowly up; the horizon inflated gradually towards the external source. For $\mathcal{M}$ greater than the values given above, the inner radii of the sequence increase again. When the disc mass reaches $1.92M$, the position of the inner rim starts to be constrained by the Lagrangian point of zero field (below which free circular motion is impossible within a certain radial range); this, however, goes up steeply to infinity at $\mathcal{M} = 2M$, i.e. the solutions of the given type with $\mathcal{M} \geq 2M$ cannot be given the freely counter-rotating interpretation. The Keplerian orbital speed was checked to be subluminal everywhere within the resulting sequence of discs, reaching roughly 0.7 at its maximum.

Only the beginning of the above sequence of stable physical discs can be realistic astrophysically because the mass $\mathcal{M}$ of the actual accretion discs is not supposed to exceed $(0.01 \div 0.1)M$. For $\mathcal{M}$ between 0 and $0.0722M$, the inner rim of our disc sequence follows the marginally stable circular geodesic of the corresponding complete spacetime (for $\mathcal{M} > 0.0722M$, the rim has to lie *above* the marginally stable orbit to avoid instability of the disc at larger radii). Even with such a minor increase of mass, the rim can go down considerably: from $r = 6M$ to $r = 4.6162M$, from $R = 6M$ to $R = 4.6644M$ and from $d_\rho = 7.1914M$ to $d_\rho = 5.4806M$.

4.2 *Oscillations and stability of self-gravitating discs*

The basic criterion for accretion-disc stability, $\ell_{,\rho} > 0$, only concerns perturbations within the disc plane. Actually, the frequency κ (measured with respect to infinity) of a small free horizontal oscillation about a circular equatorial geodesic in a stationary axisymmetric spacetime is given by[v]

$$\kappa^2 = \frac{e^{2\nu-2\lambda}}{(u^t)^3\ell}\, g_{\alpha\phi,\rho} u^\alpha g^{t\beta} u_{\beta,\rho} = \frac{e^{2\nu-2\lambda}\,\ell_{,\rho}}{(u^t)^4\rho^2 B^2}\left[\ell_{,\rho} - (u^t)^3\rho^2 B^2 \Omega_{,\rho}\right] \tag{74}$$

---

[u] Such a statement is of course coordinate-dependent. Therefore, we also used some physically more relevant measures rather then the Schwarzschild radius r only — the equatorial circumferential radius $R = \rho e^{-\nu(\rho,z=0)} = r e^{-\nu_{\rm disc}(r,\theta=\pi/2)}$ and the proper radial distance from the horizon (computed along the equatorial plane) $d_\rho = \int_0^\rho e^{(\lambda-\nu)(\rho,z=0)} {\rm d}\rho = \int_{2M}^r \frac{e^{(\lambda-\lambda_{\rm Schw}-\nu_{\rm disc})(r,\theta=\pi/2)}}{\sqrt{1-2M/r}} {\rm d}r$. Note that these transformations are non-trivial as they involve the functions $\nu_{\rm disc}$ and $\lambda_{\rm disc}$. Consequently, a given r corresponds, for different discs, to *different* R's and d_ρ's (and vice versa). However, the graphs using r and using physical radii turned out not to differ much (see the direct comparison of all the three coordinates in figure 7 of [274]).

[v] Note that we consider non-gravitating perturbations, i.e. the metric itself is not perturbed. It would be more sophisticated, and also much more difficult, to allow the perturbation to generate its own, self-consistent field. However, this would only have a second-order effect on oscillation frequencies.

([275] and references therein). These oscillations (called epicyclic) have been studied in the astrophysical literature since the 1980's, recently as the cause of low-frequency disc modes that might explain the quasi-periodic variability observed at some sources probably containing a black hole with an accretion disc (see section 2.1 and the review [157]; the history of frequency formulae is described in [275]). κ is zero at radial infinity; in the Newtonian case it increases monotonically towards the centre, whereas in relativity it has a maximum at a certain distance and then goes back to zero at the marginally stable orbit. Hence, in black-hole sources, the disc oscillations produced near the inner rim should remain trapped in a region below this maximum rather than propagate to the outer parts of the disc [158].

In [275], perturbations in vertical direction were found to be also important, in fact, to be even more dangerous than the horizontal ones for discs with a greater mass. Perturbations perpendicular to the disc plane must be treated carefully because they take place in the vicinity of the source which is singular (infinitely thin). Although the symmetry would imply $g_{\mu\nu,z} = 0$ in the equatorial plane, there appears a jump of the field in the normal direction as is usual in case of a mass (charge) layer. For the inverted first Morgan-Morgan disc, in particular, one finds that (at $\rho > b$)

$$\lim_{z\to 0^\pm} \nu_{\mathrm{disc},z} = \pm\frac{4\mathcal{M}b}{\pi\rho^4}\sqrt{\rho^2 - b^2}\,. \tag{75}$$

Even in "small" (linear) oscillations the particle spends all of the time *outside* of the equatorial plane where the field is given by (75) rather than vanishing. The equation of geodesic deviation then yields

$$\Omega_\perp^2 = {\Gamma^z}_{tt,z} + 2{\Gamma^z}_{t\phi,z}\Omega + {\Gamma^z}_{\phi\phi,z}\Omega^2$$
$$-4({\Gamma^z}_{tt} + {\Gamma^z}_{t\phi}\Omega)({\Gamma^t}_{tz} + {\Gamma^t}_{\phi z}\Omega) - 4({\Gamma^z}_{t\phi} + {\Gamma^z}_{\phi\phi}\Omega)({\Gamma^\phi}_{tz} + {\Gamma^\phi}_{\phi z}\Omega) \tag{76}$$

for the "perpendicular" frequency.

In the static case, the frequency formulae simplify to

$$\kappa^2 = \frac{e^{4\nu-2\lambda}}{1-\rho\nu_{,\rho}}(\nu_{,\rho\rho} + 4\rho\nu_{,\rho}^3 - 6\nu_{,\rho}^2 + 3\nu_{,\rho}/\rho), \tag{77}$$

$$\Omega_\perp^2 = \frac{e^{4\nu-2\lambda}}{1-\rho\nu_{,\rho}}\left[\nu_{,zz} - 4\nu_{,z}^2(1-2\rho\nu_{,\rho})\right]. \tag{78}$$

Without the external source, $\Omega_\perp^2$ reduces to the square of the Schwarzschild orbital frequency, $\Omega_\pm^2 = M(M + \sqrt{\rho^2 + M^2})^{-3}$, whereas κ^2 remains different due to the pericentre precession effect.

Requiring linear stability (also) with respect to vertical perturbations, the range of physically acceptable Lemos-Letelier superpositions within the $(\mathcal{M}, b)$-plane is narrowed further (figure 1 in [322]). The curve of the marginal vertical stability is more restrictive than that of the marginal horizontal stability for the discs with $\mathcal{M} > 0.2296M$, it even fully excludes the discs with $\mathcal{M} \geq 2M/7$. As the boundary of the delimited sector of physical discs (figure 2 in [322]), one obtains a sequence of stable discs whose inner rims lie right on, or very close to, the circular geodesics marginally stable with respect to both perturbations. We learned, in particular, that for the "realistic" part of this sequence, with relative mass below 0.07, the self-gravity makes the oscillations of the inner disc parts faster, and that the region of horizontal mode trapping gets somewhat smaller.

4.3 Redshift from observers in the disc

The redshift from a given point in the source is often mentioned as a measure of the field strength. In Weyl fields, the frequency shift between an observer moving on a circular orbit with an angular velocity Ω and an observer at rest at infinity reads

$$g \equiv \frac{f_\infty}{f_{\rm emit}} = e^\nu \sqrt{1 - e^{-4\nu}\rho^2\Omega^2}. \tag{79}$$

In our case, it is natural to compute the shift from two privileged observers within and below the disc, the static one ($\Omega = 0$) and the geodesic one ($|\Omega| = \frac{e^{2\nu}}{\rho}\sqrt{\frac{\rho\nu_{,\rho}}{1-\rho\nu_{,\rho}}}$; the observer co-rotates with the matter in terms of which the disc is interpreted). Figure 3 of [322] shows the curves obtained (for discs with different mass and inner radius) from the respective forms of equations (79), $g = e^\nu$ and $g = e^\nu\sqrt{\frac{1-2\rho\nu_{,\rho}}{1-\rho\nu_{,\rho}}}$, by varying the emitter's orbital radius.

4.4 Singularity at the rim of the first Morgan-Morgan disc

The class of Morgan-Morgan [229] static axisymmetric solutions describes the fields of a sequence of finite thin counter-rotating discs with densities

$$w^{(m)}(\rho \le b) = \frac{(2m+1)\mathcal{M}}{2\pi b^2}\left(1 - \frac{\rho^2}{b^2}\right)^{m-1/2} \qquad (m = 1, 2, \ldots). \tag{80}$$

The zeroth ($m = 0$) member being clearly singular at the rim, the main attention has been devoted to the first ($m = 1$) member. The latter has been discussed or referred to as a prototype of a simple and physically meaningful Weyl field. In addition to the papers mentioned in previous sections, the first Morgan-Morgan disc has appeared as the limiting case of a more general class of *stationary* disc solutions [14, 114], while its inverted counterpart has been considered as a possible seed for stationary black hole–disc superpositions [325] (see the next section).

The first Morgan-Morgan disc has however turned out to have a curvature singularity at the rim [271]. It is caused by too steep a decrease of density (80): the gradient

$$w^{(m)}_{,\rho} = -\frac{(4m^2-1)\mathcal{M}\rho}{2\pi b^4}\left(1 - \frac{\rho^2}{b^2}\right)^{m-3/2} \tag{81}$$

goes to infinity at $\rho \to b^-$ if $m = 0, 1$. The singularity is inherited by the annular discs, obtained by the inversion

$$\rho \to R = \frac{b^2\rho}{\rho^2 + z^2}, \qquad z \to Z = \frac{b^2 z}{\rho^2 + z^2}, \tag{82}$$

(see [322]). It was thus recommended [271] to consider "higher" ($m > 1$) members of the Morgan-Morgan counter-rotating family in superpositions. More accurately, even these discs have certain singularities at their rims — those calculated from higher derivatives of the metric. This can already be expected from a "Newtonian look" at densities (80): the m-th member of the family has infinite m-th derivative of density at the rim.

4.5 Inverting the m-th Morgan-Morgan counter-rotating disc

The potential for the whole family can be most easily written in oblate spheroidal coordinates x and $y(=\cos\theta)$, introduced by

$$\rho^2 = b^2(x^2+1)(1-y^2), \qquad z = bxy \qquad (0 \le x < \infty,\ 1 \ge y \ge -1): \tag{83}$$

$$\nu_{\rm MM}^{(m)} = -\frac{\mathcal{M}}{b}\sum_{n=0}^{m} C_{2n}^{(m)}\,\mathrm{i}Q_{2n}(\mathrm{i}x)P_{2n}(y), \tag{84}$$

where

$$C_{2n}^{(m)} = (-1)^n\frac{(4n+1)(2n)!(2m+1)!(m+n)!}{(n!)^2(m-n)!(2m+2n+1)!} \qquad (n \le m), \tag{85}$$

$P_{2n}(y)$ and $Q_{2n}(\mathrm{i}x)$ being the Legendre polynomials and Legendre functions of the second kind, respectively; note that it is often suitable to express

$$\mathrm{i}Q_{2n}(\mathrm{i}x) = P_{2n}(\mathrm{i}x)\,\mathrm{arccot}\,x - \mathrm{i}\sum_{k=1}^{2n}\frac{1}{k}P_{k-1}(\mathrm{i}x)P_{2n-k}(\mathrm{i}x). \tag{86}$$

Thanks to the invariance of the Laplace equation with respect to the Kelvin transformation, one can invert finite thin discs with respect to their outer rims, to obtain annular discs (which can in principle be superposed with some central body then). In terms of the oblate coordinates, the inversion (82) is

$$x \to X = \frac{y}{\sqrt{x^2+1-y^2}}, \qquad y \to Y = \frac{x}{\sqrt{x^2+1-y^2}}, \tag{87}$$

where X, Y are assumed to be related to R, Z in the same manner (83) as x, y are related to ρ, z. The inverted counterpart of the solution $\nu_{\rm MM}^{(m)}(x,y)$, satisfying the Laplace equation and having the correct asymptotics $\sim -\frac{\mathcal{M}}{bx}$, reads

$$\nu_{\rm disc}^{(m)} = \frac{2^{2m+1}(m!)^2}{\pi(2m+1)!}\,\frac{\nu_{\rm MM}^{(m)}(|X|,Y)}{\sqrt{x^2+1-y^2}}. \tag{88}$$

The inverted discs of the Morgan-Morgan class can now be superposed (for example) with a Schwarzschild black hole or they can be chosen as seeds in some procedure generating *stationary* generalization of such a superposition (as in the two-soliton inverse-scattering solution described in the following section).

5 Stationary thin discs around rotating black holes

In order to proceed to stationary superpositions, we employed the Belinskii-Zakharov inverse-scattering method [27, 28] — one of the generating techniques which, in principle, provide very general classes of solutions for the problem with two commuting symmetries. In [325], a real-two-soliton version of the method was applied to a general Weyl metric (31); the metric functions of this "seed" will be denoted by a hat, i.e. $\hat{\nu}$ and $\hat{\lambda}$, leaving ν and λ (and ω) for the *resulting* potentials. In Boyer-Lindquist–type coordinates (t,r,θ,ϕ), introduced by (25), the obtained metric appears as

$$\mathrm{d}s^2 = -\frac{\Delta}{\Sigma}\left(\mathcal{P}e^{\hat{\nu}}\mathrm{d}t + \mathcal{S}e^{-\hat{\nu}}\mathrm{d}\phi\right)^2 + \frac{\sin^2\theta}{\Sigma}\left(\mathcal{R}e^{\hat{\nu}}\mathrm{d}t - \mathcal{T}e^{-\hat{\nu}}\mathrm{d}\phi\right)^2$$

$$+\bar{C}^2 e^{2\hat{\lambda}-2\hat{\nu}}\left(\frac{\Sigma}{\Delta}\mathrm{d}r^2 + \Sigma\mathrm{d}\theta^2\right). \tag{89}$$

It involves two functions given by quadratures and it depends on (two) potentials describing the seed spacetime and on five independent constants; one can also choose two sign values in the derivation. If the seed is flat, one ends up with the Kerr-NUT solution. Restricting to reflectionally symmetric, asymptotically flat spacetimes containing a black hole and adjusting the coordinates in the most natural manner, the only freedom remains in the seed $(\hat{\nu}, \hat{\lambda})$ and in two constants (one puts $\bar{C}^2 = 1$, among others) which we denote M and a as in the Kerr solution. The functions present in (89) then read

$$\Delta = (r-M)^2 - k^2 = r^2 - 2Mr + a^2, \tag{90}$$

$$\Sigma = \left(r\mathcal{P} + \frac{a^2}{k}\sinh u\right)^2 + (\mathcal{R}y + a\sinh v)^2 \quad (= \mathcal{PT} + \mathcal{RS}), \tag{91}$$

$$\mathcal{P} = \cosh u - (M/k)\sinh u, \tag{92}$$

$$\mathcal{R} = a\cosh v, \tag{93}$$

$$\mathcal{S} = (1+y^2)\mathcal{R} + 2ay\sinh v - 2a\mathcal{P}e^{2\hat{\nu}}, \tag{94}$$

$$\mathcal{T} = (r^2 - a^2)\mathcal{P} + \frac{2a^2}{k}(r-M)\sinh u + 2a\mathcal{R}e^{2\hat{\nu}}, \tag{95}$$

where $k \equiv \sqrt{M^2 - a^2}$ and u and v are given by equations

$$v_{,\theta} = -\frac{\Delta u_{,r}}{k\sin\theta} = \frac{2\Delta}{\Delta + k^2\sin^2\theta}\left[(r-M)\hat{\nu}_{,r}\sin\theta + \hat{\nu}_{,\theta}y\right], \tag{96}$$

$$v_{,r} = \frac{u_{,\theta}}{k\sin\theta} = -\frac{2}{\Delta + k^2\sin^2\theta}\left[(r-M)\hat{\nu}_{,\theta}\sin\theta - \Delta\hat{\nu}_{,r}y\right], \tag{97}$$

with boundary conditions $u(y=\pm 1) = 0$, $v(y=\pm 1) = \pm 2\hat{\nu}(|y|=1)$.

A number of properties of the above metric were analyzed. The (outer) horizon is given by (the larger root of) the equation $\Delta = 0$. From equations of section 3.3, its area works out at

$$A = 4\pi \left(r_{\rm H}^2 e^{-2\hat{\nu}_{\rm H}(y=1)} + a^2 e^{2\hat{\nu}_{\rm H}(y=1)} \right) \tag{98}$$

and its surface gravitation and angular velocity with respect to infinity at

$$\kappa = \frac{4\pi k}{A} \quad \text{and} \quad \omega_{\rm H} = \frac{4\pi a}{A} \cosh 2\hat{\nu}_{\rm H}(y=1). \tag{99}$$

The Gaussian curvature of the horizon reduces to

$$\mathcal{C}_{\rm H}(y=1) = \frac{\left(r_{\rm H}^2 - a^2 e^{4\hat{\nu}_{\rm H}(1)}\right)\left[1 - 4\hat{\nu}_{{\rm H},y}(1)\right] - 2a^2}{\left(r_{\rm H}^2 + a^2 e^{4\hat{\nu}_{\rm H}(1)}\right)^2} e^{2\hat{\nu}_{\rm H}(1)} \tag{100}$$

at the axis. This can become negative for rapid rotation, namely for

$$a > 2M \frac{\sqrt{e^{4\hat{\nu}_{\rm H}(1)} + \frac{2}{1-4\hat{\nu}_{{\rm H},y}(1)}}}{1 + e^{4\hat{\nu}_{\rm H}(1)} + \frac{2}{1-4\hat{\nu}_{{\rm H},y}(1)}}. \tag{101}$$

The static limit is located where

$$\sqrt{\Delta}\mathcal{P} = \mathcal{R} \sin\theta. \tag{102}$$

Writing the metric in the Weyl-Lewis-Papapetrou form (1), the asymptotics of the gravitational potential ν and of the dragging angular velocity ω,

$$\nu = -\frac{M + \hat{M}}{r} + O(r^{-2}), \qquad \omega = \frac{2a(M + 2\hat{M})}{r^3} + O(r^{-4}), \tag{103}$$

tell us that $M + \hat{M}$ is the total mass and $(M + 2\hat{M})a$ is the total angular momentum of the solution; $\hat{M}$ denotes the mass of the seed spacetime for which we assumed the asymptotics $\hat{\nu} = -\hat{M}/r + O(r^{-2})$.

In the static case ($a = 0$), the result represents a superposition of a given seed with a Schwarzschild black hole, with u playing the role of an interaction term — $\nu = \hat{\nu} + \nu_{\rm Schw}$, $\lambda = \hat{\lambda} + \lambda_{\rm Schw} - u$. Thus one can take just the "external source" as a seed, the black hole is "supplied" by the soliton method itself. Hopefully, this applies to a general, stationary case as well. Actually, our main aim has been to learn whether the Belinskii-Zakharov technique could yield the field of a rotating black hole surrounded by an axisymmetric disc.

It seems that for small enough (though by far not negligible) values of a the solution (89)–(97) does not contain singularities on or above the horizon. However, it is quite extensive to calculate and analyse curvature invariants, even if using computer algebra. Coordinates were found in which metric is regular on the horizon (they have the meaning of the Kerr ingoing/outgoing coordinates there), but this is not enough to claim the absence of a singularity. Unfortunately, we found the horizon to be pinched rather than smooth in the equatorial plane. Indeed, the proper azimuthal circumference $2\pi(g_{\phi\phi})_{\rm H}$ of the horizon has a sharp minimum in the equatorial plane if $a > 0$ (and $\hat{M} > 0$). It is to be clarified whether this pathology is physical or the Boyer-Lindquist–type coordinates just do not cover the whole manifold. Most probably, however, there appears a supporting surface between the hole and the external disc.

5.1 *Stationary superposition cultivated from the inverted first Morgan-Morgan disc*

Some formulas were already specified to the case of a thin equatorial source in [325], in particular, those for the calculation of the energy-momentum tensor and of integral quantities characterizing the resulting spacetime. Let us mention below several properties of the metric (89)–(97) obtained for the inverted first Morgan-Morgan disc [273], i.e. for $\hat{\nu}$ given by (69) (with the original disc mass now denoted by $\hat{M}$). We saw in section 4.4 that this disc is singular at the rim. However, its other properties are quite satisfactory and the singularity is only caused by a sharp increase of density, so it is reasonable to study first this simple case and only later to embark on superpositions generated from "less-singular" seeds of inverted higher Morgan-Morgan discs (section 4.5).

On the horizon,

$$\hat{\nu} = \hat{\nu}_{\rm H} = -\frac{2\hat{M}}{\pi k^3|y|^3}\left[(k^2y^2+b^2)\arctan\frac{k|y|}{b} - bk|y|\right]. \tag{104}$$

This reduces to $\hat{\nu}_{\rm H}(y=0) = -\frac{4\hat{M}}{3\pi b}$ in the equatorial plane. The main horizon properties (98,99) only depend on the axial value $\hat{\nu}_{\rm H}(y=1)$. In [273] it is shown that with growing mass or decreasing inner radius of the disc the horizon inflates, its surface gravity weakens and its rotation slows down. The presence of the external source weakens the dependence of the horizon curvature on a/M. The existence of the sharp equatorial contraction, revealed by the behaviour of the proper horizon circumference, has been supported by plotting the isometric embedding (29).

The only undetermined quantities in the metric (89) are the functions u and v. They are fixed on the axis by the boundary conditions $u(y=\pm 1)=0$, $v(y=\pm 1) = \pm 2\hat{\nu}(|y|=1)$. On the horizon, one easily obtains $u_{\rm H}(y) = 2\hat{\nu}_{\rm H}(1) - 2\hat{\nu}_{\rm H}(y)$, $v_{\rm H}(y) = 2\hat{\nu}_{\rm H}(1)\cdot \operatorname{sign} y = \mp\text{const}$. In order that the spacetime be reflectionally symmetric, the seed potential $\hat{\nu}$ must be an even function of y. Then u is also even in y, whereas v is odd. Details of the courses of u and v are determined, according to the equations (96) and (97), by $\hat{\nu}$. In our case, $\hat{\nu}$ (69) is continuous and negative. It is maximal on the axis, having $\hat{\nu}_{,r} > 0$ and $\hat{\nu}_{,\theta} = 0$ there [while $\hat{\nu}_{,y}(y=\pm 1) \gtrless 0$]. In the equatorial plane $\hat{\nu}$ is minimal; $\hat{\nu}_{,r}$ is continuous there, being negative/positive below/above the radius $\rho = \sqrt{3/2}\,b$ ($\Leftrightarrow r = M + \sqrt{k^2+b^2}$) where $\hat{\nu}$ assumes its global minimum of $-(2/3)^{3/2}(\hat{M}/b)$. The latitudinal gradient $\hat{\nu}_{,\theta}$ is only zero below the inner radius of the disc, while it has a finite jump across the disc, given by (75):

$$\lim_{y\to 0\pm}\hat{\nu}_{,\theta} = -(r-M)\lim_{z\to 0\pm}\hat{\nu}_{,z} = \mp\frac{4\hat{M}b}{\pi\Delta^2}(r-M)\sqrt{\Delta - b^2}\,. \tag{105}$$

At infinity, $\hat{\nu} = -\hat{M}/r + O(r^{-2})$ which implies the asymptotics

$$u = \frac{\hat{M}k\sin^2\theta}{r^2} + O(r^{-3}), \qquad v = -\frac{2\hat{M}y}{r} + O(r^{-2}). \tag{106}$$

The values of u and v at general locations must be computed numerically, one can only proceed further analytically in the equatorial plane. There, equations (96) and (97) yield latitudinal derivatives

$$u_{,\theta} = -\frac{2k}{r-M}\hat{\nu}_{,\theta}, \qquad v_{,\theta} = \frac{2\Delta}{r-M}\hat{\nu}_{,r}. \tag{107}$$

Hence, u increases in latitudinal direction from zero on the axis to a maximum in the equatorial plane; this maximum is smooth/sharp ($u_{,\theta}$ is zero / has a finite jump there) below/above the inner disc radius. The global maximum of u lies where $\hat{\nu}_{,r} = 0$ in the equatorial plane (i.e. at $\rho = \sqrt{3/2}\,b$; $\hat{\nu}$ has a global minimum there). The function v is negative/positive in the upper/lower hemisphere, having a finite jump across the equatorial plane; this jump is constant — equal to $-4\hat{\nu}_{\rm H}(y=1)$ — from the horizon up to the inner disc edge, then it decreases with increasing radius. $v_{,\theta}$ is smooth across the equatorial plane, being positive/negative above/below $\rho = \sqrt{3/2}\,b$.

The radial derivatives of the equations (96) and (97) reduce to

$$u_{,r} = -\frac{2k}{r-M}\hat{\nu}_{,r}, \qquad v_{,r} = -\frac{2}{r-M}\hat{\nu}_{,\theta} \tag{108}$$

in the equatorial plane. Substituting from (71), (72) and (105), one comes to expressions which can be integrated to explicit formulae for $u(z=0, \rho<b)$, $u(z=0, \rho>b)$, $v(z=0^{\pm}, \rho<b) = \pm 2\hat{\nu}_{\rm H}(y=1)$, $v(z=0^{\pm}, \rho>b)$ (see [273]). While $u(z=0)$ increases from the horizon to a maximum at $\rho(=\sqrt{\Delta}) = \sqrt{3/2}\,b$ and then decreases to zero at radial infinity, $v(z=0^{\pm})$ increases/decreases to zero monotonically from $\rho = b$.

Using the equatorial expressions for u and v, one can, for example, graph the dependence of the equatorial radius of the static limit on $\hat{M}$, b and a. We have found that the ergosphere widens both with a/M and with $\hat{M}$. One can also plot the radial course of the light-like, dragging and geodesic angular velocities $\Omega_{\substack{\max\\ \min}}$, ω and $\Omega_{\pm}$.

5.2 Extreme limit

We assumed an under-extreme case ($k>0$) in [325] in deriving the general metric (89)–(97). However, it can be checked where the quantities go in the extreme *limit* of $k \to 0$, i.e. $a \to M$. One finds that they behave as in the extreme limit of the pure Kerr solution.

The event horizon approaches $r_{\rm H} = M$, its area, surface gravity and angular velocity reach the values

$$A = 8\pi M^2 \cosh 2\hat{\nu}_{\rm H}, \qquad \kappa = 0, \qquad \omega_{\rm H} = \frac{1}{2M}, \tag{109}$$

where $\hat{\nu}_{\rm H} = -\frac{4\hat{M}}{3\pi b}$ for the inverted first Morgan-Morgan disc. The Gaussian curvature of the horizon works out at

$$C_{\rm H} = -\frac{4\pi e^{2\hat{\nu}_{\rm H}} \cosh^3 2\hat{\nu}_{\rm H}}{A} \frac{(1+|y|)^3 e^{4\hat{\nu}_{\rm H}} + (1-|y|)^3 e^{-4\hat{\nu}_{\rm H}} - 6\sin^2\theta}{\left[1 + (|y|\cosh 2\hat{\nu}_{\rm H} + \sinh 2\hat{\nu}_{\rm H})^2\right]^3} \tag{110}$$

and reduces to

$$C_{\rm H}(y=1) = -\frac{1}{2M^2\cosh 2\hat{\nu}_{\rm H}} = -\frac{4\pi}{A}, \tag{111}$$

$$C_{\rm H}(y=0) = \frac{3-\cosh 4\hat{\nu}_{\rm H}}{M^2\cosh^4 2\hat{\nu}_{\rm H}}\, e^{2\hat{\nu}_{\rm H}} \tag{112}$$

at the axis and in the equatorial plane.

In the extreme limit, u is seen to vanish. The equatorial behaviour of u and v is important in discussion of the singularity location $\Sigma = 0$. In [325] it was shown that this singularity cannot lie above the horizon (at least) if a/M is sufficiently small. For our seed (69), however, $\Sigma(r > r_{\rm H})$ is positive in general even in the extreme limit. Actually, the first term of (91) then goes over to $[r - M(r-M)u/k]^2$. Using the expression for the equatorial maximum of $u(k \to 0)$, the root r of the above term must satisfy

$$\frac{r}{M}\left(\frac{\hat{M}}{M} - 3\frac{b^2}{M^2}\right) \geq \frac{\hat{M}}{M}. \tag{113}$$

This can only hold (for some positive r) if $\hat{M}/M > 3b^2/M^2$. Such a situation can be considered in principle, but it is, in fact, excluded astrophysically: $\hat{M}/M$ is typically claimed to be much less than 1, while a realistic value of $3b^2/M^2$ amounts to the order of ten.

The static-limit surface $r_0(\theta)$ must be found numerically as equation (102) remains rather cumbersome even in the extreme limit. Regarding the limiting forms of u and v, one can, however, infer the effect of the external disc on the equatorial value of r_0. It is given by

$$(r-M)(1-Mu/k) = M\cosh v, \qquad \text{thus} \qquad r - M > M \tag{114}$$

in the extreme limit, which means that $r_0(y=0)$ is greater than its Kerr value ($2M$).

It is quite difficult to express, in terms of the parameters $\hat{M}$; b, M and a, the contributions of the black hole and of the disc to the total mass and angular momentum of the spacetime. However, the crucial Komar formula for the angular momentum of the black hole $J_{\rm H}$, treated in equations (154)–(156) of [325], can at least be integrated in the extreme-limit case, to end up with just

$$J_{\rm H} = M^2, \quad M_{\rm H} = M, \quad \mathcal{M} = \hat{M}, \quad \mathcal{J} = 2M\hat{M}. \tag{115}$$

Another interesting feature of the $k \to 0$ limit of our superposition is the vanishing of the "external" gravitational field $\hat{\nu}_{,r}$, $\hat{\lambda}_{,r}$ on the horizon [272]. This effect of expulsion of the external (stationary axisymmetric) fields from rotating (and/or charged) black holes, analogous to the Meissner effect in (super)conductors, was observed in magnetic fields before (see the recent survey [39]).

6 Concluding remarks

In sections 4 and 5, the classes of static/stationary axisymmetric superpositions were presented for a black hole with an external thin equatorial disc. They give examples of the effects that can occur if the self-gravity of the disc is not completely negligible. The sequence of stable static discs with minimal possible radii, generated in section 4.1, does *not* (necessarily) describe the behaviour of a real accretion disc with its mass increasing (or its inner radius changing). Actually, static annular discs need not be of the given (inverted first Morgan-Morgan) type (and need not remain within this class when parameters are changed dynamically). Also, real accretion discs should be stationary rather than static. The results reviewed in the previous sections, however, allow for various generalizations. First, instead of the first Morgan-Morgan solution, singular at the rim, one can superpose an inversion of *any* member of the Morgan-Morgan family of static counter-rotating[w] finite thin discs (88). Second, a different annular disc can be employed, not belonging to the Morgan-Morgan family at all (e.g. the inverted isochrone disc of Klein [164]) — and possibly even a thick one (here we only mean the exterior, vacuum field). The stationary metric (89)–(97) applies to *any* vacuum static axisymmetric seed, too.

We will try to clarify the nature of the pathology which occurs in the equatorial plane of the metric (89)–(97). If it corresponds to a thin layer of mass rather than to a real singularity, it might only be a consequence of the fact that the Boyer-Lindquist coordinates cut off a certain part of the manifold. It is more probable, however, that this layer represents a supporting surface which bears witness that the system's stationarity is only artificial. What remains an open question in any case is the structure of the black-hole interior, altered by the presence of the external source. We hope to be able to answer it in future.

On a general level, one could conclude with the interview with Profs. W. B. Bonnor and R. Penrose which Jiří Bičák organized in 1968,[x] about mathematics, relativity and "brain drain" [30] (it was a pretty hot topic at the time when "brain strain" was to continue in Czechoslovakia, as Prof. Kuchař could confirm). In his answer to the question of in which direction to expect progress in relativity, Prof. Bonnor remarked: "It is also possible that quasars will indeed have something to do with gravitational collapse." As I noticed in the literature, at those days this was not yet a theme for a gentleman to speak about in a decent astronomical society. Although the "collapsed-matter" interpretation of the quasi-stellar radio sources [268] and the "collapsed-star" interpretation of X-ray binaries [324] were both introduced by "ApJ". (The very importance of accretion as an energy source was apparently first recognized by [323].)

Nowadays, ApJ offers "Viewing the shadow of the black hole at the Galactic center" within the next few years [108]. Nevertheless, will not a (*gravitational*) "*shine* of a black hole" be delivered sooner? Gravitational radiation may prove medicinal for rheumatism or elsewhere. Will travel agencies cite papers like [175,

[w] However artificial the counter-rotating interpretation may seem, counter-rotating stellar discs in galaxies have become an observational matter [304].

[x] It was exactly in 1968 that Prof. J. A. Wheeler coined the word "black hole", but he was only interviewed by J. B. several years later [31].

292] when offering flights intended for "gravitational basking"? In any case, the abundance of event horizons in the last years' NASA `ADS` and `astro-ph` libraries supports Prof. Bonnor's tip. Today it is in fact impossible to follow, at least in some detail, but a narrow sector of the black hole research. Not within relativity, but mainly not within astrophysics. In the bibliography below, we try to recommend those papers which should be interesting both for physicists and astrophysicists, mostly avoiding pure observational and pure mathematical references.

By far not all the aspects of ("classical" general relativistic) black holes were mentioned in the survey part of this paper. The very question of their origin was only touched in connection with the cosmic censorship, with the formation of heavy discs and with the decay of perturbations. Not a single word about thermodynamics and quantum effects, about cosmology and dark matter, about primordial black holes, about economical and ecological aspects, about Tunguska impact ...

Also, all the black holes considered in the present paper are spheroidal. Recently, *toroidal* holes have been investigated as another intriguing possibility, not only by theoreticians but also as a (rather non-orthodox) candidate for active galactic nuclei [254, 284]. Toroidal black holes can only exist provided that some of the Hawking & Ellis' assumptions are relaxed. Gannon [118] proved the possibility of a smooth toroidal horizon by dropping stationarity; the period for which a toroidal black hole can persist is, however, very limited: the hole in the torus must close up before a light ray can pass through [146]. Toroidal horizons were really observed as a temporary stage of a numerical gravitational collapse ([295] and references therein). The case of a toroidal horizon with a (ring) singularity outside (thus violating the predictability assumption) was obtained in [251]. Other possibility is to relax the asymptotic flatness and allow for a non-zero cosmological constant Λ. A metric involving (long-term stable) higher-genus horizons in an asymptotically anti–de Sitter spacetime is quite simple; it was already obtained as a result of a gravitational collapse and generalized to charged and rotating cases (see [168] and references therein). Astronomical observations indicate that $\Lambda > 0$, but it is not excluded that the stabilizing attractive effect similar to that of $\Lambda < 0$ could be provided by a host galaxy or by a massive accretion disc around the hole (!). (For many more hints, astrophysical connotations and further references, see [284].)

In connection with the Weber's putative detection of gravitational waves in 1971, R. Penrose speculated about a naked singularity in the Galactic centre [247], but at the end he stated: "This would remove the observational discrepancies, although it would be a radical explanation... In comparison black holes can now be regarded as being 'conventional'; indeed, for this reason they must be preferred a priori in any attempt at explanation." Our essay *has been* a priori 'conventional', preferring black holes, and of spheroidal topology. Well, Penrose then continued: "Yet nature does not always prefer conventional explanations, least of all in astronomy."

Acknowledgements

I thank my former Ph.D. students M. Žáček and T. Zellerin, the results reviewed in sections 4 and 5 were obtained in collaboration with them. We were encouraged by J. Bičák, V. Karas, P. Krtouš and T. Ledvinka. Support from the Czech grants GACR-202/99/0261, GACR-205/00/1685 and GAUK-141/2000 is acknowledged.

References

1. Abrahams A. M., Heiderich K. R., Shapiro S. L., Teukolsky S. A., 1992, *Vacuum initial data, singularities, and cosmic censorship*, Phys. Rev. D **46**, 2452
2. Abramo L. R., Finelli F., 2001, *Back Reaction of Gravitational Radiation on the Schwarzschild Black Hole*, Gen. Rel. Grav. **33**, 339
3. Abramowicz M. A., Kluźniak W., 2001, *A precise determination of black hole spin in GRO J1655-40*, Astron. Astrophys. **374**, L19
4. Abramowicz M. A., Bjornsson G., Pringle J. E. (eds), 1998, *Theory of Black Hole Accretion Disks* (Cambridge Univ. Press, Cambridge)
5. Abramowicz M. A., Chen X.-M., Granath M., Lasota J.-P., 1996, *Advection-dominated accretion flows around Kerr black holes*, Astrophys. J. **471**, 762
6. Abramowicz M. A., Curir A., Schwarzenberg-Czerny A., Wilson R. E., 1984, *Self-gravity and global structure of accretion discs*, Mon. Not. R. Astron. Soc. **208**, 279
7. Adams F. C., Graff D. S., Richstone D. O., 2001, *A theoretical model for the $M_{\rm bh}$-σ relation for supermassive black holes in galaxies*, Astrophys. J. **551**, L31
8. Agol E., Krolik J. H., 2000, *Magnetic stress at the marginally stable orbit: altered disk structure, radiation, and black hole spin evolution*, Astrophys. J. **528**, 161
9. Alcubierre M., Brügmann B., 2001, *Simple excision of a black hole in 3+1 numerical relativity*, Phys. Rev. D. **63**, 104006
10. Alekseev G. A., 2001, *New integral equation form of integrable reductions of Einstein equations*, Theor. Math. Phys. **129**, 1466
11. Alekseev G. A., Garcia A. A., 1996, *Schwarzschild black hole immersed in a homogeneous electromagnetic field*, Phys. Rev. D **53**, 1853
12. Aliev A. N., Gal'tsov D. V., 1989, *"Magnetized" black holes*, Sov. Phys. Usp. **32**, 75
13. Andalib S. W., Tohline J. E., Christodoulou D. M., 1997, *A survey of the principal modes of nonaxisymmetric instability in self-gravitating accretion disk models*, Astrophys. J. Suppl. **108**, 471
14. Ansorg M., 2001, *Differentially Rotating Disks of Dust: Arbitrary Rotation Law*, Gen. Rel. Grav. **33**, 309
15. Ansorg M., Kleinwächter A., Meinel R., Neugebauer G., 2002, *Dirichlet boundary value problems of the Ernst equation*, Phys. Rev. D **65**, 044006
16. Armitage P. J., Natarajan P., 1999, *Lense-Thirring precession of accretion disks around compact objects*, Astrophys. J. **525**, 909
17. Bailey I., Israel W., 1980, *Relativistic Dynamics of Extended Bodies and Polarized Media: An Eccentric Approach*, Annals Phys. **130**, 188
18. Baker J., et al., 2000, *Nonlinear and perturbative evolution of distorted black holes: Odd-parity modes*, Phys. Rev. D **62**, 127701
19. Bao G., Hadrava P., Wiita P. J., Xiong Y., 1997, *Polarization variability of*

active galactic nuclei and X-ray binaries, Astrophys. J. **487**, 142
20. Barack L., 1999, *Late time decay of scalar, electromagnetic, and gravitational perturbations outside rotating black holes*, Phys. Rev. D **61**, 024026
21. Barbour J., Pfister H. (eds), 1995, *Mach's principle — From Newton's Bucket to Quantum Gravity* (Birkhauser, Boston)
22. Bardeen J. M., 1973, *Rapidly Rotating Stars, Disks, and Black Holes*, in DeWitt C., DeWitt B. S. (eds), *Black Holes*, Les Houches 1972 (Gordon and Breach, New York), p. 241
23. Bardeen J. M., Petterson J. A., 1975, *The Lense-Thirring effect and accretion disks around Kerr black holes*, Astrophys. J. **195**, L65
24. Barrabès C., Israel W., 1991, *Thin shells in general relativity and cosmology: The lightlike limit*, Phys. Rev. D **43**, 1129
25. Becker P. A., Subramanian P., Kazanas D., 2001, *Relativistic outflows from advection-dominated accretion disks around black holes*, Astrophys. J. **552**, 209
26. Beig R., Schmidt B., 2000, *Time-Independent Gravitational Fields*, in Schmidt B. G. (ed), *Einstein's Field Equations and Their Physical Implications*, Lect. Notes Phys. 540 (Springer-Verlag, Berlin Heidelberg), p. 325
27. Belinskii V. A., Zakharov V. E., 1978, *Integration of Einstein's equations by an inverse scattering method and calculation of exact soliton solutions*, Sov. Phys. JETP **48**, 985
28. Belinskii V. A., Zakharov V. E., 1979, *Stationary gravitational solitons with axial symmetry*, Sov. Phys. JETP **50**, 1
29. Bethe H. A., Brown G. E., 1998, *Evolution of binary compact objects that merge*, Astrophys. J. **506**, 780
30. Bičák J., 1969, Interview *With Professors W. B. Bonnor and R. Penrose about Mathematics, Relativity and "Brain Drain"*, Czech. J. Phys. A **19**, 210 (in Czech)
31. Bičák J., 1978, *Art of Science — An Interview with Professor John Archibald Wheeler*, Czech. J. Phys. A **28**, 365 (in Czech)
32. Bičák J., 1979, *On the theories of the interacting perturbations of the Reissner-Nordström black hole*, Czech. J. Phys. B **29**, 945
33. Bičák J., 2000, *Selected Solutions of Einstein's Field Equations: Their Role in General Relativity and Astrophysics*, in Schmidt B. G. (ed), *Einstein's Field Equations and Their Physical Implications*, Lect. Notes Phys. 540 (Springer-Verlag, Berlin Heidelberg), p. 1
34. Bičák J., Dvořák L, 1977, *Stationary electromagnetic fields around black holes. I. General solutions and the fields of some special sources near a Schwarzschild black hole*, Czech. J. Phys. B **27**, 127
35. Bičák J., Dvořák L, 1976, *Stationary Electromagnetic Fields Around Black Holes. II. General Solutions and the Fields of Some Special Sources Near a Kerr Black Hole*, Gen. Rel. Grav. **7**, 959
36. Bičák J., Dvořák L, 1980, *Stationary electromagnetic fields around black holes. III. General solutions and the fields of current loops near the Reissner-Nordström black hole*, Phys. Rev. D **22**, 2933
37. Bičák J., Hoenselaers C., 1985, *Two equal Kerr-Newman sources in stationary*

equilibrium, Phys. Rev. D **31**, 2476
38. Bičák J., Ledvinka T., 1993, *Relativistic Disks as Sources of the Kerr Metric*, Phys. Rev. Lett. **71**, 1669
39. Bičák J., Ledvinka T., 2000, *Electromagnetic fields around black holes and Meissner effect*, Nuovo Cim. B **115**, 739
40. Blanchet L., Faye G., 2001, *Lorentzian regularization and the problem of point-like particles in general relativity*, J. Math. Phys. **42**, 4391
41. Blandford R. D., Znajek R. L., 1977, *Electromagnetic extraction of energy from Kerr black holes*, Mon. Not. R. Astron. Soc. **179**, 433
42. Bodo G., Curir A., 1992, *Models of self-gravitating accretion disks*, Astron. Astrophys. **253**, 318
43. Bogovalov S. V., 2001, *Acceleration and collimation of relativistic plasmas ejected by fast rotators*, Astron. Astrophys. **371**, 1155
44. Bogoyavlenskij O. I., 2000, *MHD model of astrophysical jets*, Phys. Lett. A **276**, 1
45. Bonnor W. B., 1992, *Physical Interpretation of Vacuum Solutions of Einstein's Equations. Part I. Time-independent solutions*, Gen. Rel. Grav. **24**, 551
46. Bonnor W. B., 2001, *The interactions between two classical spinning particles*, Class. Quantum Grav. **18**, 1381
47. Bonnor W. B., 2001, *The interactions of charged, spinning, magnetized masses*, Class. Quantum Grav. **18**, 2853 (Corrigendum: 2001, **18**, 5591)
48. Bonnor W. B., Griffiths J. B., MacCallum M. A. H., 1994, *Physical Interpretation of Vacuum Solutions of Einstein's Equations. Part II. Time-dependent solutions*, Gen. Rel. Grav. **26**, 687
49. Brady P. R., Moss I. G., Myers R. C., 1998, *Cosmic Censorship: As Strong As Ever*, Phys. Rev. Lett. **80**, 3432
50. Bretón N., Manko V. S., 1995, *A binary system of 'antisymmetric' Kerr-Newman masses*, Class. Quantum Grav. **12**, 1969
51. Bretón N., García A. A., Manko V. S., Denisova T. E., 1998, *Arbitrarily deformed Kerr-Newman black hole in an external gravitational field*, Phys. Rev. D **57**, 3382
52. Bretón N., Manko V. S., Sánchez J. A., 1998, *On the equilibrium of charged masses in general relativity: the electrostatic case*, Class. Quantum Grav. **15**, 3071
53. Brown G. E., Bethe H. A., Lee C.-H., 2001, *Selected Papers: Formation and Evolution of Black Holes in the Galaxy* (World Scientific, Singapore), in print
54. Brown G. E., et al., 2000, *A theory of gamma-ray bursts*, New Astron. **5**, 191
55. Butterworth E. M., Ipser J. R., 1976, *On the structure and stability of rapidly rotating fluid bodies in general relativity. I. The numerical method for computing structure and its application to uniformly rotating homogeneous bodies*, Astrophys. J. **204**, 200
56. Calvani M., Nobili L., 1979, *Dressing up a Kerr naked singularity*, Nuovo Cim. B **51**, 247
57. Campanelli M., Lousto C. O., 1999, *Second order invariant gravitational perturbations of a Kerr black hole*, Phys. Rev. D **59**, 124022
58. Campanelli M., Khanna G., Laguna P., Pullin J., Ryan M. P., 2001, *Perturba-*

tions of the Kerr spacetime in horizon-penetrating coordinates, Class. Quantum Grav. **18**, 1543

59. Campanelli M., Gómez R., Husa S., Winicour J., Zlochower Y., 2001, *Close limit from a null point of view: The advanced solution*, Phys. Rev. D **63**, 124013
60. Cartas-Fuentevilla R., 2000, *Adjoint operators and perturbation theory of black holes*, J. Math. Phys. **41**, 7521
61. Carter B., 1973, *Black Hole Equilibrium States*, in DeWitt C., DeWitt B. S. (eds), *Black Holes*, Les Houches 1972 (Gordon and Breach, New York), p. 57
62. Carter B., 1979, *The general theory of the mechanical, electromagnetic and thermodynamic properties of black holes*, in Hawking S. W., Israel W. (eds), *General Relativity, An Einstein centenary survey* (Cambridge Univ. Press, Cambridge), p. 294
63. Celotti A., Miller J. C., Sciama D. W., 1999, *Astrophysical evidence for the existence of black holes*, Class. Quantum Grav. **16**, A3
64. Chakrabarti S. K., 1988, *Spacetime with Self-Gravitating Thick Disc*, J. Astrophys. Astr. **9**, 49
65. Chakrabarti S. K., 1996, *Grand unification of solutions of accretion and winds around black holes and neutron stars*, Astrophys. J. **464**, 664
66. Chakrabarti S. K. (ed), 1999, *Observational Evidence for the Black Holes in the Universe* (Kluwer, Dordrecht)
67. Chandrasekhar S., 1978, *The Kerr metric and stationary axisymmetric gravitational fields*, Proc. R. Soc. Lond. A **358**, 405
68. Chandrasekhar S., 1979, *An introduction to the theory of the Kerr metric and its perturbations*, in Hawking S. W., Israel W. (eds), *General Relativity, An Einstein Centenary Survey* (Cambridge Univ. Press, Cambridge), p. 370
69. Chandrasekhar S., 1983, *The Mathematical Theory of Black Holes* (Oxford Univ. Press, New York)
70. Charlton N., Clarke C. J. S., 1990, *On the outcome of Kerr-like collapse*, Class. Quantum Grav. **7**, 743
71. Chaudhuri S., Das K. C., 1997, *Two-Soliton Solutions of Axially Symmetric Metrics*, Gen. Rel. Grav. **29**, 75
72. Chaudhuri S., Das K. C., 1997, *Axially symmetric metrics from Laplace's seed by inverse scattering method*, J. Math. Phys. **38**, 5792
73. Chinea F. J., 2000, *Self-gravitating stationary axisymmetric fluids with differential rotation*, Annalen Phys. **9**, SI-38
74. Chinea F. J., Gonzáles-Romero L. M. (eds), *Rotating Objects and Relativistic Physics*, Lect. Notes Phys. 423 (Springer-Verlag, Berlin Heidelberg)
75. Chruściel P. T., 1999, *Towards a classification of static electrovacuum spacetimes containing an asymptotically flat spacelike hypersurface with compact interior*, Class. Quantum Grav. **16**, 689
76. Chrzanowski P. L., 1975, *Vector potential and metric perturbations of a rotating black hole*, Phys. Rev. D **11**, 2042
77. Chrzanowski P. L., 1976, *Applications of metric perturbations of a rotating black hole: Distortion of the event horizon*, Phys. Rev. D **13**, 806
78. Ciufolini I., 2000, *The 1995-99 measurements of the Lense-Thirring effect using*

laser-ranged satellites, Class. Quantum Grav. **17**, 2369

79. Ciufolini I., Wheeler J. A., 1995, *Gravitation and Inertia* (Princeton Univ. Press, New Jersey)
80. Clarke C. J. S., 1993, *The Analysis of Space-Time Singularities*, Cambridge Lect. Notes Phys. 1 (Cambridge Univ. Press, Cambridge)
81. Clarke C. J. S., 1993, *A review of cosmic censorship*, Class. Quantum Grav. **10**, 1375
82. Clément G., 2000, *Generating rotating Einstein-Maxwell fields*, Annalen Phys. **9**, SI-42
83. Cosgrove C. M., 1980, *Relations between the group-theoretic and soliton-theoretic techniques for generating stationary axisymmetric gravitational solutions*, J. Math. Phys. **21**, 2417
84. Cosgrove C. M., 1981, *Bäcklund transformations in the Hauser-Ernst formalism for stationary axisymmetric spacetimes*, J. Math. Phys. **22**, 2624
85. Cosgrove C. M., 1982, *Relationship between the inverse scattering techniques of Belinskii-Zakharov and Hauser-Ernst in general relativity*, J. Math. Phys. **23**, 615
86. Cui W., Shrader C. R., Haswell C. A., Hynes R. I., 2000, *Discovery of high-frequency quasi-periodic oscillations in the black hole candidate XTE J1859+226*, Astrophys. J. **535**, L123
87. Cui W., Zhang S. N., Chen W., 1998, *Evidence for frame-dragging around spinning black holes in X-ray binaries*, Astrophys. J. **492**, L53
88. Dabrowski Y., Lasenby A. N., 2001, *Reflected iron line from a source above a Kerr black hole accretion disc*, Mon. Not. R. Astron. Soc. **321**, 605
89. Damour T., Soffel M., Xu C., 1991, *General-relativistic celestial mechanics. I. Method and definition of reference systems*, Phys. Rev. D **43**, 3273
90. Damour T., Soffel M., Xu C., 1992, *General-relativistic celestial mechanics. II. Translational equations of motion*, Phys. Rev. D **45**, 1017
91. de Andrade V. C., Blanchet L., Faye G., 2001, *Third post-Newtonian dynamics of compact binaries: Noetherian conserved quantities and equivalence between the harmonic-coordinate and ADM-Hamiltonian formalisms*, Class. Quantum Grav. **18**, 753
92. de Felice F., Yunqiang Y., 2001, *Turning a black hole into a naked singularity*, Class. Quantum Grav. **18**, 1235
93. de Felice F., Zanotti O., 2000, *Jet Dynamics in Black Hole Physics: Acceleration During Subparsec Collimation*, Gen. Rel. Grav. **32**, 1449
94. Demianski M., 1976, *Stationary Axially Symmetric Perturbations of a Rotating Black Hole*, Gen. Rel. Grav. **7**, 551
95. Dietz W., Hoenselaers C., 1982, *Stationary System of Two Masses Kept Apart by Their Gravitational Spin-Spin Interaction*, Phys. Rev. Lett. **48**, 778
96. Dietz W., Hoenselaers C., 1985, *Two-Mass Solutions of Einstein's Vacuum Equations: The Double Kerr Solution*, Annals Phys. **165**, 319
97. Di Matteo T., Psaltis D., 1999, *Quasi-periodic variability and the inner radii of thin accretion disks in Galactic black hole systems*, Astrophys. J. **526**, L101
98. Dixon W. G., 1979, *Extended bodies in general relativity; their description and motion*, in Ehlers J. (ed), *Isolated Gravitating Systems in General Relativity*,

Proc. LXVII-th Int. School of Phys. "E. Fermi" (North-Holland, Amsterdam), p. 156

99. Ehlers J., Rudolph E., 1977, *Dynamics of extended bodies in general relativity*, Gen. Rel. Grav. **8**, 197
100. Eikenberry S. S., Matthews K., Morgan E. H., Remillard R. A., Nelson R. W., 1998, *Evidence for a disk-jet interaction in the microquasar GRS 1915+105*, Astrophys. J. **494**, L61
101. Ellis G. F. R., Schmidt B. G., 1977, *Singular Space-Times*, Gen. Rel. Grav. **8**, 915
102. Elvis M., 2000, *A structure for quasars*, Astrophys. J. **545**, 63
103. Embacher F., 1984, *The Analog of Electric and Magnetic Fields in Stationary Gravitational Systems*, Found. Phys. **14**, 721
104. Ernst F. J., 1968, *New Formulation of the Axially Symmetric Gravitational Field Problem*, Phys. Rev. D **167**, 1175
105. Ernst F. J., 1968, *New Formulation of the Axially Symmetric Gravitational Field Problem. II*, Phys. Rev. D **168**, 1415
106. Esteban E. P., Ramos E., 1988, *Rotating black hole in an external electromagnetic field*, Phys. Rev. D **38**, 2963
107. Fabian A. C., Iwasawa K., Reynolds C. S., Young A. J., 2000, *Broad Iron Lines in Active Galactic Nuclei*, Publ. Astron. Soc. Pacf. **112**, 1145
108. Falcke H., Melia F., Agol E., 2000, *Viewing the shadow of the black hole at the Galactic center*, Astrophys. J. **528**, L13
109. Fendt C., Memola E., 2001, *Collimating, relativistic, magnetic jets from rotating disks. The axisymmetric field structure of relativistic jets and the example of the M 87 jet*, Astron. Astrophys. **365**, 631
110. Fernandes J. F. Q., Lun A. W. C., 1996, *Gauge invariant perturbations of black holes. I. Schwarzschild space-time*, J. Math. Phys. **37**, 836
111. Fernandes J. F. Q., Lun A. W. C., 1997, *Gauge invariant perturbations of black holes. I. Kerr space-time*, J. Math. Phys. **38**, 330
112. Fishbone L. G., Moncrief V., 1996, *Relativistic fluid disks in orbit around Kerr black holes*, Astrophys. J. **207**, 962
113. Fragile P. C., Mathews G. J., Wilson J. R., 2001, *Bardeen-Peterson effect and quasi-periodic oscillations in X-ray binaries*, Astrophys. J. **553**, 955
114. Frauendiener J., Klein C., 2001, *Exact relativistic treatment of stationary counterrotating dust disks: Physical properties*, Phys. Rev. D **63**, 084025
115. Frolov V. P., Novikov I. D., 1998, *Physics of Black Holes. Basic Concepts and New Developments* (Kluwer, Dordrecht)
116. Fukue J., Sakamoto C., 1992, *Vertical Structures of Self-Gravitating Gaseous Disks around a Central Object*, Publ. Astron. Soc. Jpn. **44**, 553
117. Fukue J., Yamamoto S., 1986, *Bipolar Jets from a Star-Torus System: Effect of Torus' Gravity*, Publ. Astron. Soc. Jpn. **38**, 895
118. Gannon D., 1976, *On the Topology of Spacelike Hypersurfaces, Singularities, and Black holes*, Gen. Rel. Grav. **7**, 219
119. Garcia M. R., et al., 2001, *New evidence for black hole event horizons from CHANDRA*, Astrophys. J. **553**, L47
120. Gebhardt K., et al., 2000, *Black hole mass estimates from reverberation map-*

ping and from spatially resolved kinematics, Astrophys. J. **543**, L5

121. Ghez A. M., Morris M., Becklin E. E., Tanner A., Kremenek T., 2000, *The accelerations of stars orbiting the Milky Way's central black hole*, Nature **407**, 349
122. Ghosh P., Abramowicz M. A., 1997, *Electromagnetic extraction of rotational energy from disc-fed black holes: the strength of the Bladford-Znajek process*, Mon. Not. R. Astron. Soc. **292**, 887
123. Gibbons G., 1980, *Non-existence of equilibrium configurations of charged black holes*, Proc. Roy. Soc. London A **372**, 535
124. Gierliński M., Maciołek-Niedźwiecki A., Ebisawa K., 2001, *Application of a relativistic accretion disc model to X-ray spectra of LMC X-1 and GRO J1655-40*, Mon. Not. R. Astron. Soc. **325**, 1253
125. Gleiser R. J., Nicasio C. O., Price R. H., Pullin J., 2000, *Gravitational radiation from Schwarzschild black holes: the second-order perturbation problem*, Phys. Rep. **325**, 41
126. González G. A., Letelier P. S., 2000, *Rotating relativistic thin disks*, Phys. Rev. D **62**, 064025
127. Gundlach C., 1998, *Nonspherical perturbations of critical collapse and cosmic censorship*, Phys. Rev. D **57**, R7075
128. Gutsunayev Ts. I., Manko V. S., 1985, *On the Gravitational Field of a Mass Possessing a Multipole Moment*, Gen. Rel. Grav. **17**, 1025
129. Gutsunaev Ts. I., Manko V. S., 1988, *On a Family of Solutions of the Einstein-Maxwell Equations*, Gen. Rel. Grav. **20**, 327
130. Hachisu I., Eriguchi Y., Nomoto K., 1986, *Fate of merging double white dwarfs*, Astrophys. J. **308**, 161
131. Harada T., Maeda H., 2001, *Convergence to a self-similar solution in general relativistic gravitational collapse*, Phys. Rev. D **63**, 084022
132. Hashimoto M., Eriguchi Y., Müller E., 1995, *Equilibrium structure of self-gravitating Keplerian disks*, Astron. Astrophys. **297**, 135
133. Hawking S. W., 1998, *Is Information Lost in Black Holes?*, in Wald R. M. (ed), *Black Holes and Relativistic Stars* (Univ. Chicago Press, Chicago), p. 221
134. Hod S., 2000, *Mode coupling in rotating gravitational collapse: Gravitational and electromagnetic perturbations*, Phys. Rev. D **61**, 064018
135. Hoenselaers C., Dietz W. (eds), 1984, *Solutions of Einstein's Equations: Techniques and Results*, Lect. Notes Phys. 205 (Springer-Verlag, Berlin Heidelberg)
136. Hubeny I., Hubeny V., 1997, *Non-LTE models and theoretical spectra of accretion disks in active galactic nuclei*, Astrophys. J. **484**, L37
137. Hubeny I., Hubeny V., 1997, *Non-LTE models and theoretical spectra of accretion disks in active galactic nuclei. II. Vertical structure of the disk*, Astrophys. J. **505**, 558
138. Hubeny I., Agol E., Blaes O., Krolik J. H., 2000, *Non-LTE models and theoretical spectra of accretion disks in active galactic nuclei. III. Integrated spectra for hydrogen-helium disks*, Astrophys. J. **533**, 710
139. Hubeny I., Blaes O., Krolik J. H., Agol E., 2001, *Non-LTE models and theoretical spectra of accretion disks in active galactic nuclei. IV. Effects of Compton scattering and metal opacities*, Astrophys. J. **559**, 680

140. Hubeny V. E., 1999, *Overcharging a black hole and cosmic censorship*, Phys. Rev. D **59**, 064013
141. Igumenshchev I. V., Abramowicz M. A., 2000, *Two-dimensional models of hydrodynamical accretion flows into black holes*, Astrophys. J. Suppl. **130**, 463
142. Igumenshchev I. V., Abramowicz M. A., Narayan R., 2000, *Numerical simulations of convective accretion flows in three dimensions*, Astrophys. J. **532**, L67
143. Islam J. N., 1985, *Rotating fields in general relativity* (Cambridge Univ. Press, Cambridge)
144. Israel W., 1998, *The Internal Structure of Black Holes*, in Wald R. M. (ed), *Black Holes and Relativistic Stars* (Univ. Chicago Press, Chicago), p. 137
145. Israelian G., Rebolo R., Basri G., Casares J., Martin E. L., 1999, *Evidence of a supernova origin for the black hole in the system GRO J1655-40*, Nature **401**, 142
146. Jacobson T., Venkataramani S., 1995, *Topology of event horizons and topological censorship*, Class. Quantum Grav. **12**, 1055
147. Jantzen R. T., Carini P., Bini D., 1992, *The Many Faces of Gravitoelectromagnetism*, Annals Phys. **215**, 1
148. Jezierski J., 1999, *Energy and Angular Momentum of the Weak Gravitational Waves on the Schwarzschild Background — Quasilocal Gauge-invariant Formulation*, Gen. Rel. Grav. **31**, 1855
149. Jhingan S., Magli G., 2000, *Black holes versus naked singularities formation in collapsing Einstein clusters*, Phys. Rev. D **61**, 124006
150. Junor W., Biretta J. A., Livio M., 1999, *Formation of the radio jet in M87 at 100 Schwarzschild radii from the central black hole*, Nature **401**, 891
151. Kaaret P., Ford E. C., Chen K., 1997, *Strong-field general relativity and quasi-periodic oscillations in X-ray binaries*, Astrophys. J. **480**, L27
152. Kaper L., van den Heuvel E. P. J., Woudt P. A. (eds), 2001, *Black Holes in Binaries and Galactic Nuclei*, Proc. ESO Workshop at Garching (Springer, Berlin Heidelberg)
153. Karas V., 1999, *Quasi-Periodic Features Due to Clumps Orbiting around a Black hole*, Publ. Astron. Soc. Jpn. **51**, 317
154. Karas V., Dovčiak M., 1997, *Region of Magnetic Dominance near a Rotating Black Hole*, Gen. Rel. Grav. **29**, 1011
155. Karas V., Czerny B., Abrassart A., Abramowicz M. A., 2000, *A cloud model of active galactic nuclei: the iron Kα line diagnostics*, Mon. Not. R. Astron. Soc. **318**, 547
156. Karas V., Lanza A., Vokrouhlický D., 1995, *Emission-line profiles from self-gravitating thin disks*, Astrophys. J. **440**, 108
157. Kato S., 2001, *Basic Properties of Thin-Disk Oscillations*, Publ. Astron. Soc. Jpn. **53**, 1
158. Kato S., Fukue J., 1980, *Trapped Radial Oscillations of Gaseous Disks around a Black Hole*, Publ. Astron. Soc. Jpn. **32**, 377
159. Kato S., Fukue J., Mineshige S., 1998, *Black-Hole Accretion Disks* (Kyoto Univ. Press, Kyoto)

160. Kegeles L. S., Cohen J. M., 1979, *Constructive procedure for perturbations of spacetimes*, Phys. Rev. D **19**, 1641
161. Khanna G., 2001, *Collision of spinning black holes in the close limit*, Phys. Rev. D **63**, 124007
162. Khanna R., Chakrabarti S. K., 1992, *Effects of a self-gravitating disc on test particle motion around a Kerr black hole*, Mon. Not. R. Astron. Soc. **259**, 1
163. King M., Pfister H., 2001, *Electromagnetic Thirring problems*, Phys. Rev. D **63**, 104004
164. Klein C., 1997, *Counter-rotating dust rings around a static black hole*, Class. Quantum Grav. **14**, 2267
165. Klein C., Richter O., 1998, *Physically realistic solutions to the Ernst equation on hyperelliptic Riemann surfaces*, Phys. Rev. D **58**, 124018
166. Klein M. J., Kox A. J., Schulmann R. (eds), 1993, *The Collected Papers of Albert Einstein*, Vol. 5 (Princeton Univ. Press, Princeton), p. 531 (in German; English transl.: 1995, p. 340)
167. Klein M. J., Kox A. J., Renn J., Schulmann R. (eds), 1995, *The Collected Papers of Albert Einstein*, Vol. 4 (Princeton Univ. Press, Princeton), p. 174 (in German; English transl.: 1996, p. 126)
168. Klemm D., Moretti V., Vanzo L., 1998, *Rotating topological black holes*, Phys. Rev. D **57**, 6127
169. Koide S., Meier D. L., Shibata K., Kudoh T., 2000, *General relativistic simulations of early jet formation in a rapidly rotating black hole magnetosphere*, Astrophys. J. **536**, 668
170. Kordas P., 1999, *Aspects of Solution-generating Techniques for Space-times with Two Commuting Killing Vectors*, Gen. Rel. Grav. **31**, 1941
171. Kormendy J., Gebhardt K., 2001, *Supermassive Black Holes in Galactic Nuclei*, in Wheeler J. C., Martel H. (eds), *Relativistic Astrophysics: 20th Texas Symposium*, (AIP, Conf. Proc. 586), p. 363
172. Korotkin D., 1997, *Some remarks on finite-gap solutions of the Ernst equation*, Phys. Lett. A **229**, 195
173. Kozłowski M., Wiita P. J., Paczyński B., 1979, *Self-gravitating accretion disk models with realistic equations of state and opacities*, Acta Astron. **29**, 157
174. Kramer D., Stephani H., Herlt E., MacCallum M. A. H., 1980, *Exact Solutions of Einstein's Field Equations* (Cambridge Univ. Press, Cambridge)
175. Krasnikov S., 2000, *Traversable wormhole*, Phys. Rev. D **62**, 084028
176. Krivan W., Price R. H., 1998, *Initial data for superposed rotating black holes*, Phys. Rev. D **58**, 104003
177. Krolik J. H., 1999, *Active Galactic Nuclei: From the Central Black Hole to the Galactic Environment* (Princeton Univ. Press, Princeton)
178. Krtouš P., 2001, *Infall of a charged shell in a Reissner-Nordström field*, in preparation
179. Kurpiewski A., Jaroszyński M., 2000, *Radiation Spectra of Advection Dominated Accretion Flows around Kerr Black Holes*, Acta Astron. **50**, 79
180. Kyriakopoulos E., *Petrov Type I, Stationary, Axisymmetric, Perfect Fluid Solution of Einstein's Equations*, Mod. Phys. Lett. A **14**, 7
181. Lanza A., 1992, *Self-gravitating thin disks around rapidly rotating black holes*,

Astrophys. J. **389**, 141

182. Laor A., 1999, *Astrophysical evidence for massive black holes*, Phys. Rep. **311**, 451
183. Leaver E. W., 1986, *Solutions to a generalized spheroidal wave equation: Teukolsky's equations in general relativity, and the two-center problem in molecular quantum mechanics*, J. Math. Phys. **27**, 1238
184. Ledvinka T., 1998, *Thin disks as sources of stationary axisymmetric electrovacuum spacetimes*, Ph.D. Thesis (Charles Univ., Prague)
185. Lee H. K., Lee C. H., van Putten M. H. P. M., 2001, *Electric charge and magnetic flux on rotating black holes in a force-free magnetosphere*, Mon. Not. R. Astron. Soc. **324**, 781
186. Lee H. K., Wijers R. A. M. J., Brown G. E., 2000, *The Blandford-Znajek process as a central engine for a gamma-ray burst*, Phys. Rep. **325**, 83
187. Lehner L., 2001, *Numerical relativity: a review*, Class. Quantum Grav. **18**, R25
188. Lemos J. P. S., Letelier P. S., 1994, *Exact general relativistic thin disks around black holes*, Phys. Rev. D **49**, 5135
189. Letelier P. S., 1989, *On soliton solutions to the vacuum Einstein equations obtained from a general seed solution*, Class. Quantum Grav. **6**, 875
190. Letelier P. S., Oliveira S. R., 1987, *Exact self-gravitating disks and rings: A solitonic approach*, J. Math. Phys. **28**, 165
191. Letelier P. S., Oliveira S. R., 1998, *Double Kerr-NUT spacetimes: spinning strings and spinning rods*, Phys. Lett. A **238**, 101
192. Levy H., 1968, *Classification of Stationary Axisymmetric Gravitational Fields*, Nuovo Cim. B **56**, 253
193. Lewin W. H. G., van Paradijs J., van den Heuvel E. P., 1995, *X-Ray Binaries*, Cambridge Astrophys. Ser. 26 (Cambridge Univ. Press, Cambridge)
194. Li L.-X., 2000, *Extracting energy from a black hole through the transition region*, Astrophys. J. **540**, L17
195. Linet B., 1977, *Stationary axisymmetric test fields on a Kerr metric*, Phys. Lett. A **60**, 395
196. Linet B., 1979, *Stationary axisymmetric electromagnetic fields in the Kerr metric*, J. Phys. A **12**, 839
197. Livio M., 1999, *Astrophysical jets: a phenomenological examination of acceleration and collimation*, Phys. Rep. **311**, 225
198. Livio M., Ogilvie G. I., Pringle J. E., 1999, *Extracting energy from black holes: the relative importance of the Blandford-Znajek mechanism*, Astrophys. J. **512**, 100
199. Lousto C. O., 2001, *Perturbative evolution of nonlinear initial data for binary black holes*, Phys. Rev. D **63**, 047504
200. Lu Y., Yu Q., 2001, *The relationship between X-ray variability and the central black hole mass*, Mon. Not. R. Astron. Soc. **324**, 653
201. Lynden-Bell D., Bičák J., Katz J., 1999, *On Accelerated Inertial Frames in Gravity and Electromagnetism*, Annals Phys. **271**, 1
202. Lynden-Bell D., Katz J., Bičák J., 1995, *Mach's principle from the relativistic constraint equations*, Mon. Not. R. Astron. Soc. **272**, 150

203. MacFadyen A. I., Woosley S. E., 1999, *Collapsars: gamma-ray bursts and explosions in "failed supernovae"*, Astrophys. J. **524**, 262
204. Machida M., Matsumoto R., Mineshige S., 2001, *Convection-dominated, Magnetized Accretion Flows into Black Holes*, Publ. Astron. Soc. Jpn. **53**, L1
205. Maciejewski W., Binney J., 2001, *Kinematics from spectroscopy with a wide slit: detecting black holes in galaxy centres*, Mon. Not. R. Astron. Soc. **323**, 831
206. Maison D., 2000, *Duality and Hidden Symmetries in Gravitational Theories*, in Schmidt B. G. (ed), *Einstein's Field Equations and Their Physical Implications*, Lect. Notes Phys. 540 (Springer-Verlag, Berlin Heidelberg), p. 273
207. Makishima K., et al., 2000, *The nature of ultraluminous compact X-ray sources in nearby spiral galaxies*, Astrophys. J. **535**, 632
208. Manko V. S., 1992, *The Exterior Gravitational Field of a Static and Stationary Mass with an Arbitrary Set of Multipole Moments*, Gen. Rel. Grav. **24**, 35
209. Manko V. S., Novikov I. D., 1992, *Generalizations of the Kerr and Kerr-Newman metrics possessing an arbitrary set of mass-multipole moments*, Class. Quantum Grav. **9**, 2477
210. Manko V. S., Ruiz E., 2001, *Exact solution of the double-Kerr equilibrium problem*, Class. Quantum Grav. **18**, L11
211. Manko V. S., Martín J., Ruiz E., 1994, *Metric of two arbitrary Kerr-Newman sources located on the symmetry axis*, J. Math. Phys. **35**, 6644
212. Manmoto T., 2000, *Advection-dominated accretion flow around a Kerr black hole*, Astrophys. J. **534**, 734
213. Mano S., Takasugi E., 1997, *Analytic Solutions of the Teukolsky Equation and Their Properties*, Prog. Theor. Phys. **97**, 213
214. Mano S., Suzuki H., Takasugi E., 1996, *Analytic Solutions of the Teukolsky Equation and Their Low Frequency Expansions*, Prog. Theor. Phys. **95**, 1079
215. Mashhoon B., 1985, *Gravitational Effects of Rotating Masses*, Found. Phys. **15**, 497
216. Masuda N., Nishida S., Eriguchi Y., 1998, *The runaway instability of self-gravitating tori with non-constant specific angular momentum around black holes*, Mon. Not. R. Astron. Soc. **297**, 1139
217. Meier D. L., 2001, *The association of jet production with geometrically thick accretion flows and black hole rotation*, Astrophys. J. **548**, L9
218. Meier D. L., Koide S., Uchida Y., 2001, *Magnetohydrodynamical Production of Relativistic Jets*, Science **291**, 84
219. Meinel R., 2000, *Gravitational field of rotating disks and black holes*, Annalen Phys. **9**, 335
220. Melia F., Bromley B. C., Liu S., Walker C. K., 2001, *Measuring the black hole spin in Sagittarius A^**, Astrophys. J. **554**, L37
221. Merloni A., Vietri M., Stella L., Bini D., 1999, *On gravitomagnetic precession around black holes*, Mon. Not. R. Astron. Soc. **304**, 155
222. Merritt D., Ferrarese L., 2001, *The $M_\bullet$-σ relation for supermassive black holes*, Astrophys. J. **547**, 140
223. Mészáros P., 2001, *Gamma-Ray Bursts: Accumulating Afterglow Implications, Progenitor Clues, and Prospects*, Science **291**, 79

224. Miller J. M., et al., 2001, *Relativistic iron emission and disk reflection in Galactic microquasar XTE J1748-288*, Astrophys. J. **546**, 1055
225. Miller J. M., et al., 2001, *High frequency quasi-periodic oscillations in the 2000 outburst of the Galactic microquasar XTE J1550-564*, Astrophys. J. **563**, 928
226. Mineshige S., Manmoto T., 1999, *Advection-dominated accretion flows: low-luminosity objects can contain supermassive black holes!*, Adv. Space Res. **23**, 1065
227. Mineshige S., Umemura M., 1996, *Self-similar, self-gravitating viscous disks*, Astrophys. J. **469**, L49
228. Mirabel I. F., Rodríguez L. F., 1999, *Sources of Relativistic Jets in the Galaxy*, Annu. Rev. Astron. Astrophys. **37**, 409
229. Morgan T., Morgan L., 1969, *The Gravitational Field of a Disk*, Phys. Rev. **183**, 1097 (Errata: 1969, **188**, 2544; 1970, D **1**, 3522)
230. Morgan E. H., Remillard R. A., Greiner J., 1997, *RXTE observations of QPOs in the black hole candidate GRS 1915+105*, Astrophys. J. **482**, 993
231. Nakamura T., Oohara K., Kojima Y., 1987, *General Relativistic Collapse to Black Holes and Gravitational Waves from Black Holes*, Prog. Theor. Phys. Suppl. **90**, 1
232. Nandi A., Manickam S. G., Rao A. R., Chakrabarti S. K., 2001, *On the source of quasi-periodic oscillations of the black hole candidate GRS 1915+105: some new observations and their interpretation*, Mon. Not. R. Astron. Soc. **324**, 267
233. Narayan R., Garcia M. R., McClintock J. E., 1997, *Advection-dominated accretion and black hole event horizons*, Astrophys. J. **478**, L79
234. Neugebauer G., 1996, *Gravitostatics and rotating bodies*, in Hall G. S., Pulham J. R. (eds), *General Relativity*, 46th Scottish Universities Summer School in Physics (IOP, London), p. 61
235. Neugebauer G., 2000, *Rotating bodies as boundary value problems*, Annalen Phys. **9**, 342
236. Neugebauer G., Meinel R., 1995, *General Relativistic Gravitational Field of a Rigidly Rotating Disk of Dust: Solution in Terms of Ultraelliptic Functions*, Phys. Rev. Lett. **75**, 3046
237. Nishida S., Eriguchi Y., 1994, *A general relativistic toroid around a black hole*, Astrophys. J. **427**, 429
238. Nordtvedt K., 1988, *Existence of Gravitomagnetic Interaction*, Int. J. Theor. Phys. **27**, 1395
239. Nordtvedt K., 1999, *30 years of lunar laser ranging and the gravitational interaction*, Class. Quantum Grav. **16**, A101
240. Ori A., 1997, *Perturbative Approach to the Inner Structure of a Rotating Black Hole*, Gen. Rel. Grav. **29**, 881
241. Ori A., 1999, *Evolution of linear gravitational and electromagnetic perturbations inside a Kerr black hole*, Phys. Rev. D **61**, 024001
242. Ostriker J., 1964, *The equilibrium of self-gravitating rings*, Astrophys. J. **140**, 1067
243. Paczyński B., 1998, *Are gamma-ray bursts in star-forming regions?*, Astrophys. J. **494**, L45

244. Papaloizou J. C., Savonije G. J., 1991, *Instabilities in self-gravitating gaseous discs*, Mon. Not. R. Astron. Soc. **248**, 353
245. Peng H., 1983, *On Calculation of Magnetic-Type Gravitation and Experiments*, Gen. Rel. Grav. **15**, 725
246. Penrose R., 1969, *Gravitational collapse: The role of general relativity*, Riv. Nuovo Cim. **1**, 252
247. Penrose R., 1972, *Black holes*, Sci. Am. **226**, 38
248. Penrose R., 1998, *The Question of Cosmic Censorship*, in Wald R. M. (ed), *Black Holes and Relativistic Stars* (Univ. Chicago Press, Chicago), p. 103
249. Perjés Z., 2000, *Rotating perfect fluid models in general relativity*, Annalen Phys. **9**, 368
250. Perry G. P., Cooperstock F. I., 1997, *Electrostatic equilibrium of two spherical charged masses in general relativity*, Class. Quantum Grav. **14**, 1329
251. Peters P. C., 1979, *Toroidal black holes?*, J. Math. Phys. **1979**, 1481
252. Pfister H., Schedel Ch., 1987, *Can gravitational attraction be compensated for by gravimagnetic effects?*, Class. Quantum Grav. **4**, 141
253. Poisson E., 1997, *Gravitational radiation from infall into a black hole: Regularization of the Teukolsky equation*, Phys. Rev. D **55**, 639
254. Pompilio F., Harun-or-Rashid S. M., Roos M., 2000, *A toroidal black hole for the AGN phenomenon*, Astron. Astrophys. **362**, 865
255. Press W. H., Teukolsky S. A., 1973, *Perturbations of a rotating black hole. II. Dynamical stability of the Kerr metric*, Astrophys. J. **185**, 649
256. Psaltis D., Belloni T., van der Klis M., 1999, *Correlation in quasi-periodic oscillation and noise frequencies among neutron star and black hole X-ray binaries*, Astrophys. J. **520**, 262
257. Psaltis D., et al., 1999, *On the magnetospheric beat-frequency and Lense-Thirring interpretations of the horizontal-branch oscillation in the Z sources*, Astrophys. J. **520**, 763
258. Punsly B., Coroniti F. V., 1990, *Relativistic winds from pulsar and black hole magnetospheres*, Astrophys. J. **350**, 518
259. Punsly B., Coroniti F. V., 1990, *Ergosphere-driven winds*, Astrophys. J. **354**, 583
260. Quevedo H., 1992, *Generating solutions of the Einstein-Maxwell equations with prescribed physical properties*, Phys. Rev. D **45**, 1174
261. Quevedo H., Mashhoon B., 1991, *Generalizations of Kerr spacetime*, Phys. Rev. D **43**, 3902
262. Rees M. J., 1998, *Astrophysical Evidence for Black holes*, in Wald R. M. (ed), *Black Holes and Relativistic Stars* (Univ. Chicago Press, Chicago), p. 79
263. Rees M. J., Begelman M. C., Blandford R. D., Phinney E. S., 1982, *Ion-supported tori and the origin of radio jets*, Nature **295**, 17
264. Riffert H., 2000, *The dynamics of a viscous gas ring around a Kerr black hole*, Astrophys. J. **529**, 119
265. Rudnicki W., 1997, *Testing Cosmic Censorship in Kerr-Like Collapse Situations*, Mod. Phys. Lett. A **12**, 2237
266. Ruszkowski M., 2000, *X-ray iron line variability for the model of an orbiting flare above a black hole accretion disc*, Mon. Not. R. Astron. Soc. **315**, 1

267. Ruszkowski M., Fabian A. C., Ross R. R., Iwasawa K., 2000, *The evidence against Comptonization as a mechanism to explain the Fe Kα line profile in MCG–6-30-15*, Mon. Not. R. Astron. Soc. **317**, L11
268. Salpeter E. E., 1964, *Accretion of interstellar matter by massive objects*, Astrophys. J. **140**, 796
269. Sarbach O., Heusler M., Brodbeck O., 2001, *Self-adjoint wave equations for dynamical perturbations of self-gravitating fields*, Phys. Rev. D **63**, 104015
270. Seidel E., 1999, *The Synergy between Numerical and Perturbative Approaches to Black Holes*, in Iyer B. R., Bhawal B. (eds), *Black Holes, Gravitational Radiation, and the Universe: Essays in Honor of C.V. Vishveshwara* (Kluwer, Boston), p. 367
271. Semerák O., 2001, *Curvature singularity around first Morgan-Morgan disc*, Class. Quantum Grav. **18**, 3589
272. Semerák O., 2002, *Expulsion of external fields from extreme horizons: example of an external gravitational field*, Czech. J. Phys. **52**, 11
273. Semerák O., 2001, *Axisymmetric thin disc around a stationary black hole*, to be submitted
274. Semerák O., Žáček M., 2000, *Static discs around a Schwarzschild black hole: I*, Class. Quantum Grav. **17**, 1613
275. Semerák O., Žáček M., 2000, *Oscillations of Static Disks around Schwarzschild Black Holes: Effect of Self-Gravitation*, Publ. Astron. Soc. Jpn. **52**, 1067
276. Semerák O., Karas V., de Felice F., 1999, *Parameters of Black Holes in Sources with Periodic Variability*, Publ. Astron. Soc. Jpn. **51**, 571
277. Semerák O., Zellerin T., Žáček M., 1999, *The structure of superposed Weyl fields*, Mon. Not. R. Astron. Soc. **308**, 691 (Erratum: 2001, **322**, 207)
278. Semerák O., Žáček M., Zellerin T., 1999, *Test-particle motion in superposed Weyl fields*, Mon. Not. R. Astron. Soc. **308**, 705
279. Shlosman I., Begelman M. C., 1987, *Self-gravitating accretion disks in active galactic nuclei*, Nature **329**, 810
280. Shore S. N., White R. L., 1982, *Self-gravitating accretion disk models for active galactic nuclei: self-consistent α-models for the broad emission-line region*, Astrophys. J. **256**, 390
281. Sibgatullin N. R., 1991, *Oscillations and Waves in Strong Gravitational and Electromagnetic Fields* (Springer-Verlag, Berlin Heidelberg)
282. Sincell M. W., 2001, *Microquasars raise megaquestions*, Science **291**, 66
283. Smarr L., 1973, *Surface Geometry of Charged Rotating Black holes*, Phys. Rev. D **7**, 289
284. Spivey R. J., 2000, *Quasars: a supermassive rotating toroidal black hole interpretation*, Mon. Not. R. Astron. Soc. **316**, 856 (Erratum: 2001, **321**, 783)
285. Stella L., Vietri M., Morsink S. M., 1999, *Correlations in the quasi-periodic oscillation frequencies of low-mass X-ray binaries and the relativistic precession model*, Astrophys. J. **524**, L63
286. Strohmayer T. E., 2001, *Discovery of a 450 Hz quasi-periodic oscillation from the microquasar GRO J1655-40 with the Rossi X-ray timing explorer*, Astrophys. J. **552**, L49

287. Strohmayer T. E., 2001, *Discovery of a second high-frequency quasi-periodic oscillation from the microquasar GRS 1915+105*, Astrophys. J. **554**, L169
288. Stuchlík Z., 1981, *Evolution of Kerr naked singularities*, Bull. Astron. Inst. Czechosl. **32**, 68
289. Šubr L., Karas V., 1999, *An orbiter crossing an accretion disc*, Astron. Astrophys. **352**, 452
290. Tanabe Y., 1979, *Generation of the Lewis solutions from the Weyl solutions*, J. Math. Phys. **20**, 1486
291. Taylor J. H. (Jr.), 1994, *Nobel Lecture: Binary pulsars and relativistic gravity*, Rev. Mod. Phys. **66**, 711
292. Teo E., 1998, *Rotating traversable wormholes*, Phys. Rev. D **58**, 024014
293. Tertychniy S. I., 1990, *Some representations of the general solution of the linearised vacuum Einstein equations for axially symmetric stationary metrics*, Class. Quantum Grav. **7**, 1345
294. Teukolsky S. A., 1973, *Perturbations of a rotating black hole. I. Fundamental equations for gravitational, electromagnetic, and neutrino-field perturbations*, Astrophys. J. **185**, 635
295. Teukolsky S. A., 1998, *Black Hole Collisions, Toroidal Black Holes, and Numerical Relativity*, in Wald R. M. (ed), *Black Holes and Relativistic Stars* (Univ. Chicago Press, Chicago), p. 123
296. Thorne K. S., 1998, *Probing Black Holes and Relativistic Stars with Gravitational Waves*, in Wald R. M. (ed), *Black Holes and Relativistic Stars* (Univ. Chicago Press, Chicago), p. 41
297. Thorne K. S., Hartle J. B., 1985, *Laws of motion and precession for black holes and other bodies*, Phys. Rev. D **31**, 1815
298. Thorne K. S., Price R. H., Macdonald D. A. (eds), 1986, *Black Holes: The Membrane Paradigm* (Yale Univ. Press, New Haven)
299. Titarchuk L., Lapidus I., Muslimov A., 1998, *Mechanisms for high-frequency quasi-periodic oscillations in neutron star and black hole binaries*, Astrophys. J. **499**, 315
300. Torres del Castillo G. F., 1990, *Generation of Perturbations by means of Decoupled Equations and their Adjoints*, Gen. Rel. Grav. **22**, 1085
301. Torres del Castillo G. F., 1996, *Gravitational and electromagnetic perturbations of the Schwarzschild solution*, J. Math. Phys. **37**, 5684
302. Torres del Castillo G. F., Cartas-Fuentevilla R., 1996, *Scattering by a Reissner-Nordström black hole*, Phys. Rev. D **54**, 4886
303. Torres del Castillo G. F., Solís-Rodríguez H. G., 1999, *Self-duality and gravitational perturbations*, J. Math. Phys. **40**, 4099
304. Tremaine S., Yu Q., 2000, *Resonant capture, counter-rotating discs, and polar rings*, Mon. Not. R. Astron. Soc. **319**, 1
305. Usui F., Nishida S., Eriguchi Y., 1998, *Emission-line profiles from self-gravitating toroids around black holes*, Mon. Not. R. Astron. Soc. **301**, 721
306. van der Klis M., 2000, *Millisecond Oscillations in X-ray Binaries*, Annu. Rev. Astron. Astrophys. **38**, 717
307. van Putten M. H. P. M., 2001, *Gamma-ray bursts: LIGO/VIRGO sources of*

gravitational radiation, Phys. Rep. **345**, 1

308. Vein P. R., 1985, *2 related families of determinantal solutions of the stationary axially-symmetric vacuum Einstein equations*, Class. Quantum Grav. **2**, 899
309. Wagh S. M., Dadhich N., 1989, *The Energetics of Black Holes in Electromagnetic Fields by the Penrose Process*, Phys. Rep. **183**, 137
310. Wagoner R. V., 1999, *Relativistic diskoseismology*, Phys. Rep. **311**, 259
311. Wagoner R. V., Silbergleit A. S., Ortega-Rodríguez M., 2001, *'Stable' quasi-periodic oscillations and black hole properties from diskoseismology*, Astrophys. J. **559**, L25
312. Wald R. M., 1984, *General Relativity* (Univ. of Chicago Press, Chicago)
313. Wald R. M., 1999, *Gravitational Collapse and Cosmic Censorship*, in Iyer B. R., Bhawal B. (eds), *Black Holes, Gravitational Radiation, and the Universe: Essays in Honor of C. V. Vishveshwara* (Kluwer, Boston), p. 69
314. Wang D.-X., 2000, *Relations between Black Hole Spin and Angular Velocity of Accreting Particles near the Horizon*, Gen. Rel. Grav. **32**, 553
315. Wheeler J. C., Martel H. (eds), 2001, *Relativistic Astrophysics: 20th Texas Symposium*, (AIP, Conf. Proc. 586)
316. Wiita P. J., 1982, *Physical properties of thick supercritical accretion disks*, Astrophys. J. **256**, 666
317. Wiita P. J., 1985, *Active galactic nuclei. I. Observations and fundamental interpretations*, Phys. Rep. **123**, 117
318. Will C. M., 1974, *Perturbation of a slowly rotating black hole by a stationary axisymmetric ring of matter. I. Equilibrium configurations*, Astrophys. J. **191**, 521
319. Will C. M., 1975, *Perturbation of a slowly rotating black hole by a stationary axisymmetric ring of matter. II. Penrose processes, circular orbits, and differential mass formulae*, Astrophys. J. **196**, 41
320. Woodward J. W., Sankaran S., Tohline J. E., 1992, *Tidal disruption of a star by a massive disk (The axisymmetric Roche problem)*, Astrophys. J. **394**, 248
321. Yorke H. W., Bodenheimer P., Laughlin G., 1993, *The formation of protostellar disks. I. 1 $M_\odot$*, Astrophys. J. **411**, 274
322. Žáček M., Semerák O., 2002, *Static discs around a Schwarzschild black hole: II*, Czech. J. Phys. **52**, 19
323. Zel'dovich Ya. B., 1964, *Fate of a star and generation of gravitational energy in accretion*, Sov. Phys. – Doklady **9**, 195
324. Zeldovich Ya. B., Guseynov O. H., 1966, *Collapsed stars in binaries*, Astrophys. J. **144**, 840
325. Zellerin T., Semerák O., 2000, *Two-soliton stationary axisymmetric sprouts from Weyl seeds*, Class. Quantum Grav. **17**, 5103
326. Zhang S. N., Cui W., Chen W., 1997, *Black hole spin in X-ray binaries: observational consequences*, Astrophys. J. **482**, L155

OPTICAL GEOMETRY, INERTIAL FORCES, AND EMBEDDING DIAGRAMS

STANISLAV HLEDÍK

Faculty of Philosophy and Science, Silesian University at Opava, Bezručovo nám. 13, 746 01 Opava, Czech Republic

E-mail: Stanislav.Hledik@fpf.slu.cz

Optical reference geometry and related inertial forces are discussed for the static, spherically or axially symmetric black-hole, and naked-singularity spacetimes as well as Kerr–Newman spacetimes. Properties of the optical reference geometry are illustrated using embedding diagrams of the equatorial plane of the spacetimes considered.

1 Introduction

Analysis of geodetical motion and studies of embedding diagrams rank among the most fundamental techniques that enable us to understand phenomena present in extremely strong gravitational fields of black holes and other compact objects. The combined gravitational and electromagnetic field of charged black holes can be characterized by properties of test particle motion that obeys the Lorentz equation.

Naked singularities, i.e., spacetime singularities not hidden behind an event horizon, represent solutions of the Einstein equations for the most exotic compact gravitational objects that could be conceivable for explaining quasars and active galactic nuclei. The Penrose conjecture of cosmic censorship [1] suggests that no naked singularity can evolve from regular initial data, however, the proof and even precise formulation of the conjecture still stands as one of the biggest challenges in general relativity. Therefore, it seems important to consider their possible astrophysical consequences. Of particular interest are those effects that could distinguish a naked singularity from black holes.

Generally, the motion of both massive test particles and photons in black-hole or naked-singularity backgrounds is governed by Carter equations [2]. Properties of the motion were discussed extensively in the past decades (see, e.g., Refs [2–8]). From the astrophysical point of view, the most important classes of test particle motion are circular motion [9] and pure radial motion [10,11]. In the case of black-hole (naked-singularity) asymptotically (anti-)de Sitter spacetimes with a nonzero cosmological constant ($\Lambda \neq 0$), the geodetical motion was discussed in a series of papers [12–18]. In the case of interior Schwarzschild–anti-de Sitter spacetimes [19], the geodetical motion was discussed in [20].

The influence of charge, spin, magnetic field or cosmological constant on the structure of black-hole or naked-singularity spacetimes can suitably be demonstrated by embedding diagrams of 2-dimensional sections of the ordinary geometry (t = const hypersurfaces) into 3-dimensional Euclidean geometry and their relation to the well known 'Schwarzschild throat' [2]. The embedding diagrams were constructed for Schwarzschild–de Sitter [15], interior Schwarzschild–de Sitter [20], Reissner–Nordström [21], Reissner–Nordström–de Sitter [18], Ernst [22], and Kerr [4] spacetimes.

Properties of the motion of test particles and photons can be properly understood in the framework of optical reference geometry allowing introduction of the concept of gravitational and inertial forces in the framework of general relativity in a natural way [23].

Of course, in accord with the spirit of general relativity, alternative approaches to the concept of inertial forces are possible (see, e.g., Refs [24, 25]); however, here we shall follow the approach of Abramowicz and his coworkers [26–34], providing a description of relativistic dynamics in accord with Newtonian intuition.

The optical geometry results from an appropriate conformal $(3+1)$ splitting, reflecting certain hidden properties of the spacetimes under consideration through its geodesic structure. The geodesics of the optical geometry related to static spacetimes coincide with trajectories of light, thus being 'optically straight' [33, 35]. Moreover, the geodesics are 'dynamically straight,' because test particles moving along them are held by a velocity-independent force [27]; they are also 'inertially straight,' because gyroscopes carried along them do not precess along the direction of motion [28].

The optical reference geometry clearly illustrates the role of a nonzero cosmological constant in both vacuum black-hole (naked-singularity) spacetimes [36, 37] and nonvacuum interior Schwarzschild–de Sitter spacetimes [20].

Some fundamental properties of the optical geometry can be appropriately demonstrated by embedding diagrams of its representative sections [21–23]. Because we are familiar with the Euclidean space, 2-dimensional sections of the optical space are usually embedded into the 3-dimensional Euclidean space. (Of course, embeddings into other suitably chosen spaces can also provide interesting information, however, we shall focus our attention on the most straightforward Euclidean case.) In spherically symmetric static spacetimes, the central planes are the most convenient for embedding—with no loss of generality one can consider the equatorial plane, choosing the coordinate system such that $\theta = \pi/2$. In Kerr–Newman backgrounds, the most representative section is the equatorial plane, which is the symmetry plane. This plane is also of great astrophysical importance, particularly in connection with the theory of accretion disks [4].

In spherically symmetric spacetimes (Schwarzschild [33], Schwarzschild–de Sitter [15], Reissner–Nordström [21], and Reissner–Nordström–de Sitter [18]), an interesting coincidence appears: the turning points of the central-plane embedding diagrams of the optical space and the photon circular orbits are located exactly at the radii where the centrifugal force, related to the optical space, vanishes and reverses sign. The same conclusion holds for the Ernst spacetime [22].

However, in the rotating black-hole and naked-singularity backgrounds, the centrifugal force does not vanish at the radii of photon circular orbits [38]. Of course, the same statement is true if these rotating backgrounds carry a nonzero electric charge. One is, therefore, interested in studying how the inertial forces, defined in the framework of the optical geometry, and the photon circular orbits are related to equatorial-plane embedding diagrams of the optical geometry of rotating, charged backgrounds [39].

The plan of the paper is the following. In Sec. 2, the optical reference geometry is introduced and the related inertial forces are defined in the general case as well as for static, spherically symmetric and stationary, axially symmetric Kerr–Newman spacetimes. The notion of the inertial forces is applied to a special class of circular orbits related to the axial Killing vector field of the spacetimes. In Sec. 3, embedding diagrams of the optical geometry are defined and their properties are discussed. The embedding diagrams are constructed for some relevant spacetimes and their ties to the properties of centrifugal force are clarified. In Sec. 4, concluding remarks are presented. Throughout the paper, geometrical units ($c = G = 1$) are used.

2 Optical Geometry and Inertial Forces

The notions of the optical reference geometry and the related inertial forces are convenient for spacetimes with symmetries, particularly for stationary (static) and axisymmetric (spherically symmetric) ones. However, they can be introduced for a general spacetime lacking any symmetry [34].

2.1 General Case

Assuming a hypersurface globally orthogonal to a timelike unit vector field n^κ and a scalar field Φ satisfying the conditions

$$n_{[\kappa}\nabla_\lambda n_{\mu]} = 0, \quad n^\kappa n_\kappa = -1, \quad \dot{n}_\lambda \equiv n^\kappa \nabla_\kappa n_\lambda = \nabla_\lambda \Phi, \tag{1}$$

the 4-velocity u^κ of a test particle of rest mass m can be uniquely decomposed as

$$u^\kappa = \gamma(n^\kappa + v\tau^\kappa). \tag{2}$$

Here, τ^κ is a unit vector orthogonal to n^κ, v is the speed, and $\gamma = (1 - v^2)^{-1/2}$.

Introducing, based on the approach of Abramowicz, Nurowski and Wex [34], a projected 3-space orthogonal to n^κ with the positive definite metric giving the so-called ordinary projected geometry,

$$h_{\kappa\lambda} = g_{\kappa\lambda} + n_\kappa n_\lambda, \tag{3}$$

and the optical geometry $\tilde{h}_{\kappa\lambda}$ by conformal rescaling

$$\tilde{h}_{\kappa\lambda} = \mathrm{e}^{-2\Phi}(g_{\kappa\lambda} + n_\kappa n_\lambda), \tag{4}$$

the projection of the 4-acceleration $a^\perp_\kappa = h^\lambda_\kappa u^\mu \nabla_\mu u_\lambda$ can be uniquely decomposed into terms proportional to the zeroth, first, and second powers of v, respectively, and the velocity change

$$\dot{v} = (\mathrm{e}^\Phi \gamma v)_{,\mu}\, u^\mu. \tag{5}$$

Thus, we arrive at a covariant definition of inertial forces analogous to the Newtonian physics [34, 40]

$$ma^\perp_\kappa = G_\kappa(v^0) + C_\kappa(v^1) + Z_\kappa(v^2) + E_\kappa(\dot{v}), \tag{6}$$

where the first term

$$G_\kappa = -m\nabla_\kappa\Phi = -m\Phi_{,\kappa} \tag{7}$$

corresponds to the gravitational force, the second term

$$C_\kappa = -m\gamma^2 v n^\lambda(\nabla_\lambda\tau_\kappa - \nabla_\kappa\tau_\lambda) \tag{8}$$

corresponds to the Coriolis–Lense–Thirring force, the third term

$$Z_\kappa = -m(\gamma v)^2 \tilde{\tau}^\lambda \tilde{\nabla}_\lambda \tilde{\tau}_\kappa \tag{9}$$

corresponds to the centrifugal force, and the last term

$$E_\kappa = -m\dot{v}\tilde{\tau}_\kappa \tag{10}$$

corresponds to the Euler force. Here, $\tilde{\tau}^\kappa$ is the unit vector along τ^κ in the optical geometry and $\tilde{\nabla}_\kappa$ is the covariant derivative with respect to the optical geometry.

2.2 Spherically Symmetric Static Spacetimes

In the simple case of static spacetimes with a field of timelike Killing vectors $\vec{n} \equiv \xi_{(t)} = \partial/\partial t$, we are dealing with space components that we will denote by Latin indices in the following. The metric coefficients of the optical reference geometry are given by formula

$$\tilde{h}_{ik} = \mathrm{e}^{-2\Phi} h_{ik}, \tag{11}$$

where h_{ik} are metric components of the 3-space ordinary geometry and

$$\mathrm{e}^{2\Phi} = -g_{\mu\nu}\xi^\mu_{(t)}\xi^\nu_{(t)} = -g_{tt}. \tag{12}$$

In the optical geometry, we can define a 3-momentum of a test particle [34, 36]

$$\tilde{p}^i = \mathrm{e}^{2\Phi} p^i, \tag{13}$$

and a 3-force acting on the particle

$$\tilde{f}_i = \mathrm{e}^{2\Phi} f_i, \tag{14}$$

where p^i, f_i are space components of the 4-momentum p^μ and the 4-force f_μ. Then the equation of motion instead of the full spacetime form

$$m f_\mu = p^\nu \nabla_\nu p_\mu \tag{15}$$

takes the following form in the optical geometry

$$m\tilde{f}_i = \tilde{p}^k \tilde{\nabla}_k \tilde{p}_i + \tfrac{1}{2} m^2 \tilde{\nabla}_i \Phi, \tag{16}$$

where $\tilde{\nabla}_k$ represents the covariant derivative with respect to the optical geometry. We can see directly that photon trajectories ($m = 0$) are geodesics of the optical geometry. The first term on the right-hand side of (16) corresponds to the centrifugal force

$$\tilde{Z}_i = \tilde{p}^k \tilde{\nabla}_k \tilde{p}_i = \tilde{p}^k \partial_k \tilde{p}_i - \tfrac{1}{2}\tilde{p}^k \tilde{p}^l \partial_i \tilde{g}_{kl}; \tag{17}$$

the second term corresponds to the gravitational force

$$-m\tilde{G}_i = \tfrac{1}{2} m^2 \tilde{\nabla}_i \Phi = \tfrac{1}{2} m^2 \partial_i \Phi. \tag{18}$$

The circular motion can be considered both in the equatorial plane and outside of it ($r = \text{const}$, $\theta \neq \pi/2 = \text{const}$). All such orbits are related to the field of the axial Killing vector $\xi_{(\phi)} = \partial/\partial\phi$. In the corresponding optical geometry, circular orbits can be defined using the Killing vector field

$$\tilde{\xi}_{(\phi)i} = \mathrm{e}^{-2\Phi}\xi_{(\phi)i}. \tag{19}$$

Radius of the circular orbit, as measured in the optical geometry, is given by the relation

$$\tilde{r} = \left(\tilde{\xi}^i_{(\phi)}\tilde{\xi}_{(\phi)i}\right)^{1/2} = r\sin\theta\,\mathrm{e}^{\Phi}, \tag{20}$$

and the unit vector tangent to the orbit is

$$\tilde{\tau}_i = \tilde{r}^{-1}\tilde{\xi}_{(\phi)i}. \tag{21}$$

The geodetical curvature $\tilde{R}$ of a circle is defined by the relation

$$\tilde{\tau}^k\tilde{\nabla}_k\tilde{\tau}_i = -\tilde{R}^{-1}\tilde{\lambda}_i, \tag{22}$$

where the unit vector $\tilde{\lambda}_i$ is the first normal to the circle. The geodesic can be expressed in the form

$$\tilde{R}^{-2} = \tilde{r}^{-2}\tilde{h}^{ik}\left(\tilde{\nabla}_i\tilde{r}\right)\left(\tilde{\nabla}_k\tilde{r}\right). \tag{23}$$

Introducing velocity v of the particle relative to the optical geometry in the standard Newtonian way

$$\tilde{p}_k = mv\tilde{\tau}_k, \tag{24}$$

the magnitude of the centrifugal force is given by the relation

$$Z = \frac{mv^2}{\tilde{R}}, \tag{25}$$

that is formally identical to the classical Newtonian relation.

As an example, let us consider circular orbits in the Reissner–Nordström–(anti-)de Sitter spacetimes with the line element

$$\begin{aligned}\mathrm{d}s^2 = &-\left(1-\frac{2M}{r}+\frac{e^2}{r^2}-\frac{\Lambda}{3}r^2\right)\mathrm{d}t^2 \\ &+\left(1-\frac{2M}{r}+\frac{Q^2}{r^2}-\frac{\Lambda}{3}r^2\right)^{-1}\mathrm{d}r^2 + r^2\left(\mathrm{d}\theta^2+\sin^2\theta\,\mathrm{d}\phi^2\right),\end{aligned} \tag{26}$$

where (t, r, θ, ϕ) are the standard Schwarzschild coordinates. Putting $M = 1$ for simplicity, we arrive at the expressions [18]

$$\tilde{R}\left(\theta=\frac{\pi}{2}\right) = r\left(1-\frac{3}{r}+\frac{2e^2}{r^2}\right)^{-1}, \tag{27}$$

$$\tilde{R}\left(\theta\neq\frac{\pi}{2}\right) = r\left[\left(1-\frac{3}{r}+\frac{2e^2}{r^2}\right)^2+\left(1-\frac{2}{r}+\frac{e^2}{r^2}-yr^2\right)\cot^2\theta\right]^{-1/2}. \tag{28}$$

The magnitude of the centrifugal force is then given by the relations

$$Z\left(\theta=\frac{\pi}{2}\right) = \frac{mv^2}{r}\left(1-\frac{3}{r}+\frac{2e^2}{r^2}\right), \tag{29}$$

$$Z\left(\theta\neq\frac{\pi}{2}\right) = \frac{mv^2}{r}\left[\left(1-\frac{3}{r}+\frac{2e^2}{r^2}\right)^2+\left(1-\frac{2}{r}+\frac{e^2}{r^2}-yr^2\right)\cot^2\theta\right]^{1/2}. \tag{30}$$

Notice that the geodetical curvature of circular orbits in the equatorial plane is independent of the cosmological parameter y while for circular orbits outside the equatorial plane, y enters into the expression for $\tilde{R}$ and the same statement holds for the magnitude of the centrifugal force.

It is important that in the equatorial plane, $\tilde{R} \to \infty$ at the radii corresponding to photon circular orbits. Since $\tilde{R} \to \infty$ there, these circular orbits are geodesics of the optical geometry and, as we can directly see from Eq. (29), the centrifugal force vanishes (and changes its sign) at the radii corresponding to photon circular geodesics.

2.3 Kerr–Newman Spacetimes

Denoting M the mass, a the specific angular momentum, and e the electric charge, the line element of the Kerr–Newman spacetime, expressed in terms of standard Boyer–Lindquist coordinates, reads

$$\mathrm{d}s^2 = -\left(1 - \frac{2Mr - e^2}{\Sigma}\right)\mathrm{d}t^2 - \frac{2a(2Mr - e^2)}{\Sigma}\sin^2\theta\,\mathrm{d}t\mathrm{d}\phi + \frac{A\sin^2\theta}{\Sigma}\mathrm{d}\phi^2 + \frac{\Sigma}{\Delta}\mathrm{d}r^2 + \Sigma\,\mathrm{d}\theta^2\,, \tag{31}$$

where

$$\Delta = r^2 - 2Mr + a^2 + e^2\,, \tag{32}$$

$$\Sigma = r^2 + a^2\cos^2\theta\,, \tag{33}$$

$$A = (r^2 + a^2)^2 - \Delta a^2 \sin^2\theta\,. \tag{34}$$

For simplicity, we put $M = 1$ in the following (equivalently, we use mass units of M). If $a^2 + e^2 \le 1$, the metric (31) represents black-hole spacetimes. The loci of their horizons, r_- (the inner one) and r_+ (the outer one), determined by real roots of $\Delta(r; a, e) = 0$, can be equivalently given by the relation

$$a^2 = a_{\mathrm{h}}^2(r; e) = r(2 - r) - e^2\,. \tag{35}$$

The Kerr–Newman spacetimes, being stationary and axially symmetric, admit two commuting Killing vector fields; the vector field η^κ is (at least asymptotically) timelike, having open trajectories, the vector field ξ^κ is spacelike, having closed trajectories. Now, the vector field n^κ relevant for construction of the ordinary projected geometry and the optical reference geometry can be expressed using these Killing vector fields [26] and corresponds to the family of locally nonrotating frames (LNRF) or zero angular momentum observers (ZAMO) introduced by Bardeen [4]. Namely,

$$\begin{gathered} n^\kappa = \mathrm{e}^{\Phi}(\eta^\kappa + \Omega_{\mathrm{LNRF}}\xi^\kappa)\,, \quad \Omega_{\mathrm{LNRF}} = -\xi^\lambda\eta_\lambda/\eta^\mu\eta_\mu\,, \\ \Phi = -\tfrac{1}{2}\ln\left(-\eta^\lambda\eta_\lambda - 2\Omega_{\mathrm{LNRF}}\xi^\lambda\eta_\lambda - \Omega_{\mathrm{LNRF}}^2\xi^\lambda\xi_\lambda\right). \end{gathered} \tag{36}$$

The LNRF vector field n^κ can be used to define the inertial forces introduced above. Assuming a circular motion with angular velocity $\Omega = \mathrm{d}\phi/\mathrm{d}t$ as measured by the stationary observers at infinity, the 4-velocity is given by

$$u^\kappa = A(\eta^\kappa + \Omega\xi^\kappa)\,, \quad A = \left(-\eta^\lambda\eta_\lambda - 2\Omega\xi^\lambda\eta_\lambda - \Omega^2\xi^\lambda\xi_\lambda\right)^{-1/2}; \tag{37}$$

τ^κ is directed along the rotational Killing vector ξ^κ now. The gravitational (7), Coriolis–Lense–Thirring (8), and centrifugal (9) forces can be written down as [39]

$$G_\kappa = -m\Phi_{,\kappa} = -m\frac{1}{2}\partial_\kappa \left[\ln\left(\frac{g_{t\phi}^2 - g_{tt}g_{\phi\phi}}{g_{\phi\phi}}\right)\right], \tag{38}$$

$$\begin{aligned} C_\kappa &= mA^2\frac{\Omega - \Omega_{\rm LNRF}}{\xi^\lambda\xi_\lambda}\left[(\xi^\lambda\xi_\lambda)(\eta^\mu\xi_\mu)_{,\kappa} - (\eta^\mu\xi_\mu)(\xi^\lambda\xi_\lambda)_{,\kappa}\right] \\ &= mA^2(\Omega - \Omega_{\rm LNRF})\sqrt{g_{\phi\phi}}\left[\partial_\kappa\left(g_{t\phi}g_{\phi\phi}^{-1/2}\right) + \Omega_{\rm LNRF}\,\partial_\kappa\sqrt{g_{\phi\phi}}\right], \end{aligned} \tag{39}$$

and

$$\begin{aligned} Z_\kappa &= \frac{1}{2}mA^2\frac{(\Omega - \Omega_{\rm LNRF})^2}{\iota^\lambda\iota_\lambda}\left[(\iota^\lambda\iota_\lambda)(\xi^\mu\xi_\mu)_{,\kappa} - (\xi^\mu\xi_\mu)(\iota^\lambda\iota_\lambda)_{,\kappa}\right] \\ &= -\frac{1}{2}mA^2(\Omega - \Omega_{\rm LNRF})^2 g_{\phi\phi}\,\partial_\kappa\left[\ln\left(\frac{g_{\phi\phi}^2}{g_{t\phi}^2 - g_{tt}g_{\phi\phi}}\right)\right], \end{aligned} \tag{40}$$

respectively; we also denote $\iota^\kappa = {\rm e}^{-\Phi}n^\kappa$. The Euler force will appear for $\Omega \neq$ const only, being determined by $\dot{\Omega} = u^\lambda\nabla_\lambda\Omega$ (see Ref. [26]). The electromagnetic forces acting in charged spacetimes were defined in the framework of the optical geometry in [40, 41]. However, here we shall concentrate on the inertial forces acting on the motion in the equatorial plane ($\theta = \pi/2$) and their relation to the embedding diagram of the equatorial plane of the optical geometry.

Clearly, by definition, the gravitational force is independent of the orbiting particle velocity. On the other hand, both the Coriolis–Lense–Thirring and the centrifugal force vanish (at any r) if $\Omega = \Omega_{\rm LNRF}$, i.e., if the orbiting particle is stationary at the LNRF located at the radius of circular orbit; this fact clearly illustrates that the LNRF are properly chosen for the definition of the optical geometry and inertial forces in accord with Newtonian intuition. Moreover, both forces vanish at some radii independently of Ω. In the case of the centrifugal force, namely this property will be imprinted into the structure of the embedding diagrams.

For the stationary circular motion, $\dot{\Omega} = 0$. It is convenient to express the inertial forces in terms of a 'Newtonian' velocity

$$\tilde{v} = \gamma v. \tag{41}$$

The velocity $\tilde{v}$ takes all real values while v ranges within $[-1, 1]$. In the stationary and axially symmetric spacetimes, there is

$$v = \tilde{\Omega}\tilde{R}, \tag{42}$$

where

$$\tilde{\Omega} = \Omega - \Omega_{\rm LNRF}, \quad \tilde{R} = \tilde{r}{\rm e}^{\Phi}, \quad \tilde{r} = (\xi^\kappa\xi_\kappa)^{1/2}. \tag{43}$$

The quantity $\tilde{R}$ is the so-called radius of gyration because $\tilde{R}^2 = \tilde{\ell}/\tilde{\Omega}$, where $\tilde{\ell} = L/E$ is the specific angular momentum, $L = U^\kappa\xi_\kappa$ is the angular momentum, and $E = -U^\kappa\eta_\kappa = \gamma{\rm e}^{\Phi}$ is the energy [26, 35]. The radius of gyration plays a very important role in the theory of rotational effects in strong gravitational fields. The direction of increasing radius of gyration gives a preferred determination of the local outward direction relevant for the dynamical effects of rotation. This direction becomes misaligned with the 'global' outward direction in strong fields [35]. The condition $\tilde{R} = \text{const}$ defines the von Zeipel cylinders in stationary spacetimes, that are related to the equipotential surfaces of equilibrium configurations of perfect fluid [42].

It is convenient to introduce gravitational acceleration and velocity-independent parts of the Coriolis and centrifugal accelerations in the direction e_κ by the relations [26]

$$\mathscr{G}(r) = e^\kappa \nabla_\kappa \Phi, \quad \mathscr{C}(r) = e^\kappa \tilde{R}\, \nabla_\kappa \Omega_{\text{LNRF}}, \quad \mathscr{Z}(r) = e^\kappa \tilde{R}^{-1} \nabla_\kappa \tilde{R}. \tag{44}$$

The acceleration necessary to keep a particle in a stationary motion with a velocity $\tilde{v}$ along a circle $r = \text{const}$ in the equatorial plane can then be expressed in a very simple form

$$a(\tilde{v}, r) = -\mathscr{G}(r) - \tilde{v}^2 \mathscr{Z}(r) + (1 + \tilde{v}^2)^{1/2} \tilde{v} \mathscr{C}(r) \tag{45}$$

that enables an effective discussion of the properties of both accelerated and geodetical motion. (Of course, only the positive root of the last term on the right-hand side of Eq. (45) has a physical meaning.)

For the stationary circular motion in the equatorial plane of the Kerr–Newman spacetimes, all three parts of the acceleration have only radial components. We obtain

$$\mathscr{G}(r) = -\frac{r^4(r - e^2) + a^2[2r(r-2)(r-e^2) - e^4] + a^4(r - e^2)}{r\Delta[(r^2 + a^2)^2 - a^2\Delta]}, \tag{46}$$

$$\mathscr{C}(r) = -\frac{2a[r(3r^2 + a^2) - e^2(2r + a^2)]}{r\sqrt{\Delta}\,[(r^2 + a^2)^2 - a^2\Delta]}, \tag{47}$$

$$\begin{aligned} \mathscr{Z}(r) = {} & \left\{ r\Delta[(r^2 + a^2)^2 - a^2\Delta] \right\}^{-1} \left\{ r^4(r^2 - 3r + 2e^2) \right. \\ & \left. + a^2[r^2(r^2 - 3r + 6) + e^2 r(3r - 7) + 2e^4] - 2a^4(r - e^2) \right\}. \end{aligned} \tag{48}$$

These definitions of the gravitational and inertial forces have really a Newtonian character since the gravitational force $G = -m\mathscr{G}(r)$ is velocity-independent, $C = m(1 + v^2)^{1/2} \tilde{v} \mathscr{C}(r)$ depends on $\tilde{v}$, and the centrifugal force $Z = -m\tilde{v}^2 \mathscr{Z}(r)$ depends on $\tilde{v}^2$. Moreover, the asymptotic behavior of these forces is consistent with 'Newtonian' intuition:

$$\mathscr{G}(r \to \infty) \sim -\frac{1}{r^2}, \quad \mathscr{C}(r \to \infty) \sim -\frac{a}{r^3}, \quad \mathscr{Z}(r \to \infty) \sim \frac{1}{r}. \tag{49}$$

It follows immediately from Eq. (45) that the circular geodetical motion of photons ($\tilde{v}^2 \to \infty$) is determined by the conditions

$$\mathscr{Z}(r) - \mathscr{C}(r) = 0, \qquad \text{for corotating orbits,} \tag{50}$$

$$\mathscr{Z}(r) + \mathscr{C}(r) = 0, \qquad \text{for counterrotating orbits.} \tag{51}$$

For ultrarelativistic particles ($\tilde{v} \gg 1$, $\tilde{v} \ll -1$), we obtain an asymptotic relation

$$a(\tilde{v}, r) \approx -\mathscr{G}(r) \pm \tfrac{1}{2}\mathscr{C}(r) - v^2[\mathscr{Z}(r) \mp \mathscr{C}(r)]. \tag{52}$$

The upper signs correspond to the corotating motion ($\tilde{v} > 0$), the lower signs to the counterrotating ($\tilde{v} < 0$) motion. The ultrarelativistic particles moving at the radius of the corotating (counterrotating) photon circular geodesic are held by the acceleration

$$a(r) = -\mathscr{G}(r) \pm \tfrac{1}{2}\mathscr{C}(r), \tag{53}$$

which is independent of the velocity, with accuracy $O(\tilde{v}^{-2})$. In static spacetimes $\mathscr{C}(r) = 0$, and at the radius of the photon circular orbit the acceleration of particles is

$$a(r) = -\mathscr{G}(r), \tag{54}$$

and is independent of $\tilde{v}$ completely. (Relation (54) is not limited to the case of ultrarelativistic orbits because $\mathscr{Z}(r) = 0$ at the radius of the photon circular geodesics in static spacetimes.)

Velocities of particles moving along the circular geodesics are determined by the relation

$$\tilde{v}_{\pm}^2 = \frac{\frac{1}{2}\mathscr{C}^2 - \mathscr{Z}\mathscr{G} \mp \frac{1}{2}\mathscr{C}(\mathscr{C}^2 - 4\mathscr{Z}\mathscr{G} + 4\mathscr{G}^2)^{1/2}}{\mathscr{Z}^2 - \mathscr{C}^2}, \tag{55}$$

where the upper (lower) signs correspond to the corotating (counterrotating) orbits, $\tilde{v} > 0$ ($\tilde{v} < 0$).

Properties of the gravitational acceleration $\mathscr{G}(r)$ and of the velocity-independent parts of the Coriolis and centrifugal acceleration $\mathscr{C}(r)$ and $\mathscr{Z}(r)$ determine properties of both the accelerated and geodetical motion in a given spacetime. It is clear that for the geodetical circular motion at a given r, the condition

$$\mathscr{C}^2(r) - 4\mathscr{Z}(r)\mathscr{G}(r) + 4\mathscr{G}^2(r) > 0 \tag{56}$$

must be satisfied. In static spacetimes $\mathscr{C}(r) = 0$, and the reality condition of the geodetical orbits is $\mathscr{G}(r)\mathscr{Z}(r) < 0$, i.e., the gravitational and centrifugal forces must point in opposite directions. If $\mathscr{C}(r) \neq 0$, the cases $|\mathscr{Z}| > |\mathscr{C}|$, $|\mathscr{Z}| = |\mathscr{C}|$, and $|\mathscr{Z}| < |\mathscr{C}|$ give qualitatively different situations. Of course, the gravitational force can play an important role too. Therefore, we have to determine the behavior of the functions $\mathscr{G}(r; a, e)$, $\mathscr{C}(r; a, e)$, and $\mathscr{Z}(r; a, e)$. We give an illustration of the behavior of these functions for black-hole (Fig. 1a) and naked-singularity (Fig. 1b) spacetimes. The asymptotic behavior of these functions is given by the relations (49). Further, it is important to know whether these functions are positive or negative.

It is important to find, where the functions $\mathscr{G}(r; a, e)$, $\mathscr{C}(r; a, e)$, and $\mathscr{Z}(r; a, e)$ change their sign [39]. All of these functions diverge and change their sign at the boundary of the region of causality violations, which is determined by the relations

$$a^2 = a^2_{\text{c.v.}}(r; e) = \frac{r^4}{e^2 - 2r - r^2}. \tag{57}$$

In Kerr spacetimes ($e^2 = 0$), this region is restricted to $r < 0$.

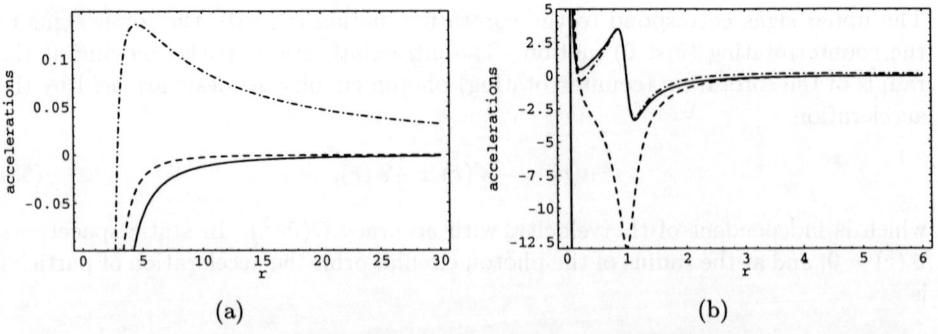

Figure 1. Behavior of the velocity-independent parts of gravitational (solid curve), Coriolis (dashed curve), and centrifugal (dashed-dotted curve) forces $\mathscr{G}(r)$, $\mathscr{C}(r)$, and $\mathscr{Z}(r)$, respectively. (a) $a^2 = 0.16$, $e^2 = 0.16$, (b) $a^2 = 0.86$, $e^2 = 0.16$. Taken from Ref. [39].

Surprisingly, the gravitational acceleration changes its sign at the zero points given by

$$a^2 = a^2_{\mathrm{g}\pm}(r;e) = \frac{e^4 - 2r(r-2)(r-e^2) \pm \sqrt{D_{\mathrm{g}}}}{2(r-e^2)}, \tag{58}$$

with

$$D_{\mathrm{g}} = 4r(r-2)(r-e^2)(2re^2 - 2r^2 - e^4) + e^8. \tag{59}$$

It holds at $r > 0$ even for the Kerr spacetimes—we arrive at a simple formula

$$a^2 = a^2_{\mathrm{g}\pm}(r) = r\left(2 - r \pm \sqrt{2-r}\right). \tag{60}$$

The Coriolis acceleration changes the sign at zero points determined by

$$a^2 = a^2_{\mathrm{c}}(r;e) = \frac{r(3r^2 - 2e^2)}{e^2 - r}. \tag{61}$$

In the Kerr spacetimes, $\mathscr{C}(r)$ does not change the sign at $r > 0$. Finally, we find that the centrifugal acceleration changes the sign at radii given by

$$a^2 = a^2_{\mathrm{z}\pm}(r;e) = \frac{r^2(r^2 - 3r + 6) + e^2 r(3r-7) + 2e^4 \pm \sqrt{D_{\mathrm{z}}}}{4(r-e^2)}, \tag{62}$$

where the discriminant is

$$\begin{aligned} D_{\mathrm{z}}(r;e) = \; & r^8 + 2r^7 - (3+2e^2)r^6 + 4(2e^2 - 9)r^5 + 3(12 + 26e^2 - e^4)r^4 \\ & - 2e^2(42 + 27e^2)r^3 + e^4(73 + 12e^2)r^2 - 28e^6 r + 4e^8. \end{aligned} \tag{63}$$

We give an example of the behavior of all the functions $a^2_{\mathrm{g}\pm}(r;e)$, $a^2_{\mathrm{c}}(r;e)$, and $a^2_{\mathrm{z}\pm}(r;e)$ in Fig. 2. Generally, the properties of stationary circular motion can be given in a straightforward manner using these functions; we shall not discuss the details here.

The centrifugal acceleration is very important since it is closely related to the radius of gyration—see Eq. (44). Therefore, it deserves a detailed study.

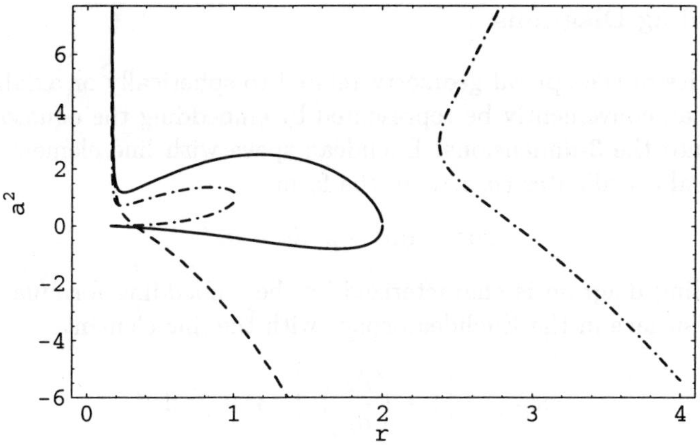

Figure 2. Behavior of the functions $a^2_{\mathrm{g}\pm}$ (solid curve), a^2_{c} (dashed curve), and $a^2_{\mathrm{z}\pm}$ (dashed-dotted curve). Plotted for a fixed charge parameter of the spacetime $e^2 = 0.16$. Taken from Ref. [39].

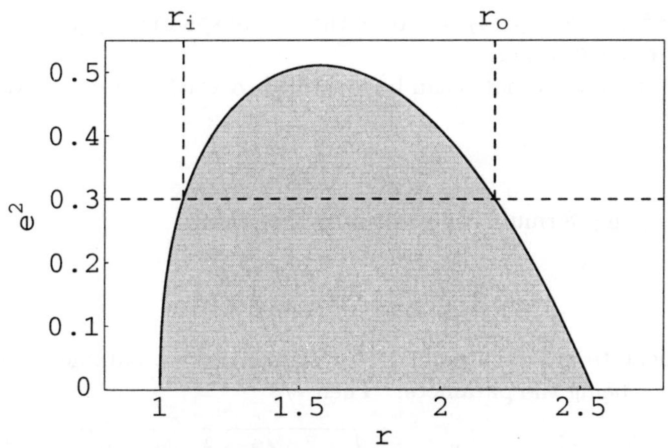

Figure 3. The function determining the reality condition $D_{\mathrm{z}}(r;e) \geq 0$ for the radii at which the centrifugal force vanishes. Inside the shaded region, $D_{\mathrm{z}}(r;e) < 0$ and the functions $a^2_{\mathrm{z}\pm}(r;e)$ are not defined. The three remaining solutions of the equation $D_{\mathrm{z}}(r;e) = 0$ are not presented here, being physically irrelevant (with $a^2 < 0$ or $e^2 < 0$). Taken from Ref. [39].

The reality condition for $a^2_{\mathrm{z}\pm}(r;e)$ is

$$D_{\mathrm{z}}(r;e) \geq 0, \tag{64}$$

which can be treated easily considering $D_{\mathrm{z}}(r;e)$ as a quartic polynom in e^2. The results are given in Fig. 3. Detailed behavior of the functions $a^2_{\mathrm{z}\pm}(r;e)$ will be determined in the following Section since it is closely related to embedding diagrams of the equatorial plane of the optical geometry.

3 Embedding Diagrams

The properties of the optical geometry related to spherically or axially symmetric spacetimes can conveniently be represented by embedding the equatorial (symmetry) plane into the 3-dimensional Euclidean space with line element expressed in the cylindrical coordinates (ρ, z, α) in the form

$$d\sigma^2 = d\rho^2 + \rho^2\, d\alpha + dz^2. \tag{65}$$

The embedding diagram is characterized by the embedding formula $z = z(\rho)$ determining a surface in the Euclidean space with the line element

$$d\ell^2_{(E)} = \left[1 + \left(\frac{dz}{d\rho}\right)^2\right] d\rho^2 + \rho^2\, d\alpha^2 \tag{66}$$

isometric to the 2-dimensional equatorial plane of the optical space determined by the line element

$$d\tilde{\ell}^2 = \tilde{h}_{rr}\, dr^2 + \tilde{h}_{\phi\phi}\, d\phi^2, \tag{67}$$

with $\tilde{h}_{rr}$ and $\tilde{h}_{\phi\phi}$ given by (4), where in the case of spherically or axially symmetric spacetimes we put $\theta = \pi/2$.

The azimuthal coordinates can be identified $(\alpha \equiv \phi)$, the radial coordinates are related via

$$\rho^2 = \tilde{h}_{\phi\phi}, \tag{68}$$

and the embedding formula is governed by the relation

$$\left(\frac{dz}{d\rho}\right)^2 = \tilde{h}_{rr}\left(\frac{dr}{d\rho}\right)^2 - 1. \tag{69}$$

It is convenient to cast the embedding formula into a parametric form $z(\rho) = z(r(\rho))$, with r being the parameter. Then

$$\frac{dz}{dr} = \pm\sqrt{\tilde{h}_{rr} - \left(\frac{d\rho}{dr}\right)^2}. \tag{70}$$

The sign in this formula is irrelevant, leading to isometric surfaces. Because

$$\frac{dz}{d\rho} = \frac{dz}{dr}\frac{dr}{d\rho}, \tag{71}$$

the turning points of the embedding diagram, giving its throats and bellies (commonly referred to as the turning points), are determined by the condition $d\rho/dr = 0$.

The embedding diagram can be constructed if the reality condition

$$\tilde{h}_{rr} - \left(\frac{d\rho}{dr}\right)^2 \geq 0 \tag{72}$$

holds. The explicit form of the reality condition will be presented and discussed below.

3.1 Schwarzschild–de Sitter Spacetimes

The line element of the optical geometry is given by the relation [15, 23]

$$d\tilde{\ell}^2_{(\mathrm{S-dS})} = \left(1 - \frac{2}{r} - yr^2\right)^{-2} dr^2 + r^2 \left(1 - \frac{2}{r} - yr^2\right)^{-1} d\phi^2 \tag{73}$$

and is to be identified with the line element (66) in order to construct the embedding diagram of the optical geometry. The azimuthal angles can directly be identified. For the radial coordinates we have to put

$$\rho = r\left(1 - \frac{2}{r} - yr^2\right)^{-1/2} . \tag{74}$$

We arrive at the formula

$$\frac{dz}{dr} = \pm \frac{r}{r - 2 - yr^3} \left[\frac{4r - 9 - yr^4}{r(r - 2 - yr^3)}\right]^{1/2} . \tag{75}$$

The embedding formula $z = z(\rho)$ can then be constructed by a numerical procedure. Further, the 'turning radii' of the embedding diagrams are given by the condition $d\rho/dr = 0$. Since

$$\frac{d\rho}{dr} = \left(1 - \frac{3}{r}\right)\left(1 - \frac{2}{r} - yr^2\right)^{-3/2} , \tag{76}$$

we can see that the turning radius determining a throat of the embedding diagram of the optical geometry is located just at $r = 3$, corresponding to the radius of the photon circular orbit; it is exactly the same result as that obtained in the pure Schwarzschild case.

As follows from Eq. (75), the embedding into the Euclidean space is possible, if the condition of embeddability

$$4r - 9 - yr^4 \geq 0 \tag{77}$$

is satisfied. For $y = 0$, we obtain the well known limit of embeddability of the optical geometry of the Schwarzschild spacetime $r_{\mathrm{e(opt)}} = 9/4$ (see Ref. [23]).

The limit of embeddability is given by the function

$$y = y_{\mathrm{e(opt)}}(r) \equiv \frac{4r - 9}{r^4} . \tag{78}$$

In the case of Schwarzschild–de Sitter spacetimes, the function $y_{\mathrm{e(opt)}}(r)$ has its maximum at $r = 3$, where $y_{\mathrm{e(opt)}} = 1/27 = y_{\mathrm{crit}}$. Therefore, for all the Schwarzschild–de Sitter spacetimes containing a static region, the embeddability of the optical geometry is restricted both from below and from above. Using a numerical procedure, the embedding diagrams are constructed for the same typical values of the parameter y (see Fig. 4). For $y \ll 1/27$, the embeddings of the optical geometry are very close to the embedding diagram of the optical geometry of the Schwarzschild spacetime. This is caused by the fact that the regions of the Schwarzschild–de Sitter spacetimes near the cosmological horizon, that have a significantly different character from that corresponding to the Schwarzschild geometry, are 'cut off' by the limit of embeddability. With y growing towards the

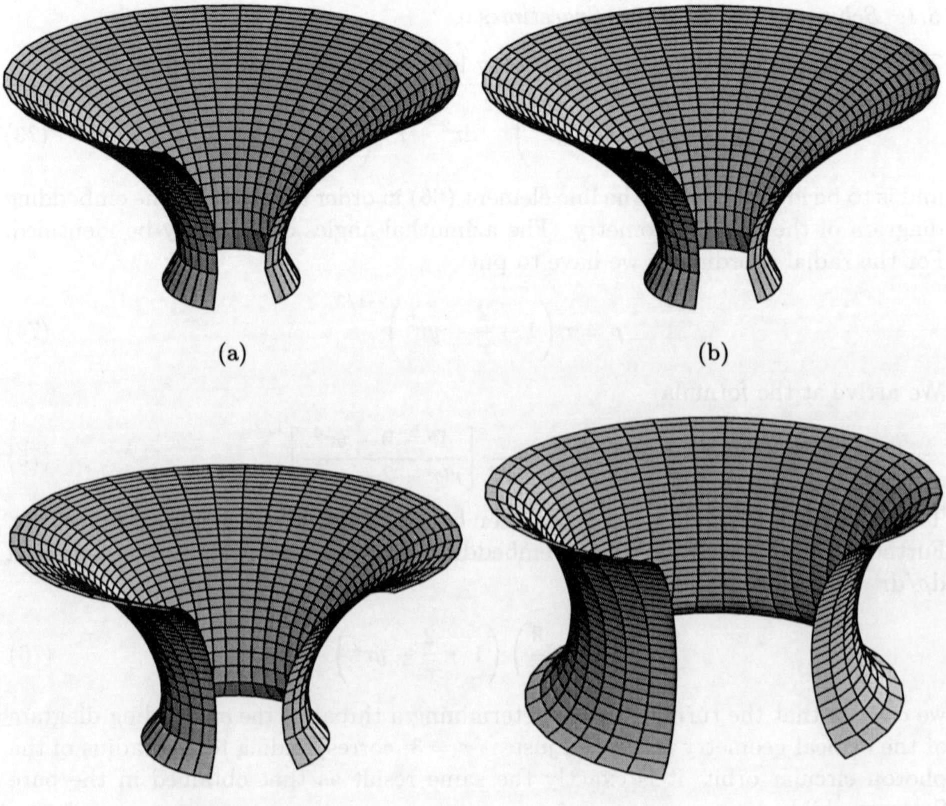

Figure 4. Embedding diagrams of the optical reference geometry of the Schwarzschild–de Sitter spacetimes. The pure Schwarzschild case is shown for comparison in (a). The diagrams are given for (b) $y = 10^{-6}$, (c) $y = 0.002$, and (d) $y = 0.03$. They are similar to the pure Schwarzschild case because the region near the cosmological horizon, being of highly different character, is 'cut off' by the limit of embeddability $y_{e(opt)}(r)$. Note that both ρ and z scales are adjusted so that all the plots occupy approximately the same area. For real proportions, see Fig. 6. Taken from Ref. [15].

critical value $y_{crit} = 1/27$, the embedding diagrams become restricted to a region symmetric around the turning radius at $r = 3$. For $y = 1/27$, the diagram degenerates into a circle at $r = 3$, similarly to the case of the embedding diagrams of the ordinary geometry.

Of course, the optical space is still well defined outside the regions of embeddability into the Euclidean space. It is useful to demonstrate its properties near the black-hole and cosmological horizons via the behavior of proper lengths along the radial direction. In the optical geometry, the proper radial length coincides with the well known Regge–Wheeler 'tortoise' coordinate:

$$r^* = \int \left(1 - \frac{2}{r} - yr^2\right)^{-1} \mathrm{d}r. \tag{79}$$

Integrating (for $0 < y < 1/27$), we arrive at

$$r^* = A \ln \frac{|r - r_\mathrm{h}|}{r + r_\mathrm{h} + r_\mathrm{c}} + B \ln \frac{|r_\mathrm{c} - r|}{r + r_\mathrm{h} + r_\mathrm{c}} + A \left(\ln \frac{r_\mathrm{c}}{r_\mathrm{h}} - \frac{1}{2} \right), \tag{80}$$

where

$$A = \frac{r_\mathrm{h}}{1 - 3yr_\mathrm{h}^2}, \quad B = \frac{r_\mathrm{c}}{1 - 3yr_\mathrm{c}^2}. \tag{81}$$

This gives a direct meaning to the 'tortoise' coordinate in the optical space. Clearly, the horizons are infinitely far away in the optical geometry; at $r \sim r_\mathrm{h}$, $r^* \to -\infty$ ($r \sim r_\mathrm{c}$, $r^* \to +\infty$) logarithmically. On the other hand, in the ordinary geometry, the horizons are located at a finite proper radial distance

$$\tilde{r} = \int \left(1 - \frac{2}{r} - yr^2 \right)^{-1/2} \mathrm{d}r; \tag{82}$$

at $r \approx r_\mathrm{h}$, $\tilde{r} \sim \sqrt{r - r_\mathrm{h}}$, while at $r \approx r_\mathrm{c}$, $\tilde{r} \approx \sqrt{r_\mathrm{c} - r}$.

The optical space extends infinitely beyond the limit of embeddability, approaching asymptotically the geometry

$$\mathrm{d}\tilde{\sigma}^2 \approx \mathrm{d}r^{*2} + \frac{r_\mathrm{h}^3 \exp(-r^*/A)}{y(r_\mathrm{c} - r_\mathrm{h})(r_\mathrm{h} - r_\mathrm{d})(2r_\mathrm{h} + r_\mathrm{c})} (\mathrm{d}\theta^2 + \sin^2\theta \, \mathrm{d}\phi^2) \tag{83}$$

for $r \to r_\mathrm{h}$, $r^* \to -\infty$ and

$$\mathrm{d}\tilde{\sigma}^2 \approx \mathrm{d}r^{*2} + \frac{r_\mathrm{c}^3 \exp(-r^*/B)}{y(r_\mathrm{c} - r_\mathrm{h})(r_\mathrm{c} - r_\mathrm{d})(2r_\mathrm{c} + r_\mathrm{h})} (\mathrm{d}\theta^2 + \sin^2\theta \, \mathrm{d}\phi^2) \tag{84}$$

for $r \to r_\mathrm{c}$, $r^* \to +\infty$. Here

$$r_\mathrm{d} = -\frac{1}{3y} \cos \frac{1}{3} \xi. \tag{85}$$

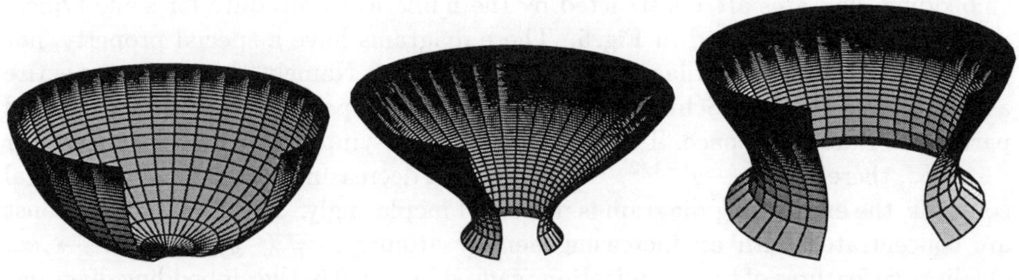

(a) (b) (c)

Figure 5. Embedding diagrams of the optical geometry of the Schwarzschild–anti-de Sitter spacetimes. They are given for (a) $y = -10^{-6}$, (b) $y = -0.002$, and (c) $y = -0.03$. Contrary to the spacetimes with $y > 0$, the diagrams cover whole asymptotic region of the optical geometry; however, it is stretched into finite region of the radial coordinate of the embedding Euclidean space. Note that both ρ and z scales are adjusted so that all the plots occupy approximately the same area. For real proportions, see Fig. 6. Taken from Ref. [15].

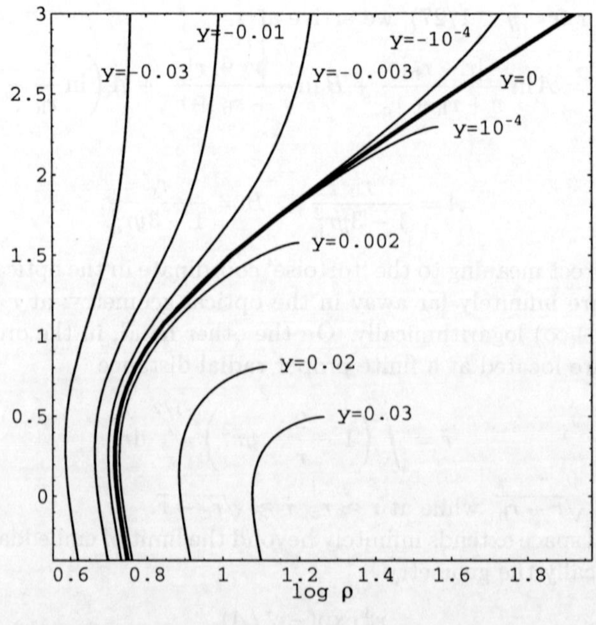

Figure 6. Qualitative features of the optical geometry embedding diagrams for the Schwarzschild–de Sitter and Schwarzschild–anti-de Sitter spacetimes in a log-log diagram. The diagrams for $y \neq 0$ 'peel off' the pure Schwarzschild diagram ($y = 0$, bold curve). All sections with $y > 0$ are complete (i.e., the maximum that can be embedded into the Euclidean space is shown), except uninteresting lower parts of the diagrams. The diagrams are indicative for modifications of the spacetime structure caused by the presence of a cosmological constant. Taken from Ref. [15].

In case of Schwarzschild–anti-de Sitter spacetimes, the limiting condition (77) restricts the embeddability of the optical geometry only from below; with $y \to -\infty$, the limit shifts to $r \to 0$ along with the radius of the black-hole horizon. The embedding diagrams are constructed by the numerical procedure for some typical values of y and illustrated in Fig. 5. These diagrams have a special property, not present in the embedding diagrams in the other cases. Namely, they cover the entire asymptotic part of the Schwarzschild–anti-de Sitter spacetime but in a restricted part of the Euclidean space. This is clear from the asymptotic behavior of $\rho(r)$. For $r \to +\infty$, there is $\rho \sim (-y)^{-1/2}$. Clearly, with a decreasing attractive cosmological constant, the embedding diagram is deformed increasingly. The circles of $r = \text{const}$ are concentrated with an increasing density around $\rho = (-y)^{-1/2}$ as $r \to \infty$. Qualitative features of the modifications caused by $y \neq 0$ in the embedding diagrams of the optical geometry are illustrated in Fig. 6.

The optical geometry is well defined under the limit of embeddability again. Now, the 'tortoise' coordinate is given by

$$r^* = D\left\{\ln|r - r_{\mathrm{h}}| - \frac{1}{2}\ln\left(r^2 + rr_{\mathrm{h}} - \frac{2}{yr_{\mathrm{h}}}\right)\right.$$

$$+\frac{6-r_\mathrm{h}}{[(6+r_\mathrm{h})(2-r_\mathrm{h})]^{1/2}}\left\{\arctan\left[\left(\frac{2-r_\mathrm{h}}{6+r_\mathrm{h}}\right)^{1/2}\left(1+\frac{2r}{r_\mathrm{h}}\right)\right]+E\right\}\Bigg\},\tag{86}$$

where

$$D=-\frac{2-r_\mathrm{h}}{2yr_\mathrm{h}(3+r_\mathrm{h})},\qquad E=-\sqrt{-y}\,[1+\ln(-4y)].\tag{87}$$

We can see again that the black-hole horizon is infinitely far away in the optical geometry. Again, the optical space extends infinitely beyond the limit of embeddability as $r\to r_\mathrm{h}$ and $r^*\to-\infty$, approaching asymptotically a geometry similar to (83).

3.2 Interior Schwarzschild–de Sitter Spacetimes with Uniform Density

The line element of the equatorial plane of the optical geometry is determined by [20]

$$\mathrm{d}\tilde{\ell}^2_{(\mathrm{S-dS,in})}=\frac{\mathrm{e}^{2\Psi(r)}}{\mathrm{e}^{2\Phi(r)}}\,\mathrm{d}r^2+\frac{r^2}{\mathrm{e}^{2\Phi(r)}}\,\mathrm{d}\phi^2,\tag{88}$$

where

$$\mathrm{e}^{\Psi(r)}=Y^{-1}(r),\tag{89}$$

$$\mathrm{e}^{\Phi(r)}=AY_1-BY(r),\tag{90}$$

$$Y(r)=\left[1-\frac{2+yx^3}{x^3}\left(\frac{r}{M}\right)^2\right]^{1/2},\tag{91}$$

$$Y_1=\mathrm{e}^{\Phi(R)}=\mathrm{e}^{-\Psi(R)}=\left(\frac{x-2-yx^3}{x}\right)^{1/2},\tag{92}$$

$$y=\tfrac{1}{3}\Lambda M^2,\quad x=\frac{R}{M},\tag{93}$$

$$\frac{a^2}{M^2}=\frac{x^3}{2+yx^3},\tag{94}$$

$$A=\frac{3}{2+yx^3},\qquad B=\frac{1-yx^3}{2+yx^3}.\tag{95}$$

Comparing this line element with the line element (66), one can see that the azimuthal coordinates can be identified ($\alpha\equiv\phi$) while the radial coordinates are related by

$$\rho=\frac{r}{\mathrm{e}^{\Phi(r)}}=\frac{a(1-Y^2(r))^{1/2}}{(AY_1-BY(r))}.\tag{96}$$

The turning points of the embedding diagram are determined by the condition $\mathrm{d}\rho/\mathrm{d}r=0$ (see, e.g., Refs [15,21]). But

$$\frac{\mathrm{d}\rho}{\mathrm{d}r}=\frac{\mathrm{d}\rho}{\mathrm{d}Y}\frac{\mathrm{d}Y}{\mathrm{d}r},\tag{97}$$

and $\mathrm{d}Y/\mathrm{d}r \neq 0$. The turning points are, therefore, given by the condition

$$\frac{\mathrm{d}\rho}{\mathrm{d}Y} = 0. \tag{98}$$

Since we have

$$\frac{\mathrm{d}\rho}{\mathrm{d}Y} = \frac{a(B - AY_1 Y)}{(1 - Y^2)^{1/2}(AY_1 - BY)^2}, \tag{99}$$

the condition

$$B - AY_1 Y = 0 \tag{100}$$

corresponding to null circular geodesics must be satisfied at the turning points [20].

Notice that for spacetimes with parameters satisfying the condition $2 + yx^3 < 0$, we have to introduce $1/\tilde{a}^2 = -1/a^2$ and $r^2 = \tilde{a}^2(Y^2 - 1)^{1/2}$. Then

$$\rho = \frac{\tilde{a}(Y^2 - 1)^{1/2}}{AY_1 - BY}, \tag{101}$$

and the turning points ($\mathrm{d}\rho/\mathrm{d}Y = 0$) are located at radii corresponding to the circular null geodesics, again.

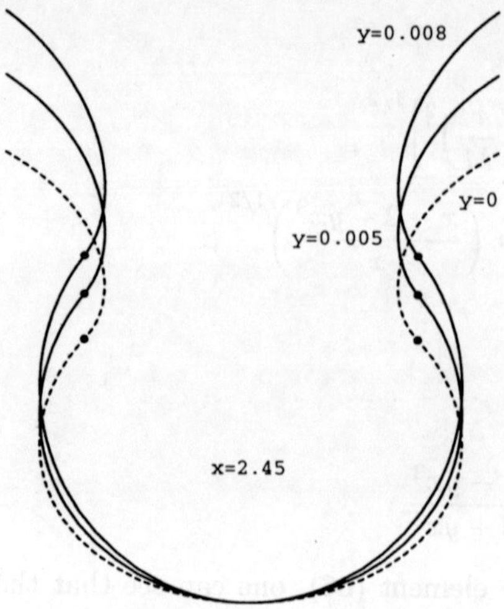

Figure 7. Axial sections of the optical reference geometry embedding diagrams for the Schwarzschild–de Sitter spacetimes, illustrating the dependence of the diagram radius on the cosmological parameter. The dashed curve represents a pure Schwarzschild spacetime, the dots mark the position of the edge of the internal spacetime (i.e., $r = R = xM$). Taken from Ref. [20].

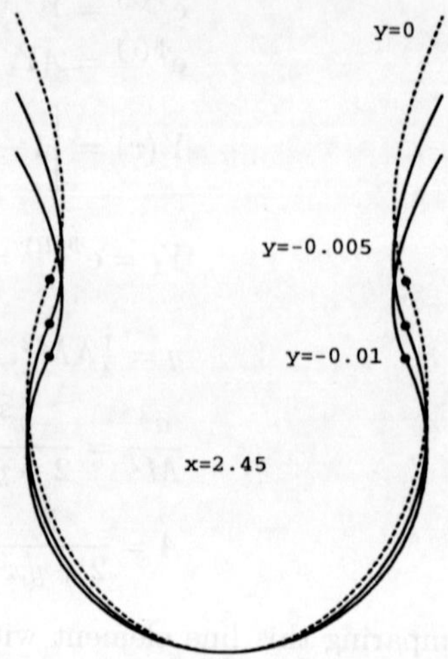

Figure 8. Axial sections of the optical reference geometry embedding diagrams for the Schwarzschild–anti-de Sitter spacetimes, illustrating the dependence of the diagram radius on the cosmological parameter. The dashed curve represents a pure Schwarzschild spacetime, the dots mark the position of the edge of the internal spacetime (i.e., $r = R = xM$). Taken from Ref. [20].

The embedding formula in a parametric form reads

$$\frac{\mathrm{d}z}{\mathrm{d}r} = \pm\frac{(1-Y^2(r))^{1/2}(A^2Y_1^2-B^2)^{1/2}}{Y(r)(AY_1-BY(r))^2}. \tag{102}$$

Writing

$$\frac{\mathrm{d}z}{\mathrm{d}r} = \pm\left(\frac{r^2}{M^2}\frac{4x-9-yx^4}{x^4}\right)^{1/2}\frac{1}{Y(AY_1-BY(r))^2}, \tag{103}$$

we can directly see that the embedding diagrams of the optical geometry can be constructed within the entire interior spacetimes for a complete range of the admitted dimensionless parameters x, y.

At the edge of the static configuration $r = R$, we obtain, using the dimensionless parameters x, y,

$$\left.\frac{\mathrm{d}z}{\mathrm{d}r}\right|_{r=R}^{\mathrm{in(op)}} = \frac{[x(4x-9-yx^4)]^{1/2}}{(x-2-yx^3)^{3/2}}; \tag{104}$$

however, using the embedding formula of the optical geometry for the external vacuum Schwarzschild–(anti-)de Sitter spacetimes (see Ref. [15]), we see that

$$\left.\frac{\mathrm{d}z}{\mathrm{d}r}\right|_{r=R}^{\mathrm{in(op)}} = \left.\frac{\mathrm{d}z}{\mathrm{d}r}\right|_{r=R}^{\mathrm{ex(op)}}, \tag{105}$$

and we conclude that the embedding diagrams are smoothly matched at $r = R$. Notice that the (inner) limit of embeddability for the optical geometry of the external, vacuum spacetimes coincides with the (inner) limit on the external radius of the static configuration.

The embedding diagrams are constructed for some typical values of x and y. They are given in Fig. 7 for $y > 0$ and in Fig. 8 for $y < 0$. If $x < 3$, there are turning points corresponding to circular null geodesics. The diagrams have an 'ellipsoidal' shape; the ellipsoids are oblate. If $x > 3$, there are no turning points in the embedding diagrams.

3.3 Reissner–Nordström–de Sitter Spacetimes

The line element of the equatorial plane of the optical geometry is given by [18]

$$\mathrm{d}\tilde{\ell}^2_{(\mathrm{R-N-dS})} = \frac{r^4}{\left(-yr^4+r^2-2r+e^2\right)^2}\,\mathrm{d}r^2 + \frac{r^4}{(-yr^4+r^2-2r+e^2)}\,\mathrm{d}\phi^2, \tag{106}$$

The azimuthal angle ϕ can again be directly identified with the azimuthal angle α in (66); for the radial coordinates we have to put

$$\rho = r\left(1-\frac{2}{r}+\frac{e^2}{r^2}-yr^2\right)^{-1/2}. \tag{107}$$

The parametric form of the embedding formula reads

$$\frac{\mathrm{d}z}{\mathrm{d}r} = \pm\frac{r\left[-yr^6+4r^3-(3e^2+9)r^2+12e^2r-4e^4\right]^{1/2}}{\left(-yr^4+r^2-2r+e^2\right)^{3/2}}. \tag{108}$$

The turning radii of the embedding diagrams are given by

$$\frac{\mathrm{d}\rho}{\mathrm{d}r} = \frac{r\left(r^2 - 3r + 2e^2\right)}{\left(-yr^4 + r^2 - 2r + e^2\right)^{3/2}} = 0. \tag{109}$$

We see directly that the turning radii are located right at $r = r_{\mathrm{ph}+}$ (and $r = r_{\mathrm{ph}-}$ in the case of naked singularities with parameters y and e chosen appropriately), corresponding to the radii of photon circular orbits. At $r = r_{\mathrm{ph}+}$, there is a throat of the embedding diagram while at $r = r_{\mathrm{ph}-}$ it has a belly. The radii of the photon circular orbits are important from the dynamical point of view. At $r > r_{\mathrm{ph}+}$, the dynamics is qualitatively Newtonian with the centrifugal force directed towards increasing r. At $r = r_{\mathrm{ph}+}$, the centrifugal force vanishes and at $r < r_{\mathrm{ph}+}$, it is directed towards decreasing r. In the field of naked singularities admitting a circular photon orbit at $r_{\mathrm{ph}-}$, the centrifugal force vanishes at $r = r_{\mathrm{ph}-}$ and change its sign again, i.e., it is directed towards increasing r at $r < r_{\mathrm{ph}-}$ (see Ref. [18] for details).

The condition of embeddability

$$4r^3 - (3e^2 + 9)r^2 + 12e^2 r - 4e^4 - yr^6 \geq 0 \tag{110}$$

can be expressed in the form

$$y \leq y_{\mathrm{e(opt)}}(r; e) \equiv \frac{4r^3 - (3e^2 + 9)r^2 + 12e^2 r - 4e^4}{r^6}, \tag{111}$$

where the function $y_{\mathrm{e(opt)}}(r; e)$ determines the limits of embeddability, i.e., the boundaries of embedding diagrams if they do not coincide with the horizons of the spacetime under consideration. The asymptotic behavior of $y_{\mathrm{e(opt)}}(r; e)$ is given by the relations $y_{\mathrm{e(opt)}}(r \to 0, e) \to -\infty$, $y_{\mathrm{e(opt)}}(r \to \infty, e) \to 0$. The zero points of the function $y_{\mathrm{e(opt)}}(r; e)$ are given by the relation

$$e^2 = e^2_{\mathrm{z(e(opt))}\pm}(r) \equiv \frac{3r(4 - r) \pm r\sqrt{r(9r - 8)}}{8}. \tag{112}$$

For $e = 0$, $y_{\mathrm{e(opt)}}(r; e = 0)$ has a zero point at $r_{\mathrm{(opt)}} = 9/4$. Since

$$\frac{\partial y_{\mathrm{e(opt)}}(r, e)}{\partial r} = \frac{-12(r - e^2)(r^2 - 3r + 2e^2)}{r^7}, \tag{113}$$

we can see that the local extrema of $y_{\mathrm{e(opt)}}(r; e)$ are located at

$$r_1 = e^2, \quad r_2 = r_{\mathrm{ph}-}(e), \quad r_3 = r_{\mathrm{ph}+}(e), \tag{114}$$

where $r_{\mathrm{ph}\pm}(e)$ determine the radii of photon circular orbits. Thus, we conclude (see Ref. [18]) that for $0 < e^2 < 1$, there is a local minimum at r_1 and local maxima at r_2, r_3, for $e^2 = 1$, r_1 and r_2 coalesce at an inflex point, for $1 < e^2 < 9/8$, there are local maxima at r_1 and r_3 and a local minimum at r_2, for $e^2 = 9/8$, r_2 and r_3 coalesce at an inflex point, and for $e^2 > 9/8$ there is only a local maximum at r_1. The properties of the embedding diagrams are thoroughly discussed in [18]; here we present typical embedding diagrams in Fig. 9 for asymptotically de Sitter spacetimes ($y > 0$) and in Fig. 10 for asymptotically anti-de Sitter spacetimes ($y < 0$). If $y > 0$, there exist two types of embedding diagrams for black-hole spacetimes

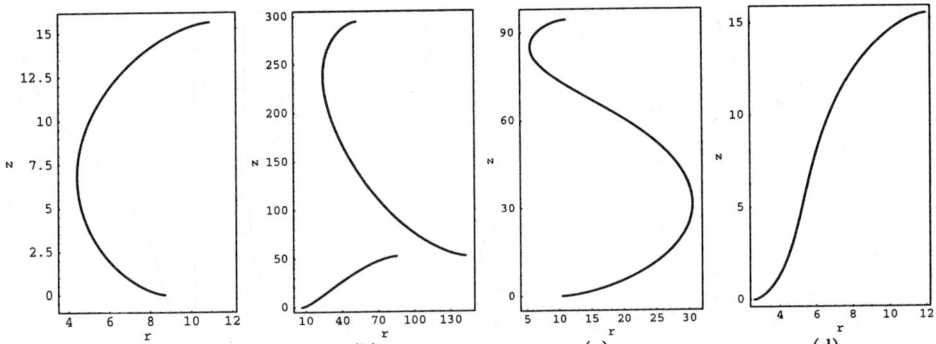

Figure 9. Typical embedding diagrams of the optical reference geometry for asymptotically de Sitter spacetimes (in axial-section form). In black-hole spacetimes, a limited region between the outer black-hole horizon and the cosmological horizon can be embedded—see (a) drawn for $y = 0.01$, $e^2 = 0.98$ and some part of the region between the singularity and the inner black-hole horizon—see (b) drawn for $y = 0.0702$, $e^2 = 1.11$. For naked-singularity spacetimes, a limited region between the singularity and the cosmological region can be embedded—see (c) drawn for $y = 0.033$, $e^2 = 1.04$ with a throat and a belly and (d) drawn for $y = 0.05$, $e^2 = 1.21$ without any turning point. Beside the types depicted, there exist cases for which no region of the optical geometry can be embedded into the Euclidean space.

and two types of embedding diagrams for naked-singularity spacetimes. If $y < 0$, there is one type of an embedding diagram for black-hole spacetimes and three types of embedding diagrams for naked-singularity spacetimes. Above the black-hole horizon, the embedding diagram always has the same character as in the corresponding Schwarzschild–(anti-)de Sitter spacetimes, i.e., it has a throat. In the naked-singularity spacetimes, the embedding diagram can have a throat and a belly but there also exist embedding diagrams with no turning point.

The well known Regge–Wheeler 'tortoise' coordinate

$$r^* = \int \left(1 - \frac{2}{r} + \frac{e^2}{r^2} - yr^2\right)^{-1} \mathrm{d}r. \tag{115}$$

enables us to understand the behavior of the optical reference geometry beyond the limits of optical embeddability [18]. At $r \sim r_{\mathrm{b}+}$, there is $r^* \sim +\ln|r - r_{\mathrm{b}+}| \to -\infty$ while at $r \sim r_{\mathrm{c}}$, there is $r^* \sim -\ln|r_{\mathrm{c}} - r| \to +\infty$. On the other hand, in the ordinary geometry, the horizons are located at a finite proper radial distance

$$\tilde{r} = \int \left(1 - \frac{2}{r} + \frac{e^2}{r^2} - yr^2\right)^{-1/2} \mathrm{d}r; \tag{116}$$

at $r \sim r_{\mathrm{b}+}$, $\tilde{r} \sim \sqrt{r - r_{\mathrm{b}+}}$, and at $r \sim r_{\mathrm{c}}$, $\tilde{r} \sim \sqrt{|r_{\mathrm{c}} - r|}$. In the Reissner–-Nordström–de Sitter spacetimes, the optical geometry extends infinitely beyond the limit of embeddability, approaching asymptotically the geometry

$$\mathrm{d}\tilde{\sigma}^2 \approx \mathrm{d}r^{*^2} + C_1 exp\left[-r^*/C(r_{\mathrm{b}+})\right] \left(\mathrm{d}\theta^2 + \sin^2\theta\, \mathrm{d}\phi^2\right) \tag{117}$$

for $r \to r_{\mathrm{b}+}$, $r^* \to -\infty$, and

$$\mathrm{d}\tilde{\sigma}^2 \approx \mathrm{d}r^{*^2} + C_2 exp\left[-r^*/C(r_{\mathrm{c}})\right] \left(\mathrm{d}\theta^2 + \sin^2\theta\, \mathrm{d}\phi^2\right) \tag{118}$$

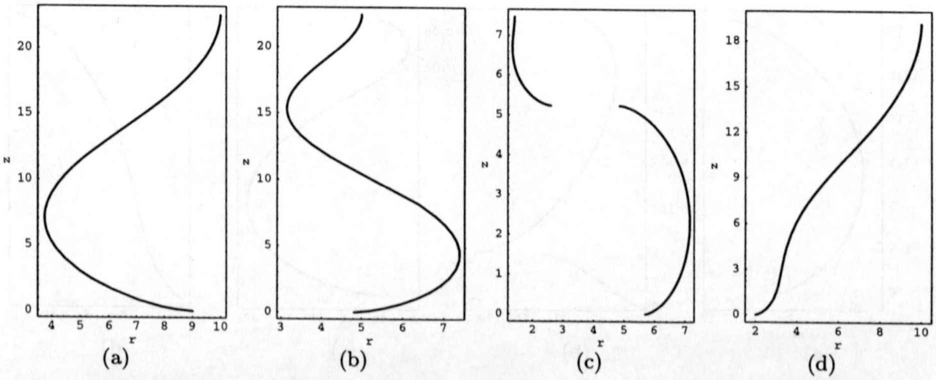

Figure 10. Typical embedding diagrams of the optical reference geometry for asymptotically anti-de Sitter spacetimes (in axial-section form). There is one type of embedding for black holes—see (a) drawn for $y = -0.01$, $e^2 = 0.98$ and three types of embedding for naked singularities—see (b) drawn for $y = -0.04$, $e^2 = 0.98$ with a throat and a belly, (c) drawn for $y = -0.5$, $e^2 = 0.795$ with two leaves, the outer one having a belly, the inner one having a throat, and (d) drawn for $y = -0.01$, $e^2 = 1.21$ without any turning point. In all these cases, the embedding is possible for $r \to \infty$, however, it is shrunk to a finite value of the embedding coordinate ρ, depending on the cosmological parameter y. Turning radii of the embedding diagrams correspond to the radii of the photon circular geodesics, that are independent of y.

for $r \to r_{\mathrm{c}}$, $r^* \to +\infty$; C_1, C_2 and $C(r_{\mathrm{b}+})$, $C(r_{\mathrm{c}})$ are constants given in terms of the spacetime parameters.

For Reissner–Nordström–anti-de Sitter black holes, the optical space has again the property that at $r \sim r_{\mathrm{b}+}$, there is $r^* \sim \ln|r - r_{\mathrm{b}+}| \to -\infty$, while for $r \to \infty$, there is $r^* \sim (-y)^{-1} \ln r \to +\infty$ and the asymptotical behavior of the optical geometry is determined by formulae similar to (117) and (118), respectively.

In the vicinity of the inner black-hole horizon, $r \to r_{\mathrm{b}-}$, there is $r^* \sim -\ln|r_{\mathrm{b}-} - r|$ and the optical geometry has the asymptotical form similar to (117). On the other hand, for $r \to 0$, there is $r^* \sim r^3/3e^2 \to 0$ independently of the value of $y \neq 0$. At $r \to 0$, the asymptotic form of the optical geometry of spacetimes with $e^2 \neq 0$ is given by

$$\mathrm{d}\tilde{\sigma}^2 \approx \mathrm{d}r^{*2} + \frac{r^4}{e^2}\left(\mathrm{d}\theta^2 + \sin^2\theta\,\mathrm{d}\phi^2\right). \tag{119}$$

In the naked singularity spacetimes, the asymptotic form of r^* and $\mathrm{d}\tilde{\sigma}^2$ at $r \to r_{\mathrm{c}}$ (for $y > 0$), and at $r \to \infty$ (for $y < 0$), respectively, is the same as in the black-hole spacetimes, i.e., $r^* \to +\infty$. But there is $r^* \sim r^3/3e^2 \to 0$ for $r \to 0$ and there is no other divergence of the 'tortoise' coordinate in the naked-singularity spacetimes. At $r \sim 0$, the optical geometry is again determined by the formula (119).

3.4 Ernst Spacetimes

In our discussion, this is a special case of a spacetime that is static and axially symmetric (not spherically symmetric). In the equatorial plane of the Ernst spacetime,

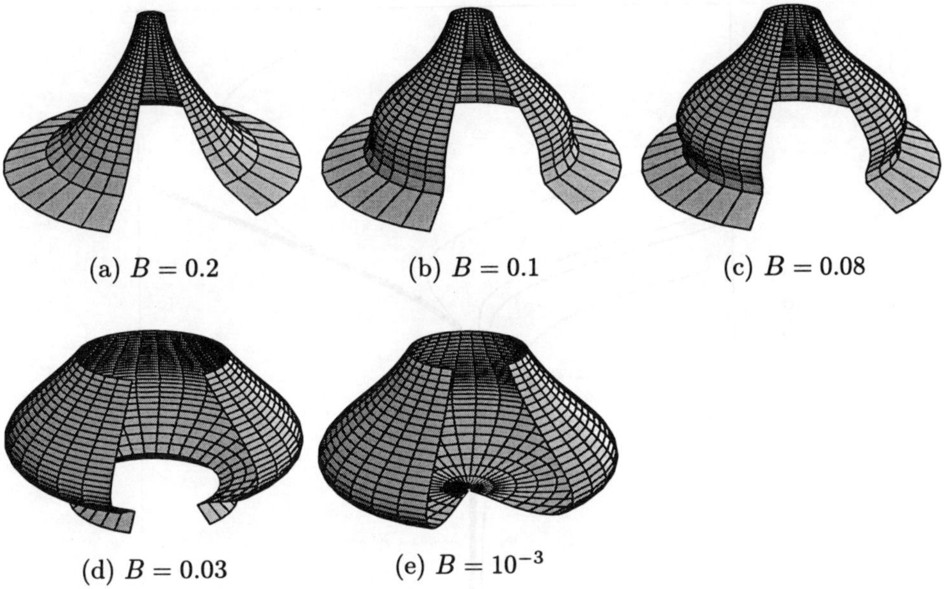

Figure 11. A series of embedding diagrams of the Ernst optical space. From the optical embeddings, one can easily see that (a), (b) for $B > B_c$ no circular photon orbits exist; (c), (d) for $B < B_c$ and (e) $B \ll B_c$ stable and unstable circular photon orbits exist. In the last case, the opening and the 'fold' in the lower part of the 'cup,' representing the unstable circular orbit are very small and beyond the image resolution. Note that the scales of the diagrams are shrunk or expanded so that all the plots occupy the same area. For real proportions, see Fig. 12. Taken from Ref. [22].

the line element of the optical geometry is given by the relation [22]

$$d\tilde{\ell}^2_{\text{Ernst}} = \frac{dr^2}{(1-2/r)^2} + \frac{r^2}{(1+B^2r^2)^4\,(1-2/r)}\,d\phi^2. \tag{120}$$

with M and B being the mass and the strength of asymptotically homogeneous magnetic field, respectively. We put $M = 1$ in the metric; see Ref. [22] for details.[a]

We identify the azimuthal coordinates of the Euclidean space (66) and Ernst optical space while for the radial coordinates we obtain relation [22]

$$\rho = \frac{r}{(1+B^2r^2)^2\sqrt{1-2/r}}. \tag{121}$$

Since

$$\frac{d\rho}{dr} = \frac{1-3/r+B^2r(5-3r)}{(1+B^2r^2)^3\,(1-2/r)^{3/2}}, \tag{122}$$

[a]The dimensionless combination of these parameters, BM, has to be very small in astrophysically realistic situations. Using, for a moment, the c.g.s. units, we find $BM \sim 2\times10^{-53}B_{\text{cgs}}M_{\text{cgs}}$ where $M = M_{\text{cgs}}G/c^2$ and $B = B_{\text{cgs}}G^{1/2}/c^2$. Then for a black hole with $M = 10^9 M_\odot$, typical for active galactic nuclei, the parameter $BM \sim 1$ could be obtained for an extremely strong magnetic field of $B_{\text{cgs}} \sim 10^{11}$ Gauss.

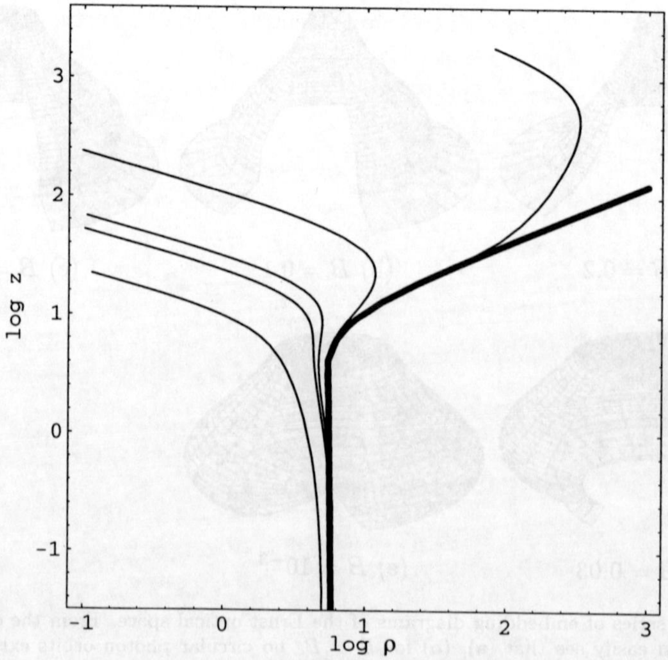

Figure 12. Behavior of the embedding function $z(\rho)$ for Ernst optical space (axial cuts of the diagrams in Fig. 11, thin curves) in comparison with the pure Schwarzschild case (bold curve), in log-log coordinates that enable us to show differences of the orders of magnitude in size. The Ernst embedding functions are given (left to right) for the following values of the parameter B: 0.2, 0.1, 0.08, 0.03, and 10^{-3}. Taken from Ref. [22].

the embedding diagrams shrink to a zero radius asymptotically. Further, there is $\mathrm{d}\rho/\mathrm{d}r = 0$ exactly at the loci of the circular photon orbits, as shown in [22]. (Note that this is a general property of spherically symmetric spacetimes [15].) The parametric form of the embedding formula reads

$$\left(\frac{\mathrm{d}z}{\mathrm{d}r}\right)^2 = \frac{(1-2/r)\,(1+B^2r^2)^6 - \left[1-3/r+B^2r(5-3r)\right]^2}{(1-2/r)^3\,(1+B^2r^2)^6}, \tag{123}$$

and the condition of embeddability

$$\left(1-\frac{2}{r}\right)(1+B^2r^2)^6 - \left[1-\frac{3}{r}+B^2r(5-3r)\right]^2 \geq 0 \tag{124}$$

can be treated numerically only. For $B = 0$, it is reduced to the well-known Schwarzschild condition $r > 9/4$. The embedding diagrams are drawn in Fig. 11 for some typical values of the parameter B, emphasizing the qualitatively different behavior. For comparison with the Schwarzschild case, we give the functions $z(\rho)$ in Fig. 12. Notice that the turning points of the embedding diagrams correspond precisely to the circular photon orbits despite the fact that the Ernst spacetime is axially, not spherically, symmetric.

In [22], it is shown that a maximum strength of the magnetic field exists

$$B_{\rm c} = \sqrt{\frac{3(\sqrt{19}-1)}{(8+\sqrt{19})^2(3+\sqrt{19})}} \approx 0.0947, \tag{125}$$

that limits the Ernst spacetimes with two circular photon orbits and two turning points of the optical geometry embedding (see Fig. 11). Two circular orbits exist, if $B < B_{\rm c}$.

3.5 Kerr–Newman Spacetimes

The line element of the equatorial plane of the optical geometry reads

$$\mathrm{d}\tilde{\ell}^2_{\rm K-N} = \frac{(r^2+a^2)^2 - a^2\Delta}{\Delta^2}\,\mathrm{d}r^2 + \frac{\left[(r^2+a^2)^2 - a^2\Delta\right]^2}{r^4\Delta}\,\mathrm{d}\phi^2. \tag{126}$$

The parametric form of the embedding formula reads

$$\frac{\mathrm{d}z}{\mathrm{d}r} = \pm\sqrt{\frac{(r^2+a^2)^2 - a^2\Delta}{\Delta^2} - \left(\frac{\mathrm{d}\rho}{\mathrm{d}r}\right)^2}\,. \tag{127}$$

The turning points of the embedding diagram, giving its throats and bellies, are determined by the condition $\mathrm{d}\rho/\mathrm{d}r = 0$, where

$$\begin{aligned}\frac{\mathrm{d}\rho}{\mathrm{d}r} = \{&r^4[r(r-3)+2e^2] + a^2[2e^4 + e^2r(3r-7) + r^2(r^2-3r+6)] \\ &- 2a^4(r-e^2)\}[r^3(r^2-2r+a^2+e^2)^{3/2}].\end{aligned} \tag{128}$$

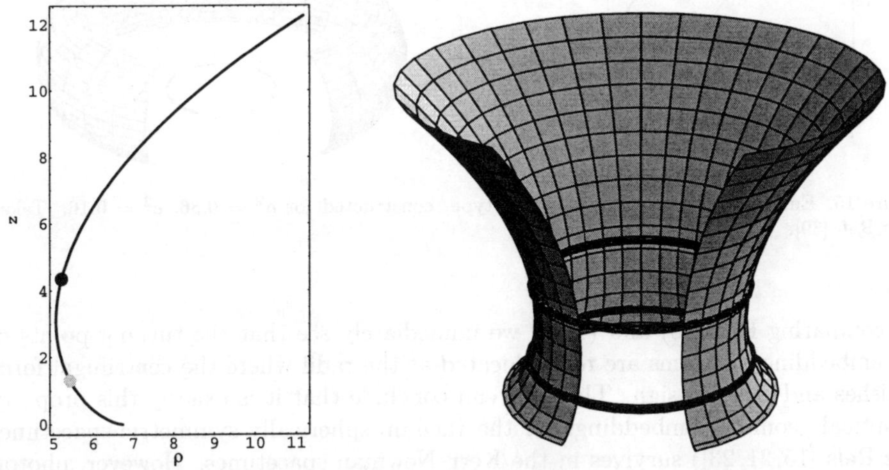

Figure 13. Embedding diagram of the Kerr–Newman black holes of the type $\mathrm{BH_1}$, constructed for $a^2 = 0.16$, $e^2 = 0.16$. The rings in the 3D diagram represent photon circular orbits. Both corotating (gray spot in 2D diagram) and counterrotating (black spot in 2D diagram) orbits are displaced from the throat of the diagram where the centrifugal force vanishes. This is a general property of the rotating backgrounds. Taken from Ref. [39].

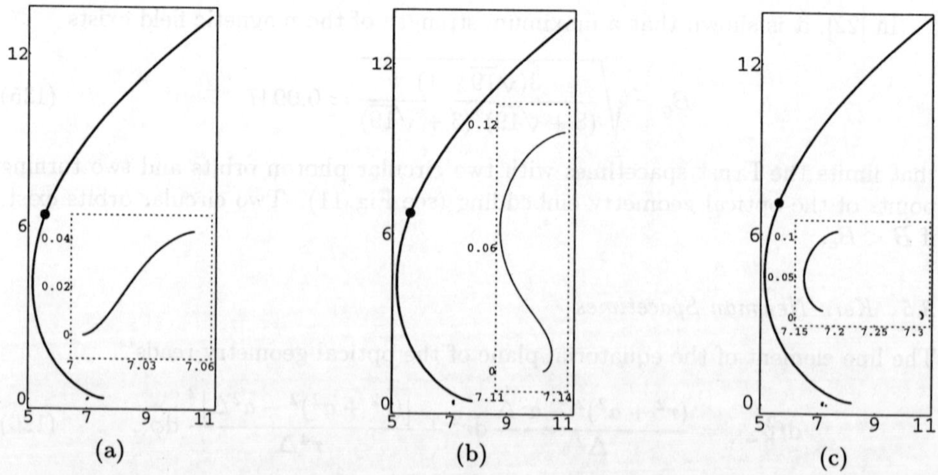

Figure 14. Embedding diagrams of the black holes classes BH_2–BH_4. (a) Class BH_2 ($a^2 = 0.7$, $e^2 = 0.16$); (b) class BH_3 ($a^2 = 0.75$, $e^2 = 0.16$); (c) class BH_4 ($a^2 = 0.81$, $e^2 = 0.16$). Taken from Ref. [39].

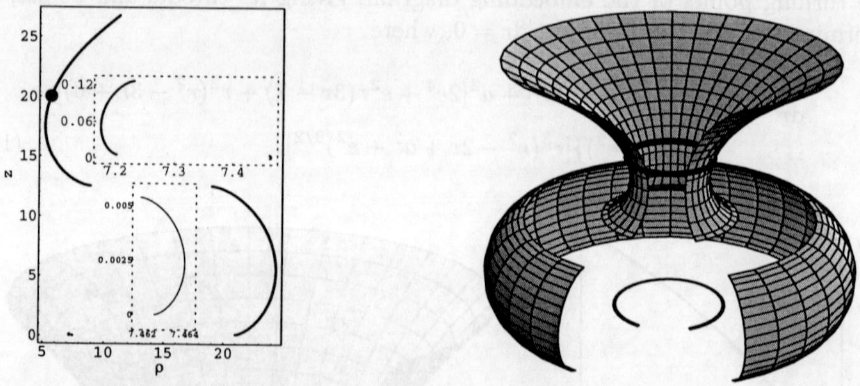

Figure 15. Embedding diagram of the NS_1 type, constructed for $a^2 = 0.86$, $e^2 = 0.16$. Taken from Ref. [39].

By comparing Eqs (48) and (128), we immediately see that the turning points of the embedding diagrams are really located at the radii where the centrifugal force vanishes and changes sign. Thus, we can conclude that it is exactly this property of optical geometry embeddings for the vacuum spherically symmetric spacetimes (see Refs [15, 21, 23]) survives in the Kerr–Newman spacetimes. However, photon circular orbits are displaced from the radii corresponding to the turning points of the embedding diagrams. Therefore, it is interesting to find all the situations where the photon circular orbits lie within the regions of optical geometry embeddability. A detailed account of these situations can be found in [39]. Here, we focus on the embeddability conditions and the turning points of the diagrams.

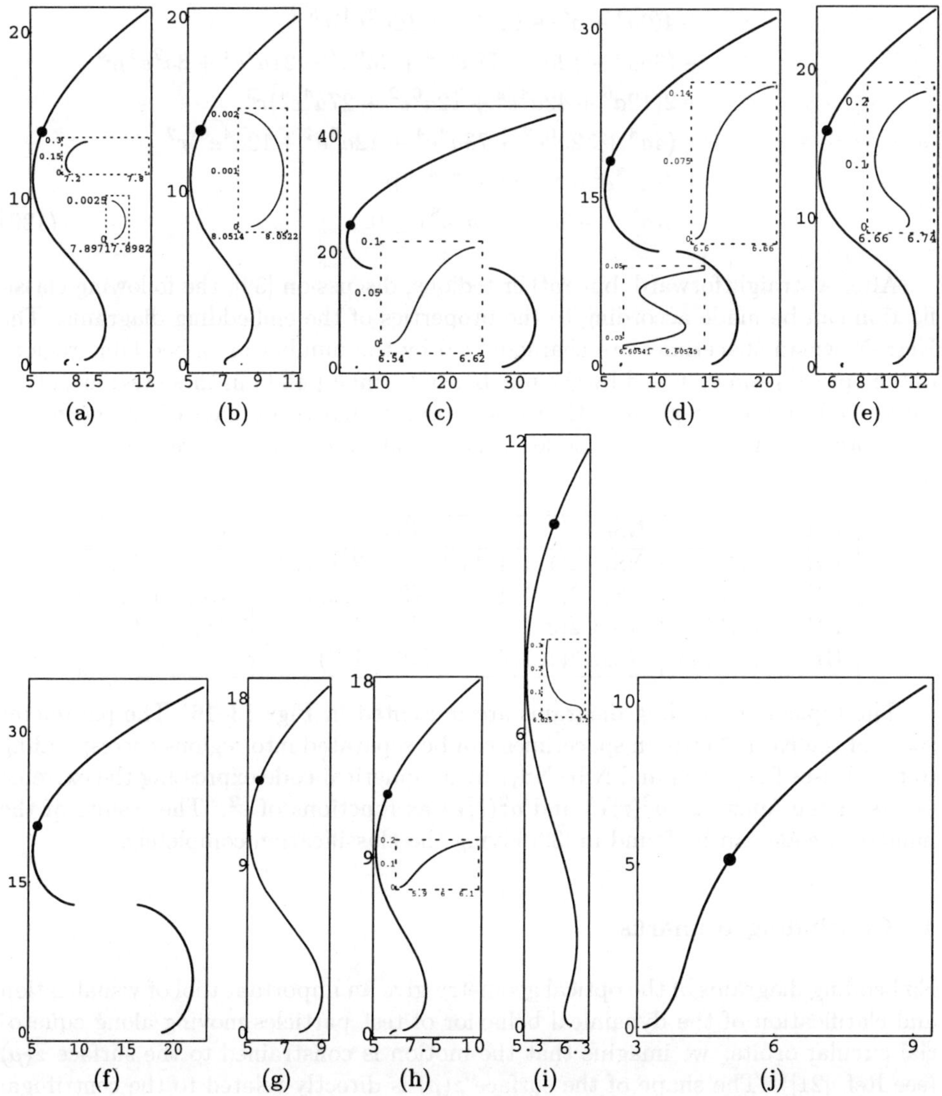

Figure 16. Embedding diagrams of the types (a) NS_2 ($a^2 = 1$, $e^2 = 0.16$), (b) NS_3 ($a^2 = 1.05$, $e^2 = 0.16$), (c) NS_4 ($a^2 = 0.82$, $e^2 = 0.19$), (d) NS_5 ($a^2 = 0.847$, $e^2 = 0.19$), (e) NS_6 ($a^2 = 0.9$, $e^2 = 0.19$), (f) NS_7 ($a^2 = 0.77$, $e^2 = 0.25$), (g) NS_8 ($a^2 = 0.9$, $e^2 = 0.25$), (h) NS_9 ($a^2 = 0.99$, $e^2 = 0.25$), (i) NS_{10} ($a^2 = 1.23$, $e^2 = 0.25$), and (j) NS_{11} ($a^2 = 1$, $e^2 = 0.9$). Taken from Ref. [39].

The embeddability condition $(\mathrm{d}z/\mathrm{d}r)^2 \geq 0$ (see (127)) implies the relation

$$\begin{aligned} E(r;a,e) = {} & 4r^{11} - 3(e^2+3)r^{10} + 12(a^2+e^2)r^9 \\ & - 2(17a^2 + 5a^2e^2 + 2e^4)r^8 + 4(9a^2 + 3a^4 + 12a^2e^2)r^7 \\ & - (33a^4 + 66a^2e^2 + 11a^4e^2 + 17a^2e^4)r^6 \end{aligned}$$

$$
\begin{aligned}
&+4(9a^4+a^6+13a^4e^2+10a^2e^4)r^5\\
&-(36a^4+12a^6+78a^4e^2+4a^6e^2+21a^4e^4+8a^2e^6)r^4\\
&+2(12a^6+42a^4e^2+12a^6e^2+27a^4e^4)r^3\\
&-(4a^8+52a^6e^2+73a^4e^4+12a^6e^4+12a^4e^6)r^2\\
&+4(2a^8e^2+9a^6e^4+7a^4e^6)r\\
&-4(a^8e^4+2a^6e^6+a^4e^8)\geq 0.
\end{aligned}
\tag{129}
$$

After a straightforward, but rather tedious, discussion [39], the following classification can be made according to the properties of the embedding diagrams. The Kerr–Newman spacetimes are characterized by the number of embeddable regions of the optical geometry and by the number of turning points in these regions, given successively for the regions with a descending radial coordinate of the geometry (presented in parentheses). Therefore, the classification can be given in the following way:

BHs	NSs	NSs	NSs
BH_1: 1(1)	NS_1: 4(1,1,1,1)	NS_5: 3(1,1,2)	NS_9: 2(2,0)
BH_2: 2(1,0)	NS_2: 3(2,1,1)	NS_6: 2(2,2)	NS_{10}: 1(4)
BH_3: 2(1,2)	NS_3: 2(3,1)	NS_7: 2(1,1)	NS_{11}: 1(0)
BH_4: 3(1,1,1)	NS_4: 3(1,1,0)	NS_8: 1(2)	

The typical embedding diagrams are presented in Figs 13–16. The parameter space of the Kerr–Newman spacetimes can be separated into regions corresponding to the classes BH_1–BH_4 and NS_1–NS_{11} by a numerical code expressing the extreme points of the functions $a^2_{\mathrm{e}}(r;e)$ and $a^2_{\mathrm{c}}(r;e)$ as functions of e^2. The results of the numerical code can be found in [39] giving the classification completely.

4 Concluding Remarks

Embedding diagrams of the optical geometry give an important tool of visualization and clarification of the dynamical behavior of test particles moving along equatorial circular orbits: we imagine that the motion is constrained to the surface $z(\rho)$ (see Ref. [21]). The shape of the surface $z(\rho)$ is directly related to the centrifugal acceleration. Within the upward sloping areas of the embedding diagram, the centrifugal acceleration points towards increasing values of r and the dynamics of test particles has an essentially Newtonian character. However, within the downward sloping areas of the embedding diagrams, the centrifugal acceleration has a radically non-Newtonian character as it points towards decreasing values of r. Such a kind of behavior appears where the diagrams have a throat or a belly. At the turning points of the diagram, the centrifugal acceleration vanishes and changes its sign, i.e., $\mathrm{d}r/\mathrm{d}\rho = \mathrm{d}z/\mathrm{d}\rho = 0$.

We can understand this connection between the centrifugal force and the embedding of the optical space in terms of the radius of gyration representing the rotational properties of rigid bodies. In Newtonian mechanics, it is defined as the radius $\tilde{R}$ of the circular orbit on which a point-like particle having the same mass

M and angular velocity Ω as a rigid body would have the same angular momentum

$$J = M\tilde{R}^2\Omega. \tag{130}$$

Defining the specific angular momentum $\ell = J/M$, we obtain

$$\tilde{R} = \sqrt{\frac{\ell}{\Omega}}. \tag{131}$$

For a point-like particle moving on $r = \text{const}$, we have $\tilde{R} = r$.

In general relativity, the radius of a circle can be given in two standard ways—namely as the circumferential radius and as the proper radial distance. A third way can be given by generalizing Eq. (131) for a point particle moving along $r = \text{const}$; we use $\ell = L/E$, where L is the angular momentum of the particle and E is its energy. In Newtonian theory, these three definitions of radius yield identical results, however, in general relativity, they are distinct. The radius of gyration $\tilde{R}$ is convenient for discussing the dynamical effects of rotation; the direction of increasing $\tilde{R}$ defines the *local outward direction* of these effects. The surfaces $\tilde{R} = \text{const}$, called von Zeipel cylinders, proved to be a very useful concept in the theory of rotating fluids in stationary, axially symmetric spacetimes. In Newtonian theory, these are ordinary straight cylinders but their shape is deformed by general-relativistic effects and their topology may be noncylindrical. There is a critical family of self-crossing von Zeipel surfaces [42].

It is crucial that in all the spacetimes considered, the following formula holds

$$\left[\tilde{h}_{\phi\phi}(\theta = \pi/2)\right]^{1/2} = \rho = \tilde{R}, \tag{132}$$

i.e., the embedding diagrams of the optical geometry equatorial plane are expressed in terms of the radius of gyration. Since in the equatorial plane, the centrifugal acceleration has a radial component

$$\mathscr{X}(r) = \tilde{R}^{-1}\partial_r\tilde{R}, \tag{133}$$

the relation of the embedding diagrams, the radius of gyration, and the centrifugal force is clear. Note that the turning points of the embedding diagrams determine both the radii where the centrifugal force changes its sign and the radii of cusps where the critical von Zeipel surfaces are self-crossing. Therefore, the embedding diagrams also reflect the properties of fluid orbiting in the spacetimes.

Notice that the black-hole backgrounds have unified character above the event horizon—there exists a throat of the embedding diagram indicating a change of sign of the centrifugal force nearby the event horizon. On the other hand, the naked-singularity spacetimes give a wide variety of behavior of the embeddings and centrifugal forces, ranging from the simple 'Newtonian' backgrounds with no change of sign of centrifugal force, to the very complicated backgrounds where the sign is changed four times in the Kerr–Newman spacetimes.

Acknowledgments

The present work was supported by the grant GAČR No. 202/99/0261, by the Committee for collaboration of the Czech Republic with CERN, and Bergen Computational Physics Laboratory project, an EU Research Infrastructure at the University of Bergen, Norway, supported by the European Community Access to Research Infrastructure Action of the Improving Human Potential Programme. The author would like to express his gratitude to Prof. Jiří Bičák for introducing into the fascinating field of relativistic astrophysics and cosmology, and Prof. Zdeněk Stuchlík for helpful advice, and numerous stimulating discussions on many of the problems of black-hole physics. The present work results from a series of papers written in collaboration with Zdeněk Stuchlík, Erlend Østgaard, Josef Juráň, Petr Slaný, and Jiří Šoltés. Contributions of all of these collegues is acknowledged.

References

1. R. Penrose. Gravitational collapse: The role of general relativity. *Nuovo Cimento B*, 1(special number):252–276, 1969.
2. C. W. Misner, K. S. Thorne, and J. A. Wheeler. *Gravitation*. Freeman, San Francisco, 1973.
3. B. Carter. Black hole equilibrium states. In C. De Witt and B. S. De Witt, editors, *Black Holes*, page 57, New York–London–Paris, 1973. Gordon and Breach.
4. J. M. Bardeen. Timelike and null geodesics in the Kerr metric. In C. De Witt and B. S. De Witt, editors, *Black Holes*, page 215, New York–London–Paris, 1973. Gordon and Breach.
5. J. Bičák and Z. Stuchlík. On the latitudinal and radial motion in the field of a rotating black hole. *Bull. Astronom. Inst. Czechoslovakia*, 27(3):129–133, 1976.
6. Z. Stuchlík. Equatorial circular orbits and the motion of the shell of dust in the field of a rotating naked singularity. *Bull. Astronom. Inst. Czechoslovakia*, 31(3):129–144, 1980.
7. I. G. Dymnikova. *Uspekhi Fiz. Nauk*, 148:393, 1986. [Sov. Phys. Usp., 29:215, 1986].
8. J. Bičák, Z. Stuchlík, and V. Balek. The motion of charged particles in the field of rotating charged black holes and naked singularities. *Bull. Astronom. Inst. Czechoslovakia*, 40(2):65–92, 1989.
9. V. Balek, J. Bičák, and Z. Stuchlík. The motion of the charged particles in the field of rotating charged black holes and naked singularities. *Bull. Astronom. Inst. Czechoslovakia*, 40(3):133–165, 1989.
10. J. Bičák and Z. Stuchlík. The fall of the shell of dust on to a rotating black hole. *Monthly Notices Roy. Astronom. Soc.*, 175(2):381–393, 1976.
11. Z. Stuchlík, J. Bičák, and V. Balek. The shell of incoherent charged matter falling onto a charged rotating black hole. *Gen. Relativity Gravitation*, 31(1):53–71, 1999.
12. Z. Stuchlík. The motion of test particles in black-hole backgrounds with non-zero cosmological constant. *Bull. Astronom. Inst. Czechoslovakia*, 34(3):129–149, 1983.
13. Z. Stuchlík. An Einstein–Strauss–de Sitter model of the universe. *Bull. Astronom. Inst. Czechoslovakia*, 35(4):205–215, 1984.
14. Z. Stuchlík. Kerr–Newman–anti-de Sitter black-hole spacetimes with a surface

of degeneracy. *Acta Phys. Slovaca*, 48(5):535–548, 1998.
15. Z. Stuchlík and S. Hledík. Some properties of the Schwarzschild–de Sitter and Schwarzschild–anti-de Sitter spacetimes. *Phys. Rev. D*, 60(4):044006 (15 pages), 1999.
16. Z. Stuchlík and S. Hledík. Equatorial photon motion in the Kerr–Newman spacetimes with a non-zero cosmological constant. *Classical Quantum Gravity*, 17(21):4541–4576, 2000.
17. Z. Stuchlík and S. Hledík. Equilibrium of a charged spinning test particle in Reissner–Nordström backgrounds with a nonzero cosmological constant. *Phys. Rev. D*, accepted in July 2001, to be published in Physical Review D, 15th of month issue, 2001.
18. Z. Stuchlík and S. Hledík. Properties of the Reissner–Nordström spacetimes with a nonzero cosmological constant. (unpublished, preprint TPA 003/Vol. 2, 2001), 2001.
19. Z. Stuchlík. Spherically symmetric static configurations of uniform density in spacetimes with a non-zero cosmological constant. *Acta Phys. Slovaca*, 50(2):219–228, 2000.
20. Z. Stuchlík, S. Hledík, J. Šoltés, and E. Østgaard. Null geodesics and embedding diagrams of the interior Schwarzschild–de Sitter spacetimes with uniform density. *Phys. Rev. D*, 64(4):044004 (17 pages), 2001.
21. S. Kristiansson, S. Sonego, and M. A. Abramowicz. Optical space of the Reissner–Nordström solutions. *Gen. Relativity Gravitation*, 30(2):275–288, 1998.
22. Z. Stuchlík and S. Hledík. Photon capture cones and embedding diagrams of the Ernst spacetime. *Classical Quantum Gravity*, 16(4):1377–1387, 1999.
23. M. A. Abramowicz, B. Carter, and J.-P. Lasota. *Gen. Relativity Gravitation*, 20:1173, 1988.
24. O. Semerák. What forces drive the relativistic motion? *Nuovo Cimento B*, 110:973, 1995.
25. R. T. Jantzen, P. Carini, and D. Bini. The many faces of gravitoelectromagnetism. *Ann. Physics*, 215(1):1–50, 1992.
26. M. A. Abramowicz, P. Nurowski, and N. Wex. Optical reference geometry for stationary and axially symmetric spacetimes. *Classical Quantum Gravity*, 12(6):1467, 1995.
27. M. A. Abramowicz. Centrifugal force – a few surprises. *Monthly Notices Roy. Astronom. Soc.*, 245(4):733–746, 1990.
28. M. A. Abramowicz. Relativity of inwards and outwards – an example. *Monthly Notices Roy. Astronom. Soc.*, 256(4):710–718, 1992.
29. M. A. Abramowicz and J. Bičák. On the interplay between relativistic gravitational, centrifugal and electric forces: a simple example. *Gen. Relativity Gravitation*, 23:941, 1991.
30. M. A. Abramowicz and J. C. Miller. Ellipticity behaviour of relativistic Maclaurin spheroids. *Monthly Notices Roy. Astronom. Soc.*, 245(4):729, 1990.
31. J. C. Miller. Relativistic gravitational collapse. In G. Ellis, A. Lanza, and J. Miller, editors, *The Renaissance of General Relativity and Cosmology*, pages 73–85, Cambridge, 1993. Cambridge University Press. A Survey to Celebrate

the 65th Birthday of Dennis Sciama.

32. J. C. Miller. In C. J. S. Clarke and R. d'Inverno, editors, *Proceedings of the meeting "Classical and numerical relativity," Southampton, December 1991*, Cambridge, 1992. Cambridge University Press.
33. M. A. Abramowicz and A. R. Prasanna. Centrifugal force reversal near a Schwarzschild black-hole. *Monthly Notices Roy. Astronom. Soc.*, 245(4):720–728, 1990.
34. M. A. Abramowicz, P. Nurowski, and N. Wex. Covariant definition of inertial forces. *Classical Quantum Gravity*, 10(10):L183, 1993.
35. M. A. Abramowicz, J. Miller, and Z. Stuchlík. Concept of radius of gyration in general relativity. *Phys. Rev. D*, 47(4):1440–1447, 1993.
36. Z. Stuchlík. *Bull. Astronom. Inst. Czechoslovakia*, 41:341, 1990.
37. Z. Stuchlík and M. Calvani. Null geodesics in black hole metrics with non-zero cosmological constant. *Gen. Relativity Gravitation*, 23:507, 1991.
38. S. Iyer and A. R. Prasanna. Centrifugal force in Kerr geometry. *Classical Quantum Gravity*, 10(1):L13–L16, 1993.
39. Z. Stuchlík, S. Hledík, and J. Juráň. Optical reference geometry of Kerr–Newman spacetimes. *Classical Quantum Gravity*, 17(14):2691–2718, 2000.
40. J. M. Aguirregabiria, A. Chamorro, K. Rajesh Nayak, J. Suinaga, and C. V. Vishveshwara. Equilibrium of a charged test particle in the Kerr–Newman spacetime: force analysis. *Classical Quantum Gravity*, 13(8):2179, 1996.
41. S. Sonego and M. A. Abramowicz. Maxwell equations and the optical geometry. *J. Math. Phys.*, 39(6):3158–3166, 1998.
42. M. Kozłowski, M. Jaroszyński, and M. A. Abramowicz. The analytic theory of fluid disks orbiting the Kerr black hole. *Astronom. and Astrophys.*, 63:209–220, 1978.

APPEARANCE OF APPROXIMATED MODELS OF ROTATING NEUTRON STARS

TOMÁŠ LEDVINKA
Institute of Theoretical Physics, Charles University, Prague
E-mail: tomas.ledvinka@mff.cuni.cz

We study the influence of the strong gravitational field of neutron stars on the light seen by distant observers. The gravitational field is approximated by the Kerr metric including thus the dragging effect caused by stellar rotation. Using this approximation, we construct pictures of compact rotating neutron stars as they would look like if light of these stars could avoid being disturbed by the environment surrounding the stars.

The effect of light deflection played an important role in the history of general relativity, however, measurements of light deflection available to astronomers do not probe strong gravitational fields even 80 years after the first observations [1]. Even though there is no way we could take a photograph of a neutron star in a foreseeable future, their appearance was studied by several authors in the static case [2]. They also discussed this effect from the perspective of its astrophysical consequences (see e.g. [3,4]).

Nevertheless, from the aesthetical point of view one may simply see the light propagation in strong gravitational fields surrounding fast rotating neutron stars and its implications for the star's appearance as a natural extension of light deflection phenomena to the realm of both draggings and redshifts. Since the author of these lines had been introduced into the subject by Jiří Bičák, the aesthetical point of view was a sufficient motivation for obtaining "sharp images" of neutron stars despite the fact that neither an unperturbed propagation of light near neutron stars nor the availability of telescopes with a sufficient resolution can be expected.

1 Neutron star models

Neutron stars considered here are quiet bodies of perfect fluid rotating with a constant angular velocity and placed in a stationary axisymmetric asymptotically flat spacetime. The existence of the Killing vector fields $\xi^\mu = (\partial/\partial t)^\mu$ and $\eta^\mu = (\partial/\partial\varphi)^\mu$ in this spacetime enables one (see e.g. [5]) to describe such a spacetime with the line element

$$ds^2 = -e^{2\nu}dt^2 + e^{2\zeta-2\nu}(d\rho^2 + \rho^2 d\vartheta^2) + B^2 e^{-2\nu}\rho^2\sin^2\vartheta\,(d\varphi - \omega dt)^2\,, \quad (1)$$

where only four metrical potentials ν, ζ, ω, and B appear. These functions must satisfy certain regularity conditions on the axis $\sin\vartheta = 0$ and at infinity $r = \infty$ so that they describe an asymptotically flat spacetime without singularities (see e.g. [6]).

The velocity of the fluid's circular motion is a linear combination of both Killing vectors

$$u^\mu = u^t(\xi^\mu + \Omega\,\eta^\mu) = u^t[1,0,0,\Omega]\,. \quad (2)$$

This velocity enters the energy-momentum tensor of perfect fluid

$$T^{\mu\nu} = (\epsilon + p)u^\mu u^\nu + pg^{\mu\nu} , \tag{3}$$

where ϵ is the energy density and p is pressure. Due to its symmetries, this source term in Einstein equations is compatible with the given line element (1) and Einstein equations yield nonlinear elliptic partial differential equations in coordinates ρ and θ for the potentials ν, ω, and B and a first order one for ζ (see e.g. [6,7]).

We will assume that both entropy density and angular rotation velocity differences can be neglected and a simple equation of state $\epsilon = \epsilon(p)$ and a rotation law $\Omega =$ constant hold.

The equations of motion for the stationary fluid simplify to the stationary version of the hydrostatic equilibrium equation [5]

$$\nabla_\mu p = (\epsilon + p)\ \nabla_\mu\ \ln u^t . \tag{4}$$

The velocity normalization factor

$$u^t = \frac{1}{\sqrt{-g_{tt} - 2\Omega g_{t\varphi} - \Omega^2 g_{\varphi\varphi}}} = \frac{e^{-\nu}}{\sqrt{1 - B^2 e^{-4\nu}\rho^2 \sin^2\vartheta(\omega - \Omega)^2}} \tag{5}$$

can be directly used to calculate the pressure since the integral of (4) reads

$$h(p) = \int_0^p \frac{dp}{\epsilon(p) + p} = \ln\left(\frac{u^t}{u^t_{polar}}\right) = \nu_{polar} - \nu - \ln\sqrt{1 - v^2}, \tag{6}$$

where the function $h(p)$ is given by the equation of state. This is a relativistic version of the Bernouli equation where the speed of fluid $v = Be^{-2\nu}\rho\sin\vartheta(\Omega - \omega)$ is measured by zero angular momentum observers. Note that the surface p=0 is given by the co-rotating equipotential

$$u^t = u^t_{polar} = 1 + z_p, \tag{7}$$

where z_p is the redshift of photons radiated from poles.

The equation of state for the neutron star matter is the crucial unknown in the construction of a neutron star model since the conditions in the center of the most massive neutron stars exceed current knowledge of nuclear physics (see e.g. the review [8]). Once an equation of state is supplied the equations can be solved using various numerical methods (see e.g. overview and references in [9] concerning both equations of state and numerical methods). Apart from the values of the metric potentials, pressure or energy density, such a solution can be described by a number of physically descriptive quantities. We will follow here [9] where Nozawa *et al.* provide the following parameters for each neutron star model:

Gravitational Mass M is the most important parameter each model provides. It can be both determined by a distant observer ($\nu \sim -M/\rho$) and (since the spacetime is stationary) also calculated as the integral over the region of non-vanishing energy-momentum tensor. Volume integrals of baryon density or energy density are also mentioned and designated the *baryon* and the *proper mass.*

Total angular momentum J determines the asymptotical behavior of $\omega \sim 2J/\rho^3$. It is also obtained as an "integral over source" this time due to the axial symmetry.

Central energy density ϵ_c is, together with the *equation of state*, crucial to the internal structure of a star (we can ignore both of them when studying light propagation above the stellar surface).
Angular velocity Ω coincides with the observed *pulsar* periods. The rotation directly deforms the shape of meridians on a star's image due to the abberation – the equatorial speed for the fastest models reaches $c/2$.
Equatorial circumferential radius R_c invariantly describes the star's size (compactness). Some models have $R_c < 4M$. Because the circular photon orbits around a Schwarzschild black hole have circumferential radius equal to $3M$ we can expect a serious distortion of its globe-like appearance for such neutron stars. The ratio of polar and equatorial surface radial coordinate ρ_p/ρ_e is used to indicate star's *oblateness*.
Polar redshift z_p is another indicator of the strength of general-relativistic effects. Using relation (7), this parameter together with Ω determine the surface $p = 0$ for given metric potentials.
Forward and *backward equatorial redshift* z^f_{eq},z^b_{eq} are the minimal and maximal redshifts of a photon leaving the equator. They indicate whether the rotation is relativistic as well as how much is the star's image distorted (see Section 3).

2 Raytracing in a stationary axisymmetric spacetime

The information about a neutron star's appearance is brought to an observer by light originating at the star's surface. We assume that because the neutron star's diameter is of order 10 km, the image that an observer sees in her telescope can be thought of as created by individual interference-free photons registered on a photographic plate. [11] Photons are radiated from the surface and after traveling along their geodesics some arrive at the observer's telescope aperture.

The fact that the spacetime under consideration is stationary and axisymmetric allows one to time-reverse the worldline of each photon and check whether and where the photon starting from the observer's camera hits the star's surface. To find out which photons taking off from the surface of the star arrive at the observer's telescope would be more complicated. Assuming the spacetime symmetries, one can clearly see that if $x^\mu(\lambda) = [t(\lambda), r(\lambda), \vartheta(\lambda), \varphi(\lambda)]^\mu$ is a null geodesic then ${}^*x^\mu(\lambda) = [-t(-\lambda), r(\lambda), \vartheta(\lambda), -\varphi(-\lambda)]^\mu$ is also a geodesic. The former starts at $t = 0$ from the observer's telescope (placed at $r = r_0$, $\vartheta = \vartheta_0$, and $\varphi = 0$) and hits the star at $t = t_1$, $\vartheta = \vartheta_1$, and $\varphi = \varphi_1$. The latter describes a photon which leaves the star at $t = -t_1$, $\vartheta = \vartheta_1$, and $\varphi = -\varphi_1$ and at $t = 0$ is registered by the observer. This is the way we obtain "photos" of compact stars.

We suppose that the observer is equipped with a *camera obscura*. First assume a flat spacetime. The axis of the camera is directed towards the star and it is aligned with the vector $-\vec{e}_r$. The position on the photographic plate of camera is described by appropriately scaled Cartesian coordinates ξ and η aligned with the basis vectors $\vec{e}_\varphi$ and $-\vec{e}_\vartheta$, respectively ($\vec{e}_r$, $\vec{e}_\vartheta$, $\vec{e}_\varphi$ being the basis vectors in spherical coordinates). Using explicit equations for a beam of rays passing (when

$\lambda = 1$) through the pinhole of camera

$$x^\mu(\lambda) = \begin{bmatrix} t \\ x \\ y \\ z \end{bmatrix}^\mu = \begin{bmatrix} \lambda\, r_0 \sqrt{1 + (\xi^2 + \eta^2)/r_0^2} \\ -\,\eta\,(\lambda - 1)\, \cos\vartheta_0 + \lambda\, r_0\, \sin\vartheta_0 \\ \xi\,(\lambda - 1) \\ \eta\,(\lambda - 1)\, \sin\vartheta_0 + \lambda\, r_0\, \cos\vartheta_0 \end{bmatrix}^\mu \tag{8}$$

one can, after transforming from the Cartesian to the spherical Minkowski coordinates, obtain the covariant components of the photon's four-momentum. Their values for a photon reaching the observer at infinity are

$$\lim_{r_0 \to \infty} p_\mu\,_{|\lambda=1} = [p_t, p_r, p_\vartheta, p_\varphi]_\mu = [-1, 1, -\eta, \xi\, \sin\vartheta_0]_\mu\,. \tag{9}$$

As long as the focal length L of the camera is small enough, equation (8) locally describes the worldline of the photon in the distant observer's local frame on the way through the camera and $[L\xi, L\eta]$ are its impact coordinates on the photographic plate although (8) is not globally a null geodesic in the spacetime (1).

We will consider worldlines of photons moving from the observer's position towards the star with an initial momentum (9). For sufficiently small ξ and η, the photon hits the star's surface at Y^μ. After inverting the sign of Y^φ and Y^t we get the event ${}^*Y^\mu$ corresponding to the latitude, longitude, and time of flight of a photon that, if radiated from surface at ${}^*Y^\mu$, ends up with coordinates $X^a = [L\xi, L\eta]^a$ on the photographic plate of observer's camera. Due to the rotation of the star, the astrographical longitude of the event ${}^*Y^\mu$ is $-Y^\varphi + \Omega\, Y^t$. We will discuss later how to cope with the fact that for an observer located at infinity the time of flight is infinite.

Apart from the point of a photon's origin at the star's surface, there is another important property of the photon registered by the observer – its redshift. Due to the equation of the surface isobar $p = 0$ (7), the redshift of a photon with impact parameters ξ and η reads

$$1 + z \;=\; \frac{E_{surf}}{E_{obs}} \;=\; \frac{u^t p_t + u^\varphi p_\varphi}{p_t} \;=\; u^t_{polar}(1 - \xi\, \Omega\, \sin\vartheta_0). \tag{10}$$

The conservation of energy and angular momentum of photons implies that the redshift of photons originating on star's surface is a linear function of ξ. This means there is no need for a color photographic film in the observer's camera. Even if the picture of star is highly distorted due to the gravitational deflection of rays, the color will be a linear function of the coordinate ξ – the lines of constant redshift are equidistantly distributed parallel lines.

There is another consequence of (10). Three redshifts z^b_{eq}, z_p, z^f_{eq} provided by numerical models contain information about the distortion of the star's image as seen by a distant equatorial ($\vartheta = \pi/2$) observer. It shows that $z^b_{eq} - z_p > z_p - z^f_{eq}$ as a result of dragging and thus the image tends to be asymmetric – the approaching and receding parts of the image have different sizes.

3 Appearance of fluid body in Kerr geometry

It is possible today to determine numerically the structure and properties of the most compact neutron stars with a high accuracy for every reasonable equation of state. A detailed comparison of the available methods can be found in [9]. The typical accuracy showed to be better than 10^{-3}. On the other hand, the uncertainty caused by the wide scale of plausible equations of state available today for describing neutron star matter is much larger.

We want to illustrate the effects related to the appearance of compact fast rotating neutron stars. We should bet on a certain equation of state, then compute its external gravitational field, solve for photon geodesics through this field, and finally find points of impact of these photons on the surface. Both the connection coefficients and surface coordinates have to be interpolated consistently with the interpolation scheme used when solving field equations to avoid numerical artifacts. In fact, metric coefficients (and their derivatives) are the least precise quantities provided by numerical models. In [9] we read: "... *More sensitive are the three redshifts and the eccentricity. It should be noted that redshift factors are local quantities which reflect metric potentials at each point. This implies that metric potentials do not have the same agreement between different numerical codes as integrated global quantities. For the eccentricity, one needs to compute the length along the surface of the star.... This causes differences up to a few percent in eccentricity for rapidly rotating models. On the other hand global quantities ... agree quite well among results of different codes.*"

This illustrates what kind of a problem we might hit while trying to directly solve the geodesic equations with potentials given as a numerical solution of field equations. In fact, a majority of geodesics we calculate when rendering the neutron star "photos" are *computed* almost *along the surface* and they are the most sensitive ones regarding the appearance of the star.

To simplify the problem, we shall try to approximate the external field with the simplest gravitational field possible – the Kerr metric. It, of course, describes the gravitational field of a black hole. We assume that some unrealistic matter, which behaves like an ordinary perfect fluid on the surface, can be made to create the Kerr metric outside of the star. This is not an outstanding approximation but closed formulae enable one to have an exact metric and exact equation for star's surface.

The line element in the Boyer-Lindquist coordinates reads [10–12]

$$\begin{aligned} ds^2 &= -\left(\frac{\Delta - a^2\sin^2\vartheta}{\Sigma}\right)dt^2 - \frac{2Mar}{\Sigma}\sin^2\vartheta\, dt\, d\varphi \\ &\quad + \frac{\Sigma}{\Delta}\,dr^2 + \left[r^2 + a^2 + \frac{2Mra^2}{\Sigma}\sin^2\vartheta\right]\sin^2\vartheta\, d\varphi^2, \end{aligned} \tag{11}$$

where

$$\Sigma = r^2 + a^2\cos^2\vartheta, \tag{12}$$

$$\Delta = r^2 + a^2 - 2\,M\,r. \tag{13}$$

The Boyer-Lindquist radial coordinate r is related to ρ via

$$r = \rho + M + \frac{M^2 - a^2}{\rho}\,. \tag{14}$$

The inner metric is not needed when studying the appearance of the star and thus the metric (11) together with the surface isobar equation (7) fully describe the problem.

There are four constants of motion for a test particle in the Kerr metric. Three of them are common to every stationary axisymmetric spacetime – the mass of the particle m, the energy $E = -p_t$, and the axial component of angular momentum $L_z = p_\varphi$. An additional constant of motion exists in the Kerr geometry (see [12])

$$\mathcal{Q} = p_\vartheta^2 + \cos^2\vartheta\left[a^2(m^2 - E^2) + \frac{L_z^2}{\sin^2\vartheta}\right] \tag{15}$$

and its existence enables one to write down first-order differential equations for test particle motion

$$\Sigma\frac{dr}{d\lambda} = \pm\sqrt{[E(r^2+a^2) - L_z a]^2 - \Delta[m^2r^2 + (L_z - aE)^2 + \mathcal{Q}]}\;, \tag{16}$$

$$\Sigma\frac{d\vartheta}{d\lambda} = \pm\sqrt{\mathcal{Q} - \cos^2\vartheta\left[a^2(m^2 - E^2) + \frac{L_z^2}{\sin^2\vartheta}\right]}\;, \tag{17}$$

$$\Sigma\frac{d\varphi}{d\lambda} = -aE + \frac{L_z}{\sin^2\vartheta} + a\frac{E(r^2+a^2) - L_z a}{\Delta}\;, \tag{18}$$

$$\Sigma\frac{dt}{d\lambda} = a(L_z - aE\sin^2\vartheta) + (r^2+a^2)\frac{E(r^2+a^2) - L_z a}{\Delta}\;. \tag{19}$$

These equations can be transformed into the form that is more appropriate when studying photons ($E = -1$, $m = 0$) starting from infinity. We introduce $u = 1/r$, $\mu = \cos\vartheta$, and a new (non-affine) parameter χ such that photons can reach $u = 0$ with a finite value of χ. We also increase the order of equations (16-17) to remove square roots. The equations of motion then take form

$$\frac{d^2u}{d\chi^2} = -(\mathcal{Q} + L_z^2 - a^2)u + 3M\left[\mathcal{Q} + (L_z - a)^2\right]u^2 - 2a^2\mathcal{Q}u^3\;, \tag{20}$$

$$\frac{d^2\mu}{d\chi^2} = -(\mathcal{Q} + L_z^2 - a^2)\mu - 2a^2\mu^3\;, \tag{21}$$

$$\frac{d\varphi}{d\chi} = \frac{L_z}{1-\mu^2} + au\frac{2M - L_z au}{1 - 2Mu + a^2u^2}\;, \tag{22}$$

$$\frac{dt}{d\chi} = a^2\mu^2 + 2M\frac{2M - L_z au}{1 - 2Mu + a^2u^2} + \left(\frac{2M}{u} + \frac{M^2}{u^2}\right)\left(1 - \frac{du}{d\chi}\Theta_h(u)\right)\;. \tag{23}$$

The initial conditions are $u(0) = 0$, $\mu(0) = \mu_0$, $\varphi(0) = 0$, $t(0) = 0$, $du/d\chi(0) = 1$, and $d\mu/d\chi(0) = \sqrt{\mathcal{Q} - \mu_0^2(\mathcal{Q} + L_z^2 - a^2) - a^2\mu_0^4}$, . The term $\Theta_h(u)$ is added to (19) to subtract the diverging time a photon needs to reach the infinity $u = 0$. It must vanish for $u > u_x$ where u_x lies outside of the star; it should be sufficiently smooth at $u = u_x$ and behave as $1 + O(u^2)$ at $u = 0$ to cancel the M^2/u^2 term in (23), for example $\Theta_h(u) = (3u/u_x + 1)(1 - u/u_x)^3$. The integral of (23) from the observer's radius $1/u_0$ to the star's surface (i.e. the time of flight) changes by a constant number $\int_{u_0}^{u_x}(2M/u + m^2/u^2)\Theta_h(u)du$. This constant becomes infinite as $u_0 \to \infty$ and cancels the diverging part of the time of flight, Y^t.

Equations (20-23) can be easily solved numerically. The numerical code checks each geodesic for the intersection with the star's surface and then draws the "photo".

4 The pictorial

We use two neutron star models from about a dozen presented in [9] to illustrate the influence of a curved spacetime in their vicinity on the star's appearance.

The A model represents a neutron star with quite a usual mass 1.4 $M_\odot$ and a very high rotation rate 1850 Hz. Its equation of state (EOS) is very soft. The B model is a highly relativistic neutron star. It has a maximal mass of 3.3 $M_\odot$ allowed by its (very stiff) EOS and it is supported against the collapse by extreme rotation 1300 Hz (without rotation the EOS of this model gives the maximal mass of 2.7 $M_\odot$). Details regarding the EOS and numerical methods can be found in [9] and its references. Summary of the relevant parameters of both models is given in the Table 1.

	Model A		Model B	
Model	Gbr	Kerr	L(L)mr	Kerr
$M/M_\odot$	1.41	–	3.32	–
R_c/M	4.65	4.68	3.96	3.62
J/M^2	0.60	0.60	0.70	0.65
ΩM	0.077	0.082	0.127	0.134
ρ_p/ρ_e	0.64	0.72	0.55	0.62
z_p	0.47	0.47	0.83	0.84
z_{eq}^f	-0.24	-0.23	-0.35	-0.33
z_{eq}^b	1.28	1.28	2.45	2.43

Table 1. Parameters of two models with a realistic EOS and their Kerr approximations.

The parameters of the Kerr field approximations for both models are also included in Table 1. The parameters of the A model can be fitted quite well using the Kerr approximation, model B shows that, as the star becomes more compact, differences arise. The most visible difference is that the surface of a real star is more oblate than that in Kerr approximation.

Appearance of these models is then presented in Figures 1-6. We plotted the standard coordinate grid co-rotating with the star. The star rotates in such a way that the righthand side is approaching. We also show six rotation phases for the B model with a hot spot on surface. The gravitational lensing is apparent.

Figures constructed in this section were also inspired by [2] where the appearance of a spherical neutron star is depicted and by the animations these authors produced in early 90's where a real neutron star model was shown rotating.

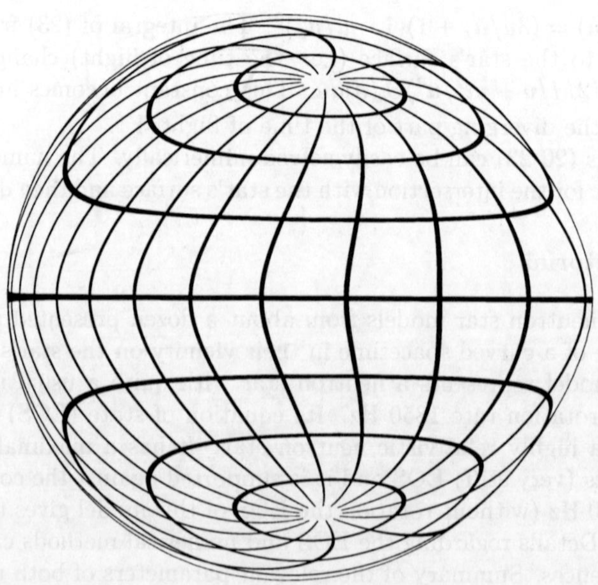

Figure 1. Model A as seen by an observer within the equatorial plane.

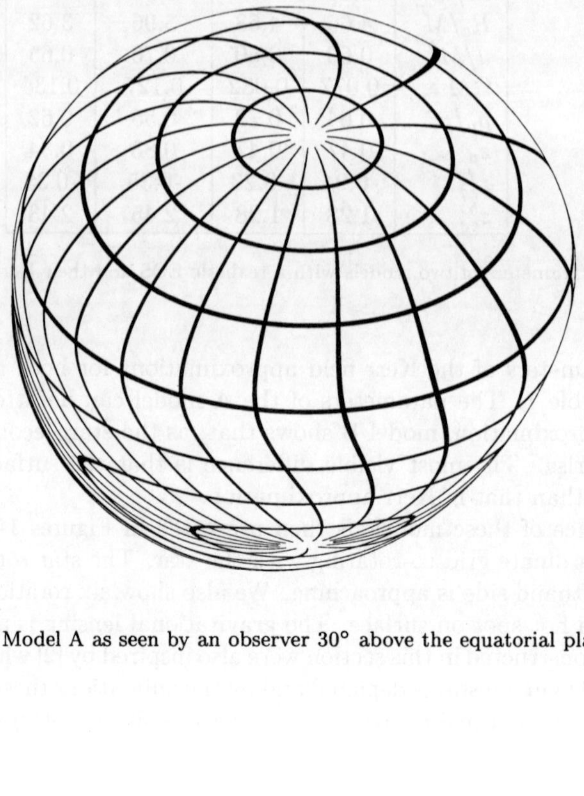

Figure 2. Model A as seen by an observer 30° above the equatorial plane.

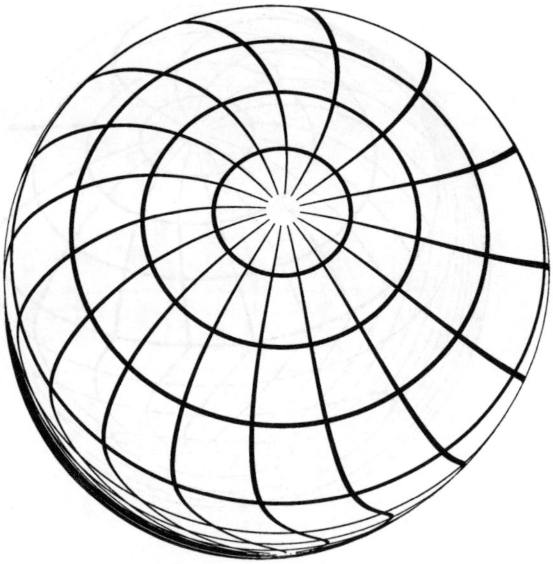

Figure 3. Model A as seen by an observer 60° above the equatorial plane.

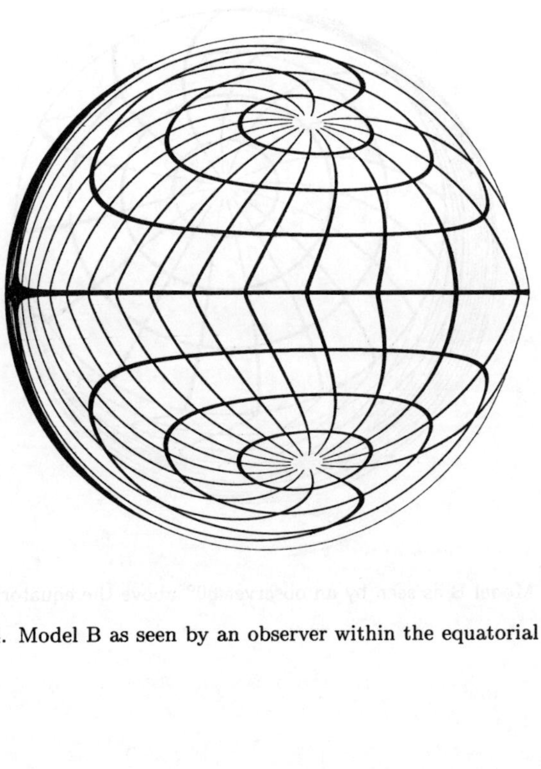

Figure 4. Model B as seen by an observer within the equatorial plane.

Figure 5. Model B as seen by an observer 30° above the equatorial plane.

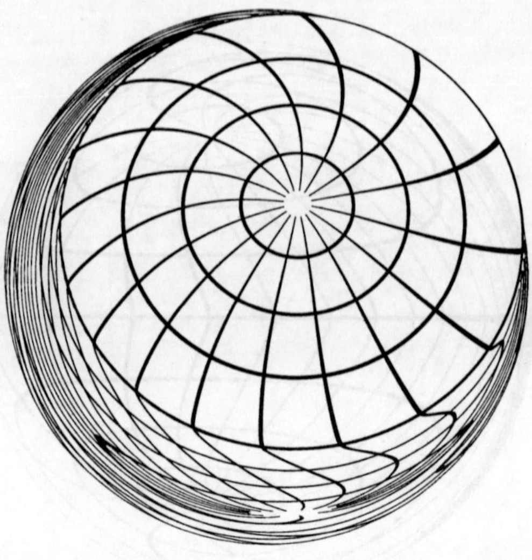

Figure 6. Model B as seen by an observer 60° above the equatorial plane.

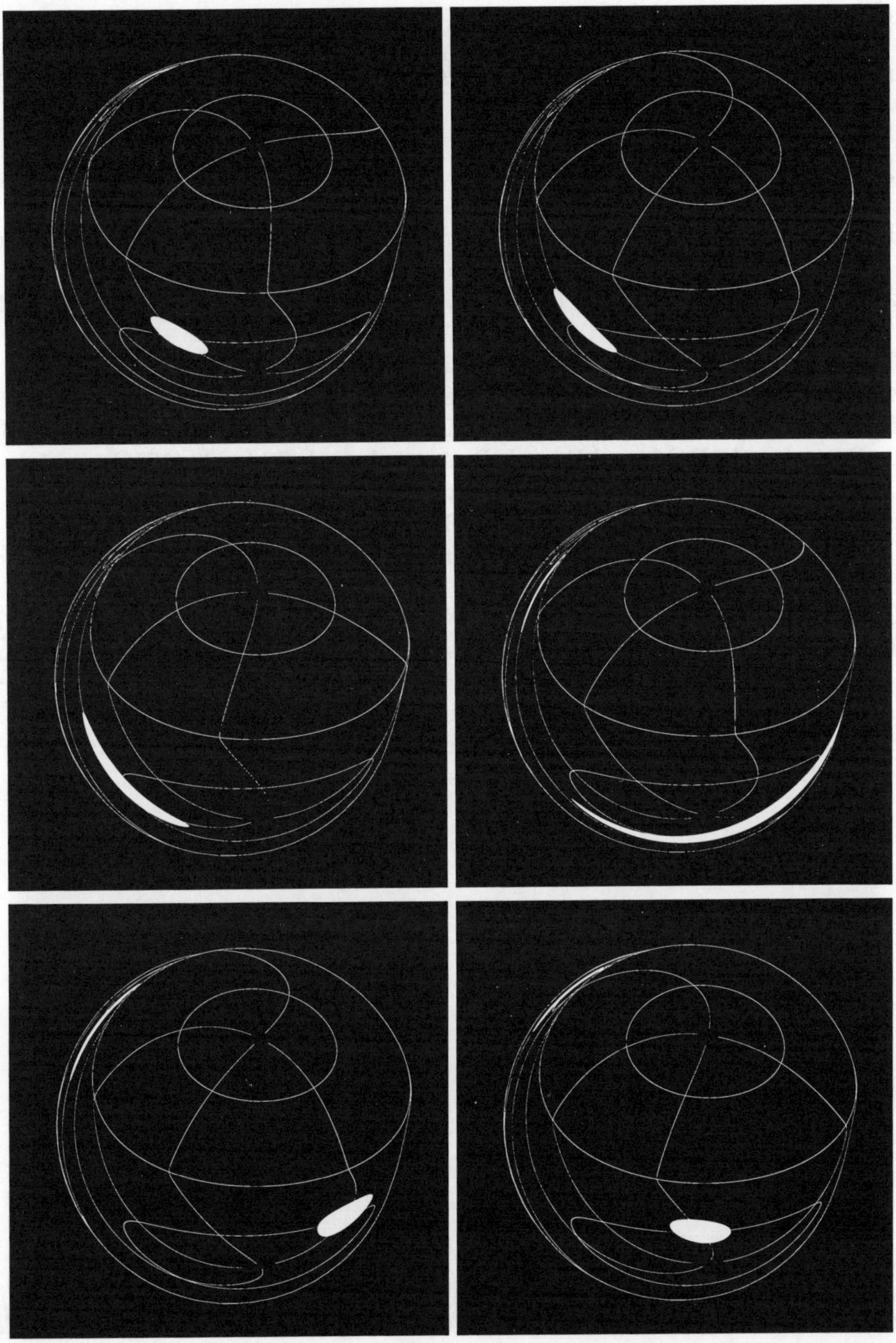

Figure 7. Six rotation phases of the B model with a hot spot on the surface as seen by an observer 30° above the equatorial plane. The second image of the hot spot appears due to the gravitational lensing, its position on receding edge of star's disk implies strong redshift.

Acknowledgments

Support from the grant No. GAČR 202/99/0261 of the Czech Republic is sincerely acknowledged.

References

1. Dyson, F. W., Eddington, A. S., and Davidson, C. A Philosophical Transactions of the Royal Society of London, Series A **220**, 291 (1920).
2. Nollert, H.-P., Ruder, H., Herold, H., Kraus, U. Astron. Astrophys. **208**, 153 (1989).
3. Riffert, H., Mészáros Astrophys. J. **325**, 207 (1988).
4. Strohmayer, T. E. Astrophys. J. **338**, 138 (1992).
5. Barden, J. M. (1972) Rapidly Rotating Stars, Disks and Black Holes, in [13].
6. Butterworth, E. M. and Ipser J. R. Astrophys. J. **204**, 200 (1976).
7. Friedman, J. L., Ipser, J. R., Parker, L.304 **115**, 1986 (.)
8. Stergioulas, N. Living Rev. Rel. **1**, 8 (1998)
9. Nozawa, T., Stergioulas, N., Gourgoulhon, E., Eriguchi, Y. Astron. Astrophys. Suppl. Ser. **132**, 431 (1998)
10. Barden, J. M. (1972) Timelike and Null geodesics in the Kerr metrics, in [13].
11. Misner, C. W., Thorne, K. S., Wheeler, J. A. Gravitation, Freeman, San Francisco, 1973.
12. Frolov, V., Novikov, I. (1998) Physics of Black Holes, Kluwer, Dordrecht.
13. Black Holes (Les Houches 1972), eds. C. DeWitt and B. S. DeWitt, Gordon and Breach, New York-London-Paris.

EXACT IMPULSIVE GRAVITATIONAL WAVES IN SPACETIMES OF CONSTANT CURVATURE

JIŘÍ PODOLSKÝ

Institute of Theoretical Physics, Charles University in Prague, V Holešovičkách 2, 180 00 Prague 8, Czech Republic.

Exact solutions exist which describe impulsive gravitational waves propagating in Minkowski, de Sitter, or anti-de Sitter universes. These may be either nonexpanding or expanding. Both cases in each background are reviewed here from a unified point of view. All the main methods for their construction are described systematically: the Penrose "cut and paste" method, explicit construction of continuous coordinates, distributional limits of sandwich waves, embedding from higher dimensions, and boosts of sources or limits of infinite accelerations. Attention is concentrated on the most interesting specific solutions. In particular, the nonexpanding impulsive waves that are generated by null multipole particles are described. These generalize the well-known Aichelburg–Sexl and Hotta–Tanaka monopole solutions. Also described are the expanding spherical impulses that are generated by snapping and colliding strings. Geodesics and some other properties of impulsive wave spacetimes are also summarized.

1 Outline and personal notes

In recent years, various aspects of exact solutions of Einstein's equations which describe impulsive waves in spaces of constant curvature have been thoroughly investigated. However, these studies are mostly scattered in a great number of particular articles published in different journals. The first purpose of this essay is to collect and unify the principal results. We attempt to provide a brief but comprehensive and self-contained review of all impulsive gravitational waves which may propagate in Minkowski, de Sitter, or anti-de Sitter universe. We hope that such a summary may be useful for further studies in classical relativity and also in quantum gravity or string theory.

Nevertheless, the main motivation for our work was to pay a tribute to Professor J. Bičák on the occasion of his 60th birthday. Jiří Bičák's lifelong devotion to the investigation of exact radiative spacetimes is famous. He has contributed to various specific topics, including those closely related to impulsive waves, which we survey here. But Professor Bičák not only wrote many contributions which will be mentioned below in the appropriate context. During the past decades he has also been an inexhaustible fountain-head of inspiration and a continuous source of encouragement — and not only to members of the Prague relativity group.

Even more personally, it was Jiří Bičák — my academic teacher — who in 1985 suggested the topic of my diploma thesis "Gravitational radiation in cosmology". Later he carefully supervised my doctoral thesis "On exact radiative space-times with cosmological constant". During all these years I have been learning fundamental issues concerning the theory of gravitational radiation from him. In particular, the geometrical and physical investigation of the Kundt and the Robinson–Trautman classes of exact gravitational waves later proved to be crucial for my understanding of possible impulses in spaces of constant curvature. Moreover, almost all of my recent works on impulsive solutions have emerged from a fruitful collaboration with Professor J. B. Griffiths from Loughborough University. And again, it was Jiří Bičák who established our mutual contacts and has been promoting and encouraging these during the last years ... It is thus only natural to dedicate this summarizing essay to him on the occasion of the exceptional anniversary of his birthday.

The present work is organized as follows. We shall first briefly describe the background spaces of constant curvature, present all (nontwisting) type N exact vacuum solutions with a cosmological constant, and "heuristically" introduce gravitational impulses as limits of corresponding sandwich waves. In the main section 3, we shall describe in detail the methods of construction of these impulsive waves — the Penrose "cut and paste" method, explicit construction of continuous coordinates, distributional limits of sandwich waves, geometrical embedding from higher dimensions, and boosts or limits of infinite acceleration of particular solutions. Using all these methods, both nonexpanding and expanding impulsive waves are constructed, as summarized at the end of section 3. The physically important particular solutions are presented in section 4, namely nonexpanding impulses generated by multipole particles, and expanding impulses generated by snapping or colliding strings. Finally, some other properties are also discussed, in particular the behaviour of geodesics.

2 Exposition

2.1 Spaces of constant curvature

The spaces of constant curvature — Minkowski, de Sitter, and anti-de Sitter spacetimes — which form backgrounds for the impulsive solutions investigated here, are the simplest exact solutions of Einstein's vacuum field equations with vanishing, positive, and negative cosmological constant Λ, respectively. Yet, these are extremely important in contemporary theoretical physics. As Jiří Bičák writes in his recent substantial essay [1] in honour of Jürgen Ehlers: "These solutions have played a crucial role in many issues in general relativity and cosmology, and most recently, they have become important prerequisites on the stage of the theoretical physics of the 'new age', including string theory and string cosmology". The de Sitter universe is the basic model of early quasi-exponential inflationary phase of expansion of the universe [2]. Moreover, according to the "cosmic no-hair" conjecture [3], it is locally the asymptotic state of many cosmological models with a positive Λ. Recently, the anti-de Sitter spacetime has also become a subject of intensive studies thanks to the Maldacena's conjecture [4] which relates string theory in asymptotically anti-de Sitter space to a non-gravitational conformal field theory on the boundary at spatial infinity.

The spacetimes of constant curvature are highly symmetric: they admit ten Killing vector fields which correspond to the maximal number of isometries possible in a 4-dimensional spacetime. Also, they are conformally flat and can thus be understood as specific portions of the Einstein static universe. Their global conformal structures are however different for different signs of Λ (see below).

It is well-known that the de Sitter and anti-de Sitter spacetimes can be represented as the 4-dimensional hyperboloid

$$-{Z_0}^2 + {Z_1}^2 + {Z_2}^2 + {Z_3}^2 + \epsilon {Z_4}^2 = \epsilon a^2 \ , \qquad a = \sqrt{3/|\Lambda|} \ , \tag{1}$$

in the flat 5-dimensional space with metric

$$\mathrm{d}s_0^2 = -\mathrm{d}{Z_0}^2 + \mathrm{d}{Z_1}^2 + \mathrm{d}{Z_2}^2 + \mathrm{d}{Z_3}^2 + \epsilon \mathrm{d}{Z_4}^2 \ , \tag{2}$$

where $\epsilon = 1$ for $\Lambda > 0$ and $\epsilon = -1$ for $\Lambda < 0$. Obviously, the de Sitter spacetime has the topology $R^1 \times S^3$ (R^1 corresponding to the time), whereas the topology of the anti-de Sitter spacetime is $S^1 \times R^3$.

Various parameterizations of the hyperboloid (1) are known which introduce suitable coordinates for the de Sitter or the anti-de Sitter spacetime. For example, the expressions

$$\begin{aligned}
Z_0 &= \tfrac{1}{\sqrt{2}}(\mathcal{U} + \mathcal{V}) / \left[1 + \tfrac{1}{6}\Lambda(\eta\bar\eta - \mathcal{U}\mathcal{V})\right] \ , \\
Z_1 &= \tfrac{1}{\sqrt{2}}(\mathcal{U} - \mathcal{V}) / \left[1 + \tfrac{1}{6}\Lambda(\eta\bar\eta - \mathcal{U}\mathcal{V})\right] \ , \\
Z_2 &= \tfrac{1}{\sqrt{2}}(\eta + \bar\eta) / \left[1 + \tfrac{1}{6}\Lambda(\eta\bar\eta - \mathcal{U}\mathcal{V})\right] \ , \\
Z_3 &= \tfrac{-\mathrm{i}}{\sqrt{2}}(\eta - \bar\eta) / \left[1 + \tfrac{1}{6}\Lambda(\eta\bar\eta - \mathcal{U}\mathcal{V})\right] \ , \\
Z_4 &= a\left[1 - \tfrac{1}{6}\Lambda(\eta\bar\eta - \mathcal{U}\mathcal{V})\right] / \left[1 + \tfrac{1}{6}\Lambda(\eta\bar\eta - \mathcal{U}\mathcal{V})\right] \ ,
\end{aligned} \tag{3}$$

i.e. inversely $\mathcal{U} = \sqrt{2}a(Z_0 + Z_1)/(Z_4 + a)$, $\mathcal{V} = \sqrt{2}a(Z_0 - Z_1)/(Z_4 + a)$, $\eta = \sqrt{2}a(Z_2 + \mathrm{i}Z_3)/(Z_4 + a)$, give the line element for all spaces of constant curvature in the conformally flat unified form

$$\mathrm{d}s_0^2 = \frac{2\,\mathrm{d}\eta\,\mathrm{d}\bar\eta - 2\,\mathrm{d}\mathcal{U}\,\mathrm{d}\mathcal{V}}{[1 + \frac{1}{6}\Lambda(\eta\bar\eta - \mathcal{U}\mathcal{V})]^2} \ . \tag{4}$$

This is de Sitter space when $\Lambda > 0$, anti-de Sitter space when $\Lambda < 0$, and Minkowski space when $\Lambda = 0$.

However, the most natural coordinates which cover the whole de Sitter hyperboloid (1) are

$$\begin{aligned}
Z_0 &= a \sinh(\tau/a) \ , \\
Z_1 &= a \cosh(\tau/a) \cos\chi \ , \\
Z_2 &= a \cosh(\tau/a) \sin\chi \sin\theta \cos\phi \ , \\
Z_3 &= a \cosh(\tau/a) \sin\chi \sin\theta \sin\phi \ , \\
Z_4 &= a \cosh(\tau/a) \sin\chi \cos\theta \ .
\end{aligned} \tag{5}$$

With these the line element (2) becomes

$$\mathrm{d}s_0^2 = -\mathrm{d}\tau^2 + a^2\cosh^2(\tau/a)\,[\,\mathrm{d}\chi^2 + \sin^2\chi\,(\mathrm{d}\theta^2 + \sin^2\theta\,\mathrm{d}\phi^2)\,] \ . \tag{6}$$

In this case, the privileged timelike observers have worldlines on which the space coordinates χ, θ and ϕ are constant. For this family, the 3-spaces $\tau =$ const. (which represent the universe at this instant) have the geometry of S^3 which contracts to a minimum size at $\tau = 0$ and then re-expand. The global conformal structure of the de Sitter spacetime becomes obvious if we introduce the conformal time η_E by

$\sin\eta_{_E} = 1/\cosh(\tau/a)$. The line element (6) then becomes $\mathrm{d}s_0^2 = \Omega^2\,\mathrm{d}s_{_E}^2$, where $\Omega = a/\sin\eta_{_E}$, and $\mathrm{d}s_{_E}^2 = -\mathrm{d}\eta_{_E}^2 + \mathrm{d}\chi^2 + \sin^2\chi\,(\mathrm{d}\theta^2 + \sin^2\theta\,\mathrm{d}\phi^2)$ is the standard metric of the Einstein static universe. Therefore, the corresponding Penrose conformal diagram covers the coordinate ranges $\chi \in [0,\pi]$ and $\eta_{_E} \in (0,\pi)$. The boundaries of $\eta_{_E}$ (i.e. $\tau = \pm\infty$) represent the conformal infinities $\mathcal{J}^\pm$. Obviously, these have a spacelike character.

The natural global coordinates for the anti-de Sitter spacetime is obtained by

$$
\begin{aligned}
Z_0 &= a\cosh(\mathrm{r})\sin(\mathrm{t}/a)\,,\\
Z_4 &= a\cosh(\mathrm{r})\cos(\mathrm{t}/a)\,,\\
Z_1 &= a\sinh(\mathrm{r})\cos\theta\,,\\
Z_2 &= a\sinh(\mathrm{r})\sin\theta\cos\phi\,,\\
Z_3 &= a\sinh(\mathrm{r})\sin\theta\sin\phi\,,
\end{aligned}
\tag{7}
$$

with which the line element (2) becomes

$$
\mathrm{d}s_0^2 = -\cosh^2\mathrm{r}\,\mathrm{d}\mathrm{t}^2 + a^2\,[\,\mathrm{d}\mathrm{r}^2 + \sinh^2\mathrm{r}\,(\mathrm{d}\theta^2 + \sin^2\theta\,\mathrm{d}\phi^2)\,]\,. \tag{8}
$$

Again, the privileged timelike (non-geodesic) observers have worldlines r, θ and ϕ constants, their proper times are given by $\tau = \mathrm{t}\cosh(\mathrm{r})$. However, the timelike slices $\mathrm{t} = \mathrm{const.}$ are 3-dimensional spaces of constant negative curvature. The conformal structure is seen by introducing $\eta_{_E} = \mathrm{t}/a$ and $\tan\chi = \sinh(\mathrm{r})$. Then $\mathrm{d}s_0^2 = \Omega^2\,\mathrm{d}s_{_E}^2$, where $\Omega = a/\cos\chi$, so that the conformal diagram of the anti-de Sitter spacetime is given by the range $\chi \in [0,\frac{\pi}{2})$, $\eta_{_E}$ arbitrary. Thus, contrary to the Minkowski or de Sitter universe, the conformal infinity $\mathcal{J}$ corresponding to $\chi = \frac{\pi}{2}$ has a timelike character in the anti-de Sitter spacetime.

Fundamental geometrical properties of the spaces of constant curvature have been described in detail in [5] and in more recent works and reviews, see e.g. [6]. Note also that the Minkowski, de Sitter, and anti-de Sitter spacetimes are stable with respect to general, nonlinear (but "weak") vacuum perturbations, as shown recently in [7], [8].

2.2 Exact type N Einstein spaces

Impulsive waves can (at least from a physical point of view) most naturally be understood as "distributional limits" of suitable families of waves with "sandwich" profiles, such that these have ever "shorter duration" but simultaneously become "stronger" in the limit. Impulsive gravitational waves can be intuitively understood in exactly this way. Incidentally, in our contributions [9], [10] with Jiří Bičák we investigated classes of exact radiative spacetimes, which may be used for the above construction. It will thus be useful and convenient for our later purposes to summarize the fundamental properties of these spacetimes here.

In particular, we studied all nontwisting Petrov-type N solutions of vacuum Einstein field equations with a cosmological constant. For type N spacetimes, the Debever–Penrose vector field k^α is quadruple and defines a privileged null congruence characterized by expansion $\Theta = \frac{1}{2}k^\alpha{}_{;\alpha}$, twist $\omega = \sqrt{\frac{1}{2}k_{[\alpha;\beta]}k^{\alpha;\beta}}$, and shear

$|\sigma| = \sqrt{\frac{1}{2}k_{(\alpha;\beta)}k^{\alpha;\beta} - \Theta^2}$. The Bianchi identities and the Kundt–Thompson theorem for type N Einstein spaces give $\sigma = 0$. Considering the nontwisting case only, we are left with two classes:

- the *Kundt* class of *nonexpanding* gravitational waves ($\Theta = 0$), cf. [11], [12], Ch. 27 in [13],
- the *Robinson–Trautman* class of *expanding* gravitational waves ($\Theta \neq 0$), cf. [14], Ch. 24 in [13].

Hereafter, we denote the above Kundt class by $KN(\Lambda)$, and the Robinson–Trautman class by $RTN(\Lambda)$.

In [9] we described these spacetimes in detail from a geometrical point of view in a unified formalism, we presented invariant subclasses, and found relations between them. All solutions of the Kundt class $KN(\Lambda)$ can be written in suitable coordinates, first given in [12], as

$$\mathrm{d}s^2 = 2\,\frac{1}{p^2}\,\mathrm{d}\xi\mathrm{d}\bar{\xi} - 2\,\frac{q^2}{p^2}\,\mathrm{d}u\mathrm{d}v + \mathrm{F}\,\mathrm{d}u^2\ , \tag{9}$$

where $p = 1 + \frac{1}{6}\Lambda\xi\bar{\xi}$, $q = (1 - \frac{1}{6}\Lambda\xi\bar{\xi})\,\alpha + \bar{\beta}\,\xi + \beta\,\bar{\xi}$, $\mathrm{F} = \kappa(q^2/p^2)v^2 - (q^2/p^2)_{,u}\,v + (q/p)\,\mathrm{H}$, $\kappa = \frac{1}{3}\Lambda\alpha^2 + 2\beta\bar{\beta}$, and $\mathrm{H} = \mathrm{H}(\xi, \bar{\xi}, u)$. The vacuum field equation is $p^2\,\mathrm{H}_{\xi\bar{\xi}} + \frac{1}{3}\Lambda\,\mathrm{H} = 0$, which has a general solution given by $\mathrm{H} = (f_{,\xi} + \bar{f}_{,\bar{\xi}}) - \frac{1}{3}\Lambda(\bar{\xi}f + \xi\bar{f})/p$, where $f(\xi, u)$ is an *arbitrary* function (otherwise (9) is a pure radiation spacetime). As shown in [12], [9], there exist the following geometrically distinct canonical subclasses characterized by specific choices of the parameters α and β:

$$KN(\Lambda)\begin{cases} \Lambda = 0 \begin{cases} \kappa = 0 : PP & \equiv KN(\Lambda = 0)[\alpha = 1, \beta = 0]\ , \\ \kappa > 0 : KN & \equiv KN(\Lambda = 0)[\alpha = 0, \beta = 1]\ , \end{cases} \\ \\ \Lambda \neq 0 \begin{cases} \kappa > 0 : KN(\Lambda)I & \equiv KN(\Lambda)[\alpha = 0, \beta = 1]\ , \\ \kappa < 0 : KN(\Lambda^-)II & \equiv KN(\Lambda < 0)[\alpha = 1, \beta = 0]\ , \\ \kappa = 0 : KN(\Lambda^-)III & \equiv KN(\Lambda < 0)[\alpha = 1, \beta = \sqrt{-\frac{\Lambda}{6}}e^{i\omega}] \\ \qquad \to KN(\Lambda^-)III_K & \equiv KN(\Lambda < 0)[\alpha = 1, \beta = \sqrt{-\frac{\Lambda}{6}}]\ . \end{cases} \end{cases}$$

For vanishing cosmological constant Λ there are two subclasses. The simpler one represents the well-known pp-waves denoted here as PP, for which $p = 1 = q$, $\mathrm{F} = f_{,\xi} + \bar{f}_{,\bar{\xi}}$. This has been investigated by many authors (for details and references see section 21.5 in [13]). The second subclass with $\Lambda = 0$ is KN which is a special class of solutions discovered by Kundt [11] (see Ch. 27 in [13]). Interestingly, there is an asymmetry in the case $\Lambda \neq 0$: there exist *three* distinct subclasses of nonexpanding waves for $\Lambda < 0$, whereas there is only *one* such subclass for $\Lambda > 0$, namely $KN(\Lambda)I$. The reason is the condition $\kappa = \frac{1}{3}\Lambda\alpha^2 + 2\beta\bar{\beta}$ which for $\Lambda > 0$ excludes the cases $\kappa < 0$ and $\kappa = 0$. Note also that there exists a special subsubclass $KN(\Lambda^-)III_K$ of the $KN(\Lambda^-)III$ subclass such that the Debever–Penrose vector is the *Killing* vector. It was explicitly demonstrated [9] that this subsubclass is identical with an interesting family of solutions found by Siklos [15], and analyzed in detail in [16].

The class of expanding Robinson–Trautman solutions $RTN(\Lambda)$ has been known for a long time [14]. In 1981, a more convenient coordinate parametrization was found [17]

$$ds^2 = 2\,v^2 \mathrm{d}\xi \mathrm{d}\bar{\xi} + 2\,v\bar{\mathrm{A}}\,\mathrm{d}\xi \mathrm{d}u + 2\,v\mathrm{A}\,\mathrm{d}\bar{\xi}\mathrm{d}u + 2\,\psi\,\mathrm{d}u\mathrm{d}v + 2\,(\mathrm{A}\bar{\mathrm{A}} + \psi \mathrm{B})\,\mathrm{d}u^2\ , \qquad (10)$$

where $\mathrm{A} = \epsilon\xi - vf$, $\mathrm{B} = -\epsilon + \frac{1}{2}v\,(f_{,\xi} + \bar{f}_{,\bar{\xi}}) + \frac{1}{6}\Lambda v^2\psi$, $\psi = 1 + \epsilon\xi\bar{\xi}$, and $\epsilon = -1, 0, +1$. Again, the solutions depend on an arbitrary function $f(\xi, u)$. The parameter ϵ is proportional to the Gaussian curvature of the 2-surfaces $u = const.$ Thus, there are nine invariant subclasses of this class since the cases $\epsilon = +1, 0, -1$ and $\Lambda > 0$, $\Lambda = 0$, $\Lambda < 0$ are independent.

In the subsequent paper [10] we presented a physical interpretation of all the nontwisting type N Einstein spaces (9), (10) which we based on the study of the deviation of geodesics [18]. It was demonstrated that there always exists an (essentially unique) orthonormal frame tied to any timelike observer. For the $KN(\Lambda)$ spacetimes this is given by

$$e^{\mu}_{(0)} = \left(\dot{v}, \dot{\xi}, \dot{\bar{\xi}}, \dot{u}\right), \quad e^{\mu}_{(1)} = -\frac{p}{\sqrt{2}}\left(\frac{2\,\mathcal{R}e\dot{\xi}}{q^2\dot{u}}, 1, 1, 0\right),$$

$$e^{\mu}_{(2)} = -\frac{p}{\sqrt{2}}\left(\frac{2\,\mathcal{I}m\dot{\xi}}{q^2\dot{u}}, i, -i, 0\right), \quad e^{\mu}_{(3)} = -\left(\dot{v} - \frac{p^2}{\dot{u}q^2}, \dot{\xi}, \dot{\bar{\xi}}, \dot{u}\right), \qquad (11)$$

and for the $RTN(\Lambda)$ solutions by

$$e^{\mu}_{(0)} = \left(\dot{v}, \dot{\xi}, \dot{\bar{\xi}}, \dot{u}\right), \quad e^{\mu}_{(1)} = \frac{1}{\sqrt{2}\,v}\left(\frac{2v}{\psi\dot{u}}\mathcal{R}e\{v\dot{\xi} + \mathrm{A}\dot{u}\}, -1, -1, 0\right),$$

$$e^{\mu}_{(2)} = \frac{1}{\sqrt{2}\,v}\left(\frac{2v}{\psi\dot{u}}\mathcal{I}m\{v\dot{\xi} + \mathrm{A}\dot{u}\}, -i, i, 0\right), \quad e^{\mu}_{(3)} = -\left(\dot{v} + \frac{1}{\psi\dot{u}}, \dot{\xi}, \dot{\bar{\xi}}, \dot{u}\right). \qquad (12)$$

An invariant form of the equation of geodesic deviation with respect to the above interpretation frame, along any timelike geodesic in the $KN(\Lambda)$ and $RTN(\Lambda)$ spacetimes, is then given by

$$\begin{aligned}
\ddot{Z}^{(1)} &= \frac{\Lambda}{3}\,Z^{(1)} - \mathcal{A}_{+}Z^{(1)} + \mathcal{A}_{\times}Z^{(2)}\ , \\
\ddot{Z}^{(2)} &= \frac{\Lambda}{3}\,Z^{(2)} + \mathcal{A}_{+}Z^{(2)} + \mathcal{A}_{\times}Z^{(1)}\ , \\
\ddot{Z}^{(3)} &= \frac{\Lambda}{3}\,Z^{(3)}\ .
\end{aligned} \qquad (13)$$

The amplitudes of the gravitational waves are $\mathcal{A}_{+} = \mathcal{R}e\,\{\mathcal{A}\}$, $\mathcal{A}_{\times} = \mathcal{I}m\,\{\mathcal{A}\}$ such that

$$\mathcal{A} = -\frac{pq}{2}\dot{u}^2\ f_{,\xi\xi\xi} \qquad (14)$$

for the $KN(\Lambda)$ spacetimes, and

$$\mathcal{A} = -\frac{\psi}{2v}\dot{u}^2\ f_{,\xi\xi\xi} \qquad (15)$$

in the $RTN(\Lambda)$ spacetimes (evaluated along the geodesics). The equations (13) express relative accelerations of nearby free test particles in terms of their actual positions. The frame components of the displacement vector $Z^{(i)} \equiv e^{(i)}_\mu Z^\mu$ determine the invariant distance between the test particles. Similarly, $\ddot{Z}^{(i)} \equiv e^{(i)}_\mu (D^2 Z^\mu / d\tau^2)$ are relative accelerations. The system (13) enables us to draw the following physical conclusions:

- The particles move isotropically one with respect to the other if $\mathcal{A} = 0$, in which case no gravitational wave is present. Indeed, both the $KN(\Lambda)$ and $RTN(\Lambda)$ spacetimes are conformally flat vacuum for $f_{,\xi\xi\xi} = 0$, and therefore Minkowski ($\Lambda = 0$), de Sitter ($\Lambda > 0$), and anti-de Sitter ($\Lambda < 0$). These maximally symmetric, homogeneous, and isotropic spacetimes are thus natural backgrounds for other "non-trivial" $KN(\Lambda)$ and $RTN(\Lambda)$ radiative solutions.

- If $f_{,\xi\xi\xi} \neq 0$ then the amplitudes $\mathcal{A}$ do not vanish. The particles are influenced by the wave, in a similar way as they are affected by a standard gravitational wave in Minkowski background. For $\Lambda \neq 0$, the influence of the wave is added to the (anti-)de Sitter isotropic expansion/contraction. Therefore, the $KN(\Lambda)$ and $RTN(\Lambda)$ metrics can be interpreted as exact gravitational waves which propagate in spaces of constant curvature.

- The wave has a transverse character as only motions in the plane $(\mathbf{e}_{(1)}, \mathbf{e}_{(2)})$ perpendicular to $\mathbf{e}_{(3)}$ are affected. The propagation direction is $\mathbf{e}_{(3)}$ and coincides with the projection of the Debever–Penrose vector on the hypersurface orthogonal to the observer's four-velocity.

- There are two polarization modes of the wave, "+" and "×", with the corresponding amplitudes $\mathcal{A}_+$ and $\mathcal{A}_\times$. Under rotations in the transverse plane they transform according to $\mathcal{A}'_+ = \cos 2\vartheta \mathcal{A}_+ - \sin 2\vartheta \mathcal{A}_\times$, $\mathcal{A}'_\times = \sin 2\vartheta \mathcal{A}_+ + \cos 2\vartheta \mathcal{A}_\times$, so that the helicity of the wave is 2, as with linearized waves on a Minkowski background.

- Curvature singularities (which might be considered as "sources" of the gravitational waves) occur where the amplitudes $\mathcal{A}$ diverge. Note that the singularities of the $RTN(\Lambda)$ spacetimes can be characterized by the non-vanishing invariant constructed recently by Bičák and Pravda [19] from the second derivatives of the Riemann tensor.

- Special classes of explicit geodesics were also discussed in [10]. For a *positive* cosmological constant, the amplitudes $\mathcal{A}$ decay exponentially fast, i.e. the *waves are damped.* The spacetimes locally approach the de Sitter universe, which is an explicit demonstration of the cosmic no-hair conjecture [3] under the presence of waves in exact model spacetimes. With Jiří Bičák we also discussed the validity of the cosmic no-hair conjecture in the Robinson–Trautman radiative spacetimes of Petrov type *II* [20].

2.3 Motivation: impulses as limits of exact sandwich waves

We have already indicated at the beginning of previous section that probably the most natural way how to introduce impulsive waves in spaces of constant curvature is to consider a distributional limit of sandwich waves of the above $KN(\Lambda)$ and $RTN(\Lambda)$ spacetimes. Indeed, all these solutions depend, in suitable coordinates (9), (10), on an arbitrary function $f(\xi, u)$ where the coordinate u is the "retarded time". Thus, an arbitrary wave-profile can be prescribed within the above families of exact radiative solutions. In particular, we can consider sandwich waves for which the wave-profiles $d_\varepsilon(u)$ of the function $f(\xi, u) \equiv f(\xi)\, d_\varepsilon(u)$ are nonvanishing only on a *finite interval* of u (of the length ε around $u = 0$). In the context of the simplest class of *pp*-waves this has been introduced already in the late 50's by Bondi, Pirani and Robinson [21].

Now, the distributional limit $d_\varepsilon(u) \to \delta(u)$, where δ is the Dirac delta distribution, results (at least formally) in a solution which clearly represents an impulsive gravitational wave localized on a single wave-front $u = 0$. Such a procedure, in which a sequence of sandwich *pp*-waves with profiles $d_\varepsilon(u)$ becomes infinitesimal in duration but unbounded in the amplitude as $\varepsilon \to 0$, was considered in [22], [23] and later elsewhere [24]. In this simplest case given by the metric (9) with $p = 1 = q$, $\mathrm{F} = \mathrm{H}(\xi, \bar{\xi})\, d_\varepsilon(u)$, one directly obtains the metric

$$\mathrm{d}s^2 = 2\,\mathrm{d}\xi\,\mathrm{d}\bar{\xi} - 2\,\mathrm{d}u\,\mathrm{d}v + \mathrm{H}\,\delta(u)\,\mathrm{d}u^2\ , \tag{16}$$

which is the well-know Brinkmann form of impulsive plane wave in Minkowski background [22], [25]. Analogously, more general nonexpanding impulsive waves of the Kundt class $KN(\Lambda)$ for $\Lambda \neq 0$, and expanding Robinson–Trautman impulsive waves of the $RTN(\Lambda)$ class can be constructed (see section 3.3 below for more details).

Mathematical problems and questions arise however with the above procedure. What is the meaning of the metric with distributional terms? What is the manifold structure? Usually, the metric coefficients are considered to be C^2 continuous functions. And even worse: in the limit of the $RTN(\Lambda)$ sandwich-wave spacetimes one obtains a metric which contains a term proportional to δ^2. The product of the Dirac distributions is a mathematically ill-defined object within the Schwartz linear distribution theory. For this, it may be necessary to invoke the more advanced Colombeau nonlinear theory of generalized functions [26].

In view of these (and other) open questions, it is of primary importance to investigate the construction of impulsive-wave spacetimes by *various* methods and approaches. These may provide a deeper understanding of the impulsive solutions based on mathematically sound definitions, discover mutual relations, investigate specific properties, and elucidate their physical interpretation.

3 Main theme: methods of construction of impulsive waves

The history of studies of impulsive waves can be divided (roughly speaking) into three periods. During the first epoch, which culminated in classic works by Lichnerowicz and others [27], principal mathematical properties of possible shock or impulsive waves in general relativity were found and investigated. It was demonstrated that these must necessarily be localized on null hypersurfaces, across which the derivatives of the metric (the second or the first) are discontinuous. Nevertheless, explicit constructions of impulsive solutions and further investigation of their properties had not been performed until the fundamental contributions by Penrose [25] and Aichelburg and Sexl [28] appeared at the beginning of 1970's. During this "Golden Era" most of the different methods of construction occured almost simultaneously. These methods, based on various approaches, are summarized in the following table:

method of construction	*basic characteristics*
"cut and paste"	general but only "formal"
continuous coordinates	general and explicit
limits of sandwich waves	general, explicit (problems for expanding)
embedding	special (nonexpanding $\Lambda \neq 0$ only), explicit
boosts/infinite acceleration	special (but physically important), explicit

The last era, which can be called the "Renaissance", started at the beginning of 1990's with a fundamental paper by Hotta and Tanaka [29]. By boosting the Schwarzschild–(anti-)de Sitter black hole to the speed of light they constructed (after previous "no-go" attempts by several authors) a nonexpanding impulse in the de Sitter and anti-de Sitter universes. Simultaneously, an interesting expanding impulsive-wave solution generated by a "snapping" cosmic string in Minkowski space was discovered and discussed by Gleiser and Pullin [30], Bičák [31], Nutku and Penrose [32], and Hogan [33], [34].

As a result of these and subsequent systematic investigations an almost complete picture has now emerged. This comprises both nonexpanding and expanding impulses in constant-curvature spaces with an arbitrary value of the cosmological constant. We will describe the classic and also more recent results here. We shall present these in the context of a unifying description of the main methods by which complete families of impulsive gravitational waves can be constructed. We will also mention the corresponding references and historical remarks in some more detail.

3.1 The "cut and paste" method

Let us start with an elegant geometrical method for constructing general nonexpanding (plane) and expanding (spherical) impulsive gravitational waves in Minkowski background. This was presented by Penrose in classic works [22], [25]. His "cut and paste" approach, which is in some respects similar to that of Israel [35], is based on cutting the spacetime manifold $\mathcal{M}$ along a suitable null hypersurface and then re-attaching the two pieces with a specific "warp". The first case leads to impulsive *pp*-waves, the second case gives expanding impulsive waves. Recently, this approach has also been extended to backgrounds of constant curvature with a nonzero cosmological constant Λ [36] - [40].

3.1.1 Nonexpanding impulsive waves

In this case, the Penrose method is based on the removal of the null hypersurface $\mathcal{U}=0$ from spacetime in the metric form (4)

$$ds_0^2 = \frac{2\,d\eta\,d\bar{\eta} - 2\,d\mathcal{U}\,d\mathcal{V}}{[1+\frac{1}{6}\Lambda(\eta\bar{\eta}-\mathcal{U}\mathcal{V})]^2}\,, \tag{17}$$

and re-attaching the halves $\mathcal{M}_-(\mathcal{U}<0)$ and $\mathcal{M}_+(\mathcal{U}>0)$ by making the identification with a "warp" in the coordinate $\mathcal{V}$ such that

$$\Big[\,\eta,\bar{\eta},\mathcal{V},\mathcal{U}=0_-\Big]_{\mathcal{M}_-} \equiv \Big[\,\eta,\bar{\eta},\mathcal{V}-H(\eta,\bar{\eta}),\mathcal{U}=0_+\Big]_{\mathcal{M}_+}, \tag{18}$$

where $H(\eta,\bar{\eta})$ is an arbitrary function of η and $\bar{\eta}$ alone. It was shown in [25] that the condition (18) automatically guarantees that the Einstein field equations are satisfied everywhere including on $\mathcal{U}=0$. However, impulsive components are introduced into the curvature tensor proportional to $\delta(\mathcal{U})$. These represent gravitational (plus possibly null-matter) impulsive waves.

In [25] Penrose considered only the Minkowski background space (4) with $\Lambda=0$, in which case the impulsive surface $\mathcal{U}=0$ is a plane. Impulsive *pp*-waves are thus obtained. However, the above method can easily be generalized to $\Lambda\neq 0$ cases. It was shown in [37] that exactly the same junction conditions (18) applied to a general background spacetime (17) introduce impulsive waves also in the de Sitter ($\Lambda>0$) or anti-de Sitter ($\Lambda<0$) universes. However, the geometries of these impulses are different as the null hypersurface $\mathcal{U}=0$, along which the spacetime is cut and pasted, is described by the 2-metric $d\sigma^2 = 2\,(1+\frac{1}{6}\,\Lambda\,\eta\,\bar{\eta}\,)^{-2}\,d\eta\,d\bar{\eta}$. This is obviously a 2-dimensional space of constant Gaussian curvature $K=\frac{1}{3}\Lambda$. When $\Lambda=0$ the impulsive wave surface is a *plane*. For $\Lambda>0$ it is a *sphere*, while for $\Lambda<0$ it is a *hyperboloid*. For cases $\Lambda\neq 0$, the geometry of these impulsive spherical and hyperboloidal waves was described in detail in [41] using various convenient coordinate representations. It was also demonstrated that the wave surfaces are indeed nonexpanding.

Let us remark that the Penrose "cut and paste" approach to construction of nonexpanding impulsive waves is closely related to the Dray and 't Hooft method of "shift function", which is prescribed on the horizon of a suitable background space [42].

3.1.2 Expanding impulsive waves

In [25] Penrose demonstrated that his "cut and paste" method can be also used to construct expanding spherical impulsive (purely) gravitational waves. However, the junction conditions are somewhat more complicated than in the nonexpanding case. Instead of starting with the familiar form of Minkowski or (anti-)de Sitter spacetime (4), it is necessary to first perform the transformation

$$\mathcal{U}=\frac{Z\bar{Z}}{p}V-U\ ,\qquad \mathcal{V}=\frac{V}{p}-\epsilon U\ ,\qquad \eta=\frac{Z}{p}V\ , \tag{19}$$

where

$$p=1+\epsilon Z\bar{Z}\ ,\qquad \epsilon=-1,0,+1\ . \tag{20}$$

With this, the constant curvature spacetimes take the following form

$$\mathrm{d}s_0^2=\frac{2\,(V/p)^2\,\mathrm{d}Z\,\mathrm{d}\bar{Z}+2\,\mathrm{d}U\,\mathrm{d}V-2\epsilon\,\mathrm{d}U^2}{\left[1+\frac{1}{6}\Lambda U(V-\epsilon U)\right]^2}\ . \tag{21}$$

In these coordinates, the hypersurface $U=0$ is a *null cone* (a sphere expanding with the speed of light) as $\eta\bar{\eta}-\mathcal{U}\mathcal{V}\equiv U(V-\epsilon U)=0$. The spacetime can now be divided into two halves $\mathcal{M}_-(U<0)$ inside the null cone, and $\mathcal{M}_+(U>0)$ outside this. Now, the Penrose junction conditions prescribe the identification

$$\Big[Z,\,\bar{Z},\,V,\,U=0_-\Big]_{\mathcal{M}_-}\equiv\Big[h(Z),\,\bar{h}(\bar{Z}),\,\frac{(1+\epsilon h\bar{h})\,V}{(1+\epsilon Z\bar{Z})|h'|},\,U=0_+\Big]_{\mathcal{M}_+} \tag{22}$$

of the points from the two re-attached parts across the impulsive sphere $U=0$. In (22) an arbitrary function $h(Z)$ introduces a specific "warp" corresponding to one of all possible impulsive *vacuum* spacetimes of this type.

In [25] Penrose described the above "cut and paste" construction for the case of expanding spherical gravitational waves in a Minkowski background. He assumed the simplest case $\Lambda=0$, $\epsilon=0$. Generalization to impulsive spherical waves in the de Sitter and anti-de Sitter universe (with $\epsilon=0$) was later found by Hogan [36]. In [34] he also introduced impulses in Minkowski spacetime with $\epsilon=+1$. The completely general form (21), (22) of the Penrose junction conditions has been recently presented in our contributions [38], [40].

Note finally that a similar (yet somewhat different) "cut and paste" approach was considered by Gleiser and Pullin [30] for constructing a specific solution which represents a spherical impulsive gravitational wave generated by a "snapping" cosmic string in Minkowski space. This particular solution will be discussed below in subsection 3.5.2.

3.2 Continuous coordinates

The Penrose "cut and paste" approach described in the previous section 3.1 is a general method. By prescribing the junction conditions (18) or (22) in (17) or (21), respectively, all nonexpanding and expanding impulsive gravitational waves in Minkowski, de Sitter, and anti-de Sitter universe can be constructed. Nevertheless, these formal identifications of points on both sides of the impulsive hypersurface do not immediately provide explicit metric forms of the complete spacetimes. It is thus of primary interest to find a suitable coordinate system for the above solutions in which the metric is explicitly continuous everywhere, including on the impulse. In the following we describe the construction of these privileged continuous coordinates.

3.2.1 Nonexpanding impulsive waves

We start again with the metric (4) which represents spaces of constant curvature. Let us perform the transformation $\mathcal{U} = U$, $\mathcal{V} = V + H + UH_{,Z}H_{,\bar{Z}}$, and $\eta = Z + UH_{,\bar{Z}}$, where $H(Z,\bar{Z})$ is an arbitrary real function. This yields the metric

$$\mathrm{d}s_0^2 = \frac{2\,|\mathrm{d}Z + U(H_{,Z\bar{Z}}\mathrm{d}Z + H_{,\bar{Z}\bar{Z}}\mathrm{d}\bar{Z})|^2 - 2\,\mathrm{d}U\mathrm{d}V}{[\,1+\frac{1}{6}\Lambda(Z\bar{Z} - UV - UG)\,]^2}\,, \tag{23}$$

where $G = H - ZH_{,Z} - \bar{Z}H_{,\bar{Z}}$. We now consider (23) for $U > 0$ and combine this with the line element (4) in which we take $\mathcal{U} = U$, $\mathcal{V} = V$, $\eta = Z$ for $U < 0$. The resulting metric can be written as

$$\mathrm{d}s^2 = \frac{2\,|\mathrm{d}Z + U\Theta(U)(H_{,Z\bar{Z}}\mathrm{d}Z + H_{,\bar{Z}\bar{Z}}\mathrm{d}\bar{Z})|^2 - 2\,\mathrm{d}U\mathrm{d}V}{[\,1+\frac{1}{6}\Lambda(Z\bar{Z} - UV - U\Theta(U)G)\,]^2}\,, \tag{24}$$

where $\Theta(U)$ is the Heaviside step function. This is continuous across the null hypersurface $U = 0$ but the discontinuity in the derivatives of the metric introduces impulsive components in the Weyl and curvature tensors proportional to the Dirac delta distribution, $\Psi_4 = (1+\frac{1}{6}\Lambda Z\bar{Z})^2 H_{,ZZ}\,\delta(U)$, $\Phi_{22} = [\,(1+\frac{1}{6}\Lambda Z\bar{Z})^3((1+\frac{1}{6}\Lambda Z\bar{Z})^{-1}H)_{,Z\bar{Z}} + \frac{1}{3}\Lambda H\,]\,\delta(U)$, in a natural tetrad [37]. The metric (24) thus explicitly describes impulsive waves in de Sitter, anti-de Sitter or Minkowski backgrounds. For $\Lambda = 0$, the line element (24) reduces to the well-known Rosen form of impulsive *pp*-waves [25], [43], [44]. Note also that the continuous coordinate system for the particular Aichelburg–Sexl solution [28] (see (52) further in the text) was found by D'Eath [45] and used for analytic investigation of ultrarelativistic black-hole encounters. This was used for studies of high-energy scattering in quantum gravity on the Planck scale [46].

The complete transformation which relates (4) and (24) can be written as

$$\begin{aligned} \mathcal{U} &= U\,, \\ \mathcal{V} &= V + H\,\Theta(U) + U\,\Theta(U)\,H_{,Z}H_{,\bar{Z}}\,, \\ \eta &= Z + U\,\Theta(U)\,H_{,\bar{Z}}\,, \end{aligned} \tag{25}$$

which is discontinuous in the coordinate $\mathcal{V}$ on $\mathcal{U} = 0$. From (25), using the fact that the coordinates U, V, Z are continuous, we obtain exactly the Penrose junction condition (18) for reattaching the two halves of the spacetime $\mathcal{M}_-$ and $\mathcal{M}_+$ with a warp. Thus, the above procedure is indeed an *explicit* Penrose's "cut and paste" construction of all nonexpanding impulsive gravitational waves.

Of course, the discontinuity in the complete transformation (25), which (formally) relates the continuous to the distributional form of impulsive solutions, causes some mathematical problems. However, it has been recently shown in [44], using the previous results [47], that (25) is in fact a *rigorous* example of a generalized coordinate transformation in the sense of Colombeau's generalized functions (at least in the case of impulsive *pp*-waves). Moreover, it is possible to interpret this change of coordinates as the distributional limit of a family of smooth transformations which is obtained by a general regularization procedure, i.e. by considering the impulse as a limiting case of sandwich waves with an *arbitrarily* regularized wave profile. These results put the formal ("physical") equivalence of both continuous and distributional forms of impulsive spacetimes on a solid ground.

3.2.2 Expanding impulsive waves

In this case, we perform a more involved transformation of the metric (4) given by

$$\begin{aligned} \mathcal{V} &= AV - DU \,, \\ \mathcal{U} &= BV - EU \,, \\ \eta &= CV - FU \,, \end{aligned} \tag{26}$$

where

$$\begin{aligned} A &= \frac{1}{p|h'|}\,, \qquad B = \frac{|h|^2}{p|h'|}\,, \qquad C = \frac{h}{p|h'|}\,, \\ D &= \frac{1}{|h'|}\left\{\frac{p}{4}\left|\frac{h''}{h'}\right|^2 + \epsilon\left[1 + \frac{Z}{2}\frac{h''}{h'} + \frac{\bar{Z}}{2}\frac{\bar{h}''}{\bar{h}'}\right]\right\}\,, \\ E &= \frac{|h|^2}{|h'|}\left\{\frac{p}{4}\left|\frac{h''}{h'} - 2\frac{h'}{h}\right|^2 + \epsilon\left[1 + \frac{Z}{2}\left(\frac{h''}{h'} - 2\frac{h'}{h}\right) + \frac{\bar{Z}}{2}\left(\frac{\bar{h}''}{\bar{h}'} - 2\frac{\bar{h}'}{\bar{h}}\right)\right]\right\}\,, \\ F &= \frac{h}{|h'|}\left\{\frac{p}{4}\left(\frac{h''}{h'} - 2\frac{h'}{h}\right)\frac{\bar{h}''}{\bar{h}'} + \epsilon\left[1 + \frac{Z}{2}\left(\frac{h''}{h'} - 2\frac{h'}{h}\right) + \frac{\bar{Z}}{2}\frac{\bar{h}''}{\bar{h}'}\right]\right\}\,. \end{aligned} \tag{27}$$

Here $h = h(Z)$ is an arbitrary function and the derivative with respect to its argument Z is denoted by a prime. With this, the metric (4) of a constant curvature space becomes

$$ds_0^2 = \frac{2\left|(V/p)\,dZ + U\,p\bar{H}\,d\bar{Z}\right|^2 + 2\,dU\,dV - 2\epsilon\,dU^2}{\left[1 + \frac{1}{6}\Lambda U(V - \epsilon U)\right]^2}\,, \tag{28}$$

where

$$H(Z) = \frac{1}{2}\left[\frac{h'''}{h'} - \frac{3}{2}\left(\frac{h''}{h'}\right)^2\right]. \tag{29}$$

Similarly as in the nonexpanding case, we may now combine the line element (28) for $U > 0$ and attach this to the metric (21) for $U < 0$ (which was obtained from (4) by the transformation (19)). The resulting complete line element can thus be written in the form

$$ds^2 = \frac{2\left|(V/p)\,dZ + U\Theta(U)\,p\bar{H}\,d\bar{Z}\right|^2 + 2\,dU\,dV - 2\epsilon\,dU^2}{\left[1 + \frac{1}{6}\Lambda U(V - \epsilon U)\right]^2}\,. \tag{30}$$

This metric which was presented for a Minkowski background in [32] - [34], with a cosmological constant in [36], and in the most general form in [38], [40] is explicitly continuous everywhere, including across the null hypersurface $U = 0$. Again, the discontinuity in the derivatives of the metric yields impulsive components in the Weyl and curvature tensors, $\Psi_4 = (p^2H/V)\,\delta(U)$, $\Phi_{22} = (p^4H\bar{H}/V^2)\,U\,\delta(U)$, using a naturally adapted tetrad. The spacetime is clearly conformally flat everywhere except on the impulsive wave surface $U = 0$. It is a vacuum solution everywhere except on the wave surface at $V = 0$, and at possible singularities of the function $p^2 H$.

By comparing the transformations (19) at $U = 0_-$ with (26), (27) at $U = 0_+$, we obtain exactly the Penrose junction conditions (22). The above procedure is thus an *explicit* "cut and paste" construction of expanding spherical pure gravitational waves in spaces of constant curvature.

Note also that new continuous coordinates which generalize (30) for the case $\Lambda = 0$ were found recently [48] as an extension of previous results for spherical shock waves [49]. These contain an additional parameter which is related to acceleration of the coordinate system.

3.3 *Limits of sandwich waves*

We have already mentioned in the introductory section 2 that impulsive waves can be understood as distributional limits of appropriate sequence of sandwich waves in a suitable family of exact radiative spacetimes. In fact, this seems to be the most intuitive way of construction of impulsive waves, although mathematical difficulties occur with this approach in the case of expanding impulses of the Robinson–Trautman type.

3.3.1 *Nonexpanding impulsive waves*

It has been demonstrated in [50] that ***all*** nonexpanding impulses in Minkowski, de Sitter or anti-de Sitter universes can simply be constructed from the Kundt class $KN(\Lambda)$ of exact type *N* solutions (9) by considering the distributional form $\mathrm{H}(\xi,\bar\xi)\,\delta(u)$ of the structural function H. These metrics can thus be written as

$$\mathrm{d}s^2 = 2\,\frac{1}{p^2}\,\mathrm{d}\xi\mathrm{d}\bar\xi - 2\,\frac{q^2}{p^2}\,\mathrm{d}u\mathrm{d}v + \left[\kappa\frac{q^2}{p^2}\,v^2 - \frac{(q^2)_{,u}}{p^2}\,v + \frac{q}{p}\,\mathrm{H}(\xi,\bar\xi)\,\delta(u)\right]\,\mathrm{d}u^2\ , \quad (31)$$

where $p = 1 + \frac{1}{6}\Lambda\xi\bar\xi$, $q = (1 - \frac{1}{6}\Lambda\xi\bar\xi)\,\alpha + \bar\beta\,\xi + \beta\,\bar\xi$, and $\kappa = \frac{1}{3}\Lambda\alpha^2 + 2\beta\bar\beta$. As indicated in section 2.2, for a general wave-profile of the function H there exist various distinct canonical subclasses of (9) characterized by specific choices of the parameters α and β. However, ***impulsive*** limits of these subclasses become (locally) ***equivalent***. Indeed, in the case of vanishing cosmological constant Λ there are two subclasses, namely the *pp*-waves *PP* and the Kundt subclass *KN*. However, the transformation

$$\mathcal{U} = (\xi+\bar\xi)(1+uv)u\ , \qquad \mathcal{V}+1 = (\xi+\bar\xi)v\ , \qquad \eta = \xi + (\xi+\bar\xi)uv\ , \qquad (32)$$

converts the impulsive *KN* metric (31) with $\alpha = 0$, $\beta = 1$ to the impulsive *PP* metric with $\alpha = 1$, $\beta = 0$. The only non-trivial impulsive gravitational waves of the form (31) in Minkowski space are thus impulsive *pp*-waves (16).

Similar results hold also for the $\Lambda \neq 0$ case. Generically, there are three distinct subclasses of nonexpanding waves $KN(\Lambda)I$, $KN(\Lambda^-)II$, and $KN(\Lambda^-)III$ (see section 2.2). It was shown in [50] that although these canonical subclasses are different for smooth profiles, they are equivalent for impulsive profiles. For example, the metric of the $KN(\Lambda^-)I$ subclass of (31) given by $\alpha = 0$, $\beta = 1$, $\Lambda < 0$ can be expressed, using the transformation $\xi = \sqrt{2}a\, e^{i\phi_1}\tanh(R_1/2)$, $v = \frac{1}{\sqrt{2}}\, a/t_1$, $u = \frac{1}{\sqrt{2}}\,(\rho_1 - t_1)/a$, as

$$\begin{aligned} \mathrm{d}s^2 = \frac{a^2}{t_1^2}\sinh^2 R_1\cos^2\phi_1\,(\mathrm{d}\rho_1^2 - \mathrm{d}t_1^2) + a^2(\mathrm{d}R_1^2 + \sinh^2 R_1 \mathrm{d}\phi_1^2) \\ + \sinh R_1\cos\phi_1\,\mathrm{H}(R_1,\phi_1)\delta(t_1-\rho_1)(\mathrm{d}t_1 - \mathrm{d}\rho_1)^2\ . \end{aligned} \tag{33}$$

Similarly, the transformation $\xi = \sqrt{2}a\, e^{i\phi_2}\tanh(R_2/2)$, $v = -a^2/\rho_2$, $u = t_2 - \rho_2$ brings the impulsive subclass $KN(\Lambda^-)II$ of the metric (31) with $\alpha = 1$, $\beta = 0$, $\Lambda < 0$ to the form

$$\begin{aligned} \mathrm{d}s^2 = \frac{a^2}{\rho_2^2}\cosh^2 R_2\,(\mathrm{d}\rho_2^2 - \mathrm{d}t_2^2) + a^2(\mathrm{d}R_2^2 + \sinh^2 R_2 \mathrm{d}\phi_2^2) \\ + \cosh R_2\,\mathrm{H}(R_2,\phi_2)\delta(t_2-\rho_2)(\mathrm{d}t_2 - \mathrm{d}\rho_2)^2\ . \end{aligned} \tag{34}$$

However, the two metrics (33) and (34) are equivalent under the transformation

$$\begin{aligned} t_1^2 &= t_2^2 - \rho_2^2(1-\tanh^2 R_2\cos^2\phi_2)\ , &\qquad \rho_1 &= \rho_2\tanh R_2\cos\phi_2\ , \\ \cosh R_1 &= \frac{t_2}{\rho_2}\cosh R_2\ , &\qquad \sin\Phi_1 &= -\frac{\sinh R_2\sin\phi_2}{\sqrt{t_2^2\rho_2^{-2}\cosh^2 R_2 - 1}}\ . \end{aligned} \tag{35}$$

In this way, by considering the above distributional limit (31) of the $KN(\Lambda)$ class (9) we obtain an explicit form of solutions which represent nonexpanding impulses. The metric (31) contains a single Dirac delta in the $\mathrm{d}u^2$ term, and obviously generalizes the well-known Brinkmann form of impulsive pp-waves (16) to which it reduces when $p = 1 = q$.

Interestingly, there exists yet another simple coordinate form of the complete family of nonexpanding impulses of the Kundt class $KN(\Lambda)$ which also contains a term with the Dirac delta. This is obtained from the continuous form of the impulsive-wave metric (24) by applying the transformation (25). If we take into account the terms which arise from the derivatives of $\Theta(\mathcal{U})$, this transformation relates (24) to

$$\mathrm{d}s^2 = \frac{2\,\mathrm{d}\eta\,\mathrm{d}\bar\eta - 2\,\mathrm{d}\mathcal{U}\,\mathrm{d}\mathcal{V} + 2H(\eta,\bar\eta)\,\delta(\mathcal{U})\,\mathrm{d}\mathcal{U}^2}{[\,1+\frac{1}{6}\Lambda(\eta\bar\eta - \mathcal{U}\mathcal{V})\,]^2}\ , \tag{36}$$

which explicitly includes the impulse located on the wavefront $\mathcal{U} = 0$. This is a gravitational wave or an impulse of null matter in any spacetime of a constant curvature (4), depending on the specific form of the function H (see [37], [51] for more details). Again, in Minkowski background this is just the Brinkmann form of a general impulsive pp-wave (16). In fact, the metric (36) is conformal to (16). (Recall in this context that Siklos [15] proved that Einstein spaces conformal to pp-waves only occur when $\Lambda < 0$. The impulsive case is clearly a counter-example to this result.)

The explicit transformation between (31) and (36) for the $KN(\Lambda)I$ subclass is

$$\mathcal{U} = \frac{(\xi+\bar{\xi})(1+uv)u}{1-\frac{1}{6}\Lambda(\xi+\bar{\xi})(1+uv)u}\,,$$
$$\mathcal{V} = \frac{(\xi+\bar{\xi})v+\frac{1}{6}\Lambda\xi\bar{\xi}}{1-\frac{1}{6}\Lambda(\xi+\bar{\xi})(1+uv)u} - 1\,, \qquad (37)$$
$$\eta = \frac{\xi+(\xi+\bar{\xi})uv}{1-\frac{1}{6}\Lambda(\xi+\bar{\xi})(1+uv)u}\,,$$

which reduces to (32) in the case $\Lambda = 0$. Similar transformations hold also for the subclasses $KN(\Lambda^-)II$ and $KN(\Lambda^-)III$. Therefore, the full family of impulsive limits (31) of nonexpanding sandwich waves of the Kundt class $KN(\Lambda)$ is indeed equivalent to the distributional form of the solutions (36), and consequently to the continuous metric (24) obtained by the explicit "cut and paste" method.

3.3.2 Expanding impulsive waves

As has been argued in [38], the family of solutions for expanding impulsive spherical gravitational waves can be considered to be an impulsive limit of the class $RTN(\Lambda)$ of vacuum Robinson–Trautman type N solutions with a cosmological constant. It is convenient to consider these solutions in the García–Plebański [17] coordinates (10). Introducing a new coordinate w by $w = \psi v$, and assuming the impulsive limit $f \equiv f(\xi)\delta(u)$, we can express the above family of expanding impulses as

$$\mathrm{d}s^2 = 2\frac{w^2}{\psi^2}\left|\,\mathrm{d}\xi - f\,\delta(u)\,\mathrm{d}u\right|^2 + 2\,\mathrm{d}u\,\mathrm{d}w + \left(\tfrac{1}{3}\Lambda w^2 - 2\epsilon\right)\mathrm{d}u^2$$
$$+w\left[(f_{,\xi}+\bar{f}_{,\bar{\xi}}) - \frac{2\epsilon}{\psi}(f\bar{\xi}+\bar{f}\xi)\right]\delta(u)\mathrm{d}u^2\,, \qquad (38)$$

where $f(\xi)$ is an arbitrary function. Similarly as in the nonexpanding case (31), the spherical impulse in the form (38) is explicitly located on the null surface $u = 0$. For $u \neq 0$ the spacetime again reduces to the background of constant curvature. However, a difficult mathematical problem occurs in this case: the metric (38) contains the *product* of the Dirac distributions in the first term. This is not a well-defined concept in the linear theory of distributions, so that the metric (38) is only formal. Nevertheless, it demonstrates that impulsive limits of Robinson–Trautman type N vacuum spacetimes are equivalent to expanding impulsive gravitational waves (30). Indeed, the transformation

$$w = \frac{\tilde{w}}{1+\frac{1}{6}\Lambda\tilde{u}(\tilde{w}-\epsilon\tilde{u})}\,, \qquad u = \int\frac{\mathrm{d}\tilde{u}}{1-\epsilon\frac{1}{6}\Lambda\tilde{u}^2}\,, \qquad (39)$$

converts (38) to

$$\mathrm{d}s^2 = \frac{1}{[1+\frac{1}{6}\Lambda\tilde{u}(\tilde{w}-\epsilon\tilde{u})]^2}\left[2\frac{\tilde{w}^2}{\psi^2}\left|\,\mathrm{d}\xi - f\,\delta(\tilde{u})\,\mathrm{d}\tilde{u}\right|^2 + 2\,\mathrm{d}\tilde{u}\,\mathrm{d}\tilde{w} - 2\epsilon\,\mathrm{d}\tilde{u}^2\right.$$
$$\left.+\tilde{w}\left[(f_{,\xi}+\bar{f}_{,\bar{\xi}}) - \frac{2\epsilon}{\psi}(f\bar{\xi}+\bar{f}\xi)\right]\delta(\tilde{u})\mathrm{d}\tilde{u}^2\right]\,. \qquad (40)$$

Dropping the tildes and performing the following discontinuous transformation

$$\begin{aligned}u &= U + \Theta(U)\,[U_{inv}(U,V,Z,\bar{Z}) - U]\,,\\ w &= V + \Theta(U)\,[V_{inv}(U,V,Z,\bar{Z}) - V]\,,\\ \xi &= Z + \Theta(U)\,[Z_{inv}(U,V,Z,\bar{Z}) - Z]\,,\end{aligned} \tag{41}$$

where the functions U_{inv}, V_{inv}, Z_{inv} are obtained by the composition of the inverse transformation to (19) with (26), we put (40) into the metric form of all expanding impulsive spherical waves expressed in the continuous coordinate system (30) (see [38] for more details). This relation is obvious for all $U \neq 0$ if we use the identity $u(w - \epsilon u) = U(V - \epsilon U)$. Keeping the distributional terms arising from Θ and its derivative, we formally obtain also the impulsive terms proportional to δ, with the identification $f \equiv Z_{inv} - Z$ (evaluated on $U = 0$).

Of course, much work is required before the metric (38) and the discontinuous transformation (41) can be fully accepted in a mathematically rigorous way. A natural tool appears to be the Colombeau theory of nonlinear generalized functions [26], in the context of which products of distribution can (in principle) be handled.

3.4 Embedding from higher dimensions

The complete class of *nonexpanding* impulsive waves in spaces of constant curvature with a *nonvanishing* cosmological constant Λ can alternatively be introduced using a convenient 5-dimensional formalism as metrics

$$\begin{aligned}ds^2 = &-dZ_0{}^2 + dZ_1{}^2 + dZ_2{}^2 + dZ_3{}^2 + \epsilon\, dZ_4{}^2\\ &+\mathrm{H}(Z_2, Z_3, Z_4)\,\delta(Z_0 + Z_1)\,(dZ_0 + dZ_1)^2\,,\end{aligned} \tag{42}$$

restricted by the constraint (1). Physically, this is an embedding of impulsive *pp*-waves (42) which propagate in 5-dimensional flat space onto the (anti-)de Sitter hyperboloid. The wave is absent when $\mathrm{H} = 0$, in which case (42) with the constraint (1) reduces to the standard form of the de Sitter or the anti-de Sitter spacetime (2). It has been demonstrated in [52] that for a nontrivial H the metric represents impulsive waves propagating in the (anti-)de Sitter universe. Each impulse is located on the null hypersurface $Z_0 + Z_1 = 0$. Considering (1), this is given by

$$Z_2{}^2 + Z_3{}^2 + \epsilon Z_4{}^2 = \epsilon a^2\,, \tag{43}$$

which is a nonexpanding 2-sphere in the de Sitter universe or a hyperboloidal 2-surface in the anti-de Sitter universe. The wave surfaces can naturally be parametrized as

$$Z_2 = a\sqrt{\epsilon(1 - z^2)}\,\cos\phi\,, \quad Z_3 = a\sqrt{\epsilon(1 - z^2)}\,\sin\phi\,, \quad Z_4 = a\,z\,. \tag{44}$$

Various 4-dimensional parametrizations of the solutions (42) on (1) can be considered. For example,

$$Z_0 = \frac{\mathcal{U} + \mathcal{V}}{\sqrt{2}\,\Omega}\,, \quad Z_1 = \frac{\mathcal{U} - \mathcal{V}}{\sqrt{2}\,\Omega}\,, \quad Z_2 + \mathrm{i}Z_3 = \frac{\sqrt{2}\,\eta}{\Omega}\,, \quad Z_4 = a\left(\frac{2}{\Omega} - 1\right), \tag{45}$$

where $\Omega = 1 + \frac{1}{6}\Lambda(\eta\bar{\eta} - \mathcal{U}\mathcal{V})$, brings the metric to the previous form (36) with $\mathrm{H} = \sqrt{2}H/(1 + \frac{1}{6}\,\Lambda\,\eta\,\bar{\eta})$. Other natural coordinates which parametrize (42) have been discussed in [41].

We have remarked in previous sections that the above metrics may describe impulsive gravitational waves and/or impulses of null matter. Purely gravitational waves occur when the vacuum field equation

$$(\Delta+\tfrac{2}{3}\Lambda)\,\mathrm{H}=0 \tag{46}$$

is satisfied [51], [53], [37], in which $\Delta\equiv\frac{1}{3}\Lambda\{\partial_z[(1-z^2)\partial_z]+(1-z^2)^{-1}\partial_\phi\partial_\phi)\}$ is the Laplacian on the impulsive surface.

3.5 Boosts and limits of infinite acceleration

Let us finally outline yet another fundamental method for the construction of particular (but physically important) impulses in spaces of constant curvature. It is based on boosting suitable, initially static sources to the speed of light which yields specific nonexpanding impulsive waves. Similarly, limits of infinite acceleration of specific sources give special expanding impulsive solutions.

3.5.1 Nonexpanding impulsive waves

It was first demonstrated in 1971 by Aichelburg and Sexl in a classic paper [28] that a specific impulsive gravitational *pp*-wave solution can be obtained by boosting the Schwarzschild black hole to the speed of light, while its mass is reduced to zero in an appropriate way. Let us start with the metric

$$\mathrm{d}s^2=-\mathrm{d}t^2+\mathrm{d}r^2+r^2(\mathrm{d}\vartheta^2+\sin^2\vartheta\,\mathrm{d}\varphi^2)+\Psi(\mathrm{d}t^2+\mathrm{d}r^2)\,, \tag{47}$$

where $\Psi=2M/r$, which is the Schwarzschild (static and spherically symmetric) solution linearized for small values of M. We introduce Cartesian coordinates $x=r\sin\vartheta\cos\varphi$, $y=r\sin\vartheta\sin\varphi$, $z=r\cos\vartheta$, and perform a boost in the x-direction,

$$t=\frac{\tilde t+\mathrm{v}\tilde x}{\sqrt{1-\mathrm{v}^2}}\,,\qquad x=\frac{\tilde x+\mathrm{v}\tilde t}{\sqrt{1-\mathrm{v}^2}}\,. \tag{48}$$

In the limit as $\mathrm{v}\to 1$, the line element (47) takes the form

$$\mathrm{d}s^2=-\mathrm{d}\tilde t^2+\mathrm{d}\tilde x^2+\mathrm{d}y^2+\mathrm{d}z^2+2(\mathrm{d}\tilde t+\mathrm{d}\tilde x)^2\lim_{\mathrm{v}\to1}\frac{\Psi}{1-\mathrm{v}^2}\,, \tag{49}$$

where $\Psi=2M\,(x^2+y^2+z^2)^{-1/2}$ and x is given by (48). To evaluate this limit, we employ the identity

$$\lim_{\mathrm{v}\to1}\frac{1}{\sqrt{1-\mathrm{v}^2}}\,\Psi(x)=\delta(\tilde t+\tilde x)\int_{-\infty}^{+\infty}\Psi(x)\,\mathrm{d}x\,, \tag{50}$$

(see [29], [41]) which is valid in the distributional sense. It is also necessary to scale the parameter M in Ψ to zero such that $8M=b_0\sqrt{1-\mathrm{v}^2}$, where b_0 is a new constant. Obviously, we obtain an impulsive *pp*-wave metric (16),

$$\mathrm{d}s^2=-\mathrm{d}\tilde t^2+\mathrm{d}\tilde x^2+\mathrm{d}y^2+\mathrm{d}z^2+\mathrm{H}\,\delta(\tilde t+\tilde x)(\mathrm{d}\tilde t+\mathrm{d}\tilde x)^2\,, \tag{51}$$

where $\mathrm{H} = \frac{1}{2}b_0\int_{-\infty}^{+\infty}(\rho^2 + x^2)^{-1/2}\,\mathrm{d}x$, and $\rho^2 = y^2 + z^2$. The divergence in the integral can be removed by the transformation $\tilde{t} - \tilde{x} \to \tilde{t} - \tilde{x} - \frac{1}{2}b_0\lim_{\mathrm{v}\to 1}\log\left(\tilde{x} + \mathrm{v}\tilde{t} - \sqrt{(\tilde{x}+\mathrm{v}\tilde{t})^2 + 1 - \mathrm{v}^2}\right)$, which gives

$$\mathrm{H} = \tfrac{1}{2}b_0\int_{-\infty}^{+\infty}\left[(\rho^2 + x^2)^{-1/2} - (1 + x^2)^{-1/2}\right]\mathrm{d}x = -b_0\,\log\rho\,. \tag{52}$$

This is the famous Aichelburg–Sexl solution [28] which represents an axially-symmetric impulsive gravitational wave in Minkowski space generated by a single null monopole particle located on $\rho = 0$. Using a similar approach, numbers of other specific impulsive waves in flat space have been obtained by boosting various spacetimes of the Kerr–Newman [54] or the Weyl family [55].

The above method can be generalized to obtain impulses in spacetimes with $\Lambda \neq 0$. This was first done in 1993 by Hotta and Tanaka [29] who boosted the Schwarzschild–de Sitter solution to obtain a nonexpanding spherical impulsive gravitational wave generated by a pair of null monopole particles in the de Sitter background. They also described an analogous solution in the anti-de Sitter universe. Their main "trick" was to consider the boost in the 5-dimensional representation of the (anti-)de Sitter spacetime (1), (2) where the boost can explicitly (and consistently) be performed. One starts with the line element

$$\begin{aligned}\mathrm{d}s^2 = &-\left(1-\epsilon\, r^2/a^2\right)\mathrm{d}t^2 + \left(1-\epsilon\, r^2/a^2\right)^{-1}\mathrm{d}r^2 + r^2(\mathrm{d}\vartheta^2 + \sin^2\vartheta\,\mathrm{d}\varphi^2)\\ &+\Psi\left[\mathrm{d}t^2 + \left(1-\epsilon\, r^2/a^2\right)^{-2}\mathrm{d}r^2\right],\end{aligned} \tag{53}$$

where $\Psi = 2M/r$. This is a perturbation of the de Sitter ($\epsilon = +1$) or anti-de Sitter ($\epsilon = -1$) space corresponding to the Schwarzschild–(anti-)de Sitter solution [56], linearized for small M. As $\Lambda \to 0$, the line element (53) reduces to the Schwarzschild-like perturbations of the Minkowski spacetime (47). For $M = 0$ the metric (53) is the (anti-)de Sitter spacetime in standard "static" coordinates. These parametrize the hyperboloid (1) by $Z_1 = r\sin\vartheta\cos\varphi$, $Z_3 = r\sin\vartheta\sin\varphi$, $Z_2 = r\cos\vartheta$, and

$$\begin{aligned}&\text{for } \epsilon = +1:\ Z_0 = \sqrt{a^2 - r^2}\,\sinh(t/a)\,, \quad Z_4 = \pm\sqrt{a^2 - r^2}\,\cosh(t/a)\,,\\ &\text{for } \epsilon = -1:\ Z_0 = \sqrt{a^2 + r^2}\,\sin(t/a)\,, \quad Z_4 = \sqrt{a^2 + r^2}\,\cos(t/a)\,.\end{aligned} \tag{54}$$

Writing (53) as $\mathrm{d}s^2 = \mathrm{d}s_0^2 + \mathrm{d}s_1^2$, where $\mathrm{d}s_0^2$ has the form (2), we may now express the perturbation as

$$\mathrm{d}s_1^2 = a^2\Psi\left[\left(\frac{Z_4\mathrm{d}Z_0 - Z_0\mathrm{d}Z_4}{{Z_4}^2 - \epsilon {Z_0}^2}\right)^2 + \frac{a^2}{r^2}\left(\frac{Z_0\mathrm{d}Z_0 - \epsilon Z_4\mathrm{d}Z_4}{{Z_4}^2 - \epsilon {Z_0}^2}\right)^2\right], \tag{55}$$

where $r = \sqrt{{Z_0}^2 + \epsilon(a^2 - {Z_4}^2)}$. Performing a boost in the Z_1-direction,

$$Z_0 = \frac{\tilde{Z}_0 + \mathrm{v}\tilde{Z}_1}{\sqrt{1-\mathrm{v}^2}}\,, \qquad Z_1 = \frac{\tilde{Z}_1 + \mathrm{v}\tilde{Z}_0}{\sqrt{1-\mathrm{v}^2}}\,, \tag{56}$$

the (anti-)de Sitter background ds_0^2 is, of course, invariant. Using again the identity (50) and rescaling the coefficient as $8M = b_0\sqrt{1-\mathrm{v}^2}$, in the limit $\mathrm{v} \to 1$ the term ds_1^2 takes the form

$$ds_1^2 = \mathrm{H}\,\delta(\tilde{Z}_0 + \tilde{Z}_1)\,(d\tilde{Z}_0 + d\tilde{Z}_1)^2 \,, \tag{57}$$

where

$$\mathrm{H}(Z_4) = \tfrac{1}{4} b_0\, a^2 \int_{-\infty}^{+\infty} \frac{\epsilon(a^2 - {Z_4}^2){Z_4}^2 + (a^2 + {Z_4}^2)x^2}{\left({Z_4}^2 - \epsilon x^2\right)^2 \left[x^2 + \epsilon(a^2 - {Z_4}^2)\right]^{3/2}}\, dx \,. \tag{58}$$

We immediately observe that the structure of the boosted metric is exactly that of (42) with the impulse located on (43). Considering the parametrization (44) of this wave surface and substituting θ for x so that $x = a\sqrt{\epsilon(1-z^2)}\cot\theta$, we explicitly evaluate the integral (58) as

$$\mathrm{H} = \tfrac{1}{4} b_0 \int_0^{\pi} \frac{z^2 + \cos^2\theta}{(z^2 - \cos^2\theta)^2} \sin^3\theta\, d\theta = b_0 \left(\frac{z}{2} \log\left|\frac{1+z}{1-z}\right| - 1 \right) . \tag{59}$$

This is the axially-symmetric Hotta–Tanaka solution [29]. Further details on boosting monopole (and also multipole) particles to the speed of light in the (anti-)de Sitter universe, the geometry of the nonexpanding wave surfaces, and discussion of various useful coordinates can be found in [41], [55]. In particular, it was demonstrated that although the impulsive wave surface is nonexpanding, for $\Lambda > 0$ this coincides with the horizon of the closed de Sitter universe. The background space contracts to a minimum size and then reexpand in such a way that the nonexpanding impulsive wave in fact propagates with the speed of light from the "north pole" of the universe across the equator to its "south pole".

3.5.2 Expanding impulsive waves

As shown above, nonexpanding impulsive waves can be obtained by boosting particular (initially static) sources to the speed of light. Interestingly, specific expanding impulsive spherical gravitational waves can also be obtained by considering exact solutions representing accelerating sources and taking these to the limit in which the acceleration becomes unbounded. It was first realized by Jiří Bičák [31] (see also [57], with recent generalizations in [58], [59]) that expanding impulses can be obtained from boost-rotation symmetric solutions which represent gravitational field of uniformly accelerating objects. In the limit of infinite acceleration, these well-known explicit exact radiative spacetimes form spherical impulsive waves attached to conical singularities.

As Bičák writes in his essay [1], "the boost-rotation symmetric solutions representing uniformly accelerated objects ... play a unique role among radiative spacetimes since they are asymptotically flat, in the sense that they admit global smooth sections of null infinity. And as the only known radiative solutions describing finite sources they can provide expressions for the Bondi mass, the news function, or the radiation patterns in explicit forms." Special solutions of this type have been studied for almost forty years, first by Bondi [60], Bonnor and Swaminarayan [61], Israel and Khan [62], and then by many others — but Jiří Bičák in particular.

In fact, the investigation of various aspects of radiative spacetimes and fields with boost-rotation symmetry is his "most typical" topic. The analysis of radiative properties of the Bonnor–Swaminarayan solutions became the basis of Bičák's Ph. D. thesis and of his first paper published abroad [63]. He realized that these belong to a wide class of boost-rotation symmetric solutions (of which, for example, the well-known C-metric is also a member) and that the boost symmetry is the *only* admissible second symmetry of axially symmetric, asymptotically flat spacetimes which admit gravitational radiation. He and his collaborators have studied the unique properties of these spacetimes, including their global structure, in a number of fundamental works, e.g. [63] - [70], [31], [57]. He has also summarized these results in many inspiring reviews [71] - [76], [1]. Altogether, the contributions concerning boost-rotation symmetric solutions represent almost a quarter of all Bičák's original publications.

The boost-rotation symmetric spacetimes can be described by the line element

$$\mathrm{d}s^2 = e^{\lambda}\mathrm{d}\rho^2 + \rho^2 e^{-\mu}\mathrm{d}\phi^2 + (\zeta^2 - \tau^2)^{-1}\left[e^{\lambda}(\zeta\mathrm{d}\zeta - \tau\mathrm{d}\tau)^2 - e^{\mu}(\zeta\mathrm{d}\tau - \tau\mathrm{d}\zeta)^2\right] , \quad (60)$$

where specific metric functions μ and λ depend only on $\zeta^2 - \tau^2$ and ρ^2. In particular, the Bonnor–Swaminarayan solution [61] generally contains five arbitrary constants m_1, m_2, A_1, A_2, and B which determine the masses and uniform accelerations of two pairs of particles, and the singularity structure on the axis of symmetry $\rho = 0$ (see e.g. [61], [72], [58]). It is possible to choose the constant B such that some sections of the axis are regular. Of particular interest are special cases described by Bičák, Hoenselaers and Schmidt [64], [65] which represent only *two* (Curzon–Chazy) particles[a] of mass m which are accelerating in opposite directions with acceleration A. For these solutions the metric functions in (60) can be written in a simple form

$$\begin{aligned} \mu &= -\frac{2m}{AR} + 4mA + B , \\ \lambda &= -\frac{m^2}{A^2R^4}\,\rho^2(\zeta^2 - \tau^2) + \frac{2mA}{R}\,(\rho^2 + \zeta^2 - \tau^2) + B , \end{aligned} \quad (61)$$

where $R = \frac{1}{2}\sqrt{(\rho^2 + \zeta^2 - \tau^2 - A^{-2})^2 + 4A^{-2}\rho^2}$. For $B = 0$, the axis $\rho = 0$ is regular between the symmetrically located particles but these are connected to infinity by two semi-infinite strings which cause their acceleration. For $B = -4mA$ there is a finite expanding strut between the particles.

The limit of infinite acceleration $A \to \infty$ of these solutions was investigated by Bičák [31], [57]. It is necessary to scale the mass parameter m to zero in such a way that the "monopole moment" $M_0 = -4mA$ remains constant. In this limit we obtain

$$\begin{aligned} \mu &= B - M_0 , \\ \lambda &= B - \mathrm{sign}\,(\,\rho^2 + \zeta^2 - \tau^2\,)\, M_0 . \end{aligned} \quad (62)$$

[a] In [58] we investigated systematically all the spacetimes with four accelerating particles. Their limits $A_i \to \infty$ are unphysical, with the only exception of a particular solution which represents a finite string/strut between the two outer particles which snaps/breaks at its midpoint and the two broken ends separate at the speed of light.

The resulting spacetime is locally flat everywhere except on the sphere $\rho^2+\zeta^2=\tau^2$. It therefore describes an expanding spherical impulsive gravitational wave generated by two particles which move apart at the speed of light in the Minkowski background and are connected to infinity by two semi-infinite strings ($B=0$), or each other by an expanding strut ($B=M_0$). Performing the transformation (see, e.g. [68]) $\rho=\frac{1}{2}(v-u)$, $\tau=\pm\frac{1}{2}(v+u)\cosh\chi$, $\zeta=\frac{1}{2}(v+u)\sinh\chi$, we put the solution (60), (62) into the standard form of boost-rotational symmetric spacetimes

$$\mathrm{d}s^2=\tfrac{1}{4}(v-u)^2\,e^{-\mu}\,\mathrm{d}\phi^2+\tfrac{1}{4}(v+u)^2\,e^{\mu}\,\mathrm{d}\chi^2-e^{\lambda}\,du\,dv\ . \tag{63}$$

The metric functions are $\mu=-M_0$, $\lambda=[\,\Theta(uv)-\Theta(-uv)\,]\,M_0$ for the two receding strings, or $\mu=0$, $\lambda=2\,\Theta(uv)M_0$ for the expanding strut. In both cases μ is a constant, but there is a discontinuity in the otherwise constant value of λ with the step $2M_0$ on the null cone $uv=0$. It is possible to perform a further transformation to coordinates in which the metric is continuous everywhere. Indeed, in the region where the functions μ and λ are constant, the transformation $U=-u\,e^{\lambda+\mu/2}$, $V=\frac{1}{2}\,v\,e^{-\mu/2}$, $\psi=\chi\,e^{\mu}$, brings (63) into the form

$$\mathrm{d}s^2=\left(V+\tfrac{1}{2}PU\right)^2\mathrm{d}\phi^2+\left(V-\tfrac{1}{2}PU\right)^2\mathrm{d}\psi^2+2\,\mathrm{d}U\,\mathrm{d}V\ , \tag{64}$$

where $P=e^{-(\lambda+\mu)}$. The solution for ***two receding strings*** can thus be written in the form (64) with

$$P=\Theta(-U)+\beta^2\,\Theta(U)\ , \tag{65}$$

where $\beta=\exp(M_0)$. The metric (64), (65) is exactly that constructed by Gleiser and Pullin [30] by their "cut and paste" method. It represents an impulsive spherical gravitational wave propagating in the Minkowski universe. Outside the wave (in the region $U>0$), there are two receding strings with deficit angles $(1-\beta)2\pi$, which can be interpreted as remnants of one cosmic string which "snapped". However, as pointed out by Bičák [31], [77], the complete solution rather describes two semi-infinite strings approaching at the speed of light and separating again at the instant at which they collide. The complementary solution for an ***expanding strut*** is also given by (64) with

$$P=\beta^{-2}\,\Theta(-U)+\Theta(U)\ . \tag{66}$$

In this case there is a string with the deficit angle $(1-\beta^{-1})2\pi$ in the locally Minkowski space inside the impulsive wave ($U<0$).

Recall that there is an alternative continuous form of the above solutions. Introducing $Z=\frac{1}{\sqrt{2}}(\psi+\mathrm{i}\,\phi)$, we convert the metric (64), (65) into the form $\mathrm{d}s^2=2\,|\,V\mathrm{d}Z+UH\mathrm{d}\bar{Z}\,|^2+2\,\mathrm{d}U\,\mathrm{d}V$, where $H=-\frac{1}{2}\beta^2$. Performing another transformation of the metric (63) in the region $u>0$, $v>0$ as

$$\begin{aligned}U&=-\frac{uv}{u+v}\exp\left[\tfrac{\lambda}{2}-\chi\,e^{(\mu-\lambda)/2}\right],\\ V&=\tfrac{1}{2}\,(u+v)\exp\left[\tfrac{\lambda}{2}+\chi\,e^{(\mu-\lambda)/2}\right],\\ Z&=\tfrac{1}{\sqrt{2}}\,\frac{v-u}{v+u}\exp\left[-\chi\,e^{(\mu-\lambda)/2}+\mathrm{i}\,\phi\,e^{-(\mu+\lambda)/2}\right],\end{aligned} \tag{67}$$

we obtain $\mathrm{d}s^2 = 2V^2\,\mathrm{d}Z\,\mathrm{d}\bar{Z} + 2\,\mathrm{d}U\,\mathrm{d}V$. These two metrics for $U > 0$ and $U < 0$ can be matched continuously across the null cone $U = 0$. In fact, this is exactly a particular case of the metric (30) for $\Lambda = 0$, $\epsilon = 0$, and constant $H = -\frac{1}{2}\beta^2$, which describes impulsive spherical wave [25], [32], [33], [38].

Recently we analyzed [58] the limit of infinite acceleration of much larger class of boost-rotationally symmetric spacetimes which generalize the Bonnor–Swaminarayan solution. These explicit solutions were found by Bičák, Hoenselaers and Schmidt [65] and represent two uniformly accelerating particles with an *arbitrary multipole structure* attached to conical singularities (as in the two cases above). These solutions can be written as [58]

$$\begin{aligned} \mu &= 2\sum_{n=0}^{\infty} M_n \frac{P_n}{(x-y)^{n+1}} + B - M\,, \\ \lambda &= -2\sum_{k,l=0}^{\infty} M_k M_l \frac{(k+1)(l+1)}{(k+l+2)} \frac{(P_k P_l - P_{k+1}P_{l+1})}{(x-y)^{k+l+2}} \\ &\quad - \left(\frac{x+y}{x-y}\right) \sum_{n=0}^{\infty} \frac{M_n}{2^n} \sum_{l=0}^{n} \left(\frac{2}{x-y}\right)^l P_l + B\,, \end{aligned} \tag{68}$$

where the constants M_n represent the multipole moments, the argument of the Legendre polynomials P_n is $\alpha = (1-xy)/(x-y)$, B is a constant,

$$M \equiv \sum_{n=0}^{\infty} \frac{M_n}{2^n}\,, \tag{69}$$

and for x, y and α in (68) one has to substitute from

$$x - y = 4A^2 R, \quad x + y = 2A^2(\rho^2 + \zeta^2 - \tau^2), \quad \alpha = \tfrac{1}{2}(\rho^2 - \zeta^2 + \tau^2 + A^{-2})/R. \tag{70}$$

The axis is regular *between* the two particles if $B = 0$. In this case there are strings connecting the particles to infinity. The alternative situation with a strut between the particles (and the axis regular *outside*) is given by $B = M$. The axis is regular *everywhere* (except at the particles) if the combination of multipole moments satisfies the condition $M = 0$ and we assume $B = 0$. In this case the particles are self-accelerating [65]. Considering only the case $n = 0$ in (68), we recover the previous solution (61) for the accelerating monopole particles, with $M = M_0 = -4mA$.

Now, we consider the null limit $A \to \infty$ of the general class of solutions (68) in which all the multipole moments M_n are kept constant. Interestingly, in this limit we again obtain (62), only the parameter M_0 is now replaced by a more general parameter M introduced in (69). For the particular values of the constant B described above this solution describes a snapping cosmic string, or an expanding strut. The ends of the strings/strut move in opposite directions with the speed of light, generating an impulsive spherical gravitational wave. In the limit, the multipole structure of initial particles *disappears* and the solution is characterised by the single constant M only. Thus, the limit $A \to \infty$ of any uniformly accelerating multipole particle is identical to that of a monopole particle $M_0 = M$, as obtained originally in [31], [57]. Of course, the metric can also be represented in the coordinates (64) or (30) with $H = -\frac{1}{2}\beta^2$, the only difference is that the parameter β has a more general form

$$\beta = \exp\left(\sum_{n=0}^{\infty} \frac{M_n}{2^n}\right) . \tag{71}$$

In the subsequent paper [59] we investigated all possible null limits $A \to \infty$ of another important explicit class of solutions with boost-rotation symmetry, namely the well-know C-metric

$$\mathrm{d}s^2 = -A^{-2}(x+y)^{-2}\left(F^{-1}\mathrm{d}y^2 + G^{-1}\mathrm{d}x^2 + G\mathrm{d}\phi^2 - F\mathrm{d}t^2\right), \tag{72}$$

where $F = -1+y^2-2mAy^3$, $G = 1-x^2-2mAx^3$. It was shown already in 1970 by Kinnersley and Walker [78] that this metric represents two uniformly accelerating *black holes*, each of mass m. The acceleration A is caused either by a strut between the black holes or by two semi-infinite strings connecting them to infinity. Radiative and asymptotic properties were investigated in [79]. Bonnor [80] found an explicit transformation of (72) into a form (60). When $mA < 1/(3\sqrt{3})$ there are four possible spacetimes according to different ranges of the coordinates x and y [81], and the corresponding explicit functions λ, μ have more complicated forms. In [59] we investigated the limits $A \to \infty$ of all these possibilities. It was demonstrated that (scaling again the mass parameter m to zero such that mA remains constant) this limit is *identical* to the metric (64) of a spherical impulsive gravitational wave generated either by a snapping string (65), or an expanding strut (66), with

$$\beta = \tfrac{1}{2}\left[1+\sqrt{3}\cot(\varphi+\tfrac{1}{3}\pi)\right] \in (0,1], \quad \text{where } \varphi = \tfrac{1}{3}\arccos(1-54\,m^2A^2) . \tag{73}$$

Note finally that it is natural to expect that the analogous null limit of infinite acceleration $A \to \infty$ of a more general C-metric, which admits a nonvanishing value of the cosmological constant Λ [82], would generate an expanding spherical impulsive wave (30) in the (anti-)de Sitter universe. However, such limit is mathematically more involved and has not yet been explicitly performed. We are currently investigating this problem. A first step has been achieved in [83] where a physical interpretation of the parameters of the solutions [82], which represent uniformly accelerating black holes in spacetimes of constant nonvanishing curvature, was presented (see also [84]). However, a deeper understanding of the global structure of these solutions is required. The most recent contribution by Bičák and Krtouš [85] may be of a great help.

3.6 Summary

Let us now summarize for convenience all the methods described above for the construction of impulsive waves in spacetimes of constant curvature, together with the main corresponding references. This is presented in the following four diagrams, which separately describe the construction of nonexpanding/expanding waves in Minkowski/(anti-)de Sitter universes.

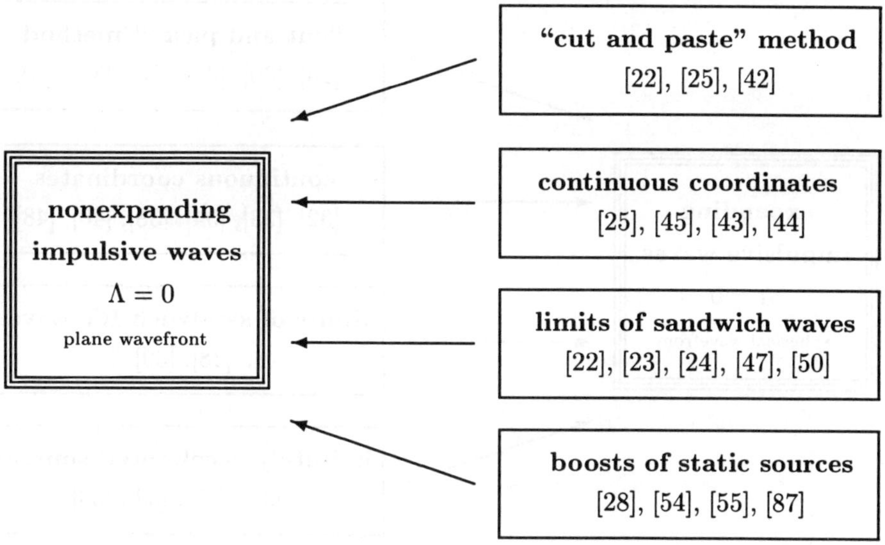

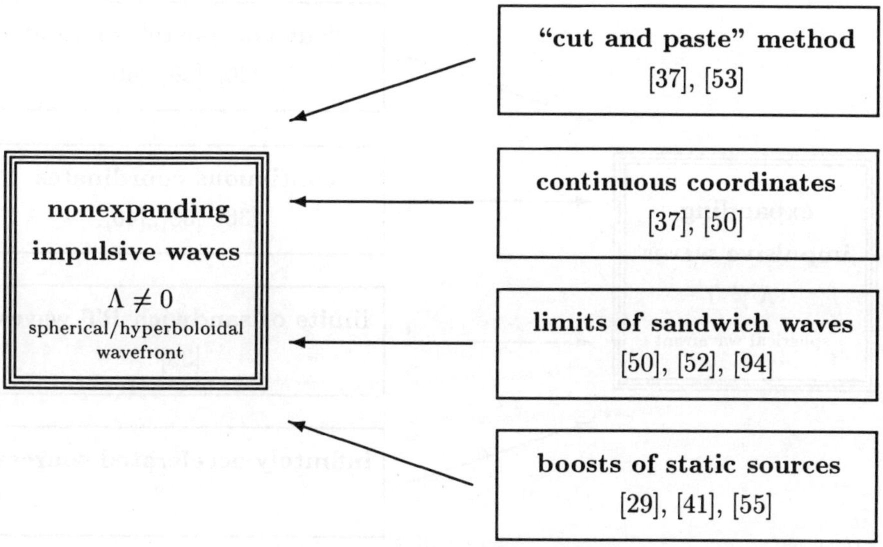

In addition, nonexpanding impulses with $\Lambda \neq 0$ can also be constructed by embedding impulsive *pp*-waves from higher dimensions on the (anti-)de Sitter hyperboloid, see the references [29], [41], [51], [52], [37].

3.6.2 Expanding impulsive waves

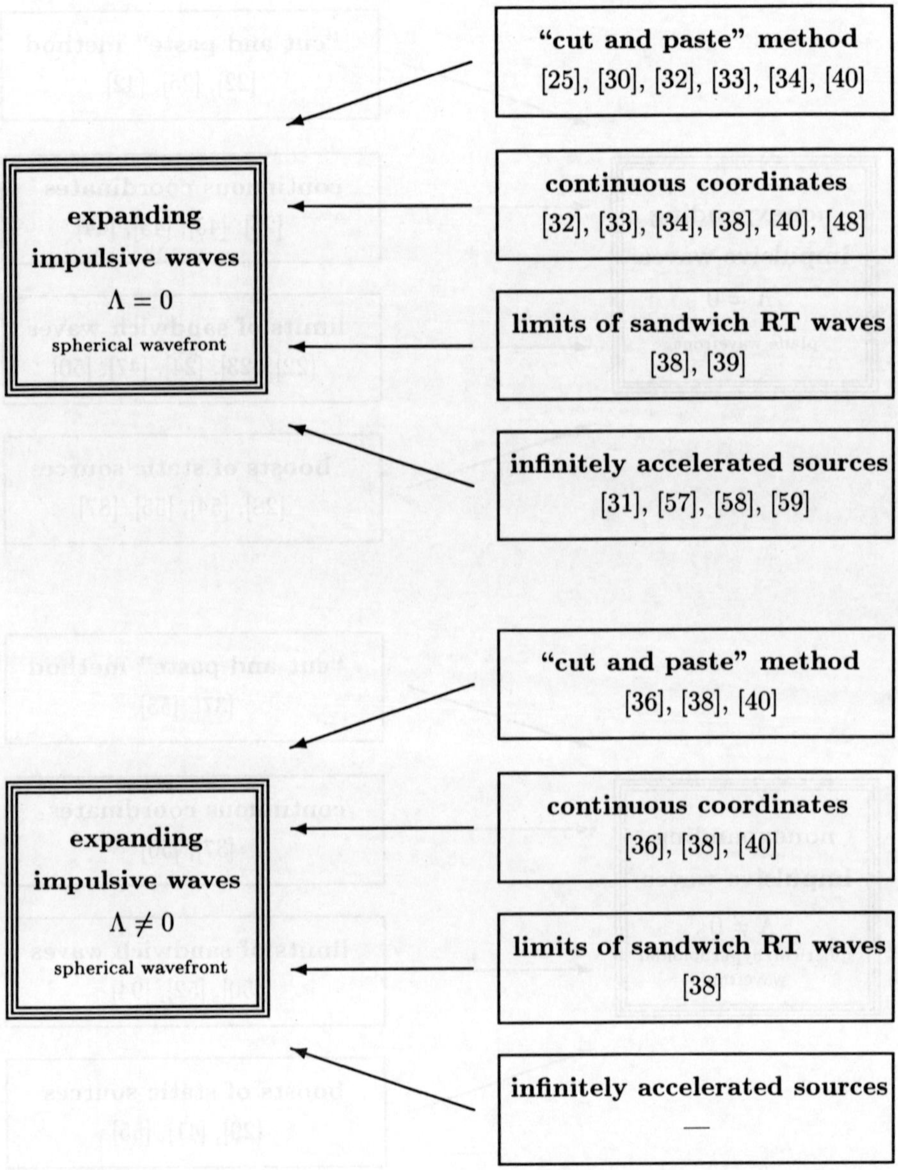

4 Particular solutions and some other properties

In the remaining part of this review some other properties of impulsive waves in spaces of constant curvature are briefly described. First, we present particular solutions of this type which are of interest from a physical point of view. Subsequently, we characterize geodesics and symmetries in these spacetimes.

4.1 Nonexpanding impulses generated by null multipole particles

It has been demonstrated above in subsection 3.5.1 that the simplest nonexpanding impulsive gravitational waves can be generated by boosting the Schwarzschild–(anti-)de Sitter black hole solutions. In the case $\Lambda = 0$ this gives the Aichelburg–Sexl solution (51), (52), and for $\Lambda \neq 0$ one obtains the Hotta–Tanaka solution (42), (59). Thus, the simplest impulsive solutions can be regarded as limits of static "monopole" solutions boosted to the speed of light. This result can be generalized: it has been shown [55] that it is possible to consider particular nonexpanding impulsive waves generated by null *multipole* particles as limits of boosted static multipole particles.

It is well-known [13], [86] that static, axially symmetric and asymptotically flat vacuum solutions which represent external field of sources with a multipole structure can be written in the Weyl coordinates as

$$ds^2 = -e^{2\psi}dt^2 + e^{-2\psi}\left[e^{2\gamma}(d\varrho^2 + dz^2) + \varrho^2 d\varphi^2\right] , \qquad (74)$$

where

$$\psi = \sum_{m=0}^{\infty} \frac{a_m}{r^{m+1}} P_m(\cos\theta) , \qquad (75)$$

$$\gamma = \sum_{m,n=0}^{\infty} \frac{(m+1)(n+1)}{m+n+2} \frac{a_m a_n}{r^{m+n+2}} \left(P_{m+1}P_{n+1} - P_m P_n\right) ,$$

$r = \sqrt{\varrho^2 + z^2}$, $\cos\theta = z/r$, and P_m are the Legendre polynomials with argument $\cos\theta$. Arbitrary constants a_m determine the mth multipole moments of the source. Performing the boost (48) where $x = \varrho\cos\varphi$, $y = \varrho\sin\varphi$, the line element (74) in the limit $\mathrm{v} \to 1$ may be written in the form (49) with $\Psi = -2\psi$. Considering again the identity (50) and rescaling the parameters a_m to zero such that $8a_m = -m\, b_m\sqrt{1-\mathrm{v}^2}$, with b_m being new constants which characterize the corresponding multipole moments of the boosted source, we obtain the impulsive pp-wave metric (51) where

$$\mathrm{H} = \sum_{m=0}^{\infty} b_m \mathrm{H}_m = \sum_{m=0}^{\infty} \tfrac{1}{2} m\, b_m \int_{-\infty}^{+\infty} \frac{1}{(\rho^2 + x^2)^{(m+1)/2}} P_m\!\left(\frac{z}{\sqrt{\rho^2 + x^2}}\right) dx \,. \qquad (76)$$

For the simplest case $a_0 \neq 0$, $a_m = 0$ for $m \geq 1$ (which corresponds to the boosted Curzon–Chazy solution) the integral gives exactly the same result (52) as the Aichelburg–Sexl solution [28] generated by a single null monopole particle. For the higher multipole components $m \geq 1$ the integral (76) can also be explicitly evaluated. As shown in [55], we obtain

$$\mathrm{H}_m(\rho,\phi)=\rho^{-m}\cos[m(\phi-\phi_m)]\,, \tag{77}$$

where $\rho^2=y^2+z^2$, $\cos(\phi-\phi_m)=z/\rho$, and ϕ_m is a constant. This term represents the mth multipole component of the exact impulsive purely gravitational *pp*-wave generated by a source of an arbitrary multipole structure [87]. Indeed, the field equations for the metric (51) with

$$\mathrm{H}=-b_0\log\rho+\sum_{m=1}^{\infty}b_m\,\rho^{-m}\cos[m(\phi-\phi_m)]\,, \tag{78}$$

correspond to a source localized on the impulsive wavefront $u\equiv\tilde{t}+\tilde{x}=0$ at $\rho=0$, which is described by $T_{uu}=J(\rho,\phi)\,\delta(u)$ where

$$J(\rho,\phi)=\tfrac{1}{8}b_0\,\delta(\rho)+\sum_{m=1}^{\infty}\tfrac{1}{8}b_m\frac{(-1)^m}{(m-1)!}\,\delta^{(m)}(\rho)\cos[m(\phi-\phi_m)]\,. \tag{79}$$

Clearly, the parameters b_m and ϕ_m represent the amplitude and phase of each multipole component.

Interestingly, there are analogous impulsive solutions also in the case $\Lambda\neq0$. It was demonstrated in [52] that nontrivial solutions of the vacuum Einstein equation (46) for the metric (42) on (1), expressed in terms of the parameters (44), can be written as

$$\mathrm{H}(z,\phi)=\sum_{m=0}^{\infty}b_m\mathrm{H}_m=\sum_{m=0}^{\infty}b_mQ_1^m(z)\cos[m(\phi-\phi_m)]\,, \tag{80}$$

where $Q_1^m(z)$ are associated Legendre functions of the second kind generated by the relation

$$Q_1^m(z)=(-\epsilon)^m|1-z^2|^{m/2}\frac{\mathrm{d}^mQ_1(z)}{\mathrm{d}z^m}\,. \tag{81}$$

The first term for $m=0$,

$$Q_1(z)\equiv Q_1^0(z)=\frac{z}{2}\log\left|\frac{1+z}{1-z}\right|-1\,, \tag{82}$$

exactly corresponds to the simplest axisymmetric Hotta–Tanaka solution (59). The higher components H_m describe nonexpanding impulsive gravitational waves in (anti-)de Sitter universe generated by null point sources with an m-pole structure, localized on the wave-front (43) at the singularities $z=\pm1$, see [52]. The general solution (80) corresponds to a source described by

$$J(z,\phi)=\sum_{m=0}^{\infty}b_m\frac{\epsilon(-1)^m}{8\pi}(1-z^2)^{m/2}\left[\delta^{(m)}(z-1)+\delta^{(m)}(-z-1)\right]\cos[m(\phi-\phi_m)]\,. \tag{83}$$

Again, each component represents a point source with an m-pole structure, where the constants b_m give the strength, and ϕ_m the orientation of each m-pole.

To complete the picture, it would be natural to regard these explicit impulsive solutions (80) in the (anti-)de Sitter universe generated by *null* multipole particles as limits of boosted *static* multipole particles, as in the case of Minkowski background. Unfortunately, no explicit exact solutions are known which describe static sources of any multipole structure in a background with $\Lambda \neq 0$. Nevertheless, we were able to argue in [55] that it is possible to obtain the correct structure of the impulsive multipole solution (42), (80) by boosting the perturbation of the (anti-)de Sitter spacetime of the form (53) if the function $\Psi(t,r,\vartheta)$ has an appropriate form. The corresponding function H in (57) is given by

$$\mathrm{H}(z,\phi) = a^3[\epsilon(1-z^2)]^{3/2} \int_0^\pi \frac{z^2+\cos^2\theta}{(z^2-\cos^2\theta)^2}\,\frac{\Psi(t,r,\vartheta)}{r^2}\,\mathrm{d}\theta\,, \tag{84}$$

in which the coordinates t, r, ϑ must be expressed using the relations

$$\cosh^2\frac{t}{a} = \frac{z^2\sin^2\theta}{z^2-\cos^2\theta}\,, \qquad r = \frac{a\sqrt{\epsilon(1-z^2)}}{\sin\theta}\,, \qquad \cos\vartheta = \sin\theta\cos\phi\,, \tag{85}$$

(for $\Lambda < 0$ the function cosh must be replaced by cos). This is a generalization of the integral (59) which yields the Hotta–Tanaka solution in the monopole case $\Psi = 2M/r$.

4.2 Expanding impulses generated by snapping and colliding strings

A physically most interesting expanding spherical impulsive gravitational wave is probably that generated by a *snapping cosmic string.* This explicit solution, described in detail above in section 3.5.2., can be written in various forms — for example using the coordinates (64) with (65), or in the form (30) with a constant value $H = -\frac{1}{2}\beta^2$ (see the transformation (67) and the related text). Alternatively, the spacetime can also be written in the form (30) with the function H given by

$$H = \frac{\frac{1}{2}\delta(1-\frac{1}{2}\delta)}{Z^2}\,, \tag{86}$$

where $\delta > 0$. This is generated from the function

$$h(Z) = Z^{1-\delta}, \tag{87}$$

by the expression (29). As shown in [32], [40], the generating function h is closely related to a geometrical interpretation of the Penrose junction conditions (22). If we evaluate the ratio $\eta/\mathcal{V}$ using (19) for $U < 0$, and using (26) for $U > 0$, which is related to the explicit construction of a continuous coordinates (30), we observe that on the impulse $U = 0$

$$\frac{\eta}{\mathcal{V}} = \begin{cases} Z & \text{for} \quad U = 0_-\,, \\ h(Z) & \text{for} \quad U = 0_+\,. \end{cases} \tag{88}$$

However, it follows from (4), (3) that $\eta/\mathcal{V} = (x + \mathrm{i}\,y)/(t - z)$ in Minkowski space, or similarly $\eta/\mathcal{V} = (Z_2 + \mathrm{i}\,Z_3)/(Z_0 - Z_1)$ in (anti-)de Sitter space. In both cases, this is exactly the relation for a stereographic correspondence between a sphere and an Argand plane by projection from the North pole onto a plane through the equator [88]. This permits us to represent the wave surface $U = 0$ either as a Riemann sphere, or as its associated complex plane parametrized by the coordinate Z. The

Penrose junction condition (22) across the wave surface then can be understood as a mapping on the complex Argand plane $Z \to h(Z)$. According to (88), this is equivalent to mapping points P_- on the "inside" of the wave surface to the identified points P_+ on the "outside".

We may assume that the spacetime $U < 0$ *inside* the impulse represented by $Z = |Z|e^{i\phi}$ covers the *complete* sphere, $\phi \in [-\pi, \pi)$. However, the range of the function $h(Z)$ does not cover the entire sphere outside the spherical impulse for $U > 0$. In particular, the complex mapping (87) covers the plane minus a wedge as $\arg h(Z) \in [-(1-\delta)\pi, (1-\delta)\pi)$. This represents Minkowski, de Sitter, or anti-de Sitter space with a deficit angle $2\pi\delta$ which may be considered to describe a snapped cosmic *string* in the region *outside* the spherical wave. The strings have a constant tension and are located along the axis $\eta = 0$.

With the help of the above geometrical insight we were able to construct even more general explicit solutions, namely expanding impulsive waves generated by two *colliding cosmic strings* [40]. In such a situation, as first suggested by Nutku and Penrose [32], two non-aligned strings initially approach each other and snap at their common point at the instant at which they collide. The remnants are four semi-infinite strings which recede from the common point of interaction with the speed of light, generating a specific spherical impulse. As shown in [40], such a solution is described by the metric (30) with the structural function of the form

$$H = \frac{\frac{1}{2}\delta(1-\frac{1}{2}\delta)}{Z^2} - \frac{\frac{1}{2}\varepsilon(1-\frac{1}{2}\varepsilon)16w_1^2(1-\delta)^2 Z^{-2\delta}}{[(w_1+\mathrm{i})Z^{1-\delta}+(w_1-\mathrm{i})]^2[(w_1-\mathrm{i})Z^{1-\delta}+(w_1+\mathrm{i})]^2}, \tag{89}$$

which is generated from

$$h(Z) = w_2 \frac{h_c^{1-\varepsilon}-1}{h_c^{1-\varepsilon}+1}, \quad \text{where} \quad h_c = -\frac{(w_1-\mathrm{i})Z^{1-\delta}+(w_1+\mathrm{i})}{(w_1+\mathrm{i})Z^{1-\delta}+(w_1-\mathrm{i})}. \tag{90}$$

Here the parameters δ and ε are the deficit angles of the first and the second pair of strings, respectively, and w_1, w_2 determine their velocities. For more details and illustrative pictures see [40].

Let us also mention that Hortaçsu and his collaborators [89] analyzed vacuum fluctuations and aspects of particle creation on the backgrounds of expanding impulsive waves.

4.3 Geodesics in impulsive spacetimes

In this final section we now mention some other properties of impulsive waves in spaces of constant curvature. In particular, we describe the behaviour of geodesics in these spacetimes.

4.3.1 Geodesics in nonexpanding impulsive waves

Geodesics in Minkowski space with plane-fronted impulsive *pp*-waves were discussed in many works, e.g. in [42], [53], [90]. However, as the corresponding geodesic and geodesic deviation equations in standard coordinates (16) contain highly singular products of distributions, an advanced framework of Colombeau algebras of generalized functions had to be employed to solve these equations in a mathematically rigorous sense [47].

In [91] we have studied the behaviour of geodesics in spacetimes which describe nonexpanding impulsive waves in the (anti-)de Sitter background. These recent results generalize those for impulsive *pp*-waves to the case of a nonvanishing cosmological constant Λ. The geodesic equations in such spacetimes can be derived conveniently using the embedding of the 5-dimensional *pp*-waves (42) onto the 4-dimensional hyperboloid (1). We obtain [91]

$$\begin{aligned} \ddot{U} &= -\tfrac{1}{3}\Lambda\, U\, e\,, \\ \ddot{V} - \tfrac{1}{2}\mathrm{H}\,\delta'(U)\,\dot{U}^2 - \mathrm{H}_{,p}\,\dot{Z}_p\,\delta(U)\,\dot{U} &= -\tfrac{1}{3}\Lambda\, V\left[e + \tfrac{1}{2}G\,\delta(U)\,\dot{U}^2\right]\,, \\ \ddot{Z}_i - \tfrac{1}{2}\mathrm{H}_{,i}\,\delta(U)\,\dot{U}^2 &= -\tfrac{1}{3}\Lambda\, Z_i\left[e + \tfrac{1}{2}G\,\delta(U)\,\dot{U}^2\right]\,, \\ \ddot{Z}_4 - \tfrac{1}{2}\epsilon\mathrm{H}_{,4}\,\delta(U)\,\dot{U}^2 &= -\tfrac{1}{3}\Lambda\, Z_4\left[e + \tfrac{1}{2}G\,\delta(U)\,\dot{U}^2\right]\,, \end{aligned} \tag{91}$$

where $p = 2, 3, 4$, $i = 2, 3$, $e = 0, -1, +1$ for null, timelike or spacelike geodesics, respectively, and $G = Z_p\,\mathrm{H}_{,p} - \mathrm{H}$ (summation convention is used). The coordinates $U \equiv \frac{1}{\sqrt{2}}(Z_0 + Z_1)$ and $V \equiv \frac{1}{\sqrt{2}}(Z_0 - Z_1)$ are introduced here for convenience, and are different from those used in the metric (24). Also, we have rescaled the structural function as $\sqrt{2}\,\mathrm{H} \to \mathrm{H}$.

The equation for U in (91) is clearly decoupled and does not involve any distributional term. The solution, which is everywhere a *smooth* function of the affine parameter τ, can be written (without loss of generality) as

$$U = \tau\,, \qquad U = a\,\dot{U}^0\sinh(\tau/a)\,, \qquad U = a\,\dot{U}^0\sin(\tau/a)\,, \tag{92}$$

for $\epsilon e = 0$, $\epsilon e < 0$ or $\epsilon e > 0$, respectively. These relations allow us to take U as the geodesic parameter. Then a general solution of the remaining four functions Z_p and V in (91) can be written as

$$\begin{aligned} Z_p(U) &= Z_p^0\sqrt{1 - \tfrac{1}{3}\Lambda e(\dot{U}^0)^{-2}U^2} \;+\; (\dot{Z}_p^0/\dot{U}^0)\,U \;+\; A_p\,\Theta(U)\,U\,, \\ V(U) &= V^0\sqrt{1 - \tfrac{1}{3}\Lambda e(\dot{U}^0)^{-2}U^2} \;+\; (\dot{V}^0/\dot{U}^0)\,U \;+\; B\,\Theta(U)\sqrt{1 - \tfrac{1}{3}\Lambda e(\dot{U}^0)^{-2}U^2} \\ &\quad + (\dot{U}^0)^{-1}(\dot{Z}_i^0 A_i + \epsilon\dot{Z}_4^0 A_4)\,\Theta(U)\,U \;+\; C\,\Theta(U)\,U\,. \end{aligned} \tag{93}$$

The coefficients are

$$\begin{aligned} A_i &= \tfrac{1}{2}\left[\mathrm{H}_{,i}(0) - \tfrac{1}{3}\Lambda\, Z_i^0 G(0)\right]\,, \quad A_4 = \tfrac{1}{2}\left[\epsilon\mathrm{H}_{,4}(0) - \tfrac{1}{3}\Lambda\, Z_4^0 G(0)\right]\,, \quad B = \tfrac{1}{2}\mathrm{H}(0)\,, \\ C &= \tfrac{1}{8}\left[\mathrm{H}_{,2}^2(0) + \mathrm{H}_{,3}^2(0) + \epsilon\mathrm{H}_{,4}^2(0) + \tfrac{1}{3}\Lambda\,\mathrm{H}^2(0) - \tfrac{1}{3}\Lambda\left(Z_p^0\,\mathrm{H}_{,p}(0)\right)^2\right]\,, \end{aligned} \tag{94}$$

in which $\mathrm{H}(0) \equiv \mathrm{H}(Z_p(0)) = \mathrm{H}(Z_p^0)$, and the constants of integration are constrained by

$$\begin{aligned} -2\,\dot{U}^0\dot{V}^0 + (\dot{Z}_2^0)^2 + (\dot{Z}_3^0)^2 + \epsilon(\dot{Z}_4^0)^2 &= e\,, \\ -2\,U^0V^0 + (Z_2^0)^2 + (Z_3^0)^2 + \epsilon(Z_4^0)^2 &= \epsilon a^2\,, \\ -U^0\dot{V}^0 - \dot{U}^0V^0 + Z_2^0\dot{Z}_2^0 + Z_3^0\dot{Z}_3^0 + \epsilon Z_4^0\dot{Z}_4^0 &= 0\,. \end{aligned} \tag{95}$$

Interestingly, these general results also describe all geodesics in impulsive *pp*-wave spacetimes in a Minkowski universe. For $\Lambda = 0$ the geodesic equations (91) exactly reduce (omitting the coordinate Z_4) to the system which has been ***rigorously*** solved by Kunzinger and Steinbauer [47] using the Colombeau algebras. Of course, the solution (93), (94) reduces to the explicit form of geodesics presented in [47] and previous works [42], [90]. The same results are also obtained from the chaotic behaviour of geodesics in nonhomogeneous sandwich *pp*-waves [92]: in the impulsive limit the chaotic motion smears and becomes regular [93].

The main advantage of the approach presented above is that for *any* Λ this yields the system (91) which includes only "weakly" singular terms. In particular, there is no square of δ, contrary to other "direct" approaches, such as that presented in [53] which used the coordinates introduced in [42].

It follows from (93) that the geodesics are continuous but refracted by the impulse in the transverse directions Z_p. However, there is a discontinuity in the longitudinal coordinate V (and its derivative) on the impulse. The jump on $U = 0$ is given by $\Delta V = B = \frac{1}{2}\mathrm{H}(0)$, which is in agreement with the Penrose junction conditions (18) in the "cut and paste" method. Further specific effects of nonexpanding impulses on privileged families of comoving observers associated with the natural coordinates (5) in the de Sitter and in the anti-de Sitter universe were discussed in [91] in detail. In particular, the behaviour of timelike and null geodesics (including their focusing) in the Hotta–Tanaka spacetime, a solution with a conformally flat pure radiation impulsive wave, and the Defrise-type impulse [94] were presented.

Note finally that symmetries of spacetimes with impulsive nonexpanding waves were studied for $\Lambda = 0$ in [95], and for $\Lambda \neq 0$ in [91]. There are always *at least three* Killing vector fields. In the metric form (36), these are null rotations generated by

$$\begin{aligned} &x\,(\partial/\partial\mathcal{V}) + \mathcal{U}\,(\partial/\partial x)\,, \qquad y\,(\partial/\partial\mathcal{V}) + \mathcal{U}\,(\partial/\partial y)\,, \\ &[1 - \tfrac{1}{12}\Lambda(x^2+y^2)]\,(\partial/\partial\mathcal{V}) - \tfrac{1}{6}\Lambda\mathcal{U}\,[x\,(\partial/\partial x) + y\,(\partial/\partial y) + \mathcal{U}\,(\partial/\partial\mathcal{U})]\,, \end{aligned} \tag{96}$$

where $\eta = \frac{1}{\sqrt{2}}(x + \mathrm{i}\,y)$. The Killing vectors (96) exactly reduce to those found previously by Aichelburg and Balasin [95] for impulsive *pp*-waves in the Minkowski background. Additional symmetries arise for specific forms of the structural function. For example, in case of the Hotta–Tanaka solution (59) the function H only depends on $z = \sqrt{1 - \epsilon a^{-2}(Z_2^2 + Z_3^2)}$. There is thus also an axial symmetry. Again, this is analogous to the axisymmetric Aichelburg–Sexl solution for the case $\Lambda = 0$ which also admits four Killing vectors [95].

4.3.2 Geodesics in expanding impulsive waves

As far as we know, an explicit form for geodesics in spacetimes with expanding spherical gravitational impulses has not been presented in the literature so far. It is possible to derive such geodesics assuming these are C^1 in the continuous coordinate system (30). With this assumption, the constants

$$\begin{aligned} &Z_i \equiv Z(\tau_i)\,, \quad V_i \equiv V(\tau_i)\,, \quad U_i \equiv U(\tau_i) = 0\,, \\ &\dot{Z}_i \equiv \dot{Z}(\tau_i)\,, \quad \dot{V}_i \equiv \dot{V}(\tau_i)\,, \quad \dot{U}_i \equiv \dot{U}(\tau_i)\,, \end{aligned} \tag{97}$$

which characterize the positions and velocities at τ_i, the instant of interaction with the impulsive wave, must have the same values when evaluated in the limits $U \to 0$ both from the region in front ($U > 0$) and behind ($U < 0$) the impulse. Let us consider here only the case $\Lambda = 0$ for the sake of simplicity. We start with a general free motion

$$\begin{aligned} t^- &= \gamma\tau \,, \\ x^- &= \dot{x}_0\,(\tau - \tau_i) + x_0 \,, \\ y^- &= \dot{y}_0\,(\tau - \tau_i) + y_0 \,, \\ z^- &= \dot{z}_0\,(\tau - \tau_i) + z_0 \,, \end{aligned} \tag{98}$$

where $\gamma = \sqrt{\dot{x}_0^2 + \dot{y}_0^2 + \dot{z}_0^2 - e}$, in the flat Minkowski region $U < 0$. It is straightforward to express the constants (97) in terms of the initial data (98) using the inverse to (19) and standard relations $\mathcal{U} = \frac{1}{\sqrt{2}}(t+z)$, $\mathcal{V} = \frac{1}{\sqrt{2}}(t-z)$, $\eta = \frac{1}{\sqrt{2}}(x + \mathrm{i}\,y)$, as

$$\begin{aligned} Z_i &= \frac{x_0 + \mathrm{i}\,y_0}{\gamma\tau_i - z_0} \,, \qquad V_i = \tfrac{1}{\sqrt{2}}\left[(1+\epsilon)\gamma\tau_i - (1-\epsilon)z_0\right] , \\ \dot{Z}_i &= \frac{\dot{x}_0 + \mathrm{i}\,\dot{y}_0}{\gamma\tau_i - z_0} - \frac{x_0 + \mathrm{i}\,y_0}{(\gamma\tau_i - z_0)^2} \\ &\quad \times \frac{[(1-\epsilon)\gamma - (1+\epsilon)\dot{z}_0](\gamma\tau_i - z_0) + 2\epsilon(x_0\dot{x}_0 + y_0\dot{y}_0)}{(1+\epsilon)\gamma\tau_i - (1-\epsilon)z_0} \,, \\ \dot{V}_i &= \tfrac{1}{\sqrt{2}}\Big\{\left[(1-\epsilon)\gamma - (1+\epsilon)\dot{z}_0\right]\left[(1-\epsilon)\gamma\tau_i - (1+\epsilon)z_0\right] \\ &\qquad + 4\epsilon(x_0\dot{x}_0 + y_0\dot{y}_0)\Big\}\left[(1+\epsilon)\gamma\tau_i - (1-\epsilon)z_0\right]^{-1} , \\ \dot{U}_i &= \sqrt{2}\,\frac{x_0\dot{x}_0 + y_0\dot{y}_0 + z_0\dot{z}_0 - \gamma^2\tau_i}{(1+\epsilon)\gamma\tau_i - (1-\epsilon)z_0} \,. \end{aligned} \tag{99}$$

Now, we express the geodesics outside the impulse in the (locally) Minkowski space $U > 0$

$$\begin{aligned} \mathcal{V}^+ &= \dot{\mathcal{V}}_0^+(\tau - \tau_i) + \mathcal{V}_0^+ \,, \\ \mathcal{U}^+ &= \dot{\mathcal{U}}_0^+(\tau - \tau_i) + \mathcal{U}_0^+ \,, \\ \eta^+ &= \dot{\eta}_0^+(\tau - \tau_i) + \eta_0^+ \,, \end{aligned} \tag{100}$$

using (26) as

$$\begin{aligned} \mathcal{V}_0^+ &= AV_i \,, \qquad \mathcal{U}_0^+ = BV_i \,, \qquad \eta_0^+ = CV_i \,, \\ \dot{\mathcal{V}}_0^+ &= A\dot{V}_i - D\dot{U}_i + (A_{,Z}\dot{Z}_i + A_{,\bar{Z}}\dot{\bar{Z}}_i)V_i \,, \\ \dot{\mathcal{U}}_0^+ &= B\dot{V}_i - E\dot{U}_i + (B_{,Z}\dot{Z}_i + B_{,\bar{Z}}\dot{\bar{Z}}_i)V_i \,, \\ \dot{\eta}_0^+ &= C\dot{V}_i - F\dot{U}_i + (C_{,Z}\dot{Z}_i + C_{,\bar{Z}}\dot{\bar{Z}}_i)V_i \,, \end{aligned} \tag{101}$$

in which the coefficients and their derivatives are given by the functions (27) evaluated at Z_i.

These general results can be used for a physical discussion of geodesics in specific impulsive solutions [96], such as the expanding spherical impulse generated by a snapping cosmic string which we have described above in sections 3.5.2 and 4.2.

5 Repetition and a few final remarks

In this essay, we have presented a brief review of exact solutions of Einstein's equations which describe impulsive waves in spaces of constant curvature. These are either gravitational and/or null matter nonexpanding impulses, or expanding spherical impulsive purely gravitational waves (attached to cosmic strings) which propagate in Minkowski, de Sitter, or anti-de Sitter universes. We have systematically and explicitly described all the main methods for their construction: the "cut and paste" method, introduction of the continuous coordinate system, considering distributional limits of sandwich waves, geometrical embedding from higher dimensions, and boosts or limits of infinite acceleration of some initially static or accelerating sources. In the further part of the contribution we concentrated on some of their properties, in particular on the behaviour of geodesics. Physically interesting special solutions were also emphasized.

We have tried to provide an overall review. Nevertheless, we are aware of the fact that the point of view presented could depend on our own approach to the subject. The essay may thus be biased, as more space was certainly devoted to a description of those aspects of impulsive spacetimes which we had personally investigated. However, we attempted to present a comprehensive list of the relevant references. If some are missing this is unintentional, and the author deeply apologizes for this.

We have dedicated the essay to Professor Bičák on the occasion of the landmark anniversary of his birthday. It has been demonstrated that he contributed significantly also to the investigation of impulsive waves. Of course, this is just one small part of his both extensive and intensive work in the field of exact radiative spacetimes.

Let me finally close this contribution by personal words once more. It was Professor Bičák who put me wise to many of the secrets, mysteries, and wonders of Einstein's theory. He has also led my first steps in the world of relativity, formed and guided my professional activities. From the very beginning I have always admired his approach to general relativity. This is based on a deep understanding of the geometry and global structure of investigated spacetimes, yet with the emphasis being placed on presenting their clear physical interpretation. I hope that I have succeeded to emulate (at least partially) his "style" here. And I also hope that Jiří Bičák will not take his pencil during the reading of this essay, making many strokes, remarks, corrections, amendments, and improvements, as he has done so many times before with my previous manuscripts.

Acknowledgements

I wish to thank very much Jerry Griffiths, my long-standing collaborator on impulsive waves, for all his suggestions and comments concerning this review. Special thanks are due to my wife for her patience and understanding. The work was supported by the grant GACR-202/99/0261 from the Czech Republic, and GAUK 141/2000 of Charles University in Prague.

References

1. Bičák J., *Selected Solutions of Einstein's Field Equations: Their Role in General Relativity and Astrophysics*, in *Einstein's Field Equations and Their Physical Implications* (Lecture Notes in Physics **540**), B.G. Schmidt (ed.) Springer, Berlin (2000) pp 1-126.
2. Guth A.H., *Inflationary universe: A possible solution to the horizon and flatness problems*, Phys. Rev. D **23** (1981) 347-356.
   Kolb E.W. and Turner M.S., *The Early Universe*, Addison-Wesley, New York (1990).
   Olive K.A., *Inflation*, Phys. Rep. **190** (1990) 307-403.
   Peebles P.J.E., *Principles of physical cosmology*, Princeton University Press, Princeton (1993).
3. Gibbons G.W. and Hawking S.W., *Cosmological event horizons, thermodynamics, and particle creation*, Phys. Rev. D **15** (1977) 2738-2751.
   Maeda K., *Inflation and Cosmic No Hair Conjecture*, in *Proc. 5th M. Grossman Meeting on Gen. Rel.*, D.G. Blair and M.J. Buckingham (eds.), World Scientific, Singapore (1989) pp 145-155.
4. Maldacena J., *The large N limit of superconformal field theories and supergravity*, Adv. Theor. Math. Phys. **2** (1998) 231-252.
   Witten E., *Anti-de Sitter Space and Holography*, Adv. Theor. Math. Phys. **2** (1998) 253-291.
   Aharony O., Gubser S.S., Maldacena J., Oouri H. and Oz Y., *Large N field theories, string theory and gravity*, Phys. Rep. **323** (2000) 183-386.
5. Penrose R., *The Structure of Space-Time*, in *Batelle Rencontres* (1967 Lectures in Mathematics and Physics), C.M. DeWitt and J.A. Wheeler (eds.) Benjamin, New York (1968) pp 121-235.
   Hawking S.W. and Ellis G.F.R., *The Large Scale Structure of Space-Time*, Cambridge University Press, Cambridge (1973).
   Misner C.W., Thorne K.S. and Wheeler J.A., *Gravitation*, Freeman, San Francisco (1973).
6. Wald R.M., *General Relativity*, The University of Chicago Press, Chicago (1984).
   d'Inverno R., *Introducing Einstein's Relativity*, Clarendon Press, Oxford (1992).
   Schmidt H.J., *On the de Sitter Space-Time – the Geometric Foundation of Inflationary Cosmology*, Fortschr. Phys. **41** (1993) 179-199.
   Podolský J., *Lorentz boosts in de Sitter and anti-de Sitter space-times*, Czech. J. Phys. **43** (1993) 1173-1176.
   Eriksen E. and Grøn Ø., *The de Sitter universe models*, Int. J. Mod. Phys. D **4** (1995) 115-159.
7. Christodoulou D. and Klainerman S., *The global nonlinear stability of the Minkowski spacetime*, Princeton University Press, Princeton (1994).
8. Friedrich H., *Asymptotic structure of space-time*, in *Recent Advances in General Relativity*, A. Janish and J. Porter (eds.) Birkhäuser, Boston (1992).

Friedrich H., *Einstein equations and conformal structure: existence of anti-de Sitter-type space-times*, J. Geom. Phys. **17** (1995) 125-184.
Friedrich H., *Einstein's Equations and Geometric Asymptotics*, in *Gravitation and Relativity: At the Turn of the Millenium* (Proc. of the GR-15 conference), N. Dadhich and J. Narlikar (eds.) Inter-University Centre for Astr. and Astrophys. Press, Pune (1998) pp 153-176.

9. Bičák J. and Podolský J., *Gravitational waves in vacuum spacetimes with cosmological constant. I. Classification and geometrical properties of nontwisting type N solutions*, J. Math. Phys. **40** (1999) 4495-4505.
10. Bičák J. and Podolský J., *Gravitational waves in vacuum spacetimes with cosmological constant. II. Deviation of geodesics and interpretation of nontwisting type N solutions*, J. Math. Phys. **40** (1999) 4506-4517.
11. Kundt W., *The Plane-fronted Gravitational Waves*, Z.Phys. **163** (1961) 77-86.
Ehlers J. and Kundt W., *Exact solutions of gravitational field equations*, in *Gravitation: an Introduction to Current Research*, L. Witten (ed.) Wiley, New York (1962) pp 49-101.
12. Ozsváth I., Robinson I. and Rózga K., *Plane-fronted gravitational and electromagnetic waves in spaces with cosmological constant*, J. Math. Phys. **26** (1985) 1755-1761.
13. Kramer D., Stephani H., MacCallum M.A.H. and Herlt E., *Exact Solutions of the Einstein's Field Equations*, Cambridge University Press, Cambridge (1980).
14. Robinson I. and Trautman A., *Spherical Gravitational Waves*, Phys. Rev. Lett. **4** (1960) 431-432.
Robinson I. and Trautman A., *Some spherical gravitational waves in general relativity*, Proc. Roy. Soc. Lond. **A265** (1962) 463-473.
15. Siklos S.T.C., *Lobatchevski Plane Gravitational Waves*, in *Galaxies, Axisymmetric Systems and Relativity*, M.A.H. MacCallum (ed.) Cambridge University Press, Cambridge (1985) pp 247-274.
16. Podolský J., *Interpretation of the Siklos solutions as exact gravitational waves in the anti-de Sitter universe*, Class. Quantum Grav. **15** (1998) 719-733.
17. García Díaz A. and Plebański J.F., *All nontwisting N's with cosmological constant*, J. Math. Phys. **22** (1981) 2655-2658.
Salazar H.I., García Díaz A. and Plebański J.F., *Symmetries of the nontwisting type-N solutions with cosmological constant*, J. Math. Phys. **24** (1983) 2191-2196.
18. Pirani F.A.E., *On the physical significance of the Riemann tensor*, Acta. Phys. Polon. **15** (1956) 389-405.
Pirani F.A.E., *Invariant formulation of gravitational radiation theory*, Phys. Rev. **150** (1957) 1089-1099.
Szekeres P., *The gravitational compass*, J. Math. Phys. **6** (1965) 1387-1391.
19. Bičák J. and Pravda V., *Curvature invariants in type N spacetime*, Class. Quantum Grav. **15** (1998) 1539-1555.
20. Bičák J. and Podolský J., *Cosmic no-hair conjecture and black-hole formation: An exact model with gravitational radiation*, Phys. Rev. D **52** (1995) 887-895.
Bičák J. and Podolský J., *Global structure of Robinson-Trautman radiative space-times with cosmological constant*, Phys. Rev. D **55** (1997) 1985-1993.

21. Bondi H., Pirani F.A.E. and Robinson I., *Gravitational waves in general relativity III. Exact plane waves*, Proc. Roy. Soc **A251** (1959) 519-533.
Bondi H. and Pirani F.A.E., *Gravitational waves in general relativity XIII. Caustic property of plane waves*, Proc. Roy. Soc **A421**, (1989) 395-410.
22. Penrose R., *Twistor Quantisation and Curved Space-Time*, Int. Jour. Theor. Phys. **1** (1968) 61-99.
23. Rindler W., *Essential Relativity*, Van Nostrand, New York (1977).
24. Podolský J. and Veselý K., *New examples of sandwich gravitational waves and their impulsive limit*, Czech. J. Phys. **48** (1998) 871-878.
Podolský J., *Smooth sandwich gravitational waves*, Czech. J. Phys. **49** (1999) 981-984.
25. Penrose R., *The geometry of impulsive gravitational waves* , in *General Relativity*, L. O'Raifeartaigh (ed.), Clarendon Press, Oxford (1972) pp 101-115.
26. Colombeau J.F., *Multiplication of Distributions*, Springer, Berlin (1992).
Oberguggenberger M., *Multiplication of Distributions and Applications to Partial Differential Equations*, Pitman Research Notes in Mathematics **259**, Longman, Harlow (1992).
Vickers J.A., *Nonlinear generalised functions in general relativity*, in *Nonlinear Theory of Generalised Functions* (Research Notes in Mathematics **401**), M. Grosser, G. Hörmann, M. Kunzinger, M. Oberguggenberger (eds.), Chapman & Hall/CRC, Boca Raton (1999) pp 275-290.
Steinbauer R., *Distributional Methods in General Relativity*, Dissertation, Universität Wien (2000).
27. Lichnerowicz A., *Thèorie relativistes de la gravitation et de l'electromagnetisme*, Masson, Paris (1955).
Synge J.L., *Relativity: the general theory*, North-Holland, Amsterdam (1960).
28. Aichelburg P.C. and Sexl R.U., *On the Gravitational Field of a Massless Particle*, Gen. Relativ. Gravit. **2** (1971) 303-312.
29. Hotta M. and Tanaka M., *Shock-wave geometry with non-vanishing cosmological constant*, Class. Quantum Grav. **10** (1993) 307-314.
30. Gleiser R. and Pullin J., *Are cosmic strings gravitationally stable topological defects?*, Class. Quantum Grav. **6** (1989) L141-L144.
31. Bičák J., *Is there a news function for an infinite cosmic string?*, Astron. Nachr. **311** (1990) 189-192.
32. Nutku Y. and Penrose R., *On Impulsive Gravitational Waves*, Twistor Newsletter No. 34, 11 May (1992) 9-12.
33. Hogan P.A., *A Spherical Impulse Gravity Wave*, Phys. Rev. Lett. **70** (1993) 117-118.
34. Hogan P.A., *Lorentz group and spherical impulsive gravity wave*, Phys. Rev. D **49** (1994) 6521-6525.
35. Israel W., *Singular Hypersurfaces and Thin Shells in General Relativity*, Nuovo Cim. B **44** (1966) 1-14.
Barrabès C. and Israel W., *Thin shells in general relativity and cosmology: The lightlike limit*, Phys. Rev. D **43** (1991) 1129-1142.
Barrabès C. and Hogan P.A., *Lightlike signals in general relativity and cosmology*, Phys. Rev. D **58** (1998) 044013.

36. Hogan P.A., *A spherical gravitational in the de Sitter universe*, Phys. Lett. A **171** (1992) 21-22.
37. Podolský J. and Griffiths J.B., *Nonexpanding impulsive gravitational waves with an arbitrary cosmological constant*, Phys. Lett. A **261** (1999) 1-4.
38. Podolský J. and Griffiths J.B., *Expanding impulsive gravitational waves*, Class. Quantum Grav. **16** (1999) 2937-2946.
39. Griffiths J.B. and Podolský J., *Exact solutions for impulsive gravitational waves*, Ann. Phys. (Leipzig) **9** (2000) SI59-SI62.
40. Podolský J. and Griffiths J.B., *The collision and snapping of cosmic strings generating spherical impulsive gravitational waves*, Class. Quantum Grav. **17** (2000) 1401-1413.
41. Podolský J. and Griffiths J.B., *Impulsive gravitational waves generated by null particles in de Sitter and anti-de Sitter backgrounds*, Phys. Rev. D **56** (1997) 4756-4767.
42. Dray T. and 't Hooft G., *The gravitational shock wave of a massless particle*, Nucl. Phys. B **253** (1985) 173-188.
43. Podolský J. and Veselý K., *Continuous coordinates for all impulsive pp-waves*, Phys. Lett. A **241** (1998) 145-147.
44. Kunzinger M. and Steinbauer R., *A note on the Penrose junction conditions*, Class. Quantum Grav. **16** (1999) 1255-1264.
45. D'Eath P.D., *High-speed black-hole encounters and gravitational radiation*, Phys. Rev. D **18** (1978) 990-1019.
    Curtis G.E., *Ultrarelativistic Black-Hole Encounters*, Gen. Relativ. Gravit. **9** (1978) 999-1008.
    D'Eath P.D. and Payne P.N., *Gravitational radiation in black-hole collisions at the speed of light. I. Perturbation treatment of the axisymmetric collision*, Phys. Rev. D **46** (1993) 658-674.
46. 't Hooft G., *Graviton dominance in ultra-high-energy scattering*, Phys. Lett. B **198** (1987) 61-63.
    Verlinde E. and Verlinde H., *High-energy scattering in quantum gravity*, Class. Quantum Grav. **10** (1993) S175-S184.
47. Steinbauer R., *Geodesics and geodesic deviation for impulsive gravitational waves*, J. Math. Phys. **39** (1998) 2201-2212.
    Kunzinger M. and Steinbauer R., *A rigorous solution concept for geodesic and geodesic deviation equations in impulsive gravitational waves*, J. Math. Phys. **40** (1999) 1479-1489.
48. Aliev A.N. and Nutku Y., *Impulsive spherical gravitational waves*, Class. Quantum Grav. **18** (2001) 891-906.
49. Nutku Y., *Spherical shock waves in general relativity*, Phys. Rev. D **44** (1991) 3164-3168.
50. Podolský J., *Non-expanding impulsive gravitational waves*, Class. Quantum Grav. **15** (1998) 3229-3239.
51. Horowitz G.T. and Itzhaki N., *Black Holes, Shock Waves, and Causality in the AdS/CFT Correspondence*, J. High Energy Phys. **2** (1999) U154-U172.

52. Podolský J. and Griffiths J.B., *Impulsive waves in de Sitter and anti-de Sitter space-times generated by null particles with an arbitrary multipole structure*, Class. Quantum Grav. **15** (1998) 453-463.
53. Sfetsos K., *On gravitational shock waves in curved spacetimes*, *Nucl. Phys.* B **436** (1995) 721-746.
54. Ferrari V. and Pendenza P., *Boosting the Kerr Metric*, Gen. Relat. Grav. **22** (1990) 1105-1117.
Loustó C.O. and Sánchez N., *The ultrarelativistic limit of the boosted Kerr-Newman geometry and the scattering of spin-$\frac{1}{2}$ particles*, Nucl. Phys. B **383** (1992) 377-394.
Balasin H. and Nachbagauer H., *The ultrarelativistic Kerr geometry and its energy-momentum tensor*, Class. Quantum Grav. **12** (1995) 707-713.
Balasin H. and Nachbagauer H., *Boosting the Kerr geometry in a arbitrary direction*, Class. Quantum Grav. **13** (1996) 731-737.
55. Podolský J. and Griffiths J.B., *Boosted static multipole particles as sources of impulsive gravitational waves*, Phys. Rev. D **58** (1998) 124024.
56. Kottler F., *Über die physikalischen Grundlagen der Einsteinschen Gravitationstheorie*, Ann. Phys. (Leipzig) **56** (1918) 410.
Lake K. and Roeder R.C., *Effects of a nonvanishing cosmological constant on a spherically symmetric vacuum manifold*, Phys. Rev. D**15** (1977) 3513-3519.
Geyer K.H., *Geometrie der Raum-Zeit der Maßbestimmung von Kottler, Weyl und Trefftz*, Astron. Nachr. **301** (1980) 135-149.
Brill D.R. and Hayward S.A., *Global structure of a black hole cosmos and its extremes* Class. Quantum Grav. **11** (1994) 359-370.
Podolský J., *The structure of the extreme Schwarzschild-de Sitter spacetime*, Gen. Relativ. Gravit. **31** (1999) 1703-1725.
57. Bičák J. and Schmidt B., *On the asymptotic structure of axisymmetric radiative spacetimes*, Class. Quantum Grav. **6** (1989) 1547-1559.
58. Podolský J. and Griffiths J.B., *Null limits of generalised Bonnor-Swaminarayan solutions*, Gen. Relativ. Gravit. **33** (2001) 83-103.
59. Podolský J. and Griffiths J.B., *Null limits of the C-metric*, Gen. Relativ. Gravit. **33** (2001) 105-110.
60. Bondi H., *Negative Mass in General Relativity*, Rev. Mod. Phys. **29** (1957) 423-428.
61. Bonnor W.B. and Swaminarayan N.S., *An Exact Solution for Uniformly Accelerated Particles in General relativity*, Z. Phys. **177** (1964) 240-256.
Bonnor W.B., *Wissenschaft. Z. der Friedrich-Schiller-Universität Jena* **15** (1966) 71.
Bonnor W.B., Griffiths J.B. and MacCallum M.A.H., *Physical Interpretation of Vacuum Solutions of Einstein's Equations. Part II. Time-dependent Solutions*, Gen. Relativ. Gravit. **26** (1994) 687-729.
62. Israel W., Khan K. A., *Collinear Particles and Bondi Dipoles in General Relativity*, Nuov. Cim. **33** (1964) 331-344.
63. Bičák J., *Gravitational radiation from uniformly accelerated particles in general relativity*, Proc. Roy. Soc. **A302** (1968) 201-224.

64. Bičák J., Hoenselaers C. and Schmidt B.G., *The solutions of the Einstein equations for uniformly accelerated particles without nodal singularities. I. Freely falling particles in external fields*, Proc. Roy. Soc. Lond. **A390** (1983) 375-409.
65. Bičák J., Hoenselaers C. and Schmidt B.G., *The solutions of the Einstein equations for uniformly accelerated particles without nodal singularities. II. Self-accelerating particles*, Proc. Roy. Soc. Lond. **A390** (1983) 411-419.
66. Bičák J. and Schmidt B.G., *Isometries compatible with gravitational radiation*, J. Math. Phys. **25** (1984) 600-606.
67. Bičák J., Reilly P. and Winicour J., *Boost-rotation symmetric gravitational null cone data*, Gen. Relativ. Gravit. **20** (1988) 171-181.
68. Bičák J. and Schmidt B.G., *Asymptotically flat radiative space-times with boost-rotation symmetry I. The general structure*, Phys. Rev. D **40** (1989) 1827-1853.
69. Bičák J. and Pravdová A., *Symmetries of asymptotically flat electrovacuum space-times and radiation*, J. Math. Phys. **39** (1998) 6011-6039.
70. Bičák J. and Pravda V., *Spinning C-metric: radiative spacetime with accelerating, rotating black holes*, Phys. Rev. **D60** (1999) 044004.
71. Bičák J., *Selected topics in the problem of energy and radiation*, in *Relativity and Gravitation*, C.G. Kuper and A. Peres (eds.) Gordon and Breach, New York (1971) pp 47-67.
72. Bičák J., *On exact radiative solutions representing finite sources*, in *Galaxies, Axisymmetric Systems and Relativity* (Essays presented to W.B. Bonnor on his 65th birthday), M.A.H. MacCallum (ed.), Cambridge University Press, Cambridge (1985) pp 91-124.
73. Bičák J., *Radiative properties of space-times with the axial and boost symmetries*, in *Gravitation and Geometry* (A volume in honour of I. Robinson), W. Rindler and A. Trautman (eds.) Bibliopolis, Naples (1987) pp 55-69.
74. Bičák J., *Exact radiative space-times*, in *Proc. 5th M. Grossman Meeting on Gen. Rel.*, D.G. Blair and M.J. Buckingham (eds.) World Scientific, Singapore (1989) pp 309-341.
75. Bičák J., *Radiative spacetimes: exact approaches*, in *Relativistic Gravitation and Gravitational Radiation* (Proceedings of the Les Houches School of Physics), J.-A. Marck and J.-P. Lasota (eds.) Cambridge University Press, Cambridge (1997) pp 67-87
76. Bičák J., *Exact radiative spacetimes: some recent developments*, Ann. Phys. (Leipzig) **9** (2000) 207-217.
77. Bičák J., *The creation and splitting of an infinite cosmis string: an exact solution* (1990) unpublished notes.
78. Kinnersley W. and Walker M., *Uniformly Accelerating Charged Mass in general Relativity*, Phys. Rev. D **2** (1970) 1359-1370.
79. Farhoosh H. and Zimmerman R.L., *Killing horizons and dragging of the inertial frame about a uniformly accelerating particle*, Phys. Rev. D **21** (1980) 317-327.
Ashtekar A. and Dray T., *On the Existence of Solutions to Einstein's Equation With Non-Zero Bondi News*, Commun. Math. Phys. **79** (1981) 581-589.

80. Bonnor W.B., *The Sources of the Vacuum C-metric*, Gen. Relat. Grav. **15** (1983) 535-551.
81. Cornish F.H.J. and Uttley W.J., *The Interpretation of the C Metric. The Vacuum Case*, Gen. Relativ. Gravit. **27** (1995) 439-454.
Pravda V. and Pravdová A., *Boost-rotation symmetric spacetimes – review*, Czech. J. Phys. **50** (2000) 333-440.
82. Plebanski J.F. and Demianski M., *Rotating, Charged, and Uniformly Accelerating Mass in general Relativity*, Ann. Phys. (NY) **98** (1976) 98-127.
Carter B., *Hamilton-Jacobi and Schrodinger Separable Solutions of Einstein's Equations*, Commun. Math. Phys. **10** (1968) 280-310.
Debever R., *On type D expanding solutions of Einstein-Maxwell equations*, Bull. Soc. Math. Belg. **23** (1971) 360-376.
83. Podolský J. and Griffiths J.B., *Uniformly accelerating black holes in a de Sitter universe*, Phys. Rev. D **63** (2001) 024006.
84. Mann R.B. and Ross S.F., *Cosmological production of charged black hole pairs*, Phys. Rev. D **52** (1995) 2254-2265.
Mann R.B., *Pair production of topological anti–de Sitter black holes*, Class. Quantum Grav. **14** (1997) L109-L114.
85. Bičák J. and Krtouš P., *Accelerated sources in de Sitter spacetime and the insufficiency of retarded fields*, Phys. Rev. D **64** (2001) 124020.
86. Quevedo H., *Multipole Moments in General Relativity — Static and Stationary Vacuum Solutions*, Fortschr. Phys. **38** (1990) 733-840.
87. Griffiths J.B. and Podolský J., *Null multipole particles as sources of pp-waves*, Phys. Lett. A **236** (1997) 8-10.
88. Penrose R. and Rindler W., *Spinors and Space-time, vol. 1*, Cambridge University Press, Cambridge (1984).
89. Hortaçsu M., *Quantum fluctuations in the field of an impulsive spherical gravitational wave*, Class. Quantum Grav. **7** (1990) L165-L169.
Özdemir N. and Hortaçsu M., *Quantum fluctuations of a spherical shock wave*, Class. Quantum Grav. **12** (1995) 1221-1227.
Bilge A.H., Hortaçsu M. and Özdemir N., *Particle Creation if a Cosmic String Snaps*, Gen. Relativ. Gravit. **28** (1996) 511-525.
Hortaçsu M., *Vacuum fluctuations for a spherical gravitational impulsive waves*, Class. Quantum Grav. **13** (1996) 2683-2691.
90. Ferrari V., Pendenza P. and Veneziano G., *Beam-Like Gravitational Waves and Their Geodesics*, Gen. Relativ. Gravit. **20** (1988) 1185-1191.
Balasin H., *Geodesics for impulsive gravitational waves and the multiplication of distributions*, Class. Quantum Grav. **14** (1997) 455-462.
Gemelli G., *Dynamical Effects of Gravitational Shock Waves*, Gen. Relativ. Gravit. **29** (1997) 1163-1180.
91. Podolský J. and Ortaggio M., *Symmetries and geodesics in (anti–)de Sitter spacetimes with non-expanding impulsive waves*, Class. Quantum Grav. **18** (2001) 2689-2706.
92. Podolský J. and Veselý K., *Chaos in pp-wave spacetimes*, Phys. Rev. D **58** (1998) 081501.
Podolský J. and Veselý K., *Chaotic motion in pp-wave spacetimes*, Class.

Quantum Grav. **15** (1998) 3505-3521.
Veselý K. and Podolský J., *Chaos in a modified Hénon-Heiles system describing geodesics in gravitational waves*, Phys. Lett. A **271** (2000) 368-376.
93. Podolský J. and Veselý K., *Smearing of chaos in sandwich pp-waves*, Class. Quantum Grav. **16** (1999) 3599-3618.
Podolský J., *Chaos in pp-wave spacetimes and its smearing in the impulsive limit*, Ann. Phys. (Leipzig) **9** (2000) SI147-SI148.
94. Defrise L., *Groupes d'isotropie et groupes de stabilité conforme dans les escapes lorentziens*, Thése: Université Libre de Bruxelles (1969).
Podolský J., *Exact non-singular waves in the anti-de Sitter universe*, Gen. Relativ. Gravit. **33** (2001) 1093-1113.
95. Aichelburg P.C. and Balasin H., *Symmetries of pp-waves with distributional profile*, Class. Quantum Grav. **13** (1996) 723-729.
Aichelburg P.C. and Balasin H., *Generalized symmetries of impulsive gravitational waves*, Class. Quantum Grav. **14** (1997) A31-A41.
96. Podolský J. and Steinbauer R., *Geodesics in expanding impulsive gravitational waves*, in preparation.

ON THE SPINNING C-METRIC

VOJTĚCH PRAVDA, ALENA PRAVDOVÁ

Mathematical Institute, Academy of Sciences, Žitná 25, 115 67 Prague 1, Czech Republic

E-mail: pravda@math.cas.cz, pravdova@math.cas.cz

Physical interpretation of some stationary and non-stationary regions of the spinning C-metric is presented. They represent different spacetime regions of a uniformly accelerated Kerr black hole. Stability of geodesics corresponding to equilibrium points in a general stationary spacetime with an additional symmetry is also studied and results are then applied to the spinning C-metric.

1 Introduction

The C-metric is a well known exact Petrov type D vacuum solution of Einstein's field equations found in the second decade of the last century [1,2]. However, it was much later when its physical interpretation was found [3,4]. Its most physical region represents a pair of Schwarzschild black holes connected by a conical singularity and uniformly accelerated in opposite directions along the axis of axial symmetry.

The C-metric is an example of boost-rotation symmetric spacetimes [5–7] corresponding to gravitational field of uniformly accelerated "particles" of various kinds. Thus for the physical interpretation it is worthwhile to express the C-metric in coordinates adapted to the boost and rotation symmetries [4].

Several generalizations of the C-metric (e.g., for charged black holes [3,8], for an external field present [9] or recently for two or more black holes accelerated in both directions along the axis of symmetry [10]) were found. Probably the most important of them is the spinning C-metric (the SC-metric) found by Plebański and Demiański [11], which is also of the Petrov type D. In the present paper, we restrict ourselves to the case with non-vanishing mass, angular momentum, and acceleration and we set electric and magnetic charges, the NUT parameter, and the cosmological constant equal to zero. Then the SC-metric in coordinates adapted to its special algebraical structure reads

$$\mathrm{d}s^2 = \frac{1}{H'}\Big[\frac{W'}{P}\mathrm{d}p^2 + \frac{P}{W'}\left(\mathrm{d}\sigma + q^2\mathrm{d}\tau'\right)^2 + \frac{W'}{Q}\mathrm{d}q^2 - \frac{Q}{W'}\left(\mathrm{d}\tau' - p^2\mathrm{d}\sigma\right)^2\Big] , \qquad (1)$$

where

$$\begin{aligned} P &= \gamma' - \varepsilon' p^2 + 2m'p^3 - \gamma' p^4 , \\ Q &= -\gamma' + \varepsilon' q^2 + 2m'q^3 + \gamma' q^4 , \\ H' &= (p+q)^2 , \\ W' &= 1 + (pq)^2 \end{aligned}$$

with γ', ε', and m' being constant.

By performing certain limiting procedures [11] that remove the acceleration or rotation parameters one obtains the Kerr solution or the C-metric, respectively.

Physical properties of the SC-metric were studied in [12–14]. Stationary regions of the SC-metric (1) and Killing horizons were identified in [13]. Then the most physical stationary region was transformed into the Weyl-Papapetrou coordinates and finally to coordinates adapted to the boost and rotation symmetries. The SC-metric is the only example of boost-rotation symmetric spacetimes with spinning sources [15] known today.

In the present paper, we start with the SC-metric in a better chosen parametrization than in (1) that makes the physical interpretation more transparent. As in [13], we first transform stationary regions of the SC-metric to the Weyl-Papapetrou form (Sec. 3) and then to coordinates adapted to the boost and rotation symmetries (Sec. 4), where non-stationary radiative regions appear. It is shown that these non-stationary regions (inside and outside the null cone of the origin) are in fact already contained in the original form of the SC-metric. This also enables us to locate $\mathcal{J}^+$ and I^+.

It is shown in [13] that the most physical stationary region $\mathcal{B}$ corresponds to a field of a uniformly accelerated Kerr black hole. Here, another stationary region is interpreted as a uniformly accelerated white hole with a ring singularity that in fact corresponds to the uniformly accelerated asymptotically flat "interior" of the Kerr solution.

In Sec. 5, geodesics and their stability are studied at first generally for an arbitrary stationary metric with another symmetry and then results are applied to the stationary region $\mathcal{B}$ of the SC-metric.

2 The spinning C-metric in the $\{\tau, x, y, \phi\}$ coordinates

Since the parameters γ' and ε' occurring in (1) do not have a straightforward connection with the angular momentum and acceleration of the black hole, we start with the SC-metric with rescaled parameters and coordinates [14]

$$\mathrm{d}s^2 = \frac{1}{H}\left[\frac{W}{F}\mathrm{d}y^2 + \frac{W}{G}\mathrm{d}x^2 + \frac{G}{W}(\mathrm{d}\phi + aAy^2\mathrm{d}\tau)^2 - \frac{F}{W}(\mathrm{d}\tau - aAx^2\mathrm{d}\phi)^2\right] , \tag{2}$$

where

$$\begin{aligned} F &= -\delta - \varepsilon y^2 - 2mAy^3 + (aA)^2\delta y^4 , \\ G &= \delta + \varepsilon x^2 - 2mAx^3 - (aA)^2\delta x^4 , \\ H &= A^2(x+y)^2 , \\ W &= 1 + (aAxy)^2 \end{aligned}$$

with m, a, A, δ, and ε being constant. The metrics (1) and (2) are related by the trivial transformation

$$p = \sqrt{aA}\, x , \quad q = \sqrt{aA}\, y , \quad \sigma = \sqrt{\frac{a}{A^3}}\, \phi , \quad \tau' = \sqrt{\frac{a}{A^3}}\, \tau ,$$

$$\gamma' = A^2\delta , \quad \varepsilon' = -\frac{A\varepsilon}{a} , \quad m' = -m\sqrt{\frac{A^3}{a^3}} .$$

Notice that for

$$\delta = 1 , \quad \varepsilon = -1 , \tag{3}$$

and $a = 0$ we get the standard form of the non-spinning C-metric. In the following, we assign the foregoing values (3) to the kinematical parameters δ and ε.

In the present paper, we study only the case when the polynomial F has four distinct real roots $y_1, \ldots, y_4$. This is satisfied if the conditions

$$m^2 > \frac{8}{9}\,a^2 \quad \text{and} \quad \alpha_1 - \alpha_2 < A^2 < \alpha_1 + \alpha_2\ , \tag{4}$$

where

$$\alpha_1 = \frac{1}{32a^6}\left(36\,a^2m^2 - 8\,a^4 - 27\,m^4\right)\ , \quad \alpha_2 = \frac{m}{32a^6}\left(9\,m^2 - 8\,a^2\right)^{3/2}\ ,$$

hold. Notice that ${\alpha_1}^2 - {\alpha_2}^2 = (a^2 - m^2)/(16a^2)$. For $m^2 > a^2$, $\alpha_1 - \alpha_2$ is negative and then A^2 has to fulfil only the upper constraint. In the limit $a \to 0$, the upper constraint for A^2, $\alpha_1 + \alpha_2$, turns out to be $1/(27m^2)$, which is the same as for the non-spinning C-metric. If y_i are the roots of F then $-y_i$ are the roots of G since $F(y) = -G(-y)$.

Similarly as in [13], it can be shown that the metric (2) has the signature $+2$ for $G > 0$ and is then stationary for $F > 0$. In the coordinates $\{\tau,\ x,\ y,\ \phi\}$, there are four stationary regions (see Fig. 1) $\mathcal{A}$, $\mathcal{B}$, $\mathcal{C} \cup \mathcal{C}'$, and $\mathcal{D} \cup \mathcal{D}'$ (as in [13], we identify $y = \infty$ and $y = -\infty$) and each of them can be transformed to the stationary Weyl-Papapetrou form (see Sec. 3).

Curvature invariants

$$R^{\alpha\beta\gamma\delta}R_{\alpha\beta\gamma\delta} = 48m^2\frac{H^3}{W^6}(W-2)(W^2-16W+16)\ , \tag{5}$$

$$\tfrac{1}{2}\epsilon^{\alpha\beta\sigma\rho}R_{\sigma\rho}{}^{\mu\nu}R_{\alpha\beta\mu\nu} = 96aAm^2xy\frac{H^3}{W^6}(3W-4)(W-4) \tag{6}$$

suggest that there are curvature singularities at points ($x = 0$, $y = \pm\infty$) and ($x = \pm\infty$, $y = 0$). The second curvature invariant also indicates that the constant a is proportional to the angular momentum of the source [16]. Both invariants vanish for $x + y = 0$. Later it will be shown that $\mathcal{J}^+$, I^+, and I^0 are located there (see Fig. 4).

Killing horizons, which are located at $y = y_i$, correspond to the black hole or acceleration horizons [13] (see Fig. 1).

3 Transformation to the Weyl-Papapetrou coordinates

As was mentioned earlier, each of the four stationary regions $\mathcal{A}$, $\mathcal{B}$, $\mathcal{C} \cup \mathcal{C}'$, and $\mathcal{D} \cup \mathcal{D}'$ can be transformed to the stationary Weyl-Papapetrou coordinates $\{\bar{t},\ \bar{\rho},\ \bar{z},\ \bar{\varphi}\}$

$$\mathrm{d}s^2 = \mathrm{e}^{-2U}[\mathrm{e}^{2\nu}(\mathrm{d}\bar{\rho}^2 + \mathrm{d}\bar{z}^2) + \bar{\rho}^2\mathrm{d}\bar{\varphi}^2] - \mathrm{e}^{2U}(\mathrm{d}\bar{t} + \bar{\omega}\mathrm{d}\bar{\varphi})^2\ , \tag{7}$$

where the metric functions U, ν, and $\bar{\omega}$ are functions of $\bar{\rho}$ and $\bar{z}$. The transformation

$$\bar{\rho}^2 = \mathcal{K}^2 FGH^{-2}\ , \tag{8}$$

$$\bar{z} = \mathcal{K}[1 + mAxy(x-y) + xy - a^2A^2x^2y^2]H^{-1}\ , \tag{9}$$

$$\tau = \kappa_1\bar{t} + \kappa_2\bar{\varphi}\ , \tag{10}$$

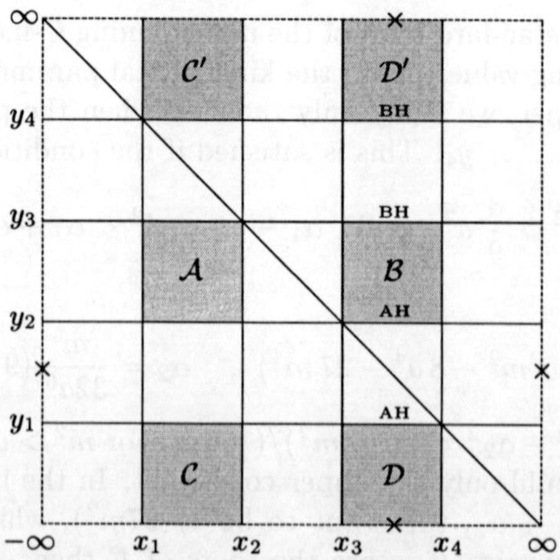

Figure 1. The structure of the SC-metric (2) in the coordinates $\{\tau,\ x,\ y,\ \phi\}$: The metric has the signature +2 for $G > 0$, i.e., in the second and fourth columns from the left, and is stationary for $F > 0$ in the shaded squares. Curvature singularities at points ($x = 0$, $y = \pm\infty$) and ($x = \pm\infty$, $y = 0$) are denoted by crosses. The black hole horizons and the acceleration horizons are labeled by BH and AH, respectively. The line $x + y = 0$, where curvature invariants (5) and (6) vanish, is also indicated.

$$\phi = \kappa_3 \bar{t} + \kappa_4 \bar{\varphi}\ , \tag{11}$$

$$\mathcal{K} = \kappa_2\kappa_3 - \kappa_1\kappa_4\ , \tag{12}$$

where $\kappa_1 \ldots \kappa_4$ are constant, can be found similarly as in [13]. It turns out that it is convenient to choose

$$\mathcal{K} = A \tag{13}$$

since then the SC-metric in the Weyl-Papapetrou coordinates with appropriately chosen constants $\kappa_1 \ldots \kappa_4$ yields the Kerr solution in the limit $A \to 0$. In the following, we assume that (13) holds.

In the present paper, τ, x, y, ϕ, $\bar{\varphi}$, κ_2, and κ_4 are dimensionless and $\bar{t}$, $\bar{\rho}$, $\bar{z}$, m, a, $1/A$, $1/\kappa_1$, and $1/\kappa_3$ have the dimension of length.

In order to express the SC-metric in the Weyl-Papapetrou form (7) we have to find the inverse transformation to (8), (9). Let us first define R_i

$$R_i \equiv \sqrt{\bar{\rho}^2 + (\bar{z} - \bar{z}_i)^2}\ , \quad i = 1 \ldots 3\ , \tag{14}$$

where $\bar{z}_i$ are the roots of the cubic equation

$$2A\bar{z}_i^3 - \bar{z}_i^2 + 2Aa^2\bar{z}_i + m^2 - a^2 = 0\ . \tag{15}$$

It can be shown that

$$R_i{}^2 = \frac{\left(\frac{A_i}{A\bar{z}_i} + \frac{A^2 m\bar{z}_i}{A_i}(x - y) - A_i xy\right)^2}{A^2(x + y)^2}\ ,$$

where

$$A_i^2 = A^2(2Aa^2\bar{z}_i + m^2 - a^2)\ .$$

Then

$$R_i = \epsilon_i \frac{\left(\frac{A_i}{A\bar{z}_i} + \frac{A^2 m \bar{z}_i}{A_i}(x-y) - A_i xy\right)}{A(x+y)} \,, \tag{16}$$

where $\epsilon_i = \pm 1$. Each of the stationary regions is characterized by a different combination of ϵ_i:

$$\begin{array}{llll} \mathcal{A}: & \epsilon_1 = -1\,, & \epsilon_2 = -1\,, & \epsilon_3 = +1\,, \\ \mathcal{B}: & \epsilon_1 = -1\,, & \epsilon_2 = +1\,, & \epsilon_3 = -1\,, \\ \mathcal{C} \cup \mathcal{C}': & \epsilon_1 = +1\,, & \epsilon_2 = +1\,, & \epsilon_3 = +1\,, \\ \mathcal{D} \cup \mathcal{D}': & \epsilon_1 = +1\,, & \epsilon_2 = -1\,, & \epsilon_3 = -1\,. \end{array}$$

Now solving Eqs. (16) for x and y, the inverse transformation can be found

$$x = \frac{F_0 + F_1}{F_2} \,, \quad y = \frac{F_0 - F_1}{F_2} \,, \tag{17}$$

where

$$F_0 = \epsilon_1 \epsilon_2 \epsilon_3 (m^2 - a^2) A^2 (\bar{z}_1 - \bar{z}_2)(\bar{z}_2 - \bar{z}_3)(\bar{z}_3 - \bar{z}_1) \,,$$
$$F_1 = \bar{z}_1 A_2 A_3 \epsilon_3 \epsilon_2 (\bar{z}_3 - \bar{z}_2) R_1 + A_1 \bar{z}_2 A_3 \epsilon_1 \epsilon_3 (\bar{z}_1 - \bar{z}_3) R_2 + A_1 A_2 \bar{z}_3 \epsilon_2 \epsilon_1 (\bar{z}_2 - \bar{z}_1) R_3 \,,$$
$$F_2 = (m^2 - a^2) A \left[\epsilon_2 \epsilon_3 (\bar{z}_2 - \bar{z}_3) A_1 R_1 + \epsilon_3 \epsilon_1 (\bar{z}_3 - \bar{z}_1) A_2 R_2 + \epsilon_1 \epsilon_2 (\bar{z}_1 - \bar{z}_2) A_3 R_3\right] .$$

From the transformation (8)–(12) follows

$$\begin{aligned} g_{\bar{\rho}\bar{\rho}} &= \frac{W}{HF({\bar{\rho}_{,y}}^2 + {\bar{z}_{,y}}^2)} \,, \\ g_{\bar{t}\bar{t}} &= \frac{-\left(\kappa_1 - a x^2 \kappa_3\, A\right)^2 F + \left(\kappa_3 + a y^2 \kappa_1\, A\right)^2 G}{HW} \,, \\ g_{\bar{\varphi}\bar{\varphi}} &= \frac{\left(\kappa_2 - a x^2 \kappa_4\, A\right)^2 F - \left(\kappa_4 + a y^2 \kappa_2\, A\right)^2 G}{HW} \,, \\ g_{\bar{t}\bar{\varphi}} &= \frac{-\left(\kappa_2 - a x^2 \kappa_4\, A\right)\left(\kappa_1 - a x^2 \kappa_3\, A\right) F + \left(\kappa_4 + a y^2 \kappa_2\, A\right)\left(\kappa_3 + a y^2 \kappa_1\, A\right) G}{HW} \end{aligned} \tag{18}$$

with x and y given by (17). The Weyl-Papapetrou metric functions are now

$$e^{2U} = -g_{\bar{t}\bar{t}} \,, \quad e^{2\nu} = -g_{\bar{\rho}\bar{\rho}} g_{\bar{t}\bar{t}} \,, \quad \bar{\omega} = g_{\bar{t}\bar{\varphi}} / g_{\bar{t}\bar{t}} \,. \tag{19}$$

Substituting Eqs. (17) and (18) into (19) we obtain the SC-metric in the Weyl-Papapetrou coordinates. Unfortunately, these expressions are very complicated and can be handled only by computer manipulations. In the non-spinning case, our approach gives again long formulas, whereas Godfrey [17] and Bonnor [4], after performing very long calculations, arrived at considerably simpler equivalent results. A similar simplification may be probably also possible in the spinning case. The spinning soliton generalization [18] of the non-spinning C-metric might be helpful in this task.

The regularity condition on the axis reads

$$\lim_{\bar{\rho}_0 \to 0} \frac{1}{2\pi} \frac{\int_0^{2\pi} \sqrt{g_{\bar{\varphi}\bar{\varphi}}(\bar{\rho}_0, \bar{z})}\, d\bar{\varphi}}{\int_0^{\bar{\rho}_0} \sqrt{g_{\bar{\rho}\bar{\rho}}(\bar{\rho}, \bar{z})}\, d\bar{\rho}} = 1 \,,$$

from which follows

$$\nu = 0\,, \tag{20}$$

$$\bar{\omega} = 0 \tag{21}$$

on the axis. When both conditions (20) and (21) are satisfied then the axis is regular. If only (21) holds then there is a conical singularity (a string or a strut); if none of these conditions is satisfied a spinning string (conical and torsion [19,20] singularity) is present and, in its vicinity, a region with closed timelike curves occurs. Vacuum Einstein's equations allow us to multiply $e^{2\nu}$ by a constant κ_5. Later we will adjust this constant to regularize some parts of the axis.

Let us now restrict ourselves to the most physically interesting stationary region $\mathcal{B}$ of the SC-metric (2), which was studied in the non-spinning and spinning cases in papers [4,7,14,21] and [13,14], respectively.

In the Weyl-Papapetrou form, the black hole horizon is located on the axis between $\bar{z}_1$ and $\bar{z}_2$, the acceleration horizon extends from $\bar{z}_3$ to ∞, and there may occur conical singularities or spinning strings on the rest of the axis depending on the values of the constants $\kappa_1 \ldots \kappa_5$ (see Fig. 2 a).

Let us now fix the appropriate values of the constants $\kappa_1 \ldots \kappa_5$:

Requiring $g_{\bar{t}\bar{\varphi}} = 0$ and $g_{\bar{\varphi}\bar{\varphi}} \neq 0$ on the acceleration horizon (where $y = y_2$, $F = 0$) leads to the condition

$$\kappa_3 + a y_2^2 A \kappa_1 = 0\,. \tag{22}$$

Demanding further that there be no torsion singularity on the axis for $\bar{z} < \bar{z}_1$ (where $G = 0$, $x = x_3 = -y_2$), i.e., $\bar{\omega}$ and thus also $g_{\bar{t}\bar{\varphi}}$ vanish there (21), we arrive at

$$\kappa_2 - a y_2^2 A \kappa_4 = 0\,. \tag{23}$$

Equivalent conditions were obtained in [13] requiring the appropriate asymptotical behaviour of the metric in the coordinates adapted to the boost and rotation symmetries discussed in Sec. 4.

From Eqs. (13), (22), and (23) we obtain

$$\kappa_4 = -\frac{A}{\kappa_1(1 + a^2 A^2 {y_2}^4)}\,. \tag{24}$$

Further requirement on the axis – to be regular for $\bar{z} < \bar{z}_1$ (20) – leads to

$$\kappa_5 = -\frac{A^2(-1 + 4y_2(mA - a^2A^2y_2) + (a^2 - m^2)A^2{y_2}^4)}{{\kappa_1}^2(1 + a^2A^2{y_2}^4)^2}\,. \tag{25}$$

By Eqs. (22)–(25) together with Eq. (31) the constants $\kappa_1 \ldots \kappa_5$ are uniquely determined.

The "angular velocity" of the black hole horizon $y = y_3$ turns out to be

$$\Omega_H \equiv -\frac{g_{\bar{t}\bar{t}}}{g_{\bar{t}\bar{\varphi}}}_{|y=y_3} = \frac{aA^2({y_3}^2 - {y_2}^2)}{1 + (aAy_2y_3)^2}\,, \tag{26}$$

where Eqs. (18), (22)–(24), and (31) were employed. Notice that the expression (26) differs from Ω_H of Letelier and Oliveira [14] since we use a different coordinate system that does not rotate asymptotically.

For $a < m$ and the acceleration $A \to 0$, Eq. (26) reduces to

$$\lim_{A\to 0} \Omega_H = \frac{a}{2m(m+\sqrt{m^2-a^2})}$$

that is equal to Ω_H for the outer horizon of the Kerr metric.

For $a < m$ and $A << 1$, the roots of (15) are

$$\bar{z}_1 = -\sqrt{m^2-a^2} + m^2 A + \mathcal{O}(A^2) \,,$$
$$\bar{z}_2 = \sqrt{m^2-a^2} + m^2 A + \mathcal{O}(A^2) \,,$$
$$\bar{z}_3 = \frac{1}{2A} - 2m^2 A + \mathcal{O}(A^2) \,.$$

In the limit $A \to 0$, there remains only the black hole horizon on the axis between $-\sqrt{m^2-a^2}$ and $+\sqrt{m^2-a^2}$ since $\bar{z}_3 \to \infty$. This indicates that in this limit the SC-metric in the Weyl-Papapetrou coordinates with the appropriate choice of $\kappa_1 \ldots \kappa_5$ approaches the Kerr solution. Indeed, the metric functions e^{2U}, $e^{2\nu}$, and $\bar{\omega}$ with $\kappa_1 \ldots \kappa_5$ given by (22)–(25) and (31) go to the corresponding metric functions of the Kerr metric in the Weyl-Papapetrou coordinates as given, e.g., in [22].

In this section, we have transformed the region $\mathcal{B}$ of the SC-metric into the Weyl-Papapetrou coordinates. It is similarly possible to transform the region $\mathcal{D} \cup \mathcal{D}'$ there, however, another choice of the constants $\kappa_1 \ldots \kappa_5$ would be appropriate in Eqs. (8)–(12). The white hole horizon appears on the axis between $\bar{z}_1$ and $\bar{z}_2$ and the acceleration horizon for $\bar{z} > \bar{z}_3$. Moreover there is a ring singularity at $\bar{z} = 0$ and $\bar{\rho} = a$ (corresponding to $x = 0$ and $y = \pm\infty$), where curvature invariants go to infinity. For the limit $A \to 0$ we obtain the region $r \in (-\infty,\ r_-)$ of the Kerr spacetime, where r is the Boyer-Lindquist coordinate. The region $\mathcal{D} \cup \mathcal{D}'$ thus corresponds to the spacetime of a uniformly accelerated white hole with a ring singularity.

4 The spinning C-metric in the coordinates adapted to the boost and rotation symmetries

Now let us transform the region $\mathcal{B}$ into the coordinates adapted to the boost and rotation symmetries $\{t,\ \rho,\ z,\ \varphi\}$. The transformation [4, 23]

$$\begin{aligned} \bar{\rho}^2 &= A^2\rho^2(z^2-t^2) \,, \\ \bar{\varphi} &= \varphi \,, \\ \bar{z} - \bar{z}_3 &= \tfrac{1}{2}A(\rho^2+t^2-z^2) \,, \\ \bar{t} &= \frac{1}{A}\operatorname{arctanh}\left(\frac{t}{z}\right) \end{aligned} \tag{27}$$

leads to the metric of boost-rotation symmetric spacetimes with spinning sources [13, 15]

$$\begin{aligned} ds^2 &= e^{\lambda}d\rho^2 + \rho^2 e^{-\mu}d\varphi^2 - 2\omega e^{\mu}(z\,dt - t\,dz)d\varphi - \omega^2 e^{\mu}(z^2-t^2)d\varphi^2 \\ &+ \frac{1}{z^2-t^2}[(e^{\lambda}z^2 - e^{\mu}t^2)dz^2 - 2zt(e^{\lambda}-e^{\mu})dz\,dt - (e^{\mu}z^2 - e^{\lambda}t^2)dt^2], \end{aligned} \tag{28}$$

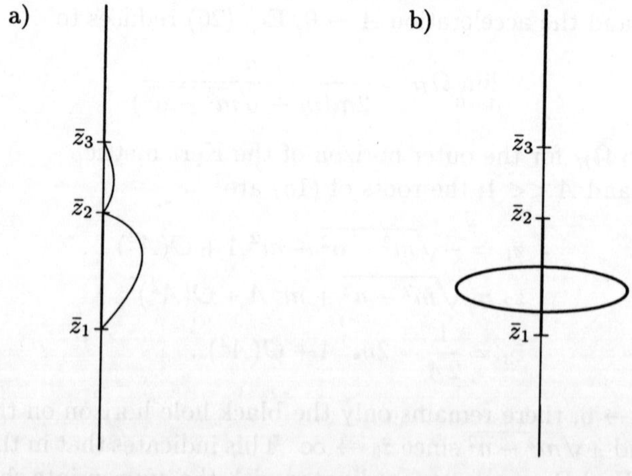

Figure 2. a) The $\mathcal{B}$ region of the SC-metric in the Weyl-Papapetrou coordinates: The black hole and acceleration horizons are located on the axis $\bar{\rho} = 0$ at $\bar{z} \in (\bar{z}_1,\ \bar{z}_2)$ and at $\bar{z} \in (\bar{z}_3,\ \infty)$, respectively. There is an ergoregion in the vicinity of the black hole horizon and a region with closed timelike curves in the neighbourhood of the spinning string between $\bar{z}_2$ and $\bar{z}_3$. b) The $\mathcal{D} \cup \mathcal{D}'$ region of the SC-metric in the Weyl-Papapetrou coordinates: The white hole and acceleration horizons are located on the axis at $\bar{z} \in (\bar{z}_1,\ \bar{z}_2)$ and $\bar{z} \in (\bar{z}_3,\ \infty)$, respectively. The ring singularity with the radius $\bar{\rho} = a$ appears at $\bar{z} = 0$.

where μ, λ, and ω – functions of ρ^2 and $z^2 - t^2$ – are given by

$$\begin{aligned} e^{\mu} &= \frac{e^{2U}}{A^2(z^2-t^2)} , \\ e^{\lambda} &= \kappa_5 \frac{A^2 e^{2\nu}}{e^{2U}} (\rho^2 + z^2 - t^2) , \\ \omega &= A\bar{\omega} . \end{aligned} \tag{29}$$

As a consequence of the transformation (27) a second black hole accelerated along the symmetry axis in the opposite direction appears (see Fig. 3).

The spacetime described by the metric (28) contains stationary and non-stationary regions separated by a so-called "roof" given by $z^2 - t^2 = 0$ (the acceleration horizon); the spacetime is stationary "bellow the roof" ($z^2 - t^2 > 0$) and radiative "above the roof" ($z^2 - t^2 < 0$), see Fig. 3. The regularity condition of the roof [13,15]

$$\lambda(\rho^2, 0) = \mu(\rho^2, 0) \tag{30}$$

yields the condition for κ_1

$$\kappa_1 = \frac{A}{\sqrt{1 + a^2 A^2 y_2{}^4}} . \tag{31}$$

The axis regularity condition [13,15]

$$\omega(0, z^2 - t^2) = 0 , \quad \lambda(0, z^2 - t^2) + \mu(0, z^2 - t^2) = 0 \tag{32}$$

is satisfied on the outer parts of the axis thanks to Eqs. (23) and (25).

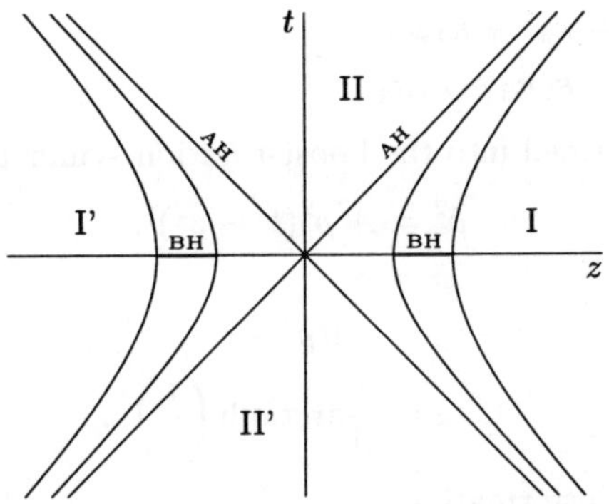

Figure 3. The stationary region $\mathcal{B}$ of the SC-metric transformed to the coordinates adapted to the boost and rotation symmetries corresponds to the region I under the roof, $z^2 - t^2 > 0$, where the roof is the acceleration horizon, denoted by AH. Under the roof, there is also another identical stationary region I' that corresponds to a second black hole accelerated along the symmetry axis in the opposite direction. The non-stationary, radiative regions II, II' appear above the roof, $z^2 - t^2 < 0$. This figure covers only the outer part of the spacetime of a uniformly accelerated Kerr black hole outside the exterior black hole horizon (BH). A similar picture, connected with $\mathcal{D} \cup \mathcal{D}'$, would describe the inner part of the spacetime of a uniformly accelerated Kerr black hole inside the interior black hole horizon.

The metric functions (29) are in fact very complicated since we have to use Eqs. (14), (17)–(19), and (27) in order to express them in coordinates $\{t, \rho, z, \varphi\}$. As the Weyl-Papapetrou metric is stationary, via the transformation (27), we get the metric functions in the stationary region bellow the roof. However, using the same expressions for the metric functions μ, λ, and ω above the roof, we obtain an analytical continuation of the metric (28) across the roof. It turns out that one has to change the sign of ϵ_3 in order to find an analytical continuation of the metric across the null cone of the origin

$$\rho^2 + z^2 - t^2 = R_3 = 0 \; . \tag{33}$$

The non-stationary region above the roof was not included in the Weyl-Papapetrou form, however, it was contained in the original metric (2) in the $\{\tau, x, y, \phi\}$ coordinates.

The region above the roof inside the null cone of the origin ($F < 0$ and $\epsilon_3 > 0$ in (16), i.e., the region β_2 in Fig. 4) can be transformed into the cylindrically symmetric metric (see [5] for the case with hypersurface orthogonal Killing vectors)

$$ds^2 = e^{-2\tilde{U}}[e^{2\tilde{\nu}}(-d\tilde{t}^2 + d\tilde{\rho}^2) + \tilde{\rho}^2 d\tilde{\varphi}^2] + e^{2\tilde{U}}(d\tilde{z} + \tilde{\omega} d\tilde{\varphi})^2 \; , \tag{34}$$

where $\tilde{U}$, $\tilde{\nu}$, and $\tilde{\omega}$ are functions of $\tilde{\rho}$ and $\tilde{t}$, by the transformation

$$\begin{aligned}
\tilde{\rho}^2 &= \mathcal{K}^2 \frac{-FG}{A^4(x+y)^4} \; , \\
\tilde{t} &= \mathcal{K} \frac{1 + mAxy(x-y) + xy - a^2A^2x^2y^2}{A^2(x+y)^2} \; , \\
\tau &= \kappa_1 \tilde{z} + \kappa_2 \tilde{\varphi} \; ,
\end{aligned}$$

$$\phi = \kappa_3 \tilde{z} + \kappa_4 \tilde{\varphi} ,$$
$$\mathcal{K} = \kappa_2 \kappa_3 - \kappa_1 \kappa_4 .$$

Then it can be transformed into the boost-rotation symmetric metric (28) by

$$\tilde{\rho}^2 = A^2 \rho^2 (t^2 - z^2) ,$$
$$\tilde{\varphi} = \varphi ,$$
$$\tilde{t} - \bar{z}_3 = \tfrac{1}{2} A(\rho^2 + t^2 - z^2) ,$$
$$\tilde{z} = \frac{1}{A} \text{arctanh} \left(\frac{t}{z} \right) .$$

Similarly by the transformation

$$\hat{t}^2 = \mathcal{K}^2 \frac{-FG}{A^4(x+y)^4} ,$$
$$\hat{\rho} = \mathcal{K} \frac{1 + mAxy(x-y) + xy - a^2A^2x^2y^2}{A^2(x+y)^2} ,$$
$$\tau = \kappa_1 \hat{z} + \kappa_2 \hat{\varphi} ,$$
$$\phi = \kappa_3 \hat{z} + \kappa_4 \hat{\varphi} ,$$
$$\mathcal{K} = \kappa_2 \kappa_3 - \kappa_1 \kappa_4 ,$$

the region above the roof outside the null cone of the origin ($F < 0$ and $\epsilon_3 < 0$ in (16), i.e., the region β_1 in Fig. 4) can be transformed into another cylindrically symmetric metric

$$\mathrm{d}s^2 = \mathrm{e}^{-2\hat{U}}[\mathrm{e}^{2\hat{\nu}}(-\mathrm{d}\hat{t}^2 + \mathrm{d}\hat{\rho}^2) + \hat{t}^2 \mathrm{d}\hat{\varphi}^2] + \mathrm{e}^{2\hat{U}}(\mathrm{d}\hat{z} + \hat{\omega}\mathrm{d}\hat{\varphi})^2 , \tag{35}$$

where the metric functions $\hat{U}$, $\hat{\nu}$, and $\hat{\omega}$ depend on $\hat{\rho}$ and $\hat{t}$. Again, it can be transformed into the form (28) by

$$\hat{t}^2 = A^2 \rho^2 (t^2 - z^2) ,$$
$$\hat{\varphi} = \varphi ,$$
$$\hat{\rho} - \bar{z}_3 = \tfrac{1}{2} A(\rho^2 + t^2 - z^2) ,$$
$$\hat{z} = \frac{1}{A} \text{arctanh} \left(\frac{t}{z} \right) .$$

Using inverse transformations from the coordinates $\{t, \rho, z, \varphi\}$ into $\{\tau, x, y, \phi\}$, we can find localization of I^0, I^+, and $\mathcal{J}^+$ as given in Fig. 4.

One can also determine the $\mathcal{J}^+$ location similarly as was done in [24] for the non-spinning C-metric. The SC-metric can be compactified by the conformal factor $\Omega = A(x+y)$. The future null infinity $\mathcal{J}^+$ is then at $\Omega = 0$, i.e., $x + y = 0$, with the induced metric

$$\mathrm{d}\tilde{s}^2_{\mathcal{J}} = G(x)(\mathrm{d}\tau^2 + \mathrm{d}\phi^2)$$

and the normal vector to $\mathcal{J}^+$

$$\tilde{n} = \tilde{\nabla}^\alpha \Omega \, \partial_{x^\alpha} = \frac{AG(x)}{1 + (aAx^2)^2} (\partial_x - \partial_y) .$$

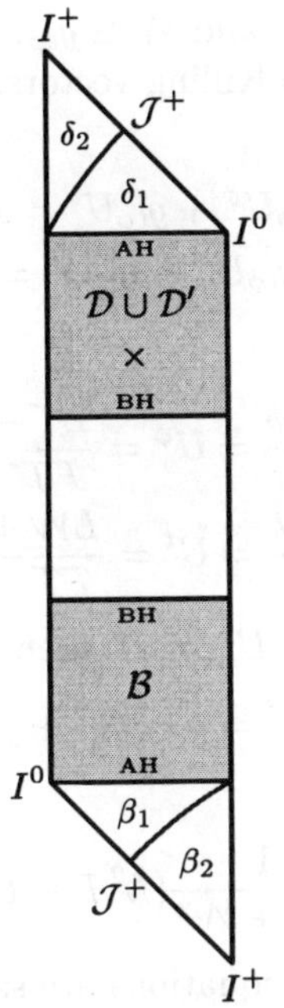

Figure 4. This figure represents the fourth column of Fig. 1, where points $y = \pm\infty$ are identified. The diagram is cut along the line $x+y=0$ where $\mathcal{J}^+$, I^+, and I^0 are located. The stationary $\mathcal{B}$ and non-stationary $\beta_1 \cup \beta_2$ regions correspond to the regions I and II in Fig. 3, respectively. The curves in both triangles represent the null cones of the origin (33) that divide them into the regions β_1 and δ_1 outside the null cone and β_2 and δ_2 inside the null cone. The stationary region $\mathcal{D} \cup \mathcal{D}'$ with the curvature singularity denoted by a cross and the triangle above it represent the inner part of the spacetime of a uniformly accelerated Kerr black hole, i.e., a uniformly accelerated white hole with a ring singularity.

5 Geodesics in the spinning C-metric

In this section, geodesics in the stationary region $\mathcal{B}$ of the SC-metric, especially their stability, are examined. As far as we know, the stability of circular orbits in stationary axisymmetric spacetimes was only studied in the case with an equatorial plane of symmetry [25, 26]. The SC-metric (and also other exact solutions, e.g., a superposition of two Schwarzschild or Kerr black holes with different parameters) does not have an equatorial plane of symmetry (see Fig. 2 a). In the following, basic theorems on the stability of solutions of ordinary non-linear differential equations [27] are employed.

Let us investigate the stability of geodesics in a general stationary spacetime with two Killing vectors, $\partial/\partial t$ and $\partial/\partial\phi$. Later we will apply the results obtained here to the SC-metric (2). We assume the metric in coordinates $\{x^0,\ x^1,\ x^2,\ x^3\} = \{t,\ X,\ Y,\ \phi\}$ to be of the form

$$\mathrm{d}s^2 = g_{XX}\mathrm{d}X^2 + g_{YY}\mathrm{d}Y^2 + g_{tt}\mathrm{d}t^2 + 2g_{t\phi}\mathrm{d}t\mathrm{d}\phi + g_{\phi\phi}\mathrm{d}\phi^2\ , \tag{36}$$

where all metric functions depend only on X and Y and g_{XX} and g_{YY} are positive.

Let us define $\mathcal{T} \equiv -g_{tt}$, $\mathcal{F} \equiv g_{\phi\phi}$, and $\mathcal{W} \equiv g_{t\phi}$.

Since the metric (36) has two Killing vectors, each freely falling particle carries two conserved quantities

$$U_\phi = L = g_{\phi\phi}U^\phi + g_{t\phi}U^t = \mathcal{F}U^\phi + \mathcal{W}U^t \; ,$$
$$U_t = -E = g_{t\phi}U^\phi + g_{tt}U^t = \mathcal{W}U^\phi - \mathcal{T}U^t \; ,$$

which imply

$$\dot{\phi} \equiv \frac{\mathrm{d}\phi}{\mathrm{d}\tau} = U^\phi = \frac{L\mathcal{T} - E\mathcal{W}}{\mathcal{F}\mathcal{T} + \mathcal{W}^2} \; ,$$
$$\dot{t} \equiv \frac{\mathrm{d}t}{\mathrm{d}\tau} = U^t = \frac{L\mathcal{W} + E\mathcal{F}}{\mathcal{F}\mathcal{T} + \mathcal{W}^2} \; . \tag{37}$$

The norm of the four-velocity $U^\alpha = \dot{x}^\alpha$ then reads

$$-1 = g_{\alpha\beta}U^\alpha U^\beta = g_{XX}\dot{X}^2 + g_{YY}\dot{Y}^2 - \mathcal{R} - 1 \tag{38}$$

with

$$\mathcal{R} \equiv -1 - \frac{1}{\mathcal{F}\mathcal{T} + \mathcal{W}^2}(L^2\mathcal{T} - E^2\mathcal{F} - 2EL\mathcal{W}) \; . \tag{39}$$

Thanks to (37), two geodesic equations are satisfied identically while the other two read

$$\ddot{X} + \mathcal{F}_1 = 0 \; , \quad \ddot{Y} + \mathcal{F}_2 = 0 \; , \tag{40}$$

where

$$\mathcal{F}_1 = \tfrac{1}{2}g^{XX}(\; g_{XX},_X \dot{X}^2 + 2g_{XX},_Y \dot{X}\dot{Y} - g_{YY},_X \dot{Y}^2 - \mathcal{R},_X\;) \; , \tag{41}$$
$$\mathcal{F}_2 = \tfrac{1}{2}g^{YY}(-g_{XX},_Y \dot{X}^2 + 2g_{YY},_X \dot{X}\dot{Y} + g_{YY},_Y \dot{Y}^2 - \mathcal{R},_Y\;) \; . \tag{42}$$

The system of two second-order differential equations (40) can be converted to four first-order differential equations

$$\dot{z}^\alpha = \mathcal{F}^\alpha\,(z^\beta) \; , \quad \mathcal{F}^\alpha = (z^2,\; -\mathcal{F}_1,\; z^4,\; -\mathcal{F}_2) \tag{43}$$

introducing new variables z^α

$$z^1 \equiv X \; , \quad z^2 \equiv \dot{X} \; , \quad z^3 \equiv Y \; , \quad z^4 \equiv \dot{Y} \; . \tag{44}$$

Stationary (equilibrium) points of the system (43), z_0^α, are given by $\mathcal{F}^\alpha(z_0^\beta) = 0$, i.e.,

$$z^2 = \dot{X} = 0 \; , \quad z^4 = \dot{Y} = 0 \; , \quad \text{i.e.,} \quad \mathcal{R}(z_0^\alpha) = 0 \; , \tag{45}$$
$$\mathcal{F}_1 = -\tfrac{1}{2}g^{XX}\mathcal{R},_X\,(z_0^\alpha) = 0 \; , \tag{46}$$
$$\mathcal{F}_2 = -\tfrac{1}{2}g^{YY}\mathcal{R},_Y\,(z_0^\alpha) = 0 \; . \tag{47}$$

One can eliminate E and L from Eqs. (46), (47) and derive the equation for the stationary points X_0, Y_0

$$(\mathcal{F},_{[X}\,\mathcal{T},_{Y]}\,)^2 - 4(\mathcal{T},_{[X}\,\mathcal{W},_{Y]}\,)(\mathcal{F},_{[X}\,\mathcal{W},_{Y]}\,) = 0 \tag{48}$$

and, using also Eq. (45), express E and L in terms of X_0 and Y_0.

Since the system (43) is autonomous, its linearization may help to determine the stability of its stationary points [27]. The linearized form of (43) in the neighbourhood of a stationary point z_0^α is

$$\dot{z}^\alpha = A^\alpha_\beta (z^\beta - z_0^\beta) ,$$

where

$$A^\alpha_\beta \equiv \frac{\partial \mathcal{F}^\alpha}{\partial z^\beta}(z_0^\gamma) = \begin{pmatrix} 0 & 1 & 0 & 0 \\ -\mathcal{F}_1^{(0)},_X & 0 & -\mathcal{F}_1^{(0)},_Y & 0 \\ 0 & 0 & 0 & 1 \\ -\mathcal{F}_2^{(0)},_X & 0 & -\mathcal{F}_2^{(0)},_Y & 0 \end{pmatrix}$$

is a constant matrix with eigenvalues

$$\lambda_\alpha = \pm\sqrt{\tfrac{1}{2}}\sqrt{-\mathcal{F}_1^{(0)},_X - \mathcal{F}_2^{(0)},_Y \pm\sqrt{(\mathcal{F}_1^{(0)},_X - \mathcal{F}_2^{(0)},_Y)^2 + 4\mathcal{F}_1^{(0)},_Y \mathcal{F}_2^{(0)},_X}} , \tag{49}$$

with $\mathcal{F}_1^{(0)},_X \equiv \mathcal{F}_1,_X (z_0^\gamma)$, etc. If all $\mathrm{Re}\lambda_\alpha$ were negative the equilibrium point of (43) would be asymptotically stable and if at least one $\mathrm{Re}\lambda_\alpha$ were positive the equilibrium point of (43) would be unstable. If $\max\{\mathrm{Re}\lambda_\alpha\}$ were zero the studied point could be stable, however, further examinations would be necessary. Since $\lambda_1 + \lambda_2 = \lambda_3 + \lambda_4 = 0$, the stationary point z_0^α is unstable if $\mathrm{Re}\lambda_\alpha \neq 0$ for any α and could be stable only if $\mathrm{Re}\lambda_\alpha = 0$ for all α, i.e., for

$$\begin{aligned} &\mathcal{F}_1^{(0)},_X \mathcal{F}_2^{(0)},_Y > \mathcal{F}_1^{(0)},_Y \mathcal{F}_2^{(0)},_X , \quad \text{i.e.,} \quad \mathcal{R}^{(0)},_{XX} \mathcal{R}^{(0)},_{YY} > \mathcal{R}^{(0)},_{XY}{}^2 , \\ &\mathcal{F}_1^{(0)},_X > 0 , \quad \mathcal{F}_2^{(0)},_Y > 0 , \qquad \text{i.e.,} \quad \mathcal{R}^{(0)},_{XX} < 0 , \quad \mathcal{R}^{(0)},_{YY} < 0 , \end{aligned} \tag{50}$$

i.e., if the function $\mathcal{R}$ has a local maximum in the equilibrium point or if

$$\begin{aligned} &\mathcal{F}_1^{(0)},_X = \mathcal{F}_2^{(0)},_Y = \mathcal{F}_1^{(0)},_Y \mathcal{F}_2^{(0)},_X = 0 , \\ \text{i.e.,} \quad &\mathcal{R}^{(0)},_{XX} = \mathcal{R}^{(0)},_{YY} = \mathcal{R}^{(0)},_{XY} = 0 . \end{aligned} \tag{51}$$

In order to determine whether the stationary points given by (48) and satisfying (50) or (51) are indeed stable, we employ the Lyapunov method. The Lyapunov function for (43) is a function V of four independent variables z^α satisfying in an open neighbourhood of an equilibrium point z_0^α the conditions

$$V(z^\alpha) > 0 \quad \text{for} \quad z^\alpha \neq z_0^\alpha , \quad V(z_0^\alpha) = 0 , \tag{52}$$

$$V' \equiv V,_{z^\alpha} \mathcal{F}^\alpha \leq 0 . \tag{53}$$

Notice that V' is the directional derivative of V in the direction of $\mathcal{F}^\alpha$, $V' = \lim_{t\to 0}[V(z^\alpha + t\mathcal{F}^\alpha) - V(z^\alpha)]/t$. If such a function exists then the equilibrium point z_0^α is stable, however, there is not a general method how to find it.

It can be shown that

$$V = g_{XX} {z_2}^2 + g_{YY} {z_4}^2 - \mathcal{R} , \tag{54}$$

which satisfies $V' = 0$ identically, is the Lyapunov function for the system (43) if $\mathcal{R}$ is negative in the neighbourhood of the equilibrium point z_0^α, i.e., if $\mathcal{R}$ has a local maximum in z_0^α since $\mathcal{R}(z_0^\alpha) = 0$. Thus, equilibrium points satisfying (50) are indeed stable.

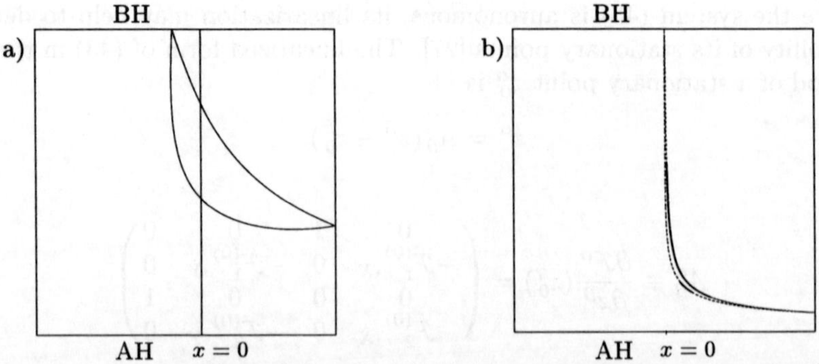

Figure 5. Stationary points in the $\mathcal{B}$ region of the SC-metric for a) $m = 1$, $a = 1/2$, $A = 1/10$; b) $m = 1/5$, $a = 1/6$, $A = 1/100$ – stable equilibrium points are highlighted, the upper and lower curves correspond to retrograde and prograde orbits, respectively.

In the special case $\mathcal{W} = 0$ with hypersurface orthogonal Killing vectors, $\mathcal{R} \sim (E^2 - V_{\rm ef}^2)/E^2$, where $V_{\rm ef}$ is the effective potential. A local maximum of $\mathcal{R}$ then corresponds to a local minimum of $V_{\rm ef}$.

To summarize: The stationary points $X = X_0$ and $Y = Y_0$ given by (48) are stable if $\mathcal{R}$ (39) has a local maximum there.

Let us apply the foregoing general considerations to the SC-metric (2). The corresponding computations are rather complicated and thus we present only the main results here.

The condition (48) for stationary points is a polynomial of the order 12 in x and y and for $a = 0$ it reduces to Eq. (10) in [28]. The stationary points x_0, y_0 in the $\mathcal{B}$ region are plotted in Fig. 5. As in the non-spinning case [28], the stationary points $x_0 = 0$ correspond to null geodesics, stationary points with $x_0 > 0$ and $x_0 < 0$ to timelike and spacelike geodesics, respectively.

In contrast to the non-spinning case, there now appear two curves in Fig. 5 – the upper and the lower one corresponding to retrograde ($L < 0$) and prograde ($L > 0$) orbits, respectively.

Using Eq. (50), one can show that for A small, there exist both retrograde and prograde stable orbits (Fig. 5 b). As was shown in [28], these geodesics correspond to co-accelerated test particles orbiting the uniformly accelerated black hole. If the parameters E and L are perturbed sufficiently the test particle falls under the black hole or acceleration horizon.

6 Conclusion

If conditions (4) are satisfied there are four stationary regions in the SC-metric. Each of them can be transformed into a different Weyl-Papapetrou metric by (8)–(12). The most physically important region $\mathcal{B}$ in the Weyl-Papapetrou coordinates represents the gravitational field of a "spinning rod" (the black hole horizon) and a semi-infinite line mass that are held in equilibrium by a spinning string. There is an ergoregion in the vicinity of the black hole horizon and a region with closed timelike curves in the neighbourhood of the spinning string (see Fig. 2 a). In the limit the acceleration $A \to 0$, there remains just the "spinning rod", which corresponds to the "exterior" of the Kerr metric. Through a further transformation to the coordinates adapted to the boost and rotation symmetries new non-stationary radiative regions appear. It turns out that these regions are already contained in the original form (2) of the SC-metric (see Fig. 4).

Another stationary region $\mathcal{D} \cup \mathcal{D}'$ corresponds to a uniformly accelerated superposition of a spinning white hole and a ring singularity, which represents a uniformly accelerated "interior" of the Kerr solution, i.e., the region bellow the inner horizon up to $r \to -\infty$. A timelike curve in the Kerr metric can start in the external, asymptotically flat region, cross two horizons, pass through the ring singularity, and emerge in another asymptotically flat white hole region, where gravity is repulsive. Similarly, in the SC-metric, a timelike curve starting in the region $\mathcal{B}$ can cross two horizons and debouch in the region $\mathcal{D} \cup \mathcal{D}'$. Then, if it is not co-accelerated it crosses the null cone of the origin and reaches I^+ (see Fig. 4).

In Sec. 5, the stability of geodesics corresponding to equilibrium points in a general stationary spacetime with two symmetries was studied. Using the Lyapunov method it can be shown that an equilibrium point is stable if the function $\mathcal{R}$ of two variables (39) has a local maximum there. It turns out that for the SC-metric there exist stable prograde and retrograde orbits corresponding to co-accelerated particles orbiting the uniformly accelerated spinning black hole.

Acknowledgments

We are grateful to Professor J. Bičák for several fruitful years of common work under his supervision, which also included the SC-metric. We thank Professors I. Babuška and I. Vrkoč for several discussions on the stability of equilibrium points and M. Žáček for a discussion on the papers [25,26]. We are also thankful to University of Texas, Austin, for hospitality. V.P. was supported by the grant GACR-202/00/P030 and A.P. by the grant GACR-202/00/P031. The work was also partially supported by the National Science Foundation under the grant NSF-Czech Rep. INT-9724783.

References

1. T. Levi-Civita, *ds^2 einsteiniani in campi newtoniani*, Rend. Acc. Lincei **27**, 343 (1918).
2. H. Weyl, *Bemerkung über die axisymmetrischen lösungen der Einsteinschen Gravitationsgleinchungen*, Ann. d. Phys **59**, 185 (1919).
3. W. Kinnersley and M. Walker, *Uniformly Accelerating Charged Mass in General Relativity*, Phys. Rev. D **2**, 1359 (1970).
4. W. B. Bonnor, *The sources of the vacuum C-metric*, Gen. Rel. Grav. **15**, 535 (1983).
5. J. Bičák and B. G. Schmidt, *Asymptotically flat radiative space-times with boost-rotation symmetry: The general structure*, Phys. Rev. D **40**, 1827 (1989).
6. J. Bičák, *Selected solutions of Einstein's field equations: their role in general*

relativity and astrophysics, in *Einstein's Field Equations and Their Physical Meaning*(Ed. B. G. Schmidt), Springer Verlag, Berlin – New York (2000).

7. V. Pravda and A. Pravdová, *Boost-rotation symmetric spacetimes – review*, Czech. J. Phys. **50**, 333 (2000); see also gr-qc/0003067.
8. F. H. J. Cornish and W. J. Uttley, *The interpretation of the C metric. The charged case when $e^2 \leq m^2$*, Gen. Rel. Grav. **27**, 735 (1995).
9. F. J. Ernst, *Generalized C-metric*, J. Math. Phys. **19**, 1986 (1978).
10. H. F. Dowker and S. N. Thambyahpillai, *Many Accelerating Black Holes*, gr-qc/0105044 (2001).
11. J. F. Plebański and M. Demiański, *Rotating, charged and uniformly accelerating mass in general relativity*, Ann. Phys. (USA) **98**, 98 (1976).
12. H. Farhoosh and R. L. Zimmerman, *Surfaces of infinite red-shift around a uniformly accelerating and rotating particle*, Phys. Rev. D **21**, 2064 (1980).
13. J. Bičák and V. Pravda, *Spinning C metric: Radiative spacetime with accelerating, rotating black holes*, Phys. Rev. D **60**, 044004 (1999).
14. P. S. Letelier and S. R. Oliveira, *Uniformly accelerated black holes*, Phys. Rev. D **64**, 064005 (2001).
15. A. Pravdová and V. Pravda, *Boost-rotation symmetric vacuum spacetimes with spinning sources*, to appear in J. Math. Phys.; see also gr-qc/0107068 (2001).
16. I. Ciufolini, *Gravitomagnetism and status of the LAGEOS III experiment*, Class. Quantum Grav. **11**, A73 (1994).
17. B. B. Godfrey, *Horizons in Weyl metrics exhibiting extra symmetries*, Gen. Rel. Grav. **3**, 3 (1972).
18. P. S. Letelier, *Static and stationary multiple soliton solutions to the Einstein equations*, J. Math. Phys. **26**, 467 (1985).
19. P. S. Letelier and S. R. Oliveira, *Double Kerr-NUT spacetimes: spinning strings and spinning rods*, Phys. Lett. A **238**, 101 (1998).
20. W. B. Bonnor, *The interactions between two classical spinning particles*, Class. Quantum Grav. **18**, 1382 (2001).
21. F. H. J. Cornish and W. J. Uttley, *The Interpretation of the C Metric. The Vacuum Case*, Gen. Rel. Grav. **27**, 439 (1995).
22. D. Kramer, H. Stephani, E. Herlt, and M. A. H. MacCallum, *Exact solutions of Einstein's field equations* (Ed. E. Schmutzer), Cambridge Univ. Press, Cambridge (1980).
23. W. B. Bonnor and N. S. Swaminarayan, *An Exact Solution for Uniformly Accelerated Particles in General Relativity*, Z. Phys. **177**, 240 (1964).
24. S. W. Hawking and S. F. Ross, *Loss of quantum coherence through scattering off virtual black holes*, Phys. Rev. D **56**, 6403 (1997).
25. J. M. Bardeen, *Stability of circular orbits in stationary, axisymmetric spacetimes*, ApJ **161**, 103 (1970).
26. O. Semerák, M. Žáček, and T. Zellerin, *Test-particle motion in superposed Weyl field*, Mon. Not. R. Astron. Soc. **308**, 705 (1999).
27. W. Walter, *Ordinary Differential Equations*, Springer-Verlag, New York (1998).
28. V. Pravda and A. Pravdová, *Co-accelerated particles in the C-metric*, Class. Quantum Grav. **18**, 1205 (2001).

TWISTING MINKOWSKI PLANE INTO A CONE

VLADIMÍR BALEK
Department of Theoretical Physics, Comenius University
Mlynská dolina, 84248 Bratislava, Slovakia
E-mail: balek@fmph.uniba.sk

Lorentzian conical surfaces of two kinds are introduced: surfaces where an angle is removed from or added to the light cone and surfaces where an angle is removed from a region ranging from the past to the future light cone. Surfaces of the first kind represent a spacetime with collapsing or exploding inertial frames and those of the second kind represent a spacetime with nonorientable time. If a surface of the second kind is glued together in a Möbius-like way along a segment of the removed angle, which is only possible if it contains a ridge, the segment acts as a time machine. All the surfaces proposed here are physically unacceptable because energy conditions are violated; the time machine, in addition, is unacceptable because it produces inconsistent histories.

1 Introduction

If we cut off a wedge-like strip of a sheet of paper and glue together the edges of the paper we get a conical surface. If we cut the paper from the edge to the center and glue a strip into it the result is again a conical surface, this time with a wavy edge. What do we get when playing in the same way with a Minkowski plane? And if we add two space dimensions, what kind of source is needed to produce such a spacetime?

Twisting a sheet of paper, when translated into mathematical terms, defines a well known object of general relativity: the cosmic string (see [1] and references therein). Indeed, the spacetime of an infinite straight string is a product of a Minkowski plane with a conical surface, or equivalently, a 4D Minkowski space with an angle cut off of the slices that are perpendicular to the string, and the sides of the angle identified. Obviously, the source of such a geometry is linear (1D): outside the tip of the removed angle there is no curvature, and hence no matter. This is reasonable in the weak field limit [2] and can be given good meaning in the exact theory, too [3]. Real strings have nonzero thickness and the spacetime they produce may be viewed as consisting of conical slices with a rounded peak (similar in shape to a shuttle-cock, when embedded into 3D) [4]. In this simplified picture, the only nonzero components of the energy-momentum tensor are the energy density proportional to the Gauss curvature of the slices, and the longitudinal stress equal to minus the energy density. (In a real string lateral tensions are also present, however, they vanish after being averaged over the cross section of the string.) To satisfy the energy conditions, the rounded peak must have nonnegative Gauss curvature everywhere. This excludes conical slices with added angle.

If one passes from a Euclidean to a Lorentzian conical surface, some implications of the change of signature may be noticed immediately. When removing an angle from a Minkowski plane, or adding an angle to it, one has to make sure that the sides of the angle are both of the same kind: timelike, null, or spacelike. A simple possibility is to remove an angle from a given quadrant of the Minkowski plane, or to add an angle from a given quadrant to the same quadrant again. The number of quadrants remains the same and the distortion is restricted to one of them. There are other possibilities, too. For example, one may let the removed angle extend from a given quadrant to the opposite one, reducing the number of quadrants to two. This possibility may be found interesting by designers of time machines because, in the resulting spacetime, the arrow of time is modified dramatically (in fact destroyed, if the surface contains no ridge to save it). Having established a conical surface, one has to show that its peak can be smoothed out so that an extended source exists for it, in principle, after two spatial dimensions are added. At this stage another new point arises. A surface in a 3D Minkowski space has an induced metric that is Lorentzian only if it crosses the light cones everywhere: for example, of the two boost-invariant hyperboloids, that of the "funnel" type (2D de Sitter universe) is Lorentzian, but that of the "bowl" type (Lobachevski plane) is not. The fact that the surfaces in question are Lorentzian everywhere puts severe restrictions on their possible embeddings. Particularly, no embedding of the "cornet" type exists for a conical surface with a sharp peak so that one cannot get rid of the peak passing to a smoothed-out surface of the "shuttle-cock" type. To complete the theory, the surface must be extended to a 4D spacetime. This may be done most simply by assuming that the extra dimensions are flat so that the 4D spacetime reduces to the spacetime of an infinite straight string with time and one space dimension interchanged. The source of such a geometry is an infinite flat wall that is created in a single moment and immediately after that turns back into a void. The string analogy tells us that the energy density as well as the lateral stress inside the wall are zero and the only nonzero components of the energy momentum tensor are two longitudinal stresses of equal magnitude. At a closer look, we find that the sign of the stresses coincides with the curvature sign of the surface we have started with. Consequently, if the curvature is nonnegative throughout the peak, the matter inside the wall obeys the weak as well as the strong energy conditions [5]. The spacetime is unphysical nevertheless because the matter inside the wall violates the energy dominance condition [5].

Since it is clear from the beginning that the spacetimes based on Lorentzian conical surfaces do not admit realistic sources, an obvious question arises why should we bother about them. Perhaps some kind of exotic matter can be invented violating the energy dominance condition in the desired way, similarly to vacuum fluctuations inside wormholes presumably violating the weak energy condition. But even if not, exploring a purely fictitious object is a kind of challenge, too, especially if it is so strange an object as this: an infinite wall appearing in a short flash just to make a vast part of spacetime "softly and suddenly vanish away" [6], or to create an equally vast part of spacetime from nothing.

In this work, properties of Lorentzian conical surfaces are investigated. In Section II, relevant facts of the Euclidean surfaces theory are summarized, in Section III, Lorentzian conical surfaces of the two kinds mentioned above are introduced and their metric properties outside the peak are analyzed, in Section IV, the geometry of the peak is studied, and in Section V, the results are discussed. The system of units is used in which $c = 1$.

2 Conical surfaces: metric, curvature, embeddings

To twist a Euclidean plane into a cone, one has to pick out an angle, remove it from the plane or add it to the plane again, and identify the sides of the wedge. The procedure is depicted in Fig. 1. Consider a region in a Euclidean plane extending from the positive x-axis to the line rotated by the angle $-\alpha$ with respect to the positive x-axis and suppose the region is removed from the plane if $\alpha > 0$ and added to the plane if $\alpha < 0$. (The angle α must be less than 2π, but it is unbounded from

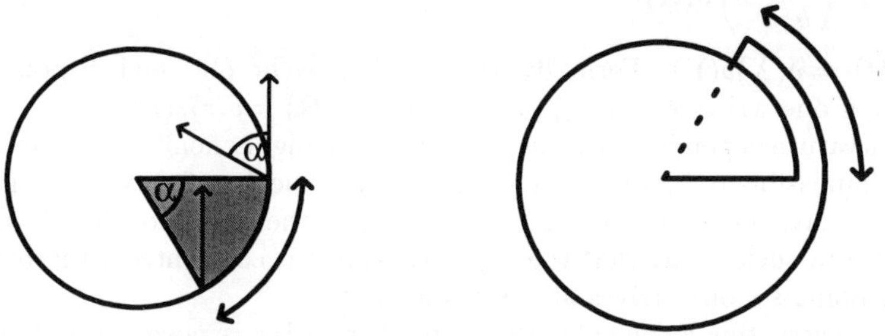

Figure 1. Conical surface with removed and added angle

below.) Outside the origin, the metric is that of the plane,

$$dl^2 = dr^2 + r^2 d\phi^2,$$

however, ϕ ranges from 0 to $2\pi b$, where $b = 1 - \alpha/(2\pi)$, rather than from 0 to 2π. The origin looks like a sharp peak when embedded into 3D and the surface possesses a δ-function-like curvature there. From the point of view of the internal geometry, the curvature manifests itself in the fact that a vector transported parallel around the peak becomes deflected. As shown in the figure above, the angle of deflection equals the angle deficit of the surface. If we wish to cover the surface by coordinates that change smoothly as we pass from one side of the removed or added angle to the other side, we must replace the ϕ-coordinate by the rescaled coordinate $\Phi = \phi/b$ ranging from 0 to 2π. Rewritten in terms of Φ, the metric is

$$dl^2 = dr^2 + b^2 r^2 d\Phi^2. \tag{1}$$

Starting from the polar coordinates (r, Φ), we may define Cartesian coordinates $X = r\cos\Phi$ and $Y = r\sin\Phi$ covering the surface smoothly everywhere.

Let us look at the curvature of a conical surface a bit closer. According to Gauss-Bonnet theorem, the integral curvature K_{int} (the surface integral of Gauss curvature) is equal to the angle of deflection of a vector transported parallel around the region of integration. If we apply this to a conical surface with a sharp peak, regarded as a limit of conical surfaces with a rounded peak, we find that the integral curvature of any part of the surface containing the peak is

$$K_{int} = \alpha. \tag{2}$$

The fact that K_{int} is constant irrespective of the choice of integration region suggests that the Gauss curvature may be expressed in terms of a δ-function. (There is a clear analogy between K_{int} and the mass of a pointlike particle.) A straightforward way how to do this is to put

$$K\sqrt{g} = K_{int}\,\delta,$$

where g is the determinant of the metric tensor and δ is a 2D δ-function at the peak. In Cartesian coordinates, $g = b^2$ at the peak, hence

$$K = 2\pi\left(\frac{1}{b} - 1\right)\delta(\mathbf{R}), \tag{3}$$

where $\delta(\mathbf{R}) = \delta(X)\delta(Y)$. Formally, this result may be obtained in polar coordinates, too, if one writes δ as $\delta(r)/(2\pi)$ and uses $\delta(\mathbf{R}) = \delta(r)/(2\pi r)$.

The curvature depends on the metric tensor in a highly nonlinear manner, therefore, it is impossible to handle the curvature of a conical surface with a sharp peak similarly as, say, the expression $\Delta \log r$. However, the theory of distributions can be extended in such a way that the expressions one is confronted with make sense and, after doing so, one arrives at Eq. (3) again [7].

When constructing a conical surface, whether with a removed or with an added angle, one may also start with an angle the sides of which are curves of the same shape rather than straight lines. Clearly, the result will be the same surface as before, smooth everywhere outside the peak. A new class of conical surfaces arises if the angle removed from the plane or added to it has sides of a different shape. The surface then contains a line of discontinuity that looks like a ridge if embedded into 3D. Two such surfaces, with the ridge represented by a toothed line, are drawn in Fig. 2. (By a toothed line one usually marks linear singularities, or edges of the space, bringing geodesics to a halt at a finite value of their affine parameter. Here, this graphic representation is used to suggest that ridges in spacetime may act like singularities on extended bodies.) Order the sides of the removed angle

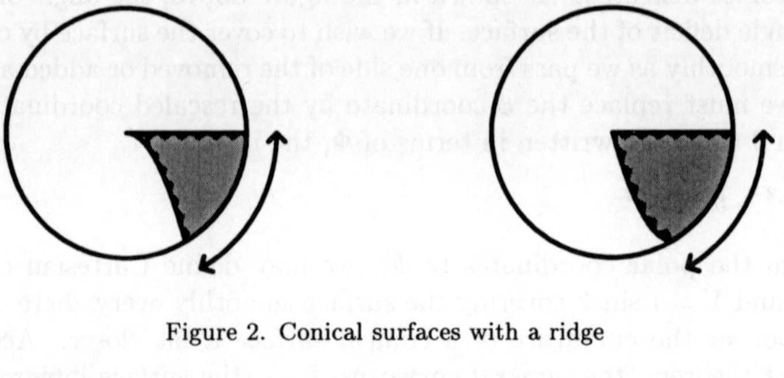

Figure 2. Conical surfaces with a ridge

counterclockwise and denote the curvature of the sides κ_1 and κ_2 with the plus sign if the center of the curvature is shifted counterclockwise with respect to the corresponding side and the minus sign otherwise. The form of the ridge is given by the jump in the curvature $\Delta\kappa = \kappa_2 - \kappa_1$: the magnitude of $\Delta\kappa$ determines the sharpness of the ridge and the sign of $\Delta\kappa$ fixes the way the ridge is bent with respect to the rest of the surface. If $\Delta\kappa > 0$, the sides of the ridge diverge from each other like on the surface on the left, while if $\Delta\kappa < 0$, the sides of the ridge converge to each other like on the surface on the right. In the former case, the ridge is bent outwards and in the latter case inwards. Note that while a surface with a nonzero $\Delta\kappa$ necessarily has a linear discontinuity, it may be free of a pointlike singularity. This is the case if the sides of the ridge are parallel at the origin, like those on the left surface in Fig. 2.

At the ridge, the curvature is proportional to a 1D δ-function. This kind of singularity is perhaps more acceptable than the singularity at the peak since it may be described within the standard theory of distributions. Choose two infinitesimal segments of the sides of the removed or added angle that merge after the sides are identified and place them on the x-axis at the distances from the origin equal to their radii of curvature. The longitudinal and transversal coordinates (l, n) in a strip containing the segment are

$$l = \rho_A \phi, \quad n = s_A(\rho_A - r), \quad A = 1, 2,$$

where ρ_1 and ρ_2 are the radii of curvature of the segments, $\rho_A = 1/|\kappa_A|$, and s_1 and s_2 are the signs of κ_1 and κ_2, respectively. Written in terms of the coordinates (l, n), the metric of the strip is

$$dl_{strip}^2 = (1 - \kappa_A n)^2 dl^2 + dn^2,$$

and the curvature

$$K = \Delta\kappa\ \delta(n). \tag{4}$$

The integral of the linear curvature is equal to the angle of line deflection from the original direction. Using this fact, we can easily check the consistency of Eq. (4) with the Gauss-Bonnet theorem.

So far we have assumed that the segments of the sides of the removed or added angle are identified if their distance from the peak is the same. If we give up this assumption we get a sort of a wrinkled ridge with adjacent regions that dilate or shrink as we pass from one to the other. Note that in this case a ridge is formed even if the sides of the removed or added angle are straight. Define dilation factors d_1 and d_2 as the ratios of distances measured on the opposite sides of the ridge to some arbitrarily defined distance measured on the ridge. Inserting the dilation factors into the definition of the longitudinal coordinate and computing the curvature in the same way as before, we obtain

$$K\sqrt{g} = -\Delta d\ \delta'(n) + \Delta(d\kappa)\ \delta(n), \tag{5}$$

where the prime denotes a derivative. While the product $K\sqrt{g}$ is well defined, K itself is not since $\sqrt{g}$ contains a jump at the ridge.

A conical surface, as suggested by its name, can be embedded into 3D as the outer surface of some infinite solid cone. Particularly, a surface with a removed angle not containing a ridge can be embedded as the outer surface of a circular cone. The induced metric is immediately transformed into the form (1) with $b = \sin\omega$, where ω is one half of the cone vertex angle, so that the angle deficit of the surface is $2\pi(1 - \sin\omega)$ (as follows from elementary considerations as well). To smooth out the metric, one may replace the peak of the surface by a spherical cup or, more generally, by any rotational surface that is round at the bottom. The resulting metric is

$$dl^2 = dr^2 + B^2 d\Phi^2, \tag{6}$$

where r ranges from some δ to certain ϵ and B is a function of r matching the function br at $r = \epsilon$ and approaching zero as $r - \delta$ when r approaches δ. (For a sphere of radius a, $\delta = a(\cot\omega - \pi/2 + \omega)$, $\epsilon = a\cot\omega$, and $B = a\sin[(r-\delta)/a]$.) The computation of Gauss curvature yields

$$K = -\frac{B''}{B}, \tag{7}$$

and after integrating the expression $K\sqrt{g} = -B''$ over r and Φ, one obtains the angle α as the Gauss-Bonnet theorem requires.

In general, a conical surface without a ridge, whether with a removed or with an added angle, looks like an expandable surface in 3D with all generating lines beginning at one point. Denote the 3D coordinate axes (ξ, η, ζ) and introduce polar coordinates (R, Φ) in the plane (ξ, η). A general conical surface is given by the equation $\zeta = \pm R/\rho$ where ρ is a nonnegative function of Φ defining the curve at which the height of the surface above the plane (ξ, η) is ± 1. Putting $r = R\rho_+/\rho$ where $\rho_+ = \sqrt{\rho^2 + 1}$, one obtains a slightly generalized expression (1) for the metric with b replaced by

$$\phi' = \frac{1}{\rho_+}\sqrt{\rho^2 + \frac{\rho'^2}{\rho_+^2}}. \tag{8}$$

The metric transforms into the form (1) if one rewrites it in terms of a new coordinate $\hat{\Phi} = b\phi(\Phi)$ and erases the hat. Given a function $\phi(\Phi)$, an embedding may not exist. We may ask, for example, whether a conical surface with a removed angle can be embedded into 3D in such a way that ϕ coincides with Φ, up to an additive constant, everywhere outside a given interval $\Delta\Phi$. Solving the variational problem, we find that the minimal value of $\Delta\phi$ is given in terms of $\Delta\Phi$ by

$$\cos\Delta\phi_{min} = \frac{\sinh\Phi_1 \sinh\Phi_2 \cos\Delta\Phi + 1}{\cosh\Phi_1 \cosh\Phi_2},$$

where Φ_1 and Φ_2 are the values of Φ at the ends of the interval $\Delta\Phi$. Consequently, $\Delta\phi$ must be greater than $\Delta\Phi$ and no embedding is possible if $\Delta\Phi$ is less than some critical angle $\Delta\Phi_{min}$ close to $\pi/2$. (The value of $\Delta\Phi_{min}$ is given by $\cos\Delta\Phi_{min}\ \cosh(2\pi - \Delta\Phi_{min}) = 1$ so that it approximately equals $\pi/2 - \epsilon$ where $\epsilon = 1/\cosh(3\pi/2) = 0.018$.)

The constraint on $\Delta\Phi$ might suggest that we cannot confine the removed angle to a sector of the embedded surface that is not wide enough. This is, however, not true. The removed angle shrinks to a line if we use embedding of the "corner" type, composed of one straight and one broken plane intersecting at a right angle, say, of the first quadrant of the plane (η, ζ) rotated by the angle α with respect to the η-axis, of the fourth quadrant of the plane (η, ζ) and of a part of the plane (ξ, ζ) containing the positive ξ-axis. In terms of the function $\rho(\Phi)$, $\rho =$ anything if

$\Phi = 0$ and $\pi/2$, and $\rho \cos \Phi = \tan \alpha$ if Φ runs from 0 to $\pi/2$. Note that a conical surface with a ridge can be embedded as a corner, too, provided the broken plane is bent. To round the peak of the corner, choose appropriate functions U and V and replace a part of the corner around the origin by the surface $\xi = U(\eta) + V(\zeta)$. The metric of the rounded peak is

$$dl^2 = du^2 + dv^2 + 2\mathcal{U}'\mathcal{V}' du\, dv, \tag{9}$$

where u and v are the lengths measured along the curves $\xi = U(\eta)$ and $\xi = V(\zeta)$ and $\mathcal{U}$ and $\mathcal{V}$ are equal to U and V considered as functions of u and v, $\mathcal{U} = U(\eta(u))$ and $\mathcal{V} = V(\zeta(v))$. This yields

$$K = \frac{\mathcal{U}''\mathcal{V}''}{(1 - \mathcal{U}'^2\mathcal{V}'^2)^2}, \tag{10}$$

which is again consistent with the Gauss-Bonnet theorem.

3 Survey of Lorentzian conical surfaces

To twist a Minkowski plane into a cone, one has to choose an angle with both sides of the same kind and then proceed in the same way as when twisting a Euclidean plane. First consider surfaces with a removed or an added angle confined to the future light cone. Two examples are shown in Fig. 3. Let the removed or added angle be located between the positive t-axis and the line boosted by the velocity $-v$ with respect to the positive t-axis, and suppose the angle is removed if $v > 0$ and added if $v < 0$. The Lorentzian analogue of polar coordinates in the future

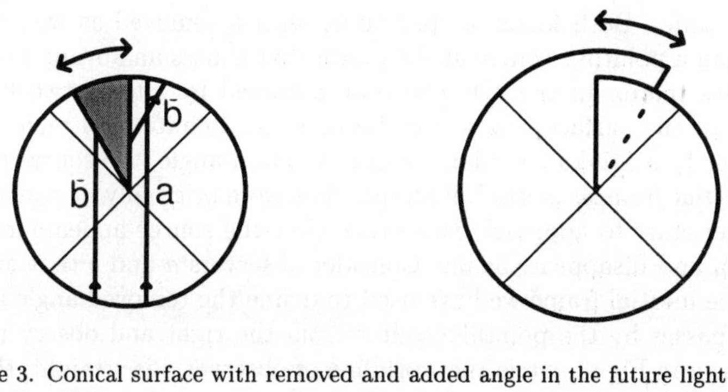

Figure 3. Conical surface with removed and added angle in the future light cone

light cone are the Milne coordinates (τ, ψ) (the proper time and the rapidity of an incoherent dust expanding from the origin) defined by $t = \tau \cosh \psi$ and $x = \tau \sinh \psi$. The Minkowski metric in these coordinates is

$$ds^2 = d\tau^2 - \tau^2 d\psi^2.$$

If an angle is removed from the light cone or added to it, the metric stays the same everywhere outside the origin but the range of the coordinate ψ changes. The sides of the removed or added angle are located at $\psi = 0$ and $\psi = -k$ where $k = \tanh v$.

Thus, instead of running smoothly from $-\infty$ to ∞ as in the original light cone, the coordinate ψ skips the interval between $-k$ and 0 if $v > 0$ and runs through that interval twice if $v < 0$. If we wish to restore the original interval of ψ, the rescaling is clearly of no help. The new coordinate has to be a nonlinear function of ψ approaching ψ rapidly enough at $\pm\infty$ or coinciding with ψ outside some interval, say

$$\Psi = \begin{cases} \psi \ \text{ if } \psi \leq -k \text{ and } \psi > M \\ \psi - k \ \text{ if } 0 < \psi \leq m \\ \psi + f(\psi) \ \text{ if } m < \psi \leq M \end{cases},$$

where f is a smooth function changing monotonically from $-k$ to 0 as ψ runs from m to M, with zero derivatives at both m and M. (If $v < 0$, the function $\Psi(\psi)$ is double-valued at $0 < \psi \leq -k$, however, the inverse function is single-valued provided $M > m - k$ and the function f is decreasing slowly enough. Note that in the limit $m = M$, the original angle between $-k$ and 0 is closed and a new angle between $m - k$ and m is opened.) Rewritten in terms of the Ψ-coordinate, the metric within the future light cone is

$$ds^2 = d\tau^2 - \tau^2\psi'^2 d\Psi^2, \tag{11}$$

where ψ is regarded as a function of Ψ. The entire surface is covered smoothly by the Lorentzian coordinates (T, X) given by $T = \tau\cosh\Psi$ and $X = \tau\sinh\Psi$ inside the future light cone and coinciding with (t, x) elsewhere.

Let us look at the way the distortion of the light cone affects metric properties of the spacetime. Both kinds of spacetime, with a removed as well as an added angle, contain a pointlike source at the origin that causes an abrupt change of metric properties there. In the 4D spacetime obtained by adding two flat spacelike dimensions to the surface, the source becomes an infinite wall mentioned in the introduction. In a Minkowski plane with a removed angle, the source causes a *collapse of inertial frames*: inertial observers that were originally at rest with respect to each other start to approach each other after the source appears for a moment among them and disappears again. Consider observers a and b that are originally at rest in the inertial frame we have used to define the removed angle and suppose observer a passes by the pointlike source from the right and observer b from the left. As shown in Fig. 3, while the worldline of observer a is straight, the worldline of observer b breaks down on one side of the removed angle and proceeds in a new direction on the other side. Let the observers compare their clocks, exchanging light signals when they are far from each other and looking at the dials when they meet, and suppose they measure their actual distance, too. To predict the results, nothing but the knowledge of the elements of the Minkowski geometry plus a little algebra is needed. Denote the distances of the observers from the pointlike source, measured by comoving rods, l_a and l_b. The proper time of the observers at their encounter, counted from the instant they pass by the source, is

$$\tau_{a0} = \frac{l_a\cosh k + l_b}{\sinh k}, \quad \tau_{b0} = \frac{l_a + l_b\cosh k}{\sinh k},$$

so that observer a finds his clock shifted with respect to the clock of observer b by

$$\Delta\tau_0 = \tanh\frac{k}{2}\,(l_a - l_b). \tag{12}$$

If the observers compare the ticking of their clocks by sending successive light signals to their counterpart, both find the other clock accelerated e^k times before the encounter and slowed down e^k times after the encounter. This is just the Doppler shift corresponding to velocities v and $-v$. By sending a light signal to the other observer and receiving it again, both observers can measure the synchronization time $\Delta\tau_{syn}$ (the shift of the other observer's clock with respect to their own clock) as well as the distance l to the other observer. The results are summarized in Table 1. The three columns of the table refer to the period before the pointlike source starts to affect the light signals, the period after the first signal is affected, and the period after both signals are affected. The regimes in the first and third periods may be interpreted simply: while in the first period the observers are at rest with respect to each other and the clocks they are equipped with are synchronized, in the third period they are moving with relative velocities $\pm v$ and their clocks are shifted by $\Delta\tau_0$ if compared in an inertial frame in which the observers move with opposite velocities.

Table 1. Synchronization times and distances on a Lorentzian conical surface

τ_b	$< -l_a$	$\in (-l_a, e^{-k}l_a)$	$> e^{-k}l_a$
$\Delta\tau_{syn}$	0	$\tanh\frac{k}{2}\,(\tau_b + l_a)$	$\tanh k\,(-\tanh\frac{k}{2}\,\tau_b + l_a)$
l	$l_a + l_b$	$-\tanh\frac{k}{2}\,\tau_b + \frac{e^{-k/2}l_a}{\cosh(k/2)} + l_b$	$-\tanh k\,\tau_b + \frac{l_a}{\cosh k} + l_b$

In a Minkowski plane with an added angle, the singularity leads to an *explosion of inertial frames*: inertial observers depart from each other, leaving an expanding empty region between them. The ticking of the clocks is Doppler-shifted again and the measurements of distances and synchronization times proceed in three regimes similar to those in a Minkowski plane with a removed angle, with the results obtained by changing the sign of k in Table 1.

A Lorentzian surface with a distorted light cone may contain a ridge, like the surfaces depicted in Fig. 4. This is the case if the sides of the removed or added angle cannot be boosted into each other. In the 4D spacetime obtained by adding two flat spacelike dimensions to the surface, the ridge becomes an infinite flat wall accelerating at different rates on the opposite sides. Consider a ridge formed from a removed angle with the sides diverging from each other. When passing through the ridge, extended bodies are subject to the action of tidal forces: a cloud of dust scatters and a rigid body is strained by forces pulling it from the centre to the edge. If the sides of the removed angle are curved strongly enough, the tidal forces

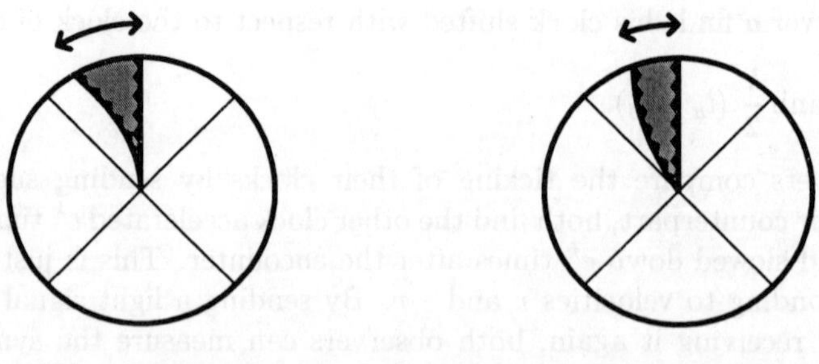

Figure 4. Conical surfaces with a ridge in the future light cone

overcome the molecular forces holding the body together and the body falls apart. The ridge then acts like a timelike singularity: if an observer decides to look at it closer, he gets crushed. Exploration of the ridge is thus similar to an expedition into a black hole, except that the observer can send signals to distant observers until the very last moment. However, since the left hand side of the removed angle departs from the time axis to infinity, the distant observers on the left get eventually crushed, too.

Let us now introduce surfaces with identified light cones. If an angle extending from the past to the future light cone is cut off of a Minkowski plane, only two of the four quadrants of the Minkowski plane survive: the light cone containing the past as well as the future of the origin and a spacelike region outside the light cone. Two surfaces of that kind are shown in Fig. 5, both with the angle deficit π, one of them smooth and one containing a ridge. The metric inside the identified light

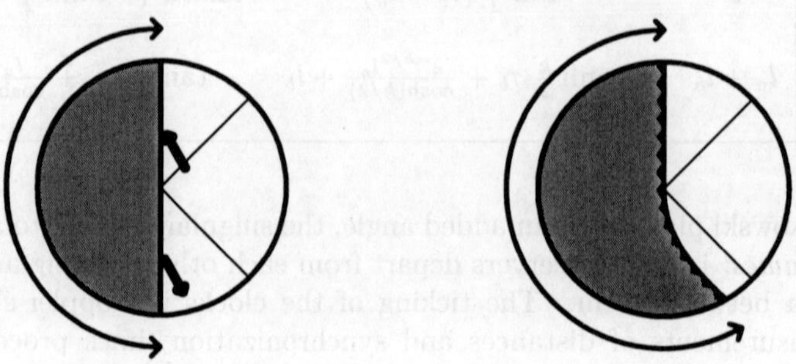

Figure 5. Conical surfaces with the past and future light cones identified.

cones is again of the form (11), however, it cannot be extended into the spacelike region as before. The coordinates covering smoothly the entire surface exist in this case, too, but they are Euclidean rather than Lorentzian.

Identifying the light cones provides us with a *nonorientable time*: an observer moving to the left and forward in time returns moving to the right and backward in time. He can meet his younger self and watch as he retreats (both ways) in time. If there is a strongly curved ridge across the identified light cones, the observer gets crushed before having such an extraordinary experience, however, time remains nonorientable for light signals and pointlike particles. One way out is to postulate that the spacelike region outside the identified light cones is forbidden for physical objects to enter. This reduces the spacetime to the 2D Milne universe. A less trivial possibility is to introduce a time-warping ridge along which the time arrow turns upside down as we pass from one side to the other. Such a ridge may be considered to be a kind of a wrinkled ridge that exists on Lorentzian surfaces only: while at an ordinary wrinkled ridge the metric tensor experiences a jump and the vector field defining the arrow of time is either continuous (on surfaces with a distorted light cone) or does not exist (on surfaces with identified light cones), at a time-warping ridge, the vector field defining the arrow of time jumps, too. If we identify the light cones by a time-warping ridge, we get rid of the nonorientable time, leaving closed timelike worldlines in the spacetime but changing their shape. Light signals and pointlike particles crossing the ridge shift back in time first and keep moving forward in time afterwards. Extended bodies must be torn apart when crossing the ridge due to the huge curvatures inside it, although the net tidal forces may be small or even zero (if the sides of the removed angle are straight). An observer thus can send himself a message from the future, but he cannot travel into the past.

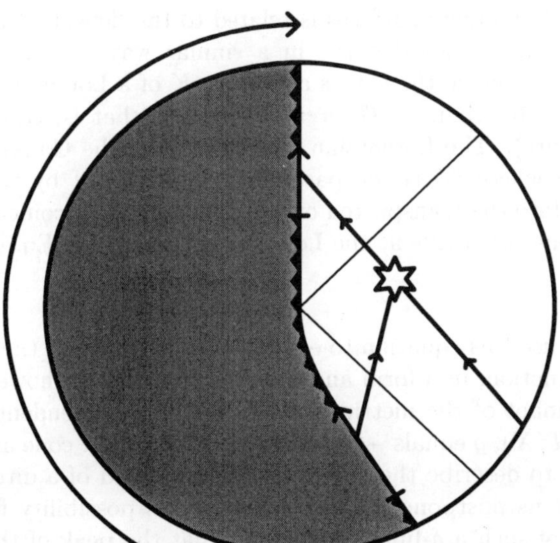

Figure 6. Time machine

If there is a time-warping ridge in the spacetime with identified light cones, we may straighten a segment of it and glue the surface along that segment in a Möbius-like way so that the upper point of the segment located on the upper side of the removed angle coincides with the upper point of the segment located on the lower side of the removed angle. This procedure, depicted in Fig. 6, yields a *time machine.* If an observer finds his way into the permeable segment, he can undertake a journey into the past and see his younger self, similarly as in a spacetime with no ridge. However, now they grow older together after they meet. The journey proceeds like in a wormhole with one important difference: unlike an observer crossing a wormhole, an observer returning in time through identified light cones emerges on the same side he has left. The same is true about any material object. A billiard ball thus collides with itself frontally after being transported back in time and cannot get into the time machine in the first place. Collisions of this kind are also unavoidable after extra dimensions are added and the ridge becomes an infinite wall. A ball heading directly towards the permeable shell in the wall cannot be hit by the later copy of itself from the side and get to the shell obliquely, since after crossing the shell it would recede from the place of collision. The time machine is therefore unacceptable even in an approach admitting multiple histories [8]: the temporal behavior of balls moving against the permeable shell is inconsistent so that for a large class of initial conditions the multiplicity of possible histories is zero.

4 Smoothing out the Lorentzian peak

The curvature of Lorentzian surfaces is related to the deflection of a vector transported parallel along a closed curve in a similar way as the curvature of Euclidean surfaces. Introduce the Gauss curvature K of a Lorentzian surface defined in terms of metric by Gauss's *Theorema egregium* (that is, equal to one half of the scalar curvature). The Lorentzian modification of the Gauss-Bonnet theorem reads $K_{int} = \delta\psi$ where $\delta\psi$ is the parameter of the boost by which the original vector is rotated into the transported one. Applying this to conical surfaces with a distorted light cone, we arrive at the Lorentzian analogue of Eq. (2),

$$K_{int} = k. \tag{13}$$

We would like to use this equation to express the curvature at the Lorentzian peak in terms of a δ-function, in a form analogous to (3). This is, however, not possible since the determinant of the metric tensor is direction-dependent at the peak: in the coordinates (T, X), g equals $-\psi'$ within the future light cone and -1 elsewhere. We may attempt to describe the curvature by some kind of a direction-dependent δ-function but let us postpone the discussion of this possibility for a while, since the introduction of such a δ-function requires that the peak of the surface is first rounded.

As mentioned in the introduction, in 4D spacetimes obtained by adding two flat spacelike dimensions to a Lorentzian surface, the only nonzero components of the energy momentum tensor are two lateral stresses of equal magnitude, the sign of which coincides with the sign of K. (They equal $K/(8\pi G)$.) On a Minkowski plane with a removed angle, K_{int} is positive so that K may be positive everywhere after the peak is smoothed out, while on a Minkowski plane with an added angle, K_{int} is negative so that K must be negative somewhere no matter what the smoothing looks like. Consequently, the source of the spacetime with collapsing inertial frames may satisfy the weak as well as the strong energy conditions while the source of the spacetime with exploding inertial frames violates these conditions inevitably.

If the original and the transported vectors point into the opposite light cones, a natural candidate for the modified Gauss-Bonnet theorem is $K_{int} = \delta\hat{\psi}$ where $\delta\hat{\psi}$ is the parameter of the boost by which the reflected vector is rotated into the transported one. However, if we try to fix K_{int} by the same argument as on a surface with the transported vector pointing into the same light cone as the original one, we fail. The argument is based on the fact that the Gauss curvature satisfies, by definition, the equation $K_{int} = \delta\psi$ in infinitesimal regions. Thus, to obtain K_{int} in a finite region, one has to sum up $\delta\psi$'s on facets of an infinitely fine-grained lattice and since the contributions of internal links cancel, one ends up with the total $\delta\psi$. In the argument, it is tacitly assumed that a field of dyads is defined throughout the region where K_{int} is computed, otherwise it would be impossible to assign definite values to $\Delta\psi$'s at the links of the lattice. On surfaces like those with identified light cones, no field of dyads exists, however. If we attempt to construct it we find that the dyads collapse into single vectors along the null lines beginning at the peak. On the other hand, we may use a field of dyads containing a jump along some sequence of links, provided the relative orientation of the dyads on the opposite sides of the links is the same everywhere. If we introduce reflected dyads along the sides of the removed angle we obtain $K_{int} = \delta\hat{\psi}$ as desired. However, another relative orientation of dyads, leading to another equation, would be equally acceptable, so that, in fact, nothing can be said about K_{int}. This is clearly due to the singularity of the dyad field at the peak. Again, we postpone further discussion until we get some idea how to round the peak of the surfaces in question.

The curvature of a Lorentzian ridge is given by Eq. (4) or (5) again, provided we identify κ_1 and κ_2 with minus the proper acceleration of the ridge sides, and d_1 and d_2 with the ratio of the proper time measured at the ridge sides to the coordinate time measured at the ridge. Obviously, we must replace g by $-g$ in (5), too. The sign of $\Delta\kappa$ is correlated with the form of the ridge in the same way as on Euclidean surfaces: $\Delta\kappa > 0$ means that the sides of the ridge diverge from each other and $\Delta\kappa < 0$ means that they converge to each other. Thus a regular (unwrinkled) ridge with diverging sides obeys energy conditions, while a regular ridge with converging sides does not. A wrinkled ridge violates energy conditions irrespective of the form of its sides since the sign of δ' is ambiguous. Finally, a time-warping ridge is presumably equally unable to satisfy energy conditions as a wrinkled ridge: the curvature connected with the reflection of the arrow of time is zero thus, after rounding the ridge, one must obtain curvature that is half positive and half negative.

A straightforward way to round the peak of a surface with a distorted future light cone is to pick out some function B depending on both τ and Ψ and write down the metric inside the future light cone in a form analogical to (6),

$$ds^2 = d\tau^2 - B^2 d\Psi^2, \tag{14}$$

leaving the metric within the remaining part of the surface unchanged. The function B is supposed to interpolate between $\tau\psi'$ at $\tau > \epsilon$ and $\tau - \delta$ at $\tau \sim \delta$. The curvature is identical to that of Eq. (7), provided the prime is understood as a partial derivative with respect to τ. (The fact that B also depends on Ψ does not affect K.) If B is chosen concave everywhere, which is presumably always possible for surfaces with a removed angle, the source of the 4D geometry fulfills the weak as well as the strong energy conditions. Inside the future light cone, the directional curvature is

$$K_{int}(\Psi) \equiv \int_{\delta}^{\epsilon} K\sqrt{-g}\, d\tau = 1 - \psi',$$

which suggests that the curvature of the sharp peak is

$$K = (\Psi' - 1)\,\frac{\delta(\tau)}{\tau}. \tag{15}$$

The expression on the right hand side is a direction-dependent δ-function with the support at the origin, located entirely within the angle $m < \psi < M$ of the future light cone.

The smoothing procedure proposed above has to be corrected in one important point: one cannot restrict the smoothing to a single quadrant of the distorted Minkowski plane, similarly as one cannot round the peak of a Euclidean cone leaving a part of it untouched. Radial lines in the distorted light cone are cut short on their way to the origin, hence the radial lines in the remaining quadrants have to be cut short, too. (On a Lorentzian plane, "to cut a path short" means "to walk longer", therefore radial lines should meet at smaller values of their affine parameters, if these parameters grow in the direction away from the origin. Particularly, the value of δ should be negative rather than positive as on Euclidean surfaces.) To match the radial lines, we must smooth out the surface in a star-shaped region with four infinitely prolonged tips along the null lines, not just in an infinite strip adjacent to the null lines inside the distorted light cone. The additional deformation of the surface should affect neither $K_{int}(\Psi)$ of the rounded peak, nor K of the sharp peak. However, it affects the nature of the source: after the remaining quadrants are deformed, the 4D geometry is generated by two colliding walls rather than by two walls appearing at some instant at the same place and departing from each other afterwards.

A surface with a distorted light cone may be also smoothed out inside a compact region, similarly to Euclidean conical surfaces. The smoothed-out metric may be constructed, this time explicitly, by embedding the surface into 3D. Introduce a 3D Minkowski space with spacelike axes ξ and η and timelike axis ζ and define the conical surface by $\zeta = \pm R/\rho$ as before. The surface breaks up into two parts separated by null lines: the timelike region at $\rho < 1$ and the spacelike region at $\rho > 1$. In the latter region, the surface reduces to the 2D Rindler space with the synchronized time χ given by an equation similar to (8),

$$\chi' = \frac{1}{\rho_-}\sqrt{\frac{\rho'^2}{\rho_-^2} - \rho^2}, \tag{16}$$

where $\rho_- = \sqrt{\rho^2 - 1}$. As seen from this equation, the metric in the spacelike region has the right signature only if the curve $R = \rho(\Phi)$ fits into radially oriented 2D cones with the vertex angle equal to $2\arctan(1/\rho_-)$. This excludes smooth closed curves, hence a conical surface must cross the plane (ξ, η) and/or have corners. Particularly, we may represent surfaces with a distorted future light cone as corners introduced in Section II. By rounding the peak of a corner in the same way as in Section II, we arrive at Eq. (10) with a reversed sign in the denominator. This time, the curved part of the surface is not an infinite star but a finite rectangle and, therefore, the source of the geometry in 4D is a wall appearing for a short time and disappearing again. For a Minkowski plane with a removed angle, the broken side of the corner points inwards, hence the sides of the rectangle may be chosen to bend inwards, too. As a result, both functions $\mathcal{U}$ and $\mathcal{V}$ are convex, the curvature is positive everywhere, and the source of the 4D geometry satisfies the weak as well as the strong energy conditions. The directional curvature is smeared over all quadrants, thus differing from the directional curvature obtained earlier and the curvature of the sharp peak is not given by Eq. (15) anymore. The δ-function must depend on the direction somehow; it must at least contain some kind of a sign indicating which quadrant is distorted. This dependence, however, does not seem to be localizable in a unique way (similar to the gravitational energy in general relativity: any connection?).

The last topic to be discussed is the rounding of the peak of a surface with identified light cones. We cannot embed such a surface into 3D since after the generating lines on the opposite sides of the surface leave the 3D light cone, they cannot be matched without spoiling the signature. We can neither use a deformed metric of the type (14) inside the light cone; in addition, we cannot use an analogous metric outside the light cone since two quadrants of a Minkowski plane far from the origin are not enough to produce a complete Minkowski plane at the origin. An alternative still exists: we may glue two surfaces at their peaks by a sort of wormhole, rather than round the peak of a single surface. Consider a pair of Lorentzian conical surfaces with an angle deficit π at the peak. To glue the surfaces together, introduce polar coordinates (r, ϕ) defined by $t = r\cos\phi$ and $x = r\sin\phi$, with the angle ϕ ranging from 0 to π and multiply the metric coefficient g_{rr} by a smooth function $\zeta(r)$ decreasing from infinity to 1 as r runs from some δ (the neck of the wormhole) to certain ϵ (the mouth of the wormhole). The metric is

$$ds^2 = \cos 2\phi\, (\zeta dr^2 - r^2 d\phi^2) - 2r\, \sin 2\phi\, dr\, d\phi, \tag{17}$$

and the curvature

$$K = -\cos 2\phi\, \frac{r\zeta' + 4\zeta(\zeta - 1)}{2r\, (\zeta \cos^2 2\phi + \sin^2 2\phi)^2}. \tag{18}$$

The curvature is half positive and half negative so that in the 4D version of the geometry, the weak energy condition is violated. The peaks are not, in fact, smoothed out completely since the curvature diverges at $\phi = \pi/4$ and $3\pi/4$ at the neck of the wormhole. As we cross these points in the direction of increasing ϕ, the light cone turns back at a right angle, changing the direction with respect to the neck of the wormhole from perpendicular to parallel at $\phi = \pi/4$ and from parallel to perpendicular at $\phi = 3\pi/4$. The angle of the light cone rotation with respect to the radial lines starting from the singular points is, at both sheets of the surface, $-3\pi/2$ so that the singular points are conical, with the full angle around them equal to 3π. (At both points, one half of the Minkowski plane is added to the surface.) The conical nature of the singular points can be demonstrated most simply if we choose

$\zeta = \delta/[4(r-\delta)]$ since then the curvature at the neck of the wormhole is zero everywhere except at the singular points. Introduce coordinates $\mu = \pm 2^{1/3}\delta^{1/6}(r-\delta)^{1/2}$ and $\nu = \mp 2^{1/3}\delta^{2/3}\Delta\phi$ in the neighbourhood of the singular points, with the two signs in the former expression corresponding to the two sheets of the surface and the two signs in the latter expression corresponding to the two singular points. The metric in the coordinates (μ, ν) is

$$ds^2 = \nu\,(d\mu^2 - d\nu^2) + 2\mu\, d\mu\, d\nu,$$

and it reduces to the Minkowski metric if we put $(\mu, \nu) = \rho^2/2\,(\sin 2\chi, \cos 2\chi)$ and $(t, x) = \rho^3/3\,(\sin 3\chi, \cos 3\chi)$. The angle χ ranges from 0 to π, thus the full angle at the origin is 3π.

Two singular points at the neck of the wormhole shrink to one if we deform the cosine in g_{rr} and $g_{\phi\phi}$ to some function $\mathcal{C}(r, \phi)$ that equals $\cos 2\phi$ if $r \geq \epsilon$, increases monotonically from -1 to zero at $\phi = \pi/2$ as r runs from ϵ to δ, and approaches 1 as r approaches δ, say,

$$\mathcal{C} = \frac{1+u}{2}\cos(\pi \mp |2\phi - \pi|^u) + \frac{1-u}{2},$$

where the upper and the lower signs correspond to $\phi < \pi/2$ and $\phi > \pi/2$ respectively and $u = (r-\delta)/(\epsilon-\delta)$. The singular point at $\phi = \pi/2$ is conical again, with the full angle around it equal to 4π. (The surface consists of two joined Minkowski planes there.)

The integral curvature of the wormhole with two singular points at the neck, computed outside an infinitesimal strip connecting the singular points, is zero. As for the contribution of the strip, it is reasonable to suppose that if a vector coincides with itself after being transported around some region, the integral curvature of that region is zero. Unlike on Euclidean surfaces, the rotation of the vector by an integer multiple of 2π does not count, since there is no dyad field with respect to which it could be determined. This suggests that K_{int} on the strip, as well as on the entire surface, is zero. The same result is obtained if we start with a limit-case wormhole with one singular point at the neck. Throughout the wormhole, we may introduce a dyad field that turns upside down along a straight line passing from one sheet of the surface to the other, say, along the time axis in the coordinates (t, x). Using the argument presented at the beginning of this section, we find that the value of K_{int} of the entire surface, except the facet containing the singular point, is zero. However, a vector coincides with itself if transported parallel around the singular facet, hence K_{int} remains zero also after this facet is taken into account. (On a Euclidean version of the surface, the metric is regular everywhere but the dyad field may be again chosen with a jump along a line crossing the wormhole and a singular point at the neck of the wormhole. The singular facet then contributes -2π to K_{int}.) The result may be generalized to any surface confined between two boundaries at which vectors change orientation after being parallel transported. Introducing a dyad field with a constant jump along some line connecting the boundaries, we find that K_{int} equals $\delta\hat{\psi}_1 + \delta\hat{\psi}_2$ where $\delta\hat{\psi}_1$ and $\delta\hat{\psi}_2$ are the values of $\delta\hat{\psi}$ at the boundaries. Particularly, if two surfaces with identified light cones are glued together, the value of K_{int} of the resulting surface is

$$K_{int} = \hat{k}_1 + \hat{k}_2, \tag{19}$$

where $\hat{k}$ is the parameter fixing the size of the removed angle, defined as the parameter of the boost by which the reflected future side of the removed angle rotates into the past side.

We may try to deform the metric (17) in such a way that $g_{\phi\phi}$ becomes negative along the neck everywhere so that the whole neck will be spacelike and there will be no need for the light cones to turn over abruptly. However, when doing so, we inevitable make both $g_{\phi\phi}$ and $g_{r\phi}$ vanish at a certain point between $\phi = \pi/4$ and $3\pi/4$. The net effect is that we push the couple of singularities from the neck into the bulk of the wormhole, placing them at the opposite sides of the neck. This suggests the validity of a *singularity theorem for conical spacetimes*: if the light cone turns upside down in the direction of circulation at the boundary, or boundaries, of a compact region in a 2D spacetime, there must be a singularity somewhere within that region.

5 Conclusions

As we have seen, exploring Lorentzian conical surfaces gives rise to a whole range of unusual properties of a spacetime, such as the collapse and explosion of inertial frames, disintegration of observers by a wall accelerated at a different rate on its opposing sides, motion of physical objects backward in time after having crossed a plane where an infinite wall appeared for a moment and disappeared again, and shift of physical objects back in time as they pass through a shell inside an infinite impenetrable wall. The first two items on the list, concerning spacetimes with a distorted future light cone, are compatible with the general notion of the way gravity sets stage for physical processes, however, the other two, concerning spacetimes with identified light cones, contradict this notion strongly. We have already noticed that a time machine based on identified light cones is unacceptable even from the point of view admitting such machines in principle. The motion backward in time is equally, if not more, pathological. A special kind of such motion was considered by Feynman when he proposed a mixture of retarded and advanced potentials with equal strengths, but in this theory, the ability to move back in time was restricted to the propagation of interaction and Feynman abandoned it shortly afterward anyway.

The spacetimes proposed here, whatever objections may be raised against them, would still deserve a second thought if the theory provided realistic sources for them. As stressed in the introduction, this is not the case. All these spacetimes violate the energy dominance condition and all the spacetimes but those with a removed angle inside the future light cone, possibly containing a ridge with diverging sides, violate the weak energy condition as well. Thus, the theory reduces to pure kinematics with no dynamics supporting it. Such a theory may be motivated by a wish to explore all possibilities that are allowed kinematically, including the impossible ones. What if under some circumstances that we cannot imagine at the moment the impossible becomes possible? When studying Lorentzian wormholes, one has already allowed for the violation of the weak energy condition, referring to a possible effect of vacuum fluctuations of matter fields. The energy dominance condition is a stronger version of the weak energy condition, implemented to guarantee the conservation of matter. In the spacetimes considered here, it is violated because we have used the simplest possible extension to 4D, adding two flat dimensions to the original surface. In such spacetimes, the energy dominance condition does not hold due to a somewhat accidental fact that the Einstein tensor is zero in 2D. We may use more sophisticated extensions, such as a product of a Lorentzian conical surface with a 2-sphere. However, the energy dominance condition would presumably be violated in such spacetimes, too, since a 4D spacetime must inherit the peak where matter is not conserved from the Lorentzian surface it comes from. Note that in a conical spacetime with a pointlike source obtained by multiplying a surface with a removed angle in the future light cone by a 2-sphere, energy conditions must be violated even more severely than in a conical spacetime with a wall-shaped source since the boundaries of the removed region form a wrinkled ridge with an indefinite, rather than zero, energy density. The fact that matter in the spacetimes proposed here is not conserved may be regarded as a shortcoming that by itself makes the theory liable to refutation. Still, one may speculate that the matter appears and disappears again due to some kind of tunneling of a matter field "from nothing to nothing". When inside a barrier, the field might act effectively as classical matter with an indefinite energy density. A barrier for tunneling via an infinite wall should be infinitely high, prohibiting possible creation of the wall, but a pointlike singularity may be produced, in principle, inside a *finite* barrier, hence with a finite probability.

Acknowledgements

I am grateful to Jiří Bičák who directed my worldline to general relativity. In this work, I used much of what I have learned from him about strange things that may happen to space and time.

References

1. A. Vilenkin and E. P. S. Shellard: *Cosmic strings and other topological effects*, Cambridge University Press, Cambridge (1995).
2. A. Vilenkin, *Phys. Rev.* D **23**, 852 (1981).
3. D. D. Sokolov and A.A.Starobinskii, *Sov. Phys. Dokl.* **22**, 312 (1977).
4. J. R. Gott III, *Ap. J.* **288**, 422 (1985); W. A. Hiscock, *Phys. Rev.* D **31**, 3288 (1985); B. Linet, *Gen. Rel. Grav.* **17**, 1109 (1985).
5. S. W. Hawking and G. R. F. Ellis: *The large scale structure of space-time*, Cambridge University Press, Cambridge (1973).
6. L. Carroll: *The Hunting Of The Snark - An Agony In Eight Fits*, in *The Complete Stories and Poems of Lewis Carroll* (Children's Leisure Products Ltd, David Dale House, New Lanark, 2001).
7. C. J. S. Clarke, J. A. Vickers and J. P. Wilson, *Class. Quantum Grav.* **13**, 2485 (1996); F. Dahia and C. Romero, *Mod. Phys. Lett.* **A14**, 1879 (1999).
8. F. Echeverria, G. Klinkhammer and K. S. Thorne, *Phys. Rev.* D **44**, 1077 (1991).

THE SEMICLASSICAL EINSTEIN EQUATIONS

TOMÁŠ KOPF

Matematický ústav Slezské univerzity v Opavě, Bezručovo náměstí 13, 74601 Opava, Czech Republic

The semiclassical Einstein equations are reviewed with special attention to the stress-energy tensor appearing in them. A functional approach to the renormalization of the stress energy tensor is suggested.

1 The quantum challenge to general relativity

General relativity is quite true with its remarkably coherent mathematical structure justified by an amazing collection of observations [1].

As a theory of nature, it does not stand alone though: Quantum field theory is also quite true with its as yet remarkably incoherent mathematical structure justified by an even more amazing collection of experiments [2].

But the two good truths do not fit together easily as has been known for the better part of the last century.

The problem is that *quantized* matter of quantum field theory cannot consistently interact with the *classical* spacetime geometry of general relativity while, based on the success of Einstein's equations, it is certain that an interaction between matter and spacetime exists.

It is commonly accepted, that a resolution of this problem can be achieved by seeing general relativity as the classical approximation to a quantum theory and there have been bold attempts to find such a fundamental quantum gravity theory, in particular the string theory and non-perturbative quantum gravity.

While work on fundamental theories is still in progress, there is an intermediate possibility of staying close to the two experimentally verified theories and discussing their interplay in approximations that do not require a full understanding of quantum gravity. Such considerations can give us hints on essential parts of the fundamental theory, in particular on its physical interpretation.

One such approximation is the quantum field theory on a curved background. This theory is considered here.

Outline

Quantum field theory on a curved background will be reviewed in Section 2. This section will also include a discussion on conceptual problems with back reaction of the field on spacetime.

The technical problems of back reaction including a reasonable definition of the stress-energy tensor are given in Section 3. A functional approach to these questions is advocated in Section 4. The limits of the present status of the functional approach are discussed in Conclusion.

As there is a vast body of literature on this subject, we will restrict ourselves here to a small and necessarily incomplete selection of citations only. A number of technical issues that are only mentioned here can be found in the books [3–5].

2 Quantum field theory on a curved background

On a fixed curved background manifold, a quantum field theory may be obtained by the quantization of a classical field theory.

As it turns out, the complications of an interacting field theory are not necessary to arrive at a substantial part of the emerging problems and will therefore be avoided here. Restricting thus to a free field theory, the classical space of solutions of the field equation (the phase space) is linear and can be quantized straightforwardly by imposing canonical commutation or anticommutation relations on linear observables.

Another potential difficulty to be brushed aside is the question of how to deal with spacetimes containing closed timelike curves (see [6] which addresses this problem). The background spacetime will always be assumed to be globally hyperbolic. This makes a definition of localized fields as solutions of the field equation more feasible. They can be given by localized initial data in two ways: Either by localizing the support of the usual initial data for the hyperbolic equations of motion within a spacelike Cauchy surface or (more generally) by associating with a function f with localized spacetime support a solution $(E_+ - E_-)f$ through the advanced Green's function E_+ and the retarded Green's function E_-. The well-definedness of the Green's functions E_+ and E_- is secured by the imposed assumption of global hyperbolicity. Within the framework of these two schemes, one can define field operators $\phi(\vec{x})$ localized at a spatial point $\vec{x}$ of a Cauchy surface Σ or field operators $\phi(x)$ localized at a spacetime point x as distributions with values in the field algebra $\mathcal{A}$.

A further simplification from the functional point of view is gained if the spatial Cauchy surfaces are required to be compact.

After cutting down so drastically on possible complications, there is an easy-to-obtained, well defined field algebra $\mathcal{A}$.

But still, a substantial problem remains: There is no vacuum state. In a generically curved background, there is, in contrast to the Minkowski spacetime, no group of spacetime isometries to single out a vacuum and the somewhat more general determination of a vacuum based on the splitting of solutions into positive and negative energy parts breaks down (unless the spacetime is static), since time evolution will mix the positive and negative energy solutions and thus what would seem at one time to be the vacuum would not be the vacuum at another time. Without a preferred vacuum there is also no preferred particle interpretation. To make matters worse, vacua determined at different times by an unsuitable description (e.g., as the ground states of instantaneous Hamiltonians) will, in general, determine Fock representations that are unitarily inequivalent, i.e., belong to different folia of states.

In an amazing historical process, a solution to the problem of the correct vacuum in quantum field theory has crystallized within a period of over thirty years. It contains the following ingredients:

1. It simply has to be accepted that there is no preferred vacuum in general and any experimental notion of a particle has to be supplemented by an explanation on how it is measured [7]. Of course, in many interesting situations, the

existence of special spacetime symmetries or of asymptotically flat regions may still provide for a preferred and useful vacuum. But this is not to be seen as something fundamental.

2. Only a special subset of states of the field algebra $\mathcal{A}$ is to be considered physical. These states $\omega : \mathcal{A} \to \mathbb{C}$ are characterized by satisfying the so-called Hadamard condition on their two-point function $\omega(\phi(x_1)\phi(x_2))$, which will be discussed in the next section. These states have a well-defined expectation value of the stress-energy tensor and under a restriction to a compact space they can all be given by density matrices in a single representation of the field algebra on a Hilbert space, the physical Hilbert space. Thus there is only a preferred Hilbert space instead of a preferred vacuum.

So far, the discussion was restricted to a fixed spacetime prescribed in advance. But the results will be essential in a more complicated situation of the following section concerned with the semiclassical Einstein equations.

3 The semiclassical Einstein equations and the expectation value of the stress-energy tensor

This section is concerned with the back reaction of a quantum field on a classically described spacetime. This mixture of quantum and classical concepts is called the semi-classical approximation.

Remark 1. The physical relevance of a quantum field given on a prescribed space-time background, as described in the previous section, can be justified by giving the field the status of a weak test field. The geometry of spacetime is not influenced by the quantum field at all and will not enter into any quantum correlations with the matter field.

This changes radically, once the back reaction is allowed. At any time, the quantum field may develop a superposition of states correlated with different spacetime geometries. E.g., significant mass may be swung into different positions, depending on the outcome of a quantum experiment on the field, producing thus correlations of the field with macroscopically different spacetime geometries and leading to a breakdown of the semiclassical picture [8]. Thus, the semiclassical approximation is useful for special states only and avoids the earlier mentioned problems of inconsistency of quantum fields back reacting to a classical geometry by refusing to deal with troubling situations where quantum geometry becomes essential.

One may even wonder whether there is any chance of the semiclassical description being useful at all. The answer, hiddenly illustrating the lack of a *theoretical* understanding of this issue, is the *practical* observation that a classical spacetime together with quantum fields seem to be around us to a high degree of precision.

The classicality of this spacetime which allows to make contact with the semiclassical approximation is hardly an accident but rather seems to be enforced by decoherence-generating mechanisms that would have to be seen from the point of view of a closed quantum system. However, such a justification seems not to be fully developed yet.

Under the assumption that a given state is appropriate for the semiclassical approximation, the classical geometry, represented in the Einstein equations by the Einstein tensor $G_{\mu\nu}$ and by a cosmological term $\Lambda g_{\mu\nu}$, can be coupled to the expectation value $\langle T_{\mu\nu}\rangle$ of the stress-energy tensor $T_{\mu\nu}$:

$$G_{\mu\nu} + \Lambda g_\mu \nu = 8\pi \langle T_{\mu\nu}\rangle. \tag{1}$$

This is the semiclassical Einstein equation. Unfortunately though, the expectation value $\langle T_{\mu\nu}\rangle$ of the stress-energy tensor $T_{\mu\nu}$ is singular and, as it stands, Equation (1) does not make sense. To save it, the singular part of the stress-energy tensor has to be removed — the stress energy tensor has to be renormalized. This can be done systematically and covariantly under the assumption that the stress-energy tensor has a certain prescribed singularity structure as expressed by the Hadamard condition. The Hadamard condition has been known in the form of a rather involved requirement but has been recently given a simple, local reformulation due to Radzikowski [9] and based on the work of Hörmander and Duistermaat [10]. Since it rests on the notion of a wave front set, the following explanation will be given first:

Explanation 1. *The* wave front set *of a distribution is the set of all directions in the cotangent space over all points x of the spacetime manifold in which its Fourier transform (after localization by a function with support within a small neighborhood around x) does not decay at infinity faster than any polynomial.*

The wave front set refines the singular support of a distribution by providing in addition to the points x in which the distribution is singular the set of co-directions over those points in which the distribution is singular.

The wave front set of a two-point function satisfying a hyperbolic equation in both entries only consists of pairs of points that can be joined by light geodesics and of pairs of co-directions in these points being cotangent to that geodesic up to a sign parallel translates of each other.

This leaves only two possible wave front set directions for a future directed light geodesic (for future directed propagation) joining the first point of the two-point function with the second: The direction cotangent with respect to the geodesic and of positive energy and the opposite direction, i.e., the direction cotangent to the geodesic and of negative energy.

The Hadamard condition can then be given as a generalization of the positive energy condition of quantum field theory on Minkowski space and can also be seen as a generalized definition of the Feynman Green's function with the appropriate conditions not required from the two-point function as such but from its wave front set only:

Definition 1. *A state ω on the quantum field with local field observables $\phi(x)$ is* Hadamard *if its two-point function $\omega(\phi(x_1)\phi(x_2))$ has in its wave front set only positive frequencies propagating forward in time and negative frequencies propagating backward in time.*

The two-point function of any Hadamard state can then be renormalized by subtracting a prescribed singularity. The subtracted part has to be constructed on the basis of the geometry of spacetime only, without direct reference to the state of the quantum field, so that a number of conditions are satisfied [11]:

1. Covariance.
2. Causality with respect to changing the geometry of spacetime in the past or in the future.
3. Matrix elements for orthogonal states should give the correct results without any renormalization.
4. The recovery of results obtained in the special case of quantum field theory on Minkowski space.

A number of technical tools to achieve this are available [3, 4] .

Both the construction of Hadamard states and their renormalization are somewhat involved, making it difficult to deal even with spacetimes with a high degree of imposed symmetries. Considerable attention in this respect has been given to Robertson-Walker spacetimes [12, 13].

4 A functional approach to the renormalization of the stress-energy tensor

There are certain parallels of the renormalization of the stress-energy tensor with the regularization of traces. A well defined local part has to be taken out of an otherwise non-local object. It is therefore natural to ask, whether there exists a functional approach to the renormalization of the stress-energy tensor in the same way as there are functional approaches to the regularization of traces.

There is a second reason to approach the situation from a functional point of view:

It may happen that the semiclassical approximation will be valid for the description of a classical spacetime together with a quantum field outside of the event representing a purely quantum measurement. The result may be what would be understood within the many-world interpretation as a splitting of spacetime into branches. The classical idea of a spacetime manifold is clearly somewhat compromised by the splitting.

What remains unchanged even during the purely quantum part of evolution though is the field algebra of the quantum field. Within the field algebra, spacetime is present in some way as a choice of localized observables. These may be given directly through field operators or through local field algebras within the Haag-Kastler framework [14].

The splitting of spacetime into branches would then amount to different choices of local observables running side by side. Under the assumption that, in addition to all the simplifications introduced above (the fields are free and evolve on a globally hyperbolic manifold with compact spatial Cauchy hypersurfaces), the considered states are pure and quasifree (i.e., given entirely by their two-point functions), all structures are reflected without loss of information on the one-particle Hilbert space which may be constructed from the phase space — the classical space of solutions of the field equations. This allows to simplify the necessary treatment considerably.

The following data might be given:

- **A description of the spacetime as encoded in the classical phase space of the field.** This can be done by representing coordinates on spacetime hypersurfaces in the phase space. Additional structures have to be provided to describe the hypersurfaces properly. This can be achieved using the spectral geometry data of A. Connes [15] and binding the hypersurfaces, understood as evolving within a Hamiltonian framework [16], together through their causal relations [17] and through a time vector. For requirements on these structures and for an example, see [18].

- A fixed projector P splitting the phase space into positive and negative frequencies. This projector should be unique up to the addition of smoothing operators. Which operators are smoothing is to be determined by the Hamiltonian $H(t)$.

- Smooth time evolution. The Hamiltonian is required to commute with the projector P at all times, up to smoothing operators.

- The Hamiltonian evolution is to be such as to allow the construction of the geometric local singular part of the two-point function to be used in a renormalization prescription of the stress-energy tensor.

Remark 2. The projector P determines the state of the quantum field. The condition that it commutes with the Hamiltonian up to smoothing operators allows for the mixing of positive and negative energy solutions and, if a scattering picture is available, for the creation of particles due to curvature in such a way that the Hadamard condition is not violated by the evolution. The last requirement is to be understood as a safety measure and should be replaced in the future.

5 Conclusion

The semiclassical Einstein equations may be useful even in cases where the evolution of matter and geometry leads to the splitting of the spacetime into branches.

A possible way to describe such effects is to encode the spacetime into the matter quantum field algebra. Compatibility conditions between the positive-negative energy splitting and the Hamiltonian evolution of the geometry provide interesting functional conditions and require further examination.

The usefulness of the semiclassical approximation depends significantly on whether semiclassical states commonly appear in nature. Their presence can possibly be facilitated by decoherence generating mechanisms based on the locality of interactions and the post-interaction propagation in the field theory involved. Here, we made simplifying assumption that the matter field under consideration was free, and the model is therefore bound to miss this important source of proper interpretation.

6 Acknowledgements

The author is thankful for support from grants MSM: J10/98:192400002 and GAČR 201/00/0724

References

1. C. M. Will: *The Confrontation between General Relativity and Experiment*, Living Reviews in Relativity, Albert-Einstein-Institut, Max-Planck-Institut für Gravitationsphysik, Potsdam.
2. D. E. Groom et al: *The review of particle physics*, Europ. Phys. Jour. C15 (2000) 1.
3. N. D. Birrell and P. C. W. Davies: *Quantum Fields in Curved Space*, Cambridge University Press, Cambridge, 1982.
4. S. A. Fulling: *Aspects of Quantum Field Theory in Curved Space-Time*, Cambridge University Press, Cambridge, 1988.
5. R. M. Wald: *Quantum Field Theory in Curved Spacetime and Black Hole Thermodynamics*, University of Chicago Press, Chicago and London, 1994.
6. U. Yurtsever: Class. Quantum Grav. **11** (1994) 999.
7. P. Hájíček: Phys. Rev. **D 15** (1977) 2757.
8. D. N. Page and C. D. Geilker: Phys. Rev. Lett. **47** (1981) 979.
9. M. J. Radzikowski: Commun. Math. Phys. **179** (1996) 529.
10. J. J. Duistermaat and L. Hörmander: Acta Math. **128** (1972) 183.
11. R. M. Wald: Comm. Math. Phys. **54** (1977) 1.
12. L. Parker: Phys. Rev. **183** (1969) 1057.
13. C. Lueders and J. E. Roberts: Commun. Math. Phys. **134** (1990) 29.
14. R. Haag and D. Kastler: J. Math. Phys. **5** (1964) 848.
15. A. Connes: Commun. Math. Phys. **182** (1996) 155.
16. E. Hawkins: Commun. Math. Phys. **187** (1997) 471.
17. T. Kopf: Int. J. Mod. Phys. **A13** (1998) 2693.
18. T. Kopf and M. Paschke: J. Math. Phys. (to appear); math-ph/0012012.

BOUNDARY QUANTUM MECHANICS

PAVEL KRTOUŠ
Institute of Theoretical Physics,
Faculty of Mathematics and Physics, Charles University,
V Holešovičkách 2, 180 00 Prague 8,
Czech Republic
e-mail: Pavel.Krtous@mff.cuni.cz

A reformulation of a physical theory in which measurements at the initial and final moments of time are treated independently is discussed, both on the classical and quantum levels. Methods of the standard quantum mechanics are used to quantize *boundary phase space* to obtain *boundary quantum mechanics* — a theory that does not depend on the distinction between the initial and final moments of time, a theory that can be formulated without reference to the causal structure. As a supplementary material, the geometrical description of quantization of a general (e.g. curved) configuration space is presented.

Introduction

Motivation

In this work we formulate *boundary quantum mechanics* — a modification of the standard quantum mechanics where states at the initial and final moments of time are treated independently.

Our primary motivation for building the boundary quantum mechanics was an observation that in a field theory one usually has to define similar structures (such as e.g., Fock bases, vacuum states, coherent states, field and momentum observables) both for the initial and final moments of time and these structures only differ in the time at which they are defined. An effort to simplify handling of different structures at the initial and final time led to a unified picture in which both time boundaries play a completely equivalent role.

The main idea is to treat both the initial and final states of a physical system in an independent way, as if they would be states of different systems. The Hilbert space of boundary quantum mechanics is thus given as a tensor product of the initial and final Hilbert spaces. The dynamics of the boundary quantum mechanics is given by specifying a special *physical state* $|phys\rangle$ that contains all information about dynamical correlations.

Surprisingly, the resulting theory does not need to distinguish between the initial and final states — it can be formulated in a way in which we do not need to split the boundary of a (space)time domain on which we study the system to the initial and final parts. The consequence of this fact is that we can use the formalism of boundary quantum mechanics in situations when we do not have a reasonable causal structure which would allow us to identify the "initial" and "final" moments of time, particularly, we can use the formalism in the Euclidian form of a theory.

The material presented here is based on some parts of Ref. [1]. The work [1] was concerned mainly with a study of the relation between quantizations of a scalar field theory and relativistic particle theory. Boundary quantum mechanics was used for an alternative description of the scalar field theory. The general framework leading to boundary quantum mechanics was scattered in several places in Ref. [1]. Here, the material is presented in a modified and more compact form with an emphasis on the construction of the general formulation of the boundary mechanics.

After boundary quantum mechanics was constructed and used in Ref. [1] the author has found out that similar ideas have been pursued in the context of quantum gravity [2, 3], however, in these works the idea of treating the different moments of time independently has been extended much farther — here "all moments of time" are treated independently, hence, the constructed Hilbert space of the theory is given by some kind of a continuous tensor product of Hilbert spaces for each time.

Plan of work

The explanation of the formalism is divided to three chapters. In Chapter 1 a classical analogue of boundary quantum mechanics is constructed, namely the boundary phase space is defined and its connection to standard phase spaces is studied. This chapter formulates the theory on a very general level. It allows to define the boundary phase space without a reference to the causal structure. However, these details are not necessary for the following chapters — if one is not interested in the causal (in)dependence of the boundary phase space, it is sufficient to understand the boundary phase space on the level explained in the overview and summary of Chapter 1. The way of treating the general dynamical theory has been mainly inspired by Ref. [4].

The material presented in Chapter 2 is actually independent of the main subject — of boundary quantum mechanics. In this chapter we present a geometrical formulation of quantization for a system with a phase space with cotangent bundle structure $\mathbf{T}^*\boldsymbol{\mathcal{V}}$ over some configuration space $\boldsymbol{\mathcal{V}}$. This is a generalization of the standard quantum mechanical methods of quantization of the position and momentum variables to the situation when the space of "positions" $\boldsymbol{\mathcal{V}}$ does not have a linear structure — when it is a general manifold. Only a construction of "position" and "momentum" observables and their position representation is presented here, dynamical questions are not discussed.

Finally, boundary quantum mechanics is formulated in Chapter 3. First, the theory is built on the basis of the usual quantum mechanics. Afterwards, it is shown that boundary quantum mechanics can be constructed immediately from the boundary phase space without reference to quantum mechanics at a given time. At the end the issue of dynamics is discussed.

The main text is supplemented by three appendices in which some of the geometrical notions are reviewed. Appendix C defines the notion of densities of a general (complex) weight on a manifold. Appendices A, B contain a geometrical formultaion of symplectic geometry. Most of this material is well known (see, e.g., [5, 6]) and it is included mainly to fix the notation and remind the reader of properties of different introduced objects. However, let us note that Appendix B also defines *covariant partial derivatives* on tangent and cotangent bundles — a notion which, to the author's knowledge, is not defined elsewhere.

Notation

We use *abstract indices* to denote the tensor structure of different tensor objects (see e.g. [7]). They indicate which space the object is from and allow us to write down a contraction in the tensor algebra by the usual repetition of the indices. We distinguish tensors with abstract indices from their coordinate representation. We use bold letters for the abstract indices and normal letters for coordinate indices (but you can hardly find them here). Hence, choosing a base $e_a^{\boldsymbol{a}}$ and dual base $\epsilon_{\boldsymbol{a}}^a$, $a = 1, 2, \ldots$, we can write $A_{\boldsymbol{b}\ldots}^{\boldsymbol{a}\cdots} = A_{b\ldots}^{a\cdots}\, e_a^{\boldsymbol{a}}\, \epsilon_{\boldsymbol{b}}^b \ldots$, and $A_{b\ldots}^{a\cdots} = A_{\boldsymbol{b}\ldots}^{\boldsymbol{a}\cdots}\, \epsilon_{\boldsymbol{a}}^a e_b^{\boldsymbol{b}} \ldots$. Here $A_{\boldsymbol{b}\ldots}^{\boldsymbol{a}\cdots}$ is a tensor object and $A_{b\ldots}^{a\cdots}$ is a "bunch" of numbers depending on the base.

Because it can be tiresome to write indices all the time (but it is sometimes inescapable), we drop them if it is clear what structure the object has. (In fact, we view the abstract indices as some kind of a "dress" of the tensor object which serves to specify tensor operations.) We also use an alternative notation for contraction using an infix operator *dot* (we use different dots for different spaces), i.e., for example, $a \cdot \omega = \omega \cdot a = a^{\boldsymbol{n}} \omega_{\boldsymbol{n}}$ or $(a \cdot g)_{\boldsymbol{n}} = a^{\boldsymbol{m}} g_{\boldsymbol{mn}}$.

1 Boundary, canonical, and covariant phase spaces

Overview

The main goal of this chapter is to define a boundary phase space — a kinematical area of the classical counterpart of boundary quantum mechanics. We will start our construction on a very general level and we will see that the boundary phase space can be introduced for a very broad class of theories. However, after this general introduction we turn to a more specific theory to grasp the meaning of the boundary phase space and to understand its relation to standard phase spaces used in physics. We represent it as a cotangent bundle over the boundary value space and, at the end, we introduce special types of observables that will be quantized in the next chapter.

Before turning to a discussion of a general situation let us note that the basic idea lying behind the boundary phase space is very simple. The boundary phase space is a space of canonical data ("values" and "momenta") at both the initial and final time with a symplectic structure induced from canonical phase spaces at the initial and final moment of time. The main nontrivial output of the general discussion below is a construction of the boundary phase space without reference to the causal structure, without necessity of splitting the boundary data to the initial and final parts.

The space of histories and the action

A physical theory can be specified by a *space of elementary histories* $\mathcal{H}$ and a dynamical structure on it. Elementary histories represent a wide class of potentially imaginable evolutions of the system, not necessarily realized in the nature.

Examples of the spaces of histories are the space of all possible trajectories in the spacetime (theory of a relativistic particle), the space of all possible field configuration on the spacetime (field theory), the space of all possible connections on a spacetime (gauge field theory) or the space of all maps from one manifold to another (target) manifold (strings, membranes, ...). The space of histories of a general nonrelativistic system is a space of trajectories in a configuration space $\mathcal{V}_o$ — in space of "positions".

We restrict ourselves to theories that are local on some *inner manifold* N and we pay attention to this dependence. Generally, histories of such local theories can be represented as sections of some fibre bundle over the inner manifold. We use $\boldsymbol{x}, \boldsymbol{y}, \ldots$ as tensor indices for objects from tangent tensor spaces $\mathbf{T}\mathcal{H}$ and the dot $\bullet$ for contraction in these spaces. Let us note that "vectors" from $\mathbf{T}\mathcal{H}$ are again sections of some bundles over the inner manifold, i.e., they are essentially functions (or distributions), and their tensor indices $\boldsymbol{x}, \ldots$ denote also the inner manifold dependence. The contraction $\bullet$ thus includes integration over the inner manifold.

Almost all known theories can be reformulated in this way. For example, elementary histories of a general nonrelativistic system can be viewed as mapping from a one dimensional inner manifold N ("time-line" manifold) to the so-called *target manifold* $\mathcal{V}_o$, i.e., as sections of the trivial fibre bundle $N \times \mathcal{V}_o$ over N. The realization of a field theory is even more straightforward — the inner manifold is spacetime and histories are sections of some bundle over it. We will restrict ourselves mainly to these two cases. Typical examples are a particle in a curved space and the scalar field theory (see [1] for a discussion of the latter case).

We assume that we are able to restrict the theory to any domain Ω in the inner manifold N. It means that we are able to speak about space of histories $\mathcal{H}[\Omega]$ on the domain Ω. We will see that the domain of dependence plays an important role in the dynamics.

On the general level, we admit any sufficiently bounded domain Ω with a smooth boundary $\partial\Omega$. We need to deal with a *bounded* domain to assure that the action functional is well defined on a sufficiently wide set of histories. Generally, if the domain is compact, the action is defined for all smooth histories. However, we can also allow domains that are not compact "in some insignificant directions". A typical example is a sandwich domain in a globally hyperbolic spacetime between two non-intersecting non-compact Cauchy surfaces. Such a domain is unbounded in the spatial direction and this fact has to be compensated by a restriction of the set of histories to those that fall-off sufficiently fast at spatial infinity. We cannot do the same thing in the temporal direction because we would exclude physically interesting histories — specifically, the solutions of the classical equations of motion. In case of a nonrelativistic system the domain Ω is simply a compact interval in the one dimensional inner manifold N.

The localization of histories on the domain Ω also gives us a localization of elements of the tangent spaces $\mathbf{T}\mathcal{H}$, i.e., we can speak about a space $\mathbf{T}\mathcal{H}[\Omega]$. We call these tangent vectors *linearized histories.* As we said, the tangent space at h can be represented as a vector bundle over the inner manifold (or over the domain Ω).

The dynamics of the system is given by a domain-dependent *action* $S[\Omega]$

$$S[\Omega] \,:\, \mathcal{H} \to \mathbb{R}\,. \tag{1.1}$$

Let us note that we cannot generalize the action to a functional $S[N]$ on the whole inner manifold — it would be infinite for most physically interesting histories.

The action is local, i.e., for any Ω

$$S[\Omega](\mathrm{h}_1) = S[\Omega](\mathrm{h}_2) \qquad \text{if } \mathrm{h}_1 = \mathrm{h}_2 \text{ on } \Omega\,, \tag{1.2}$$

and additive under smooth joining of domains, i.e.,

$$S[\Omega](\mathrm{h}) = S[(\Omega_1](\mathrm{h}) + S[\Omega_2](\mathrm{h})\,, \tag{1.3}$$

where $\Omega = \Omega_1 \cup \Omega_2$ is a domain and $\Omega_1 \cap \Omega_2$ is a submanifold without boundary which is a subset of both $\partial\Omega_1$ and $\partial\Omega_2$.

The equation of motion

In general, we work with smooth histories and smooth domains with a boundary, unless stated otherwise. We assume sufficient smoothness of the action but we skip the discussion of this issue.

However, we explicitly assume that the action is essentially of the first-order. In short, this means that the action leads to second-order equations of motion. On a general level, this can be formulated by a condition that the variation of the action (keeping the domain Ω fixed) can be written in the following way

$$\mathrm{d}S[\Omega](\mathrm{h}) = \chi[\Omega]\delta S(\mathrm{h}) - \mathfrak{P}[\partial\Omega](\mathrm{h}) . \tag{1.4}$$

This relation represents an "integration by parts" usually employed in the variation of the action. A description of individual terms follows.

We use the gradient operator $\mathrm{d}_{\boldsymbol{x}}$ on the space of histories $\boldsymbol{\mathcal{H}}$ to denote the variation. It is defined by the usual relation

$$\delta\mathrm{h} \bullet \mathrm{d}S[\Omega](\mathrm{h}) = \frac{d}{d\varepsilon} S[\Omega](\mathrm{h}_\varepsilon)\Big|_{\varepsilon=0} , \tag{1.5}$$

where h_ε is a curve in $\boldsymbol{\mathcal{H}}$ with a tangent vector $\delta\mathrm{h}^{\boldsymbol{x}}$ and, as mentioned above, the contraction $\bullet$ also includes integration over the domain Ω in the inner manifold.

$\delta S_{\boldsymbol{x}}$ represents the *variation of the action* on the entire inner manifold.[a] It is a 1-form on the space of histories $\boldsymbol{\mathcal{H}}$ and we will refer to its tensor index $\boldsymbol{x}$ as to a *variational argument*. We require that $\delta S(\mathrm{h})$ contain at most second-order inner space derivatives of the history h and it is smooth in the variational argument. Thanks to this smoothness, multiplication by $\chi[\Omega]$ is well defined.

$\chi[\Omega]$ is the characteristic function of the domain Ω (i.e., $\chi[\Omega] = 1$ inside Ω and zero outside). We will also use a bi-distribution $(\chi[\Omega]\delta)$ which projects ("cuts-off") smooth functions to functions with a support on Ω (cf. Appendix C).

$\mathfrak{P}_{\boldsymbol{x}}[\partial\Omega]$ is the *generalized momentum on the boundary* $\partial\Omega$. It is localized on the boundary in both its history dependence and in the variational argument (the momentum is a 1-form on $\boldsymbol{\mathcal{H}}$ and by variational argument we again mean the $\boldsymbol{x}$-dependence). We require that it can contain at most the first derivative in the direction normal to the boundary and cannot contain any normal derivatives in the variational argument. Hence, it can be represented as a distribution on boundary values of the linearized histories. Later (cf. Eqs. (1.13) and (1.14)) we clarify a relation of the generalized momentum to the usual definition of the momentum in the Lagrangian formalism.

The *classical equations of motion* are given by the condition

$$\delta S(\mathrm{h}) = 0 , \tag{1.6}$$

and we denote the space of their solutions by $\boldsymbol{\mathcal{S}}$ — it is the *space of classical solutions.*

The *linearized equation of motion* selects the linearized histories tangent to $\boldsymbol{\mathcal{S}}$. It has the form

$$\overset{\leftrightarrow}{\delta^2 S}(\mathrm{h}) \bullet \delta\mathrm{h} = 0 , \tag{1.7}$$

where $\delta\mathrm{h}$ is a linearized history (tangent vector to $\boldsymbol{\mathcal{H}}$) at a classical history h and the *second variation of the action* $\overset{\leftrightarrow}{\delta^2 S}_{\boldsymbol{xy}}$, given by

$$\overset{\leftrightarrow}{\delta^2 S}_{\boldsymbol{xy}}(\mathrm{h}) = \mathfrak{D}_{\boldsymbol{y}}\delta S_{\boldsymbol{x}}(\mathrm{h}) , \tag{1.8}$$

---

[a] The symbol "δ" does not represent any operation here; it should be understood as a part of the symbol δS.

is a derivative of the form δS using an ultralocal connection $\mathfrak{D}$. It is easy to check that for a classical history h, thanks to (1.6), the second variation of the action $\overleftrightarrow{\delta^2 S}(\mathrm{h})$ does not depend on the choice of the connection $\mathfrak{D}$. We define $\overleftarrow{\delta^2 S}_{\boldsymbol{xy}} = \overrightarrow{\delta^2 S}_{\boldsymbol{yx}}$ and

$$\overrightarrow{\delta^2 S}[\Omega] = (\chi[\Omega]\delta) \bullet \overleftrightarrow{\delta^2 S}\,, \quad \overleftarrow{\delta^2 S}[\Omega] = \overleftrightarrow{\delta^2 S} \bullet (\chi[\Omega]\delta)\,, \tag{1.9}$$

satisfying again $\overleftarrow{\delta^2 S}_{\boldsymbol{xy}}[\Omega] = \overrightarrow{\delta^2 S}_{\boldsymbol{yx}}[\Omega]$.

We assume that the equation of motion has a well-defined boundary problem on the domain Ω, i.e., we assume the existence of a unique solution from $\boldsymbol{\mathcal{S}}$ for a given restriction of a history to the boundary. Moreover, we assume that the linearized equation of motion has a well-formulated Dirichlet and Neumann boundary problem, i.e., there is a unique solution to the linearized equation of motion for a given linearized value on the boundary or linearized momentum on the boundary. This requires some generality of the action — for example we exclude a massless scalar field. More serious is the restriction that we must also exclude theories with local symmetries — see [4] for some details on this case.

When working with the manifold $\boldsymbol{\mathcal{S}}$, we use $\mathrm{A}, \mathrm{B}, \dots$ for tangent tensor indices.

Boundary phase space

Next we define the *boundary symplectic structure* $\mathrm{d}\mathfrak{P}[\partial\Omega]$ to be the external derivative of the generalized momentum 1-form $\mathfrak{P}[\partial\Omega]$, which turns out to be the Wronskian of the second variation of the action (see (1.8)),

$$\mathrm{d}\mathfrak{P}[\partial\Omega] = \overleftarrow{\delta^2 S}[\Omega] - \overrightarrow{\delta^2 S}[\Omega]\,. \tag{1.10}$$

We say that two histories are canonically equivalent on the boundary if they have the same restriction on the boundary and the same momentum $\mathfrak{P}$. We call the quotient of the space $\boldsymbol{\mathcal{H}}$ with respect to this equivalence the *boundary phase space* $\boldsymbol{\mathcal{B}}[\partial\Omega]$. A point from the boundary phase space thus represents values and momenta on the entire boundary $\partial\Omega$, i.e., at *both* the initial and final time.

We use $\mathrm{A}, \mathrm{B}, \dots$ as tensor indices for tensors from the tangent spaces $\mathbf{T}\,\boldsymbol{\mathcal{B}}[\partial\Omega]$.

It is straightforward to check that vectors tangent to the orbits of equivalence are degenerate directions of the boundary symplectic form $\mathrm{d}\mathfrak{P}[\partial\Omega]$ and therefore we can define its action on the space $\boldsymbol{\mathcal{B}}[\partial\Omega]$. We will require that the form $\mathrm{d}\mathfrak{P}[\partial\Omega]$ is non-degenerate on the boundary phase space. Because the external derivative of this form is zero, it is indeed a symplectic form in the sense of Appendix A, and it gives a symplectic space structure to the space $\boldsymbol{\mathcal{B}}[\partial\Omega]$, thus justifying the name boundary *phase* space.

The space $\boldsymbol{\mathcal{S}}$ is a submanifold of $\boldsymbol{\mathcal{H}}$ and, therefore, it defines a submanifold of the space $\boldsymbol{\mathcal{B}}[\partial\Omega]$, which we denote by the same letter $\boldsymbol{\mathcal{S}}$.

Lagrangian density

Until now, we have been developing the formalism on a very general level. In the following we restrict to theories with the action given by

$$S[\Omega](\mathrm{h}) = \int_\Omega L(\mathrm{h}, \mathrm{Dh}) \,, \tag{1.11}$$

where L is the *Lagrangian density* — density on the inner manifold N — ultralocally dependent on the value of the history and "velocities", i.e., inner space derivatives of the history.[b]

In general, we need some additional structure on the fibre bundle $\mathcal{H}$ to define the "velocity" (an inner space derivative of the history) — we need, for example, a connection on the bundle. There can exist a natural connection (e.g. if we can identify fibres of the bundle) or a choice of the connection can be equivalent to a specification of an external field (a gauge field of Yang-Mills theories). In the following we will have in mind mainly a general nonrelativistic system, where we do not need any additional structure — the velocity can be simply understood as a derivative of the trajectory $\mathrm{h} : N \to \mathcal{V}_o$ with respect to the "proper time" (a preferred coordinate on N).

In this simple case the variation of the action and an integration by parts gives us the decomposition (1.4) with δS and $\mathfrak{P}[\partial\Omega]$ as follows:

$$\delta S(\mathrm{h}) = \frac{\nabla L}{\partial \mathrm{h}}(\mathrm{h}, \dot{\mathrm{h}}) - \left(\frac{\partial L}{\partial \dot{\mathrm{h}}}(\mathrm{h}, \dot{\mathrm{h}})\right)^{\cdot} \,, \tag{1.12}$$

$$\mathfrak{P}[\partial\Omega](\mathrm{h}) = \frac{\partial L}{\partial \dot{\mathrm{h}}}(\mathrm{h}, \dot{\mathrm{h}})\, \delta[\partial\Omega] \,. \tag{1.13}$$

Here $\delta[\partial\Omega]$ is a delta function localized on the boundary $\partial\Omega$, i.e., with a support at the end points of the time interval Ω. A similar expression for the general case when the inner manifold is not one dimensional and a thorough discussion of the scalar field case can be found in [1].

Space of boundary values

We call the restriction of the history to the boundary $\partial\Omega$ the *boundary value* of the history. We emphasize that the boundary values do not include any inner space derivatives of the histories in the direction normal to the boundary. The *space of all boundary values* $\mathcal{V}[\partial\Omega]$ can be represented as a fibre bundle with the boundary $\partial\Omega$ as the base manifold. We use $\mathtt{x}, \mathtt{y}, \ldots$ as tensor indices for tensors from the tangent spaces $\mathbf{T}\,\mathcal{V}[\partial\Omega]$ and the dot $\cdot$ for contraction in these spaces. In general, the boundary $\partial\Omega$ is a manifold and the contraction $\cdot$ includes integration over the boundary.

Let us note that the choice of notation is adjusted to the scalar field theory. In this case the inner manifold is spacetime, the target manifold are real numbers $\mathbb{R}$,

[b] The formalism developed until now is more general — it covers for example the case of the Einstein-Hilbert action for gravity for which the Lagrangian density contains second spacetime derivatives of the metric. But this dependence is degenerate and it is possible to satisfy the above conditions if a proper boundary term is chosen.

the space of histories $\mathcal{H}$ is the space of functions on spacetime, and the indices $\mathbf{x}, \ldots$ used for vectors from $\mathbf{T}\mathcal{H}$ actually represent points in spacetime. Similarly, the space $\mathcal{V}[\partial\Omega]$ is the space of functions on the boundary $\partial\Omega$ of a spacetime domain Ω and indices $\mathrm{x}, \ldots$ represent points on the boundary. In case of the target manifold being not so simple, all these indices also carry information about a direction in the target manifold.

Another simple case is the nonrelativistic system where the inner manifold is one dimensional. In this case the linearized history $\delta\mathrm{h}^{\mathbf{x}}$ represents a time dependent target-manifold-vector-valued function, i.e., the index $\mathbf{x}$ represents a time variable and a direction in the target manifold. The boundary $\partial\Omega$ consists only of two points from N and the boundary value x of a trajectory h is a pair of end points $\mathrm{x} = [\mathrm{x_f}, \mathrm{x_i}]$. The space of boundary values $\mathcal{V}[\partial\Omega]$ is thus isomorfic to $\mathcal{V}_\mathrm{o} \times \mathcal{V}_\mathrm{o}$ and the tangent space $\mathbf{T}\,\mathcal{V}[\partial\Omega]$ to a direct sum $\mathbf{T}\,\mathcal{V}_\mathrm{o} \oplus \mathbf{T}\,\mathcal{V}_\mathrm{o}$. Therefore, the tensor indices $\mathrm{x}, \ldots$ correspond to pairs of directions in the target manifold and the contraction $\cdot$ reduces to a contraction over finite dimensional vector spaces.

We denote the projection from $\mathcal{H}$ to $\mathcal{V}[\partial\Omega]$ by $\mathbf{x}[\partial\Omega]$. We already said that the condition that the generalized momentum does not contain inner space derivatives normal to the boundary in its variational argument ensures that $\mathfrak{P}_{\mathbf{x}}[\partial\Omega]$ can be realized as a distribution (in the $\mathbf{x}$ argument) on the boundary values od the linearized histories — we do not need any other information about a linearized history $\delta\mathrm{h}^{\mathbf{x}}$ except its value on the boundary to compute $\delta\mathrm{h}^{\mathbf{x}}\mathfrak{P}_{\mathbf{x}}[\partial\Omega]$. Translation between linearized histories and its boundary values is, of course, the differential $\mathrm{D}\mathbf{x}[\partial\Omega]$ of the projection $\mathbf{x}[\partial\Omega]$. Hence, we can represent the generalized momentum in the following form:

$$\mathfrak{P}_{\mathbf{x}}[\partial\Omega] = \mathbf{p}_{\mathrm{x}}[\partial\Omega]\,\mathrm{D}^{\mathrm{x}}_{\mathbf{x}}\mathbf{x}[\partial\Omega]\,. \tag{1.14}$$

Here $\mathbf{p}[\partial\Omega](\mathrm{h})$ is from the cotangent bundle $\mathbf{T}^\star_{\mathbf{x}(\mathrm{h})}\mathcal{V}[\partial\Omega]$. The differential $\mathrm{D}^{\mathrm{x}}_{\mathbf{x}}\mathbf{x}[\partial\Omega]$ understood as a distribution (in the $\mathbf{x}$ argument) is actually a delta function with a support on the boundary (multiplied by a (finite dimensional) unit tensor on the target manifold).

$\mathbf{x}[\partial\Omega](\mathrm{h})$ and $\mathbf{p}[\partial\Omega](\mathrm{h})$ represent the values and the momenta of the history h at *both* the initial and final time. The meaning of $\mathbf{x}[\partial\Omega](\mathrm{h})$ is clear from its definition; the meaning of $\mathbf{p}[\partial\Omega](\mathrm{h})$ can be seen — at least in case of a one dimensional inner manifold — by comparing Eq. (1.14) and Eq. (1.13). The same interpretation holds in the general case, too (see [1] for the details of the scalar field case). In the following, we drop the boundary dependence of $\mathbf{x}$ and $\mathbf{p}$.

Next we define the *classical history* $\bar{\mathrm{h}}(\mathrm{x})$ with a given boundary value x

$$\delta S(\bar{\mathrm{h}}(\mathrm{x})) = 0\,, \quad \mathbf{x}(\bar{\mathrm{h}}(\mathrm{x})) = \mathrm{x} \tag{1.15}$$

and the classical action

$$\bar{S}[\Omega](\mathrm{x}) = S[\Omega](\bar{\mathrm{h}}(\mathrm{x}))\,. \tag{1.16}$$

We use $\bar{\mathrm{h}}$ also for the induced map from the space $\mathcal{V}[\partial\Omega]$ to the boundary phase space $\mathcal{B}[\partial\Omega]$, and $\mathbf{x}$ and $\mathbf{p}$ for the induced maps from the boundary phase space $\mathcal{B}[\partial\Omega]$ to the spaces $\mathcal{V}[\partial\Omega]$ and $\mathbf{T}^\star\,\mathcal{V}[\partial\Omega]$. This suggests that we can represent the boundary phase space $\mathcal{B}[\partial\Omega]$ as a cotangent bundle $\mathbf{T}^\star\,\mathcal{V}[\partial\Omega]$. Indeed, the canonical symplectic structure of the cotangent bundle (B.7) does coincide with $d\mathfrak{P}[\partial\Omega]$:

$$\nabla_{\mathsf{A}} \mathbf{p}_{\mathsf{x}} \wedge \mathrm{D}^{\mathsf{x}}_{\mathsf{B}} \mathbf{x} = \mathrm{d}_{\mathsf{A}} (\mathbf{p}_{\mathsf{x}} \mathrm{D}^{\mathsf{x}}_{\mathsf{B}} \mathbf{x}) = \mathrm{d}_{\mathsf{A}} \mathfrak{P}_{\mathsf{B}} \,. \tag{1.17}$$

The space $\boldsymbol{S}$ as a submanifold of $\boldsymbol{B}[\partial\Omega]$ can be characterized using the condition

$$\mathbf{p} = -\mathrm{d}\bar{S}(\mathbf{x}) \,, \tag{1.18}$$

which follows from

$$\begin{aligned} \mathrm{d}_{\mathbf{x}} \bar{S} &= \mathrm{D}^{\boldsymbol{x}}_{\mathbf{x}} \bar{\mathrm{h}}\, \mathrm{d}_{\boldsymbol{x}} S(\bar{\mathrm{h}}) = \mathrm{D}^{\boldsymbol{x}}_{\mathbf{x}} \bar{\mathrm{h}}\, \big(\delta S_{\boldsymbol{x}}[\Omega](\bar{\mathrm{h}}) - \mathfrak{P}_{\boldsymbol{x}}[\partial\Omega](\bar{\mathrm{h}})\big) = \\ &= -\mathrm{D}^{\boldsymbol{x}}_{\mathbf{x}} \bar{\mathrm{h}}\, \mathrm{D}^{\mathbf{y}}_{\boldsymbol{x}} \mathbf{x}(\bar{\mathrm{h}})\, \mathbf{p}_{\mathbf{y}}(\bar{\mathrm{h}}) = -\mathbf{p}_{\mathbf{x}}(\bar{\mathrm{h}}) \,. \end{aligned} \tag{1.19}$$

Linearization of Eq. (1.18) gives

$$\mathrm{D}^{\mathsf{A}}_{\mathbf{x}} \bar{\mathrm{h}} = \frac{\nabla^{\mathsf{A}}_{\mathbf{x}}}{\partial \mathbf{x}} - \big(\nabla_{\mathbf{x}} \mathrm{d}_{\mathbf{y}} \bar{S}\big) \frac{\partial^{\mathsf{A}}}{\partial \mathbf{p}_{\mathbf{y}}} \tag{1.20}$$

(see Appendix B for definitions of objects used here, especially Eq. (B.6)).

Causal structure

Until now we have not needed any time flow in the underlying inner manifold N. It could be spacetime, an inner sheet of a string, or a one dimensional time-line — in all these cases we do have some kind of time flow. However, the formalism also works in a more general situation. We can use it, for example, for the Euclidian form of the theory, where we do not have any time direction. Now, we will use the time flow for the first time and we will add an additional causal structure that will allow us to define concepts such as canonical and covariant phase spaces.

All what we will use is an assumtion that the boundary of the domain can be split into two disjoint parts without a boundary

$$\begin{aligned} \partial\Omega &= \partial\Omega_{\mathrm{f}} \cup \partial\Omega_{\mathrm{i}} = -\Sigma_{\mathrm{f}} \cup \Sigma_{\mathrm{i}} \,, \\ \partial\Omega_{\mathrm{f}} &= -\Sigma_{\mathrm{f}} \,, \quad \partial\Omega_{\mathrm{i}} = \Sigma_{\mathrm{i}} \,, \end{aligned} \tag{1.21}$$

each of them carrying a full set of data (see the condition below). Here the minus sign suggests an opposite choice of the normal direction orientation for one part of the boundary. Clearly, we have in mind two Cauchy hypersurfaces that define a sandwich domain in a globally hyperbolic spacetime, or two end points of the interval in the one dimensional inner manifold N in the case of a non-relativistic system. The decomposition in (1.21) allows us to write

$$\begin{aligned} \boldsymbol{\mathcal{V}}[\partial\Omega] &= \boldsymbol{\mathcal{V}}[\Sigma_{\mathrm{f}}] \times \boldsymbol{\mathcal{V}}[\Sigma_{\mathrm{i}}] \,, \\ \boldsymbol{\mathcal{B}}[\partial\Omega] &= -\boldsymbol{\mathcal{B}}[\Sigma_{\mathrm{f}}] \oplus \boldsymbol{\mathcal{B}}[\Sigma_{\mathrm{i}}] \,, \\ \mathbf{T}\,\boldsymbol{\mathcal{V}}[\partial\Omega] &= \mathbf{T}\,\boldsymbol{\mathcal{V}}[\Sigma_{\mathrm{f}}] \oplus \mathbf{T}\,\boldsymbol{\mathcal{V}}[\Sigma_{\mathrm{i}}] \,, \end{aligned} \tag{1.22}$$

and we will use shorthands $\boldsymbol{\mathcal{V}}$, $\boldsymbol{\mathcal{V}}_{\mathrm{f}}$, $\boldsymbol{\mathcal{V}}_{\mathrm{i}}$ and $\boldsymbol{\mathcal{B}}$, $\boldsymbol{\mathcal{B}}_{\mathrm{f}}$, $\boldsymbol{\mathcal{B}}_{\mathrm{i}}$.

In case of field theory, the space $\boldsymbol{\mathcal{V}}_{\mathrm{f}}$ (or $\boldsymbol{\mathcal{V}}_{\mathrm{i}}$, respectively) represents the space of field configurations on the final (or the initial) Cauchy hypersurface. In case of particle theory, spaces $\boldsymbol{\mathcal{V}}_{\mathrm{f}}$, $\boldsymbol{\mathcal{V}}_{\mathrm{i}}$ represent positions of the particle at the final or initial time, i.e., $\mathcal{V}_f = \mathcal{V}_i = \mathcal{V}_o$. Similarly $\mathcal{B}_f$ (or $\mathcal{B}_i$) represent canonical data (values and momenta) at the final (or initial) time.

We require that both parts contain a full set of boundary data — there should exist a unique classical history for a given element from $\mathcal{B}_f$ or $\mathcal{B}_i$.

Thanks to the locality, we can decompose the symplectic structure $d\mathfrak{P}[\partial\Omega]$ as

$$d\mathfrak{P}[\partial\Omega] = -d\mathfrak{P}[\Sigma_f] + d\mathfrak{P}[\Sigma_i]\,. \tag{1.23}$$

$d\mathfrak{P}[\Sigma_f]$ and $d\mathfrak{P}[\Sigma_i]$ play the role of the symplectic structure on $\mathcal{B}_f$ and $\mathcal{B}_i$. We will call these spaces *canonical phase spaces*. The minus sign in the relations (1.22) reflects the relation of these spaces as symplectic spaces.

The canonical phase spaces can be again represented as cotangent bundles $\mathbf{T}\,\mathcal{V}_f$ and $\mathbf{T}\,\mathcal{V}_i$ through the maps $\mathbf{x}_f$, $\mathbf{p}_f$ and $\mathbf{x}_i$, $\mathbf{p}_i$. Let us note that $\mathbf{p}_f$ takes into account the opposite orientation of the normal direction to Σ_f and $\partial\Omega_f$, so that

$$\mathbf{p} = -\mathbf{p}_f \oplus \mathbf{p}_i\,. \tag{1.24}$$

Covariant phase space

Finally, we can also give a phase space structure to the space of classical histories $\mathcal{S}$. First we note that, thanks to (1.10), solutions $\xi_1, \xi_2 \in \mathbf{T}\,\mathcal{S}$ of linearized equations of motion (1.7) satisfy

$$\xi_1 \bullet d\mathfrak{P}[\partial\Omega] \bullet \xi_2 = 0\,. \tag{1.25}$$

Therefore, it follows from (1.23) that $d\mathfrak{P}[\Sigma_f]$ and $d\mathfrak{P}[\Sigma_i]$ have the same restriction $\overleftrightarrow{\omega}$ on the space $\mathcal{S}$.

$$\xi_1^A \overleftrightarrow{\omega}_{AB} \xi_2^B = \xi_1 \bullet d\mathfrak{P}[\Sigma_f] \bullet \xi_2 = \xi_1 \bullet d\mathfrak{P}[\Sigma_i] \bullet \xi_2\,. \tag{1.26}$$

In the same way, we check that the same expression for $\overleftrightarrow{\omega}$ holds for any future-oriented Cauchy hypersurface Σ. It means that we have equipped the space of classical histories $\mathcal{S}$ with the symplectic structure $\overleftrightarrow{\omega}$. We will call this space the *covariant phase space*. From Eq. (1.17) follows that the $\mathcal{V}_f$, $\mathcal{V}_i$ and $\mathbf{T}^\star\,\mathcal{V}_f$, $\mathbf{T}^\star\,\mathcal{V}_i$-valued observables $\mathbf{x}_f$, $\mathbf{x}_i$ and $\mathbf{p}_f$, $\mathbf{p}_i$ are canonically conjugate on this space (cf. Appendix A):

$$\overleftrightarrow{\omega}_{AB} = \nabla_A \mathbf{p}_{f\mathbf{x}} \wedge D^{\mathbf{x}}_B \mathbf{x}_f = \nabla_A \mathbf{p}_{i\mathbf{x}} \wedge D^{\mathbf{x}}_B \mathbf{x}_i\,. \tag{1.27}$$

We can invert the symplectic form $\overleftrightarrow{\omega}$ to get $\overleftrightarrow{\omega}^{-1}$:

$$\overleftrightarrow{\omega}^{-1\,AM} \overleftrightarrow{\omega}_{BM} = \overleftrightarrow{\omega}^{-1\,MA} \overleftrightarrow{\omega}_{MB} = \delta^A_B\,. \tag{1.28}$$

If we view $\mathcal{S}$ as a subspace of the boundary phase space $\mathcal{B}$ we can understand $\overleftrightarrow{\omega}^{-1}$ as a tensor from $\mathbf{T}^2\,\mathcal{B}$ tangent to $\mathcal{S}$ in both indices. Because the differential $D\bar{h}$ of the map $\bar{h} : \mathcal{V} \to \mathcal{B}$ (it is a restriction of the map (1.15) to $\mathcal{B}$) plays the role of a projector of vectors from $\mathbf{T}\mathcal{V}$ to vectors from $\mathbf{T}\mathcal{B}$ tangent to $\mathcal{S}$, we can write

$$\overleftrightarrow{\omega}^{-1} = D\bar{h} \cdot g_c \cdot D\bar{h} \tag{1.29}$$

with an antisymmetric tensor $g_c \in \mathbf{T}^2\,\mathcal{V}$. Using (1.20), (1.28), and (1.27) we get

$$g_c \cdot (d_f d_i \bar{S} - d_i d_f \bar{S}) = \delta_{\mathcal{V}}\,. \tag{1.30}$$

This means that

$$\begin{aligned} g_c = g_{\mathrm{if}} - g_{\mathrm{fi}} \,, \quad g_{\mathrm{if}}^{\mathbf{yx}} = g_{\mathrm{fi}}^{\mathbf{xy}} \,, \\ g_{\mathrm{if}} \cdot \mathrm{d_f d_i} \bar{S} = \delta_{\boldsymbol{\mathcal{V}}_{\mathrm{i}}} \,, \quad g_{\mathrm{fi}} \cdot \mathrm{d_i d_f} \bar{S} = \delta_{\boldsymbol{\mathcal{V}}_{\mathrm{f}}} \,, \end{aligned} \tag{1.31}$$

and, using Eq. (1.20) again, we get

$$\begin{aligned} \overset{\leftrightsquigarrow}{\omega}^{-1} &= \left(\frac{\nabla_{\mathbf{x}}}{\partial \mathbf{x}} - \frac{\partial}{\partial \mathbf{p_u}}(\nabla_{\mathbf{u}} \mathrm{d}_{\mathbf{x}} \bar{S})\right) g_c^{\mathbf{xy}} \left(\frac{\nabla_{\mathbf{y}}}{\partial \mathbf{x}} - (\nabla_{\mathbf{y}} \mathrm{d}_{\mathbf{v}} \bar{S}) \frac{\partial}{\partial \mathbf{p_v}}\right) = \\ &= \left(\frac{\partial}{\partial \mathbf{p_{fu}}} \frac{\nabla_{\mathbf{u}}}{\partial \mathbf{x_f}} - \frac{\nabla_{\mathbf{u}}}{\partial \mathbf{x_f}} \frac{\partial}{\partial \mathbf{p_{fu}}}\right) + \left(\frac{\partial}{\partial \mathbf{p_{iu}}} \frac{\nabla_{\mathbf{u}}}{\partial \mathbf{x_i}} - \frac{\nabla_{\mathbf{u}}}{\partial \mathbf{x_i}} \frac{\partial}{\partial \mathbf{p_{iu}}}\right) + \\ &\quad + \left(\frac{\nabla_{\mathbf{x}}}{\partial \mathbf{x}}\, g_c^{\mathbf{xy}}\, \frac{\nabla_{\mathbf{y}}}{\partial \mathbf{x}}\right) + \\ &\quad + \left(\frac{\partial}{\partial \mathbf{p_{fx}}} \Big((\nabla_{\mathbf{fx}} \mathrm{d}_{\mathbf{fu}} \bar{S}) g_{\mathrm{fi}}^{\mathbf{uv}} (\nabla_{\mathbf{iv}} \mathrm{d}_{\mathbf{iy}} \bar{S}) - \mathrm{d}_{\mathbf{fx}} \mathrm{d}_{\mathbf{iy}} \bar{S} \Big) \frac{\partial}{\partial \mathbf{p_{iy}}} \right. \\ &\qquad \left. - \frac{\partial}{\partial \mathbf{p_{ix}}} \Big((\nabla_{\mathbf{ix}} \mathrm{d}_{\mathbf{iu}} \bar{S}) g_{\mathrm{if}}^{\mathbf{uv}} (\nabla_{\mathbf{fv}} \mathrm{d}_{\mathbf{fy}} \bar{S}) - \mathrm{d}_{\mathbf{ix}} \mathrm{d}_{\mathbf{fy}} \bar{S} \Big) \frac{\partial}{\partial \mathbf{p_{fy}}} \right) + \\ &\quad + \left(\frac{\nabla_{\mathbf{x}}}{\partial \mathbf{x_i}} g_{\mathrm{if}}^{\mathbf{xy}} (\nabla_{\mathbf{fy}} \mathrm{d}_{\mathbf{fu}} \bar{S}) \frac{\partial}{\partial \mathbf{p_{fu}}} - \frac{\partial}{\partial \mathbf{p_{fu}}} (\nabla_{\mathbf{fu}} \mathrm{d}_{\mathbf{fy}} \bar{S}) g_{\mathrm{fi}}^{\mathbf{yx}} \frac{\nabla_{\mathbf{x}}}{\partial \mathbf{x_i}}\right) + \\ &\quad + \left(\frac{\nabla_{\mathbf{x}}}{\partial \mathbf{x_f}} g_{\mathrm{fi}}^{\mathbf{xy}} (\nabla_{\mathbf{iy}} \mathrm{d}_{\mathbf{iu}} \bar{S}) \frac{\partial}{\partial \mathbf{p_{iu}}} - \frac{\partial}{\partial \mathbf{p_{iu}}} (\nabla_{\mathbf{iu}} \mathrm{d}_{\mathbf{iy}} \bar{S}) g_{\mathrm{if}}^{\mathbf{yx}} \frac{\nabla_{\mathbf{x}}}{\partial \mathbf{x_f}}\right) . \end{aligned} \tag{1.32}$$

Here, ∇ is any covariant derivative on the value space $\boldsymbol{\mathcal{V}}$, ∇_{f} and ∇_{i} are its restrictions to $\boldsymbol{\mathcal{V}}_{\mathrm{f}}$ and $\boldsymbol{\mathcal{V}}_{\mathrm{i}}$, and $\mathrm{d_f}$, $\mathrm{d_i}$ are gradients on $\boldsymbol{\mathcal{V}}_{\mathrm{f}}$ and $\boldsymbol{\mathcal{V}}_{\mathrm{i}}$.

Or, if we view $\boldsymbol{\mathcal{S}}$ as a subspace of the space of histories $\boldsymbol{\mathcal{H}}$ we can understand $\overset{\leftrightsquigarrow}{\omega}^{-1}$ as a tangent tensor from $\mathbf{T}^2\boldsymbol{\mathcal{H}}$ that satisfies the linear equation of motion in both indices. We call this representation the *causal Green function* G_c

$$G_c^{xy} = \mathrm{D}_{\mathbf{x}}^{x} \bar{\mathrm{h}}\, g_c^{\mathbf{xy}}\, \mathrm{D}_{\mathbf{y}}^{y} \bar{\mathrm{h}} \,, \tag{1.33}$$

where we now understand h as a map from $\boldsymbol{\mathcal{V}}$ to $\boldsymbol{\mathcal{H}}$. In the space $\mathbf{T}\boldsymbol{\mathcal{H}}$, Eq. (1.28) takes the form

$$G_c \bullet \mathrm{d}\mathfrak{P}[\Sigma] = -D_C[\Sigma] \,, \tag{1.34}$$

where $D_C[\Sigma]$ is a *Cauchy projector* of a history on the linearized classical history with the same value and momentum on the surface Σ. It is, of course, an identity on $\mathbf{T}\boldsymbol{\mathcal{S}}$.

Similarly to Eq. (1.33) we can introduce G_{if} and G_{fi}, which turn out to be the advanced and retarded Green functions. g_c, g_{if}, and g_{fi} are thus the corresponding Green functions evaluated on the boundary $\partial\Omega$.

Poisson brackets

The *Poisson brackets* of two observables on a phase space are defined by (A.5). We can compare Poisson brackets in the sense of different phase spaces. Clearly, any observable on $\boldsymbol{\mathcal{H}}$ generates an observable on $\boldsymbol{\mathcal{S}}$ and we can define

$$\{A,B\}_{\boldsymbol{\mathcal{S}}} = \mathrm{d}_x A\, G_c^{xy}\, \mathrm{d}_y B \quad \text{on} \quad \boldsymbol{\mathcal{S}} \,. \tag{1.35}$$

For observables depending only on the boundary values and momenta — i.e., for observables on $\boldsymbol{\mathcal{B}}$ — we can define the Poisson brackets in the sense of the boundary phase space

$$\{A,B\}_{\boldsymbol{\mathcal{B}}} = \mathrm{d}_{\mathsf{A}} A \; \mathrm{d}\mathfrak{P}^{-1\,\mathsf{AB}} \; \mathrm{d}_{\mathsf{B}} B = \frac{\partial A}{\partial \mathbf{p_x}} \frac{\nabla_{\mathbf{x}} B}{\partial \mathbf{x}} - \frac{\nabla_{\mathbf{x}} A}{\partial \mathbf{x}} \frac{\partial B}{\partial \mathbf{p_x}} \,. \tag{1.36}$$

With the help of (1.32) we find that the covariant Poisson brackets for such observables are given by

$$\begin{aligned}
\{A,B\}_{\boldsymbol{\mathcal{S}}} &= \mathrm{d}_{\mathsf{A}} A \; \overleftrightarrow{\omega}^{-1\,\mathsf{AB}} \; \mathrm{d}_{\mathsf{B}} B = \\
&= \Big(\frac{\partial A}{\partial \mathbf{p_{fu}}} \frac{\nabla_{\mathbf{u}} B}{\partial \mathbf{x_f}} - \frac{\nabla_{\mathbf{u}} A}{\partial \mathbf{x_f}} \frac{\partial A}{\partial \mathbf{p_{fu}}} \Big) + \Big(\frac{\partial A}{\partial \mathbf{p_{iu}}} \frac{\nabla_{\mathbf{u}} B}{\partial \mathbf{x_i}} - \frac{\nabla_{\mathbf{u}} A}{\partial \mathbf{x_i}} \frac{\partial B}{\partial \mathbf{p_{iu}}} \Big) + \\
&\quad + \Big(\frac{\nabla_{\mathbf{x}} A}{\partial \mathbf{x}} \, g_c^{\mathbf{xy}} \, \frac{\nabla_{\mathbf{y}} B}{\partial \mathbf{x}} \Big) + \\
&\quad + \Big(\frac{\partial A}{\partial \mathbf{p_{fx}}} \Big((\nabla_{\mathbf{fx}} \mathrm{d}_{\mathbf{fu}} \bar{S}) g_{\mathrm{fi}}^{\mathbf{uv}} (\nabla_{\mathbf{iv}} \mathrm{d}_{\mathbf{iy}} \bar{S}) - \mathrm{d}_{\mathbf{fx}} \mathrm{d}_{\mathbf{iy}} \bar{S} \Big) \frac{\partial B}{\partial \mathbf{p_{iy}}} \\
&\qquad - \frac{\partial A}{\partial \mathbf{p_{ix}}} \Big((\nabla_{\mathbf{ix}} \mathrm{d}_{\mathbf{iu}} \bar{S}) g_{\mathrm{if}}^{\mathbf{uv}} (\nabla_{\mathbf{fv}} \mathrm{d}_{\mathbf{fy}} \bar{S}) - \mathrm{d}_{\mathbf{ix}} \mathrm{d}_{\mathbf{fy}} \bar{S} \Big) \frac{\partial B}{\partial \mathbf{p_{fy}}} \Big) + \\
&\quad + \Big(\frac{\nabla_{\mathbf{x}} A}{\partial \mathbf{x_i}} g_{\mathrm{if}}^{\mathbf{xy}} (\nabla_{\mathbf{fy}} \mathrm{d}_{\mathbf{fu}} \bar{S}) \frac{\partial B}{\partial \mathbf{p_{fu}}} - \frac{\partial A}{\partial \mathbf{p_{fu}}} (\nabla_{\mathbf{fu}} \mathrm{d}_{\mathbf{fy}} \bar{S}) g_{\mathrm{fi}}^{\mathbf{yx}} \frac{\nabla_{\mathbf{x}} B}{\partial \mathbf{x_i}} \Big) + \\
&\quad + \Big(\frac{\nabla_{\mathbf{x}} A}{\partial \mathbf{x_f}} g_{\mathrm{fi}}^{\mathbf{xy}} (\nabla_{\mathbf{iy}} \mathrm{d}_{\mathbf{iu}} \bar{S}) \frac{\partial B}{\partial \mathbf{p_{iu}}} - \frac{\partial A}{\partial \mathbf{p_{iu}}} (\nabla_{\mathbf{iu}} \mathrm{d}_{\mathbf{iy}} \bar{S}) g_{\mathrm{if}}^{\mathbf{yx}} \frac{\nabla_{\mathbf{x}} B}{\partial \mathbf{x_f}} \Big) \,.
\end{aligned} \tag{1.37}$$

Moreover, for observables localized only on Σ_{f} or Σ_{i} we have

$$\begin{aligned}
\{A_{\mathrm{f}}, B_{\mathrm{f}}\}_{\boldsymbol{\mathcal{B}}_{\mathrm{f}}} &= -\{A_{\mathrm{f}}, B_{\mathrm{f}}\}_{\boldsymbol{\mathcal{B}}} = \{A_{\mathrm{f}}, B_{\mathrm{f}}\}_{\boldsymbol{\mathcal{S}}} \,, \\
\{A_{\mathrm{i}}, B_{\mathrm{i}}\}_{\boldsymbol{\mathcal{B}}_{\mathrm{i}}} &= \{A_{\mathrm{i}}, B_{\mathrm{i}}\}_{\boldsymbol{\mathcal{B}}} = \{A_{\mathrm{i}}, B_{\mathrm{i}}\}_{\boldsymbol{\mathcal{S}}} \,.
\end{aligned} \tag{1.38}$$

Observables at most linear in momenta

During the quantization we will be interested in a special kind of observables on the boundary phase space $\boldsymbol{\mathcal{B}}$ (or, in general, on any phase space with the cotangent bundle structure). We will investigate observables dependent only on the value and observables linear in the momentum. We define the following observables for a function f and a vector field a on $\boldsymbol{\mathcal{V}}$

$$\mathsf{F}_f = f(\mathbf{x}) \,, \tag{1.39}$$

$$\mathsf{G}_a = a^{\mathbf{x}}(\mathbf{x})\, \mathbf{p_x} \,. \tag{1.40}$$

The Poisson brackets of these observables are

$$\begin{aligned}
\{\mathsf{F}_{f_1}, \mathsf{F}_{f_2}\}_{\boldsymbol{\mathcal{B}}} &= 0 \,, \\
\{\mathsf{F}_f, \mathsf{G}_a\}_{\boldsymbol{\mathcal{B}}} &= -\mathsf{F}_{a\cdot df} \,, \\
\{\mathsf{G}_{a_1}, \mathsf{G}_{a_2}\}_{\boldsymbol{\mathcal{B}}} &= \mathsf{G}_{[a_1,a_2]} \,.
\end{aligned} \tag{1.41}$$

Here $[a_1, a_2]$ is the Lie bracket of the vector fields a_1, a_2. In the sense of the covariant phase space we have

$$\{\mathsf{F}_{f_1}, \mathsf{F}_{f_2}\}_{\boldsymbol{\mathcal{S}}} = \mathsf{F}_{\mathrm{d}f_1\cdot g_c\cdot \mathrm{d}f_2} \quad \text{on} \quad \boldsymbol{\mathcal{S}} \tag{1.42}$$

and from the condition (1.18) we get

$$\mathsf{G}_a = \mathsf{F}_{a \cdot \mathrm{d}\bar{S}} \quad \text{on} \quad \boldsymbol{\mathcal{S}} \,. \tag{1.43}$$

Summary

Canonical phase spaces $\boldsymbol{\mathcal{B}}_{\mathrm{f}}$, $\boldsymbol{\mathcal{B}}_{\mathrm{i}}$ or the covariant phase space $\boldsymbol{\mathcal{S}}$ are commonly used phase spaces of the classical theory. The dynamical evolution is described as a canonical transformation of $\boldsymbol{\mathcal{B}}_{\mathrm{i}}$ to $\boldsymbol{\mathcal{B}}_{\mathrm{f}}$ (the Schrödinger picture of the classical theory) or as an evolution of observables on $\boldsymbol{\mathcal{S}}$ (Heisenberg picture on the classical level).

If we could use experimental devices localized only on the boundary of the investigated domain (i.e., if we could perform experiments only at the initial and final moments of time), the boundary phase space $\boldsymbol{\mathcal{B}}$ would be sufficient for a description of our system. It is enough to investigate observables defined using canonical variables at the beginning and at the end. The dynamics of the system is hidden in the definition of the special subspace — of the physical phase space $\boldsymbol{\mathcal{S}}$ that tells us the relation between the initial value and momentum and the final value and momentum for a physical solution of the equation of motion. In other words, the space $\boldsymbol{\mathcal{B}}$ represents all possible values of canonical observables at the initial and the final time without knowledge of the equations of motion. The subspace $\boldsymbol{\mathcal{S}}$ represents values of canonical observables correlated via the dynamical development of the system. The advantage of the boundary phase space $\boldsymbol{\mathcal{B}}$ is that we do not need any causal structure to define it. Hence, we can construct it even for a Euclidian theory. This advantage will be even more appealing in the quantum case.

2 Configuration quantization

Overview

In this chapter we describe a construction of quantum observables at most linear in momenta for a general system with a phase space $\mathcal{G}$. This chapter is completely independent of the previous one. The starting point is a phase space with a cotangent bundle structure, i.e., with the phase space built over a general configuration space $\mathcal{V}$. We will introduce a special type of quantum observables, that are generalizations of the classical position and momentum observables of the standard quantum mechanics on a linear configuration space, and we will find their position representation. The main idea is to understand the momentum observable as a shift operator along some "direction" in the configuration space. On a general manifold, however, we have to be careful, because the "direction" is specified by a vector field that is, in general, position dependent; and, of course, the position and momentum observables don't commute.

Ideas of quantization

Quantization is a heuristic procedure of constructing a quantum theory for a given classical theory. Let us have a classical system with a kinematical area given by a phase space $\mathcal{G}$ and a symplectic structure ω. Observables are functions on $\mathcal{G}$, and the Poisson brackets are given by Eq. (A.5). Quantization tells us how to assign to (at least some) classical observables quantum observables — operators on a quantum Hilbert space $\mathcal{H}$. We will use letters with a hat to denote quantum observables and we denote $\mathcal{O}$ the algebra they obey. They should satisfy the same algebraic relations as the classical observables and the commutation relations generated by Poisson brackets. Specifically, if the quantum versions of classical observables A, B, and $C = \{A, B\}$ are $\hat{\mathrm{A}}, \hat{\mathrm{B}}, \hat{\mathrm{C}}$, they should be related by

$$[\hat{\mathrm{A}}, \hat{\mathrm{B}}] = -i\hat{\mathrm{C}} \,, \tag{2.1}$$

$[\hat{\mathrm{A}}, \hat{\mathrm{B}}] = \hat{\mathrm{A}}\hat{\mathrm{B}} - \hat{\mathrm{B}}\hat{\mathrm{A}}$ being the commutator of operators. It is well known that the procedure described above cannot be carried out for all classical observables. Because quantum observables do not commute we have an "ordering problem" for observables given by a product of non-commuting observables.

The usual quantization procedure tries to quantize a specific class of classical observables and construct physically interesting observables from them. Even in this case the operator ordering ambiguity is encountered. But we have to expect this — a quantum theory is not fully determined by the classical counterpart.

We will discuss quantization for a general example of a classical theory given on a phase space $\mathcal{G}$ that has a cotangent bundle structure $\mathbf{T}^* \mathcal{V}$. The mathematical structure of such a phase space is reviewed in Appendices A and B. The main goal is not to solve a dynamical problem here, but to formulate the quantization procedure in a geometrically covariant way. We will define quantum observables of the position and momentum in the case of a general configuration space $\mathcal{V}$ (i.e., without the usual assumption of linearity of the configuration space). This procedure will be applied to quantization of the boundary phase space in the next section.

Let us note that the procedure described below has a well-defined meaning for a finite-dimensional configuration space $\mathcal{V}$. Of course, this is not the case of, e.g., the scalar field theory, where the configuration space is the space of functions on the space manifold Σ. Technical problems with a generalization to an infinite dimensional case is one of the reasons why this kind of quantization is not usually employed to quantize a field theory, but see [1] and the end of this section for an additional discussion. However, in the following we will consider mainly a finite dimensional non-relativistic system, for example a particle in a curved space.

Algebra of observables

At the end of the last chapter we introduced two special kinds of observables on the phase space with a cotangent bundle structure. We defined the observables F_f depending only on the "position" (Eq. (1.39)) and the observables G_a linear in momentum (Eq. (1.40)). Their Poisson brackets are given in Eq. (1.41). Now, we will quantize these observables. First, we formulate more carefully what conditions we are imposing on the quantum versions of these observables.

We are looking for maps $\hat{\mathsf{F}}$ and $\hat{\mathsf{G}}$ from the test functions and test vector fields on the configuration space $\mathcal{V}$ to the space of quantum observables $\mathcal{O}$

$$\begin{aligned} \hat{\mathsf{F}}_f \in \mathcal{O} \quad &\text{for} \quad f \in \mathfrak{F}\mathcal{V} , \\ \hat{\mathsf{G}}_a \in \mathcal{O} \quad &\text{for} \quad a \in \mathfrak{T}\mathcal{V} , \end{aligned} \tag{2.2}$$

that should be hermitian (for real f and a)

$$\hat{\mathsf{F}}_f^\dagger = \hat{\mathsf{F}}_f , \quad \hat{\mathsf{G}}_a^\dagger = \hat{\mathsf{G}}_a , \tag{2.3}$$

and should satisfy commutation relations motivated by the Poisson brackets (1.41)

$$[\hat{\mathsf{F}}_{f_1}, \hat{\mathsf{F}}_{f_2}] = 0 , \tag{2.4}$$

$$[\hat{\mathsf{F}}_f, \hat{\mathsf{G}}_a] = i\hat{\mathsf{F}}_{a\cdot df} , \tag{2.5}$$

$$[\hat{\mathsf{G}}_{a_1}, \hat{\mathsf{G}}_{a_2}] = -i\hat{\mathsf{G}}_{[a_1,a_2]} . \tag{2.6}$$

Next, we have to formulate the condition that the quantum observables satisfy "the same algebraic relations as the classical ones". For the observables $\hat{\mathsf{F}}_f$ this is straightforward because they commute — we require that we be able to take out any algebraic operation $g(\dots)$ from the argument of the observable and perform the same operation on the observables:

$$\hat{\mathsf{F}}_{g(f_1,f_2,\dots)} = g(\hat{\mathsf{F}}_{f_1}, \hat{\mathsf{F}}_{f_2}, \dots) . \tag{2.7}$$

With the $\hat{\mathsf{G}}_a$ we have to be more careful — they do not commute with each other and with the $\hat{\mathsf{F}}_f$ observables. But we are interested in observables linear in momentum, so we need to investigate only the vector-field dependence of the $\hat{\mathsf{G}}_a$. We require

$$\hat{\mathsf{G}}_{a_1+\alpha a_2} = \hat{\mathsf{G}}_{a_1} + \alpha\hat{\mathsf{G}}_{a_2} \quad \text{for} \quad \alpha \in \mathbb{R} , \tag{2.8}$$

$$\hat{\mathsf{G}}_{fa} = \hat{\mathsf{F}}_{f^{\frac{1}{2}-i\gamma}} \hat{\mathsf{G}}_a \hat{\mathsf{F}}_{f^{\frac{1}{2}+i\gamma}} \quad \text{for} \quad \gamma \in \mathbb{R} . \tag{2.9}$$

The last condition is equivalent to the classical expression $\mathsf{G}_{fa} = \mathsf{F}_f\mathsf{G}_a$, but it specifies the ordering of the quantum theory. We have chosen the exponents of the function f in the special forms $(\frac{1}{2} - i\gamma)$ and $(\frac{1}{2} + i\gamma)$ to satisfy the hermiticity condition (2.3) with a real constant γ. The ordering condition can be rewritten as

$$\hat{\mathsf{G}}_{fa} = (\frac{1}{2} - i\gamma)\,\hat{\mathsf{F}}_f\hat{\mathsf{G}}_a + (\frac{1}{2} + i\gamma)\,\hat{\mathsf{G}}_a\hat{\mathsf{F}}_f = \frac{1}{2}(\hat{\mathsf{F}}_f\hat{\mathsf{G}}_a + \hat{\mathsf{G}}_a\hat{\mathsf{F}}_f) + \gamma\,\hat{\mathsf{F}}_{a\cdot df}\,. \tag{2.10}$$

Finally, we require that the position observables $\hat{\mathsf{F}}_f$ form a complete set of commuting observables

$$\forall f \quad [\hat{\mathsf{F}}_f, \hat{\mathrm{A}}] = 0 \qquad \Rightarrow \qquad \exists g \quad \hat{\mathrm{A}} = \hat{\mathsf{F}}_g\,. \tag{2.11}$$

Let us note that from these conditions it follows that

$$\hat{\mathsf{F}}_\alpha = \alpha\hat{\mathbb{1}} \quad \text{for} \quad \alpha \in \mathbb{R}\,, \tag{2.12}$$

$$\forall f, a \quad [\hat{\mathsf{F}}_f, \hat{\mathrm{A}}] = 0 \quad [\hat{\mathsf{G}}_a, \hat{\mathrm{A}}] = 0 \qquad \Rightarrow \qquad \exists \alpha \in \mathbb{R} \quad \hat{\mathrm{A}} = \alpha\hat{\mathbb{1}}\,. \tag{2.13}$$

Position representation

Next, we construct a *position base* in $\mathcal{H}$ on which the action of the operators $\hat{\mathsf{F}}_f$ and $\hat{\mathsf{G}}_a$ has a simple representation. Or, in other words, we find a realization of the operators $\hat{\mathsf{F}}_f$ and $\hat{\mathsf{G}}_a$, which satisfies the conditions formulated above, as operators on the space of densities on the configuration manifold $\boldsymbol{\mathcal{V}}$.

We have, of course, immediately a candidate for such a base. The observables $\hat{\mathsf{F}}_f$ form a complete set of commuting observables, so there exists a base of eigenvectors labeled by the position in the configuration space $\boldsymbol{\mathcal{V}}$ such that

$$\hat{\mathsf{F}}_f|pos : \mathrm{x}\rangle = f(\mathrm{x})|pos : \mathrm{x}\rangle\,. \tag{2.14}$$

Strictly speaking, $|pos : \mathrm{x}\rangle$ are not vectors from the Hilbert space $\mathcal{H}$ but generalized vectors that can be defined, for example, by the condition that the projectors to these states form an operator-valued density on the configuration space. It will be convenient later to give these vectors the character of a density of some (generally complex) weight $\alpha \in \mathbb{C}$ (see Appendix C for short review of densities on a manifold). I.e., we assume

$$|pos : \mathrm{x}\rangle \in \mathcal{H} \otimes \tilde{\mathbb{C}}^{\alpha}_{\mathrm{x}}\,\boldsymbol{\mathcal{V}}\,. \tag{2.15}$$

The base of eigenvectors is orthogonal, but we need to be careful when writing down a normalization condition. Because of the distributional character of the vectors we can normalize them only to a delta-distribution. And to get the right normalization we need to choose a volume element μ on the value space. With such a volume element we can write the orthonormality relation

$$\langle pos : \mathrm{x}|pos : \mathrm{y}\rangle = (\mu^{2\,\mathrm{Re}\,\alpha - 1}\delta)(\mathrm{x}|\mathrm{y})\,. \tag{2.16}$$

The completeness relation is

$$\hat{\mathbb{1}} = \int_{\boldsymbol{\mathcal{V}}} |pos : .\rangle\langle pos : .|\,\mu^{1-2\,\mathrm{Re}\,\alpha}\,. \tag{2.17}$$

Let us note that the conditions above do not fix the base $|pos : \mathrm{x}\rangle$ uniquely — they fix the base up to an x-dependent phase factor. We will deal with this ambiguity below.

For any vector $|state\rangle \in \mathcal{H}$ we can define a *wave function* — a density of the weight α^* on the configuration space $\mathcal{V}$

$$\Psi_{|state\rangle}(\mathrm{x}) = \langle pos : \mathrm{x}|state\rangle \ . \tag{2.18}$$

The density μ defines the scalar product on this space which is an isomorphic to the product on the quantum space

$$\langle st1|st2\rangle = (\Psi_{|st1\rangle}, \Psi_{|st2\rangle})_\mu = \int_{\mathcal{V}} \Psi^*_{|st1\rangle} \Psi_{|st2\rangle}\, \mu^{1-2\,\mathrm{Re}\,\alpha} \ , \tag{2.19}$$

and it induces the hermitian conjugation ‡ on operators acting on densities of weight α

$$(A^\ddagger \psi_1, \psi_2)_\mu = (\psi_1, A\psi_2)_\mu \ . \tag{2.20}$$

We define the *position representation* of a quantum observable $\hat{\mathsf{A}}$ as an operator $\check{A}$ on wave functions

$$\check{A}\, \Psi_{|state\rangle} = \Psi_{\hat{\mathsf{A}}|state\rangle} \ . \tag{2.21}$$

Clearly,

$$\check{\mathsf{F}}_f = f\delta \ , \tag{2.22}$$

which can be also written as

$$\hat{\mathsf{F}}_f = \int_{\mathcal{V}} f\, |pos : .\rangle\langle pos : .|\, \mu^{1-2\,\mathrm{Re}\,\alpha} \ . \tag{2.23}$$

Phase fixing

Now we proceed to find the position representation of the momentum observables $\hat{\mathsf{G}}_a$. In this section we show that there exists a unique choice of the weight α, namely $\alpha = (\frac{1}{2} - i\gamma)$, and of the phase of the position base $|pos : \mathrm{x}\rangle$, for which

$$\check{\mathsf{G}}_a = -i\mathcal{L}_a \ , \tag{2.24}$$

where $\mathcal{L}_a$ is the Lie derivative along the vector field a acting to the right.

First we define the *position shift operator* along a vector field a on the configuration space $\mathcal{V}$ as

$$\hat{\mathsf{U}}_a(\varepsilon) = \exp(-i\varepsilon\hat{\mathsf{G}}_a) \ . \tag{2.25}$$

The commutation relation (2.5) gives us

$$\hat{\mathsf{U}}_a \hat{\mathsf{F}}_f \hat{\mathsf{U}}_a^{-1} = \hat{\mathsf{F}}_{\mathbf{u}_a^\star f} \ . \tag{2.26}$$

Here $\mathbf{u}_a(\varepsilon)$ is a diffeomorphism on $\mathcal{V}$ induced by the vector field a

$$\frac{d}{d\varepsilon}\mathbf{u}_a = a \ , \quad \mathbf{u}_a(\varepsilon_1 + \varepsilon_2) = \mathbf{u}_a(\varepsilon_1)\, \mathbf{u}_a(\varepsilon_2) \ , \tag{2.27}$$

and $\mathbf{u}_a^\star$ is a map induced by the diffeomorphism on objects defined on the configuration space. The equation (2.26) gives us

$$\begin{aligned} &\hat{\mathsf{F}}_f\,\hat{\mathsf{U}}_a|pos:\mathrm{x}\rangle = \hat{\mathsf{U}}_a\hat{\mathsf{F}}_{\mathbf{u}_a^{\star\,-1}f}|pos:\mathrm{x}\rangle = f(\mathbf{u}_a\mathrm{x})\,\hat{\mathsf{U}}_a|pos:\mathrm{x}\rangle \qquad \Rightarrow \\ &\qquad \Rightarrow \qquad \hat{\mathsf{U}}_a|pos:\mathrm{x}\rangle \quad \text{is proportional to} \quad |pos:\mathbf{u}_a\mathrm{x}\rangle\;. \end{aligned} \tag{2.28}$$

Here we have to be careful about the proportionality coefficient because vectors $|pos:\mathrm{x}\rangle$ and $|pos:\mathbf{u}_a\mathrm{x}\rangle$ are also densities in different points x and $\mathbf{u}_a\mathrm{x}$. Because $\hat{\mathsf{U}}_a(\varepsilon)$ forms a commuting one-dimensional group for $\varepsilon\in\mathbb{R}$ we can write the proportionality relation as follows

$$\mathbf{u}_a^\star\Big(\hat{\mathsf{U}}_a\Psi_a(\mathrm{x})|pos:\mathrm{x}\rangle\Big) = \Psi_a(\mathbf{u}_a\mathrm{x})|pos:\mathbf{u}_a\mathrm{x}\rangle\;, \tag{2.29}$$

where Ψ_a is a density on $\boldsymbol{\mathcal{V}}$ that is defined up to a density invariant under the action of the diffeomorphism $\mathbf{u}_a$.

Next we prove that Ψ_a can be chosen as a density of weight $(\frac{1}{2}-\alpha)$ in the form

$$\Psi_a = \mu^{\frac{1}{2}-\alpha}\exp(-i\phi_a)\;, \tag{2.30}$$

where ϕ_a is a real function on $\boldsymbol{\mathcal{V}}$ defined up to a function constant on the orbits of $\mathbf{u}_a$.

We write Ψ_a as a density of the weight $(\frac{1}{2}-\alpha)$ in the form

$$\Psi_a = \mu^{\frac{1}{2}-\alpha}\rho_a\exp(-i\phi_a)\;, \tag{2.31}$$

with ρ_a, ϕ_a real functions. Remembering Eq. (2.25), the differential form of equation (2.29), gives[c]

$$\hat{\mathsf{G}}_a\Big(\mu^{\frac{1}{2}-\alpha}\rho_a\exp(-i\phi_a)|pos:.\rangle\Big) = i\mathcal{L}_a\Big(\mu^{\frac{1}{2}-\alpha}\rho_a\exp(-i\phi_a)|pos:.\rangle\Big)\;. \tag{2.32}$$

From this, it follows that the position representation of the $\hat{\mathsf{G}}_a$ observables:

$$\check{\mathsf{G}}_a = -i\mathcal{L}_a - i\Big((\frac{1}{2}-\alpha^*)\frac{1}{\mu}\mathcal{L}_a\mu + \frac{1}{\rho_a}\mathcal{L}_a\rho_a + ia\cdot\mathrm{d}\phi_a\Big)\delta\;. \tag{2.33}$$

The definition (2.20) gives us

$$\mathcal{L}_a^\ddagger = -\mathcal{L}_a + (2\,\mathrm{Re}\,\alpha - 1)\frac{1}{\mu}(\mathcal{L}_a\mu)\delta\;. \tag{2.34}$$

The hermiticity condition (2.3) implies $\check{\mathsf{G}}_a = \check{\mathsf{G}}_a^\ddagger$. Substituting to this condition we obtain

$$\frac{1}{\rho_a}\mathcal{L}_a\rho_a = 0\;, \tag{2.35}$$

which means ρ_a should be constant, e.g., $\rho_a = 1$. It proves the statement (2.30).

---

[c]Let us recall that the Lie derivative of an object A on the manifold is defined by

$$(\mathcal{L}_a A)(\mathrm{x}) = -\frac{d}{d\varepsilon}(\mathbf{u}_a^\star A)(\mathrm{x})\Big|_{\varepsilon=0} = \frac{d}{d\varepsilon}\mathbf{u}_a^{\star\,-1}A(\mathbf{u}_a\mathrm{x})\Big|_{\varepsilon=0}\;.$$

Finally we show that the function ϕ_a in equation (2.30) has the form

$$\phi_a = \varphi + \tilde{\phi}_a \, , \tag{2.36}$$
$$\tilde{\phi}_a(\mathbf{u}_a(\varepsilon)\mathrm{x}) - \tilde{\phi}_a(\mathrm{x}) = \gamma \int_0^{\varepsilon} \frac{1}{\mu}(\mathcal{L}_a\mu)\, d\varepsilon \, ,$$

where φ is a real function independent of the vector field a.

Let us define a function

$$\lambda_a = a \cdot \mathrm{d}\phi_a - \gamma \frac{1}{\mu} \mathcal{L}_a \mu \, . \tag{2.37}$$

This allows us to write $\check{\mathsf{G}}_a$ as

$$\check{\mathsf{G}}_a = \check{\mathsf{G}}'_a + \lambda_a \delta \, , \tag{2.38}$$
$$\check{\mathsf{G}}'_a = -i\mathcal{L}_a - i(\frac{1}{2} + i\gamma - \alpha^*)\frac{1}{\mu}(\mathcal{L}_a\mu)\delta \, . \tag{2.39}$$

It is easy to check that the operators $\check{\mathsf{G}}'_a$ have the same properties as the operators $\check{\mathsf{G}}_a$. Using the consequences of the properties (2.8), (2.9), and (2.6), we get conditions on λ_a:

$$\lambda_{a_1 + f a_2} = \lambda_{a_1} + f \lambda_{a_2} \, , \tag{2.40}$$
$$\lambda_{[a_1, a_2]} = a \cdot \mathrm{d}\lambda_{a_2} - a_2 \cdot \mathrm{d}\lambda_{a_1} \, .$$

The first condition implies that $\lambda_a = a \cdot \lambda$ for some form λ on the configuration space $\mathcal{V}$ and the second condition implies that this form is closed: $\mathrm{d}\lambda = 0$. So, if we ignore topological problems (see some of the comments below), we can rewrite relation (2.37) as

$$a \cdot \mathrm{d}\phi_a = a \cdot \mathrm{d}\varphi + \gamma \frac{1}{\mu} \mathcal{L}_a \mu \tag{2.41}$$

for a real function φ. Integrating along orbits of $\mathbf{u}_a$, we get the desired expression (2.36).

If we redefine our position base by the phase factor $\exp(-i\varphi)$, we obtain the position representation for the momentum observables $\check{\mathsf{G}}_a$ in the form (2.39). We see that if we choose the density weight of our position base to be

$$\alpha = \frac{1}{2} - i\gamma \, , \tag{2.42}$$

the position representation reduces to the simple form (2.24). This also allows us to write the action of $\hat{\mathsf{G}}_a$ in the position base:

$$\hat{\mathsf{G}}_a |pos : .\rangle = i\mathcal{L}_a |pos : .\rangle \, . \tag{2.43}$$

Let us note that for this choice of α we do not need any volume element on $\mathcal{V}$ because the normalization and completeness conditions (2.16) and (2.17) reduce to

$$\langle pos : \mathrm{x} | pos : \mathrm{y} \rangle = \delta(\mathrm{x}|\mathrm{y}) \, , \quad \hat{\mathbb{1}} = \int_{\mathcal{V}} |pos : .\rangle\langle pos : .| \, . \tag{2.44}$$

Uniqueness of the quantization

Now we ask the question whether all realizations of our quantization of the basic observables F_f and G_a are unitarily equivalent. Let us assume we have two quantum versions of the basic observables $\hat{\mathsf{F}}_f$, $\hat{\mathsf{G}}_a$ and $\hat{\mathsf{F}}'_f$, $\hat{\mathsf{G}}'_a$ that both satisfy all the conditions formulated above. Clearly, we can construct a unitary operator mapping the position base of the first pair to the position base of the second pair, except that we do not require a proper phase fixing. Such a unitary operator essentially identifies the position observables but not necessarily the momentum observables.

Therefore, we will investigate the relation between the observables $\hat{\mathsf{G}}_a$ and $\hat{\mathsf{G}}'_a$ that both, together with the common position observables $\hat{\mathsf{F}}_f$, satisfy all the conditions above. The commutation relation (2.5) together with the completeness condition (2.11) gives

$$\hat{\mathsf{G}}'_a - \hat{\mathsf{G}}_a = \hat{\mathsf{F}}_{\lambda_a} \tag{2.45}$$

for an a-dependent real function λ_a. The linearity (2.8) and the ordering condition (2.9) together with the property (2.7) give

$$\lambda_{a_1+fa_2} = \lambda_{a_1} + f\lambda_{a_2} \qquad \Rightarrow \qquad \lambda_a = \lambda \cdot a \tag{2.46}$$

for some form λ on $\mathcal{V}$. The commutation relation (2.6) implies

$$[a_1, a_2] \cdot \lambda = a_1 \cdot \mathrm{d}(a_2 \cdot \lambda) - a_2 \cdot \mathrm{d}(a_1 \cdot \lambda) \qquad \Rightarrow \qquad \mathrm{d}\lambda = 0\,. \tag{2.47}$$

If the configuration space $\mathcal{V}$ is sufficiently topologically trivial (precisely, if the first cohomology group is trivial), it follows from the last equation that the form λ is a gradient of some function φ. In this case we can write

$$\hat{\mathsf{G}}'_a = \exp(-i\hat{\mathsf{F}}_\varphi)\hat{\mathsf{G}}_a \exp(i\hat{\mathsf{F}}_\varphi)\,, \tag{2.48}$$

and we see that $\hat{\mathsf{G}}'_a$ and $\hat{\mathsf{G}}_a$ are unitary equivalent. If the configuration space is not topologically trivial and closed forms are not the same as exact forms, we can have unitarily inequivalent realizations of the basic quantum variables. We ignored this possibility when constructing the position base and we will not investigate it further here.

Relation between different orderings

Here we will investigate a relation between different orderings of the momentum observables. Let us assume that ${}^{\gamma_1}\hat{\mathsf{G}}$ and ${}^{\gamma_2}\hat{\mathsf{G}}$ together with the position observables $\hat{\mathsf{F}}_f$ satisfy the above conditions with the parameter γ_1 or γ_2 in the ordering condition (2.9). Similary to the previous discussion we get

$${}^{\gamma_1}\hat{\mathsf{G}}_a - {}^{\gamma_2}\hat{\mathsf{G}}_a = \hat{\mathsf{F}}_{\tilde{\lambda}_a}\,, \tag{2.49}$$

$$\tilde{\lambda}_{a_1+\alpha a_2} = \tilde{\lambda}_{a_1} + \alpha\tilde{\lambda}_{a_2} \quad \text{for} \quad \alpha \in \mathbb{R}\,, \tag{2.50}$$

$$\tilde{\lambda}_{fa} = f\tilde{\lambda}_a + (\gamma_1 - \gamma_2)a \cdot \mathrm{d}f\,. \tag{2.51}$$

If we write the function $\tilde{\lambda}_a$ using a density μ on $\mathcal{V}$ as

$$\tilde{\lambda}_a = (\gamma_1 - \gamma_2)\frac{1}{\mu}\mathcal{L}_a\mu + \lambda_a\,, \tag{2.52}$$

we find that λ_a has the same properties as in the previous section, i.e., it represents the freedom of the quantization of G_a with the given ordering parameter. Because we have discussed it already, we ignore it now. Thus, we found that different orderings of the momentum observable can be written in the form

$$ {}^{\gamma}\hat{\mathsf{G}}_a = {}^{0}\hat{\mathsf{G}}_a + \gamma \hat{\mathsf{F}}_{\mu^{-1}\mathcal{L}_a\mu} \,. \tag{2.53} $$

It is easy to check that the μ dependence for a fixed γ is of the form discussed in the previous section.

Observables quadratic in momentum

Until now we only discussed quantization of observables independent of momentum and linear in momentum. We saw that these observables were sufficient for the construction of the natural base in the quantum Hilbert space $\mathcal{H}$. But they are not usually sufficient for the construction of the dynamics of the theory. A typical Hamiltonian is quadratic in momenta. Therefore, it is necessary to address the issue of quantization of such observables. I.e., we want to quantize classical observables of the form

$$ \mathsf{K}_k(\mathrm{x}, \mathrm{p}) = \mathrm{p} \cdot k^{-1}(\mathrm{x}) \cdot \mathrm{p} \,, \tag{2.54} $$

where k is a metric on $\boldsymbol{\mathcal{V}}$ and we have restricted ourself to the case of a non-degenerate k.

We can formulate conditions similar to those above for $\hat{\mathsf{F}}_f$ and $\hat{\mathsf{G}}_a$. With a suitable choice of simplicity and covariance requirements it is possible to show that there is a one-dimensional freedom in the ordering for the quadratic observables (labeled by a parameter $\xi \in \mathbb{R}$) and that the position representation $\check{\mathsf{K}}_k$ of the quadratic observable is

$$ \check{\mathsf{K}}_k = \mathrm{L}_k + \xi \mathcal{R}_k \,, \tag{2.55} $$

where L_k and $\mathcal{R}_k$ are the Laplace operator and scalar curvature of the metric k.

Linear theory

The situation simplifies significantly if the configuration space $\boldsymbol{\mathcal{V}}$ is linear. This is usually assumed in simple quantum mechanical models and it is essential in the quantum field theory. For a linear theory, the formalism reduces to the standard canonical quantization. We will shortly illustrate the correspondence with the usual formalism using a notation borrowed from the the scalar field theory. However, these notes apply for any linear theory.

In the scalar field theory, the configuration space $\boldsymbol{\mathcal{V}}[\Sigma]$ represents the space of all field configurations on a Cauchy hypersurface Σ and the phase space $\boldsymbol{\mathcal{B}}[\Sigma]$ has the structure of a cotangent bundle $\mathbf{T}^\star \boldsymbol{\mathcal{V}}[\Sigma]$ that can be identified with the dual space $\boldsymbol{\mathcal{V}}[\Sigma]^\star$. The dual can be realized as the space of densities $\tilde{\boldsymbol{\mathcal{V}}}[\Sigma]$ of the weight 1 on the hypersurface Σ. This brings good and bad news. As we said, the value space $\boldsymbol{\mathcal{V}}[\Sigma]$ is linear, but it is infinite-dimensional. Let's ignore the infinite dimension and do some formal manipulation first.

Linearity allows us to define the observables of value and momentum $\hat{\varphi}$ and $\hat{\pi}$ (roughly speaking operators $\hat{\mathrm{x}}$ and $\hat{\mathrm{p}}$ in the previous notations). These are, of course, not well defined objects in the general non-linear case but in the case of a linear space $\boldsymbol{\mathcal{V}}$ it is possible to define them as objects from the spaces $\boldsymbol{\mathcal{V}} \otimes \mathcal{O}$ and $\tilde{\boldsymbol{\mathcal{V}}} \otimes \mathcal{O}$. They are connected with the general observables $\hat{\mathsf{F}}_f$, $\hat{\mathsf{G}}_a$ of the previous sections as

$$\begin{aligned} \hat{\mathsf{F}}_f &= f(\hat{\varphi}) , \\ \hat{\mathsf{G}}_a &= (\frac{1}{2} - i\gamma)\, a(\hat{\varphi}) \cdot \hat{\pi} + (\frac{1}{2} + i\gamma)\, \hat{\pi} \cdot a(\hat{\varphi}) , \end{aligned} \tag{2.56}$$

where we used the natural identification of the tangent space $\mathbf{T}\boldsymbol{\mathcal{V}}$ with $\boldsymbol{\mathcal{V}}$ itself. The value and momentum observables satisfy the canonical commutation relations

$$[\hat{\varphi}, \hat{\pi}] = i\delta_{\boldsymbol{\mathcal{V}}} \hat{\mathbb{1}} . \tag{2.57}$$

We can construct the *value base* normalized to a "constant measure" $\boldsymbol{\Omega}$ on the configuration space $\boldsymbol{\mathcal{V}}$

$$\begin{aligned} \hat{\varphi}|val:\varphi\rangle &= \varphi|val:\varphi\rangle \quad \text{for} \quad \varphi \in \boldsymbol{\mathcal{V}} , \\ \hat{\pi}|val:\varphi\rangle &= i\, \mathrm{d}|val:\varphi\rangle , \\ \langle val:\varphi_1|val:\varphi_2\rangle &= (\boldsymbol{\Omega}^{-1}\delta)(\varphi_1|\varphi_2) , \\ \hat{\mathbb{1}} &= \int\limits_{\varphi\in\boldsymbol{\mathcal{V}}} \boldsymbol{\Omega}\, |val:\varphi\rangle\langle val:\varphi| . \end{aligned} \tag{2.58}$$

The wave functional for a state $|state\rangle$ has the form

$$\Psi_{|state\rangle}(\varphi) = \langle val:\varphi|state\rangle , \tag{2.59}$$

the scalar product on wave functions is

$$\langle st1|st2\rangle = \int\limits_{\boldsymbol{\mathcal{V}}} \Psi^*_{|st1\rangle} \Psi_{|st2\rangle}\, \boldsymbol{\Omega} , \tag{2.60}$$

and the value representations of the observables $\hat{\varphi}$ and $\hat{\pi}$ are

$$\begin{aligned} \check{\varphi}\, \Psi_{|state\rangle}(\varphi) &= \varphi\, \Psi_{|state\rangle}(\varphi) , \\ \check{\pi}\, \Psi_{|state\rangle}(\varphi) &= -i\, \mathrm{d}\Psi_{|state\rangle}(\varphi) . \end{aligned} \tag{2.61}$$

The problem is that in the infinite-dimensional case there is no such thing as a "constant measure" $\boldsymbol{\Omega}$ on the space $\boldsymbol{\mathcal{V}}$ and we do not have a Hilbert space generated by a scalar product on wave functions. The solution to these technical difficulties lies in a restriction of the possible wave functions to those that are falling off sufficiently fast so the functional integral (2.60) has a meaning even if the measure $\boldsymbol{\Omega}$ itself does not have one. Such wave functions have to be suppressed, for example, by a Gaussian exponent — so the integral (2.60) turns into a Gaussian integration that is well defined even for infinite-dimensional spaces.

3 Boundary Quantum Mechanics

Introduction

In this chapter we will finally develop *boundary quantum mechanics* — a variation of quantum mechanics based on quantization of the boundary phase space. Here we present a general formulation of the method, an application to scalar field theory can be found in [1].

Let us start with the quantum theory described in the previous section. We quantized the observables F_f and G_a on a phase space with the cotangent bundle structure, we found quantum operators $\hat{\mathsf{F}}_f$ and $\hat{\mathsf{G}}_a$, and we constructed the special position base $|pos : \mathrm{x}\rangle$. We did not formulate the dynamical part of the theory; we will touch this issue now.

We are interested in a situation when we study the system only at the "beginning" and at the "end". More precisely, we are interested in observables with a support only on the boundary of the domain Ω. It means that in the case of the field theory, when the inner manifold is a globally hyperbolic spacetime, observables are localized only on the initial and final hypersurfaces of the sandwich domain $\Omega = \langle\Sigma_\mathrm{f}, \Sigma_\mathrm{i}\rangle$. In case of a one-dimensional inner space manifold, the boundary of Ω reduces to the initial and final moment of time. In these cases we can quantize the initial and final canonical phase spaces $\boldsymbol{\mathcal{B}}_\mathrm{i} = \boldsymbol{\mathcal{B}}[\Sigma_\mathrm{i}]$ and $\boldsymbol{\mathcal{B}}_\mathrm{f} = \boldsymbol{\mathcal{B}}[\Sigma_\mathrm{f}]$ — i.e. we construct the quantum observables $\hat{\mathsf{F}}_{\mathrm{f}\,f}$, $\hat{\mathsf{G}}_{\mathrm{f}\,a}$ and $\hat{\mathsf{F}}_{\mathrm{i}\,f}$, $\hat{\mathsf{G}}_{\mathrm{i}\,a}$ on a quantum space $\mathcal{H}$ satisfying the conditions (2.2–2.11) with the ordering parameters $\gamma_\mathrm{f} = -\gamma_\mathrm{i}$, and the corresponding position bases[d] $|\mathrm{f}\,pos : \mathrm{x}_\mathrm{f}\rangle$ and $|\mathrm{i}\,pos : \mathrm{x}_\mathrm{i}\rangle$ in the space $\mathcal{H}$. The dynamics reduces to the investigation of relations between these two sets of observables or relations between objects generated by them, for example, of the position bases. We can state we solved the dynamical problem if we find in-out transition amplitudes $\langle \mathrm{f}\,pos : \mathrm{x}_\mathrm{f} | \mathrm{i}\,pos : \mathrm{x}_\mathrm{i}\rangle$ for all $\mathrm{x}_\mathrm{f} \in \boldsymbol{\mathcal{V}}[\Sigma_\mathrm{f}]$ and $\mathrm{x}_\mathrm{i} \in \boldsymbol{\mathcal{V}}[\Sigma_\mathrm{i}]$.

We do not attempt to find the transition amplitudes in a general situation. Instead, we reformulate this setting in a slightly different language.

Construction of the boundary quantum space

We represented the quantization of both $\boldsymbol{\mathcal{B}}_\mathrm{f}$ and $\boldsymbol{\mathcal{B}}_\mathrm{i}$ on a common quantum Hilbert space $\mathcal{H}$. Let us construct another representation on the *boundary quantum space*

$$\mathcal{H}_{\boldsymbol{\mathcal{B}}} = \mathcal{H}^\dagger \otimes \mathcal{H} \tag{3.1}$$

(the space of tensor products of covector and vector elements of the form $\langle\mathrm{f}| \otimes |\mathrm{i}\rangle$, i.e., essentially operators on $\mathcal{H}$). We use the notation $|state)$ for vectors from $\mathcal{H}_{\boldsymbol{\mathcal{B}}}$, and we use the accent $\hat{}$ to denote observables on this space.

---

[d] We use the proper normalization of the position bases and choose the density weight of the base equal to $(\frac{1}{2} - i\gamma_\mathrm{f})$ and $(\frac{1}{2} - i\gamma_\mathrm{i})$, where γ_f and γ_i are the order parameters as defined in (2.9). I.e. we do not need to choose any volume element on $\boldsymbol{\mathcal{V}}[\Sigma_\mathrm{f}]$ or $\boldsymbol{\mathcal{V}}[\Sigma_\mathrm{i}]$ and it makes sense to write, for example,

$$\hat{\mathsf{F}}_{\mathrm{f}\,f} = \int\limits_{\mathrm{x}_\mathrm{f} \in \boldsymbol{\mathcal{V}}[\Sigma_\mathrm{f}]} |\mathrm{f}\,pos : \mathrm{x}_\mathrm{f}\rangle\langle \mathrm{f}\,pos : \mathrm{x}_\mathrm{f}| \;.$$

We interpret the boundary quantum space in the following way: we assign a vector $|\mathrm{fi})$ from $\mathcal{H}_{\mathcal{B}}$ to any pair of f-dependent and i-dependent vectors $|\mathrm{f}\rangle$ and $|\mathrm{i}\rangle$ by

$$|\mathrm{fi}) = \langle \mathrm{f}| \otimes |\mathrm{i}\rangle \,. \tag{3.2}$$

Here, the description f-*dependent vector* suggests that the vector is identified using quantum observables on Σ_f; but, of course, it can be any vector from $\mathcal{H}$. Particularly, we define a vector in $\mathcal{H}_{\mathcal{B}}$ for any $\mathrm{x} = [\mathrm{x}_\mathrm{f}, \mathrm{x}_\mathrm{i}] \in \mathcal{V}[\partial\Omega] = \mathcal{V}[\Sigma_\mathrm{f}] \times \mathcal{V}[\Sigma_\mathrm{i}]$:

$$|pos : \mathrm{x}) = \langle \mathrm{f}\, pos : x_\mathrm{f}| \otimes |\mathrm{i}\, pos : \mathrm{x}_\mathrm{i}\rangle \,. \tag{3.3}$$

This allows us to "lift" any f-dependent or i-dependent operator $\hat{\mathrm{A}}_\mathrm{f}$ or $\hat{\mathrm{A}}_\mathrm{i}$ on $\mathcal{H}$ to an operator on the boundary quantum space $\mathcal{H}_{\mathcal{B}}$

$$\begin{aligned} \hat{\mathrm{A}}_\mathrm{f} &= \hat{\mathrm{A}}_\mathrm{f}^\dagger \otimes \hat{\mathbb{1}}_\mathrm{i} \qquad & \hat{\mathrm{A}}_\mathrm{f}|\mathrm{fi}) &= \langle \mathrm{f}|\hat{\mathrm{A}}_\mathrm{f}^\dagger \otimes |\mathrm{i}\rangle \,, \\ \hat{\mathrm{A}}_\mathrm{i} &= \hat{\mathbb{1}}_\mathrm{f} \otimes \hat{\mathrm{A}}_\mathrm{f} \qquad & \hat{\mathrm{A}}_\mathrm{i}|\mathrm{fi}) &= \langle \mathrm{f}| \otimes \hat{\mathrm{A}}_\mathrm{i}|\mathrm{i}\rangle \,. \end{aligned} \tag{3.4}$$

Using these definitions, we find

$$[\hat{\mathrm{A}}_\mathrm{f}, \hat{\mathrm{A}}_\mathrm{i}] = 0 \,. \tag{3.5}$$

We apply this method to construct the observables $\hat{\mathsf{F}}_{\mathrm{f}\, f_\mathrm{f}}$, $\hat{\mathsf{G}}_{\mathrm{f}\, a_\mathrm{f}}$ and $\hat{\mathsf{F}}_{\mathrm{i}\, f_\mathrm{i}}$, $\hat{\mathsf{G}}_{\mathrm{i}\, a_\mathrm{i}}$ for any functions and vector fields f_f, a_f or f_i, a_i on the value spaces $\mathcal{V}[\Sigma_\mathrm{f}]$ or $\mathcal{V}[\Sigma_\mathrm{i}]$. Next we want to construct a generalization of these observables $\hat{\mathsf{F}}_f$ and $\hat{\mathsf{G}}_a$ for any function f and vector field a on the value space $\mathcal{V}[\partial\Omega] = \mathcal{V}[\Sigma_\mathrm{f}] \times \mathcal{V}[\Sigma_\mathrm{i}]$. Thanks to (2.4) for both $\hat{\mathsf{F}}_{\mathrm{f}\, f}$ and $\hat{\mathsf{F}}_{\mathrm{i}\, f}$ and to equation (3.5), we do not have ordering problems with $\hat{\mathsf{F}}_f$ for $f(\mathrm{x}) = f(\mathrm{x}_\mathrm{f}, \mathrm{x}_\mathrm{i})$

$$\begin{aligned} \hat{\mathsf{F}}_f &= \int\limits_{\substack{\mathrm{x}_\mathrm{f} \in \mathcal{V}[\Sigma_\mathrm{f}] \\ \mathrm{x}_\mathrm{i} \in \mathcal{V}[\Sigma_\mathrm{i}]}} |\mathrm{f}\, pos : \mathrm{x}_\mathrm{f}\rangle\langle \mathrm{f}\, pos : \mathrm{x}_\mathrm{f}| \otimes |\mathrm{i}\, pos : \mathrm{x}_\mathrm{i}\rangle\langle \mathrm{i}\, pos : \mathrm{x}_\mathrm{i}|\; f(\mathrm{x}_\mathrm{f}, \mathrm{x}_\mathrm{i}) = \\ &= \int\limits_{\mathrm{x} \in \mathcal{V}[\partial\Omega]} |pos : \mathrm{x})(pos : \mathrm{x}|\; f(\mathrm{x}) \,. \end{aligned} \tag{3.6}$$

Similarly, for any vector field $a(\mathrm{x}) = a_\mathrm{f}(\mathrm{x}_\mathrm{f}) \oplus a_\mathrm{i}(\mathrm{x}_\mathrm{i})$ on $\mathcal{V}[\partial\Omega]$ where $a_\mathrm{f} \in \mathfrak{T}\,\mathcal{V}[\Sigma_\mathrm{f}]$ and $a_\mathrm{i} \in \mathfrak{T}\,\mathcal{V}[\Sigma_\mathrm{i}]$, motivated by (1.24), we can write

$$\hat{\mathsf{G}}_a = -\hat{\mathsf{G}}_{\mathrm{f}\, a_\mathrm{f}} + \hat{\mathsf{G}}_{\mathrm{i}\, a_\mathrm{i}} \,. \tag{3.7}$$

It is straightforward to check that $\hat{\mathsf{F}}_f$, $\hat{\mathsf{G}}_a$ are quantizations of the observables F_f, G_a on the boundary phase space $\mathcal{B}[\partial\Omega]$; i.e. they satisfy (2.2-2.11) with the ordering parameter $\gamma = \gamma_\mathrm{i} = \gamma_\mathrm{f}$. Note that a different orientation of the boundary $\partial\Omega$ and the final hypersurface Σ_f, which translates to the different sign of the symplectic structures (1.23) and to the definition of the momenta (1.24), is compensated by the covector representation of f-dependent vectors.

Let us summarize: quantization of the basic observables on the final and initial hypersurfaces Σ_f and Σ_i induces quantization of the basic observables F_f, G_a on the entire boundary $\partial\Omega$. Hence, we are able to formulate the "kinematics" of the theory using quantization of the boundary phase space $\mathcal{B}[\partial\Omega]$ — which we call *boundary quantum mechanics.* The boundary quantum space $\mathcal{H}_{\mathcal{B}}$ represents all possible quantum states at the beginning and at the end chosen independently of the real evolution of the system. Essentially, we are treating the initial and final experiments as experiments on independent systems. States in $\mathcal{H}_{\mathcal{B}}$ represent outputs of measurements understood in this way.

Before we turn to the dynamics let us list some properties of the space $\mathcal{H}_{\mathcal{B}} = \mathcal{H}^\dagger \otimes \mathcal{H}$. For vectors and operators in the "product" form

$$\begin{aligned} |\mathrm{fi}) &= \langle \mathrm{f}| \otimes |\mathrm{i}\rangle , \\ \hat{\mathrm{O}}_{\mathrm{fi}} &= \hat{\mathrm{O}}_{\mathrm{f}}^\dagger \otimes \hat{\mathrm{O}}_{\mathrm{i}} , \end{aligned} \tag{3.8}$$

we can write

$$\begin{gathered} (\mathrm{fi}\, st1|\mathrm{fi}\, st2) = \langle \mathrm{f}\, st2|\mathrm{f}\, st1\rangle\, \langle \mathrm{i}\, st1|\mathrm{i}\, st2\rangle = \mathrm{Tr}_{\mathcal{H}}\big((|\mathrm{f}\, st2\rangle\langle \mathrm{f}\, st1|)^\dagger\, (|\mathrm{i}\, st2\rangle\langle \mathrm{i}\, st1|)\big) , \\ (\mathrm{fi}| = |\mathrm{fi})^\dagger = |\mathrm{f}\rangle \otimes \langle \mathrm{i}| , \quad \hat{\mathrm{A}}_{\mathrm{fi}}\hat{\mathrm{B}}_{\mathrm{fi}} = (\hat{\mathrm{B}}_{\mathrm{f}}^\dagger \hat{\mathrm{A}}_{\mathrm{f}}^\dagger) \otimes (\hat{\mathrm{B}}_{\mathrm{i}}\hat{\mathrm{A}}_{\mathrm{i}}) , \\ \hat{\mathrm{O}}_{\mathrm{fi}}|\mathrm{fi}) = \langle \mathrm{f}|\hat{\mathrm{O}}_{\mathrm{f}}^\dagger \otimes \hat{\mathrm{O}}_{\mathrm{i}}|\mathrm{i}\rangle , \quad \mathrm{Tr}_{\mathcal{H}_{\mathcal{B}}} \hat{\mathrm{O}}_{\mathrm{fi}} = \mathrm{Tr}_{\mathcal{H}}\, \hat{\mathrm{O}}_{\mathrm{f}}^\dagger\; \mathrm{Tr}_{\mathcal{H}}\, \hat{\mathrm{O}}_{\mathrm{i}} . \end{gathered} \tag{3.9}$$

Dynamics in the boundary quantum space

Of course, much more interesting is the dynamical part of a theory. We have to ask the question whether we are able to translate the dynamically interesting quantities to the language of boundary quantum mechanics. As we outlined, the dynamical information is hidden in the in-out transition amplitudes $\langle \mathrm{f}|\mathrm{i}\rangle$. Such an amplitude can be written as

$$\langle \mathrm{f}|\mathrm{i}\rangle = \mathrm{Tr}_{\mathcal{H}}\big(\hat{\mathbb{1}}^\dagger\, |\mathrm{i}\rangle\langle \mathrm{f}|\big) = (phys|\mathrm{fi}) , \tag{3.10}$$

where $|\mathrm{fi})$ is as in (3.8) and the *physical state* $|phys)$ is given by

$$|phys) = \sum_{\mathrm{k}} \langle \mathrm{k}| \otimes |\mathrm{k}\rangle = \hat{\mathbb{1}} \tag{3.11}$$

for a complete orthonormal base $|\mathrm{k}\rangle$ in $\mathcal{H}$.

This means that there exists a preferred physical state $|phys)$ in the boundary quantum space $\mathcal{H}_{\mathcal{B}}$ that determines the dynamics of the theory. Specifically, if we set up some initial and final experiments that determine the quantum state $|state) \in \mathcal{H}_{\mathcal{B}}$, the physical transition amplitude corresponding to this state is given by

$$A(state) = (phys|state) . \tag{3.12}$$

For example, for the position base $|pos : \mathrm{x})$ we get

$$A_{pos}(\mathrm{x}) = A_{pos}(\mathrm{x_f}|\mathrm{x_i}) = (phys|pos : \mathrm{x}) = \langle \mathrm{f}\, pos : \mathrm{x_f}|\mathrm{i}\, pos : \mathrm{x_i}\rangle . \tag{3.13}$$

We will call $A_{pos}(\mathrm{x})$ the *position transition amplitude.*

The physical state $|phys)$ is actually an entangled quantum state that carries all information about the dynamical correlations between the initial and final moment of time without reference to particular initial or final conditions.

Boundary quantum mechanics

In the previous subsections we constructed the boundary quantum space and observables on it using quantization based on the initial and final phase spaces. But it is clear that we can skip the splitting of the boundary into two pieces and quantize directly the basic observables F_f, G_a on the boundary phase space $\boldsymbol{\mathcal{B}}[\partial\Omega]$. It is a phase space with a cotangent bundle structure, so we can apply the general formalism and obtain the quantum observables $\hat{\mathsf{F}}_f$ and $\hat{\mathsf{G}}_a$. We can also construct the position base $|pos : \mathrm{x})$. And we do not need any causal information for this; we do not need any global time flow on the underlying inner manifold or a causal decomposition of the boundary.

It means that we can build boundary quantum mechanics even in situations where we do not have any natural splitting of the boundary into two pieces, for example in Euclidian theories. Therefore, we could call boundary quantum mechanics also *time-symmetric quantum mechanics.*

However, in this setting we have to find the physical state $|phys)$ without reference to the initial and final causal decomposition.[e] Again, we cannot expect an answer on a general level — this is a question equivalent to solving the quantum evolution. But we suggest methods determining the physical state — we formulate a dynamical equation for boundary quantum mechanics.

The classical evolution in the boundary phase space $\boldsymbol{\mathcal{B}}[\partial\Omega]$ is determined by a specification of the physical phase space $\boldsymbol{\mathcal{S}}$ as a subspace of $\boldsymbol{\mathcal{B}}[\partial\Omega]$. It can be done, for example, via condition (1.18), or, expressed using observables F_f, G_a, by the condition

$$\mathsf{G}_a + \mathsf{F}_{a\cdot \mathrm{d}\bar{S}} = 0 \qquad \text{for all} \quad a \in \mathfrak{T}\boldsymbol{\mathcal{V}}[\partial\Omega]\,. \tag{3.14}$$

Hence, on the classical level we specified physical states by imposing constraints in the phase space. In the usual quantum mechanics one quantizes the constrained subspace $\boldsymbol{\mathcal{S}}$. In boundary quantum mechanics we quantize the entire boundary phase space, but we have to impose conditions on the physical state inspired by the classical constraints

$$\big(\hat{\mathsf{G}}_a + \hat{\mathsf{F}}_{a\cdot \mathrm{d}\bar{S}}\big)|phys) = 0 \tag{3.15}$$

for, at least, some vector fields a on the value space $\boldsymbol{\mathcal{V}}[\partial\Omega]$.

We cannot expect the condition above to be satisfied for all vector fields a due to the noncommutativity of the position and momentum observables ($\hat{\mathsf{G}}_a$ is some ordering of the "$a(\mathbf{x})\cdot\mathbf{p}$" observable). We will see that such a strong requirement would be inconsistent. It will turn out that the choice of the class of vector fields for which the condition (3.15) is required to hold (i.e. the choice of a preferred operator ordering) is equivalent to the solution of the dynamical problem. So, if we have some preferred vector fields, it can provide us with a method for finding the transition amplitudes we are looking for.

[e] More precisely, in the Heisenberg picture, which we are using, the physical state $|phys)$ is a fixed dynamically independent state and the dynamics is hidden in the relations of the basic observables $\hat{\mathsf{F}}_f$, $\hat{\mathsf{G}}_a$ to this state. But we will be a bit vague and speak about a determination of the physical state $|phys)$ because it is more intuitive and does not influence any computation.

Let us look at the position representation of the constraint conditions. Using (2.14) and (2.43) we find

$$(phys|\big(\hat{\mathsf{G}}_a + \hat{\mathsf{F}}_{a\cdot \mathrm{d}\bar{S}}\big)|pos:.) = \big(i\mathcal{L}_a + a\cdot \mathrm{d}\bar{S}\big)(phys|pos:.) = 0\,. \tag{3.16}$$

If we represent the position transition amplitude as

$$A_{pos}(\mathrm{x}) = \mathfrak{a}(\mathrm{x})\,\exp\big(i\bar{S}(\mathrm{x})\big) \tag{3.17}$$

with $\mathfrak{a}(\mathrm{x})$ a density of the weight $(\frac{1}{2} - i\gamma)$, the condition above translates to

$$\mathcal{L}_a\mathfrak{a} = 0\,. \tag{3.18}$$

This confirms that the constraint conditions cannot be satisfied for all vector fields a — it would require $\mathfrak{a} = 0$.

If a linearly complete set of vector fields for which the constraint conditions should be satisfied is specified, the density $\mathfrak{a}$ is determined completely up to a constant multiplicative factor (this remaining freedom of choice is, of course, equivalent to the choice of a global normalization and phase factor). In case of a linear non-interacting theory,[f] such a set of vectors exists — if we require the condition (3.15) be satisfied for globally parallel vector fields[g] we obtain that $\mathfrak{a}$ is a constant and we recover the known fact that the quantum amplitudes for a non-interacting theory are given by the classical action through (3.17).

In a general case, the density $\mathfrak{a}$ contains all quantum corrections to the transition amplitude A_{pos}. It indicates that a choice of the "right" vector fields in the case of a general interacting theory can be difficult or even impossible. Therefore, we turn to another way of specifying the physical state $|phys)$ or physical amplitudes $A(state)$.

Path integral

There exists another approach to the quantization. This is the path integral quantization, which gives essentially a prescription for the position transition amplitude on the basis of a completely different calculation — through a sum of elementary amplitudes over all possible histories with fixed boundary values,

$$\langle \mathrm{f}\,pos:\mathrm{x_f}|\mathrm{i}\,pos:\mathrm{x_i}\rangle = A_{pos}(\mathrm{x_f}|\mathrm{x_i}) = \int\limits_{\substack{\mathrm{h}\in\mathcal{H}\\ \mathrm{x(h)}=[\mathrm{x_f},\mathrm{x_i}]}} \mathfrak{M}^F(\mathrm{h})\,\exp\big(iS(\mathrm{h})\big)\,. \tag{3.19}$$

However, this amplitude is exactly the wave function of the physical state $|phys)$, i.e., the dynamics of boundary quantum mechanics can be specified by the path integral

$$(phys| = \int\limits_{\mathrm{h}\in\mathcal{H}} \mathfrak{M}^F(\mathrm{h})\,\exp\big(iS(\mathrm{h})\big)\,(pos:\mathrm{x(h)}|\,. \tag{3.20}$$

---

[f] Motivated by the field theory, by *non-interacting* theory we mean a Hamiltonian quadratic in momenta and positions.

[g] The notion of global parallelism is defined thanks to the linearity.

Let us add some comments about the advantages and problems of this approach. The integral over the space of histories faces serious problems due to the infinite dimension of the space of histories and the oscillatory character of the integrand. A technical solution usually leads to the computation of some Euclidian equivalent of the integral, which is usually better defined, followed by some "Wick rotation" — a transformation from the Euclidian to the physical theory. We can view this "Euclidian business" as a mere technical detour without a physical interpretation. Only after computing the integral we are able to identify the results with the transition amplitudes of quantum mechanics. In the usual framework, we do not even know what would be the quantum mechanics of the Euclidian formulation of the theory — the usual quantum mechanics essentially uses the causal structure to define the initial and final states.

However, the formalism developed above gives us a hope for another option. We can formulate boundary quantum mechanics even for a Euclidian version of the theory — it does not need the causal structure, it can be formulated without splitting of the boundary phase space to the initial and final part. Therefore we could make the connection with the path integral already in the Euclidian formulation

$$(\boldsymbol{phys}|\boldsymbol{pos} : \mathrm{x}) = A_{\boldsymbol{pos}}(\mathrm{x}) = \int\limits_{\substack{\mathrm{h}\in\mathcal{H} \\ \mathrm{x}(\mathrm{h})=\mathrm{x}}} \mathfrak{M}^F(\mathrm{h})\, \exp\bigl(-I(\mathrm{h})\bigr)\,, \tag{3.21}$$

$I(\mathrm{h})$ being the Euclidian action.[h] For the physical version of the theory, this reduces to the relations (3.20) above. In the Euclidian case, this relation would give an interpretation for the path integral amplitude in terms of transition amplitudes of boundary quantum mechanics.

Unfortunately, the situation is not so straightforward. Even in case of a linear non-interacting theory, the form of boundary quantum mechanics formulated above applied to the Euclidian version of the theory does not correspond exactly to the Euclidian path integral. The problem is hidden in the method of quantization we used. The translation of the Poisson brackets of the classical theory to the commutators of the quantum theory

$$\{\,,\,\} \rightarrow i[\,,\,] \tag{3.22}$$

intrinsically contains reference to the physical signature. The imaginary unit in this translation can be traced to be the cause of the imaginary unit in the exponent of the transition amplitude (3.17) computed in boundary quantum mechanics independently of the physical or Euclidian version of the theory. For the Euclidian theory it would be more appropriate to construct "Euclidian quantum mechanics" based on the commutation relation generated by the rule

$$\{\,,\,\} \rightarrow [\,,\,]\,, \tag{3.23}$$

with observables represented, probably, on a real Hilbert space. We do not attempt to build such quantum mechanics here. However, escaping the necessity of a causal structure in the usual quantum mechanics through the method of quantization of the boundary phase space is the first step towards Euclidian quantum mechanics.

---

[h] $I(\mathrm{h})$ is real for "Euclidian" histories. For physical histories we have $iS(\mathrm{h}) = -I(\mathrm{h})$.

A Symplectic geometry

This appendix is a review of the standard geometrical formulation of the symplectic geometry (see, e.g., [5, 6]).

The phase space is a manifold $\mathcal{G}$ of an even dimension $2n$ with a *symplectic form* ω that satisfies

$$\begin{aligned} &\omega^{\mathsf{T}} = -\omega\,, \quad \omega \in \mathfrak{T}^0_2\,\mathcal{G}\,, \\ &\omega \quad \text{is non-degenerate}\,, \\ &\omega \quad \text{is closed (i.e. } \mathrm{d}\omega = 0)\,. \end{aligned} \tag{A.1}$$

We can invert it[i]

$$\omega^{-1} \circ \omega = -\delta_{\mathcal{G}} \tag{A.2}$$

and define a *canonical vector field* associated with a function H on $\mathcal{G}$

$$X_H = (\mathrm{d}H) \circ \omega^{-1}\,. \tag{A.3}$$

This canonical vector field generates a canonical transformation on $\mathcal{G}$ that does not change the symplectic structure:

$$\mathcal{L}_{X_H}\omega = 0\,. \tag{A.4}$$

We can define the *Poisson brackets* of functions on $\mathcal{G}$ as

$$\{A, B\} = X_A \circ \mathrm{d}B = X_A \circ \omega \circ X_B = (\mathrm{d}A) \circ \omega^{-1} \circ (\mathrm{d}B)\,. \tag{A.5}$$

We have

$$\frac{d}{dt}B \overset{\text{def}}{=} \mathcal{L}_{X_H}B = \{H, B\}\,, \tag{A.6}$$

$$[X_A, X_B] = X_{\{A,B\}}\,, \tag{A.7}$$

where $[,]$ are the Lie brackets on vector fields. The symplectic structure also induces a volume element on $\mathcal{G}$

$$d\Gamma = (2\pi)^{-n}\frac{1}{n!}\,\underbrace{|\omega \wedge \omega \wedge \cdots \wedge \omega|}_{n \text{ times}} = \left(\mathrm{Det}\,\frac{\omega}{2\pi}\right)^{\frac{1}{2}}. \tag{A.8}$$

Finally, if we choose coordinates (x^a, p_a) for $a = 1, \ldots, n$ such that

$$\omega = \mathrm{d}p_a \wedge \mathrm{d}x^a\,, \tag{A.9}$$

we get

$$\{x^a, p_b\} = -\delta^a_b \tag{A.10}$$

and (x^a, p_a) are *canonical coordinates*.

---

[i]The dot $\circ$ indicates the contraction in $\mathbf{T}\mathcal{G}$.

B Tangent and cotangent bundle geometry

In this appendix we discuss the geometry of tangent bundles of a manifold $\boldsymbol{\mathcal{V}}$. We define "partial derivatives" of observables on these spaces in a covariant way. We also show that the cotangent bundle has the structure of a symplectic manifold. We use $\mathsf{a}, \mathsf{b}, \ldots$ as indices for tensors on the manifold $\boldsymbol{\mathcal{V}}$ and indices $\mathsf{A}, \mathsf{B}, \ldots$ for tensors from tangent spaces of the cotangent bundle $\boldsymbol{\mathcal{G}} = \mathbf{T}^\star \boldsymbol{\mathcal{V}}$.

Functions such as the Lagrangian $L(x,v)$ and the Hamiltonian $H(x,p)$ are functions on the tangent and cotangent bundle, respectively, of a configuration space $\boldsymbol{\mathcal{V}}$. Because velocities v (or momenta p) are vectors (covectors) from different fibers for different position x ($\mathbf{T}_x \boldsymbol{\mathcal{V}} \neq \mathbf{T}_y \boldsymbol{\mathcal{V}}$ for $x \neq y$), we have to be careful to use a partial derivative with respect of the position x. There is no problem with the definition

$$\frac{\partial L}{\partial v^{\mathbf{a}}}(x,v) \;:\qquad \delta v^{\mathbf{n}} \frac{\partial L}{\partial v^{\mathbf{n}}}(x,v) = \frac{d}{d\varepsilon} L(x, v + \varepsilon \delta v)|_{\varepsilon=0} \tag{B.1}$$

— a *partial derivative with x constant* — but to define the *derivative with v constant* we need a connection ∇ on $\boldsymbol{\mathcal{V}}$

$$\frac{\nabla_{\mathbf{a}} L}{\partial x}(x,v) \;:\qquad \delta x^{\mathbf{n}} \frac{\nabla_{\mathbf{n}} L}{\partial x}(x,v) = \frac{d}{d\varepsilon} L(x_\varepsilon, v_\varepsilon)|_{\varepsilon=0}\,, \tag{B.2}$$

where x_ε is a curve starting from x in a direction δx and v_ε is the parallel transport of v along x_ε in the sense of the connection ∇ (i.e. $\frac{\nabla}{d\varepsilon} v_\varepsilon = 0$). Similarly, for a function on the cotangent bundle,

$$\begin{aligned} \frac{\partial H}{\partial p_{\mathbf{a}}}(x,p) \;&:\qquad \delta p_{\mathbf{n}} \frac{\partial H}{\partial p_{\mathbf{n}}}(x,p) = \frac{d}{d\varepsilon} H(x, p + \varepsilon \delta p)|_{\varepsilon=0}\,,\\ \frac{\nabla_{\mathbf{a}} H}{\partial x}(x,p) \;&:\qquad \delta x^{\mathbf{n}} \frac{\nabla_{\mathbf{n}} H}{\partial x}(x,p) = \frac{d}{d\varepsilon} H(x_\varepsilon, p_\varepsilon)|_{\varepsilon=0}\,, \end{aligned} \tag{B.3}$$

where again $\frac{\nabla}{d\varepsilon} p_\varepsilon = 0$.

We want to show that the cotangent bundle $\boldsymbol{\mathcal{G}} = \mathbf{T}^\star \boldsymbol{\mathcal{V}}$ has the structure of a phase space. It will be useful to define a covariant generalization of "coordinate" vector fields and forms

$$\begin{aligned} &\frac{\partial^{\mathsf{A}}}{\partial p_{\mathbf{a}}} \qquad \text{a vector field on } \boldsymbol{\mathcal{G}} \text{ for which } \quad \frac{\partial^{\mathsf{A}}}{\partial p_{\mathbf{a}}} \mathrm{d}_{\mathsf{A}} H = \frac{\partial H}{\partial p_{\mathbf{a}}}\,,\\ &\frac{\nabla^{\mathsf{A}}_{\mathbf{a}}}{\partial x} \qquad \text{a vector field on } \boldsymbol{\mathcal{G}} \text{ for which } \quad \frac{\nabla^{\mathsf{A}}_{\mathbf{a}}}{\partial x} \mathrm{d}_{\mathsf{A}} H = \frac{\nabla_{\mathbf{a}} H}{\partial x}\,. \end{aligned} \tag{B.4}$$

$\frac{\partial}{\partial p}$ is actually the natural identification of the vector space $\mathbf{T}^\star_x \boldsymbol{\mathcal{V}}$ with its tangent space $\mathbf{T}(\mathbf{T}^\star_x \boldsymbol{\mathcal{V}})$ and $\frac{\nabla}{\partial x}$ is the horizontal shift of the connection ∇. Form fields dual to these vector fields

$$\begin{aligned} &\mathrm{D}^{\mathbf{a}}_{\mathsf{A}} x\,, \quad \text{differential of the bundle projection } x : \mathbf{T}\boldsymbol{\mathcal{V}} \to \boldsymbol{\mathcal{V}},\ p|_x \to x\,,\\ &\nabla_{\mathsf{A}} p_{\mathbf{a}}\,, \end{aligned} \tag{B.5}$$

are defined by

$$\begin{gathered} \frac{\nabla^{\mathsf{N}}_{\mathbf{a}}}{\partial x} \mathrm{D}^{\mathbf{b}}_{\mathsf{N}} x = \delta^{\mathbf{b}}_{\mathbf{a}}\,, \quad \frac{\partial^{\mathsf{N}}}{\partial p_{\mathbf{b}}} \nabla_{\mathsf{N}} p_{\mathbf{a}} = \delta^{\mathbf{b}}_{\mathbf{a}}\,, \quad \frac{\nabla^{\mathsf{N}}_{\mathbf{a}}}{\partial x} \nabla_{\mathsf{N}} p_{\mathbf{b}} = 0\,, \quad \frac{\partial^{\mathsf{N}}}{\partial p_{\mathbf{a}}} \mathrm{D}^{\mathbf{b}}_{\mathsf{N}} x = 0\,,\\ \frac{\nabla^{\mathsf{A}}_{\mathbf{n}}}{\partial x} \mathrm{D}^{\mathbf{n}}_{\mathsf{B}} x + \frac{\partial^{\mathsf{A}}}{\partial p_{\mathbf{n}}} \nabla_{\mathsf{B}} p_{\mathbf{n}} = \delta^{\mathsf{A}}_{\mathsf{B}}\,. \end{gathered} \tag{B.6}$$

Now we can write down the canonical cotangent bundle symplectic form

$$\omega_{\mathsf{AB}} = \nabla_{\mathsf{A}} p_{\mathbf{a}} \wedge \mathrm{D}^{\mathbf{a}}_{\mathsf{B}} x = \nabla_{\mathsf{A}} p_{\mathbf{a}}\, \mathrm{D}^{\mathbf{a}}_{\mathsf{B}} x - \mathrm{D}^{\mathbf{a}}_{\mathsf{A}} x\, \nabla_{\mathsf{B}} p_{\mathbf{a}}\,, \tag{B.7}$$

$$\omega^{-1\,\mathsf{AB}} = \frac{\partial^{\mathsf{A}}}{\partial p_{\mathbf{n}}} \frac{\nabla^{\mathsf{B}}_{\mathbf{n}}}{\partial x} - \frac{\nabla^{\mathsf{A}}_{\mathbf{n}}}{\partial x} \frac{\partial^{\mathsf{B}}}{\partial p_{\mathbf{n}}} \tag{B.8}$$

and we can even explicitly write the symplectic potential

$$\omega_{\mathsf{AB}} = \mathrm{d}_{\mathsf{A}} \theta_{\mathsf{B}}\,, \quad \theta_{\mathsf{A}} = p_{\mathbf{n}} \mathrm{D}^{\mathbf{n}}_{\mathsf{A}} x\,. \tag{B.9}$$

The canonical vector fields and Poisson brackets are

$$X^{\mathsf{A}}_{F} = \frac{\partial F}{\partial p_{\mathbf{n}}} \frac{\nabla^{\mathsf{A}}_{\mathbf{n}}}{\partial x} - \frac{\nabla_{\mathbf{n}} F}{\partial x} \frac{\partial^{\mathsf{A}}}{\partial p_{\mathbf{n}}}\,, \tag{B.10}$$

$$\{A, B\} = \frac{\partial A}{\partial p_{\mathbf{n}}} \frac{\nabla_{\mathbf{n}} B}{\partial x} - \frac{\nabla_{\mathbf{n}} A}{\partial x} \frac{\partial B}{\partial p_{\mathbf{n}}}\,. \tag{B.11}$$

If we change the connection to another one,

$$\tilde{\nabla} = \nabla \oplus \Gamma\,,$$
$$\tilde{\nabla}_{\mathbf{a}} a^{\mathbf{b}} = \nabla_{\mathbf{a}} a^{\mathbf{b}} + \Gamma^{\mathbf{b}}_{\mathbf{an}} a^{\mathbf{n}}\,, \quad \tilde{\nabla}_{\mathbf{a}} p_{\mathbf{b}} = \nabla_{\mathbf{a}} p_{\mathbf{b}} - \Gamma^{\mathbf{n}}_{\mathbf{ab}} p_{\mathbf{n}}\,, \tag{B.12}$$

we get

$$\frac{\tilde{\nabla}^{\mathsf{A}}_{\mathbf{a}}}{\partial x}(x,p) = \frac{\nabla^{\mathsf{A}}_{\mathbf{a}}}{\partial x}(x,p) + p_{\mathbf{r}} \Gamma^{\mathbf{r}}{}_{\mathbf{as}}(x) \frac{\partial^{\mathsf{A}}}{\partial p_{\mathbf{s}}}(x,p)\,,$$
$$\tilde{\nabla}_{\mathsf{A}} p_{\mathbf{a}}(x,p) = \nabla_{\mathsf{A}} p_{\mathbf{a}}(x,p) - p_{\mathbf{r}} \Gamma^{\mathbf{r}}_{\mathbf{as}}(x) \mathrm{D}^{\mathbf{s}}_{\mathsf{A}} x(x,p)\,. \tag{B.13}$$

By straightforward calculations, we can check that the quantities ω, θ, X_F, and $\{\,,\}$ do not depend on the choice of the connection.

Finally, coordinates x^{a} on $\boldsymbol{\mathcal{V}}$ generate coordinates $(x^{\mathrm{a}}, p_{\mathrm{a}})$ on $\boldsymbol{\mathcal{G}}$ by

$$p_{\mathrm{a}} = p_{\mathbf{a}} \frac{\partial^{\mathbf{a}}}{\partial x^{\mathrm{a}}} \tag{B.14}$$

and they define the coordinate connection ∂ on $\boldsymbol{\mathcal{V}}$ for which

$$\partial \mathrm{d}x^{\mathrm{a}} = 0\,, \quad \partial \frac{\partial}{\partial x^{\mathrm{a}}} = 0 \quad \text{for } \mathrm{a} = 1, 2, \ldots, n\,. \tag{B.15}$$

Using this connection and expressing everything in coordinates, we get the standard relations [5,6]

$$\omega_{\mathsf{AB}} = \mathrm{d}_{\mathsf{A}} p_{\mathrm{n}} \wedge \mathrm{d}_{\mathsf{B}} x^{\mathrm{n}}\,, \quad \theta_{\mathsf{A}} = p_{\mathrm{n}} \mathrm{d}_{\mathsf{A}} x^{\mathrm{n}}\,, \tag{B.16}$$

$$X^{\mathsf{A}}_{F} = \frac{\partial F}{\partial p_{\mathrm{n}}} \frac{\partial^{\mathsf{A}}}{\partial x^{\mathrm{n}}} - \frac{\partial F}{\partial x^{\mathrm{n}}} \frac{\partial^{\mathsf{A}}}{\partial p_{\mathrm{n}}}\,, \tag{B.17}$$

$$\{A, B\} = \frac{\partial A}{\partial p_{\mathrm{n}}} \frac{\partial B}{\partial x^{\mathrm{n}}} - \frac{\partial A}{\partial x^{\mathrm{n}}} \frac{\partial B}{\partial p_{\mathrm{n}}}\,, \tag{B.18}$$

$$\{x^{\mathrm{a}}, p_{\mathrm{b}}\} = -\delta^{\mathrm{a}}_{\mathrm{b}}\,. \tag{B.19}$$

C Densities on a Manifold

In this appendix we shortly review the definition of densities on a manifold and some operations with them.

On any manifold M we can define a vector bundle of tangent densities of a weight α that we denote $\tilde{\mathbb{R}}^\alpha M$ and $\tilde{\mathbb{C}}^\alpha M$ if the densities are real and complex, respectively. The space of sections we denote $\tilde{\mathfrak{F}}^\alpha M$. The standard fiber of these bundles is the vector space of real or complex numbers. As for any tangent bundle, the density bundle can be defined by a *coordinate map* from the space of frames of the tangent vector space $\mathbf{T}\,M$ to the standard fiber $\mathbb{R}$ (or $\mathbb{C}$). The coordinate map has to be a representation of the linear group acting on the space of frames. The coordinate map $e_a \to \mu[e_a]$ tells us by what factor is a density μ different from the coordinate density ϵ given by the base e_a

$$\mu[e_a] = \mu\,\epsilon^{-1} \;, \quad \epsilon[e_a] = 1 \;. \tag{C.1}$$

For densities of weight α, the coordinate map is a representation of the linear group of the following type:

$$\mu[A^b_a e_b] = |\det A|^\alpha \; \mu[e_a] \;. \tag{C.2}$$

Clearly, we can define complex densities even for a complex weight.

Besides the linear operation, we can also define the multiplication and constant powers of densities. These operations map densities of some weight to densities of a different weight. Let us note that a complex conjugation maps the densities of weight α to densities of weight α^* and therefore "the absolute value" $|\mu| = (\mu\mu^*)^{\frac{1}{2}}$ belongs to densities of weight $\operatorname{Re}\alpha$.

Densities of weight 1 are called *volume elements* because they can be integrated. Let μ be a volume element, then we define locally

$$\int\limits_{\Omega} \mu = \int\limits_{x^a(\Omega)} \mu\Big[\frac{\partial}{\partial x^a}\Big] \; d^n x \;, \tag{C.3}$$

where x^a are arbitrary coordinates. On the right hand side the usual coordinate integration is understood. Extension to domains not covered by one coordinate system is done by standard methods [8]. The consistency of this definition (independence of a choice of coordinates) follows from the transformation properties (C.2) with $\alpha = 1$.

A metric g or a symplectic form ω (see Appendix A) define canonical volume elements:

$$\mathfrak{g}^{\frac{1}{2}} = |\operatorname{Det} g|^{\frac{1}{2}} \;, \quad d\Gamma = \left|\operatorname{Det} \frac{\overleftrightarrow{\omega}}{2\pi}\right|^{\frac{1}{2}} \;, \tag{C.4}$$

where the operation $\operatorname{Det} : \mathbf{T}^0_2\, M \to \tilde{\mathbb{R}}^1 M$ is defined by

$$(\operatorname{Det} g)[e_a] = \det g_{ab} \;. \tag{C.5}$$

We can also define a map from the space of totally antisymmetric forms to the space of volume elements $\sigma \to |\sigma|$, such that

$$|\sigma|\,[e_a] = |\sigma_{1\ldots n}| \;, \tag{C.6}$$

n beeing the dimension of the manifold M. It is easy to check that (C.2) is satisfied.

Finally, let us note that the delta distribution $\delta(x|y)$ can be understood as a bi-distribution with a density character in both arguments and of weights α, β such that $\alpha + \beta = 1$, i.e., it is a functional on test densities μ, ν of weights $1 - \alpha$ and $1 - \beta$:

$$\int \mu(x)\nu(y)\delta(x|y) = \int \mu\nu \,. \tag{C.7}$$

If we want to define the "ordinary" delta function that is not a density in any of its arguments, we have to normalize it to some volume element[j] μ. We call such a distribution $(\mu^{-1}\delta)$. It acts on test densities φ, ψ of weight 1:

$$\int \varphi(x)\psi(y)\,(\mu^{-1}\delta)(x|y) = \int \varphi\psi\mu^{-1} \,. \tag{C.8}$$

Similarly, for any smooth density f we can define a bi-distribution $(f\delta)$.

References

1. P. Krtouš. *Relationships between Scalar Field and Relativistic Particle Quantizations.* PhD thesis, University of Alberta, Edmonton, Alberta, Canada, September 1997.
2. C. J. Isham, N. Linden, K. Savvidou, and S. Schreckenberg. Continuous time and consistent histories. *J. Math. Phys.*, 39:1818–1834, 1998. e-print: quant-ph/9711031.
3. K. Savvidou. *Continuous Time in Consistent Histories.* PhD thesis, Imperial College, London, UK, 1999. e-print: gr-qc/9912076.
4. B. S. DeWitt. *Dynamical Theory of Groups and Fields.* Gordon and Brench, New York, 1965.
5. V. I. Arnol'd. *Mathematical methods of classical mechanics.* Springer, New York, 1989.
6. T. Frankel. *The Geometry of Physics, An Introduction.* Cambridge University Press, Cambridge, 1997.
7. R. Penrose and W. Rindler. *Spinors and Space-time.* Cambridge University Press, Cambridge, 1984.
8. S. W. Hawking and G. F. R. Ellis. *The Large Scale Structure of Space-time.* Cambridge University Press, Cambridge, 1973.